Gene F. Mazenko

Nonequilibrium Statistical Mechanics

Related Titles

Mazenko, G. F.

Fluctuations, Order, and Defects

540 pages
2003
Hardcover
ISBN 0-471-32840-5

Mazenko, G. F.

Equilibrium Statistical Mechanics

630 pages
2000
Hardcover
ISBN 0-471-32839-1

Reichl, L. E.

A Modern Course in Statistical Physics

842 pages with 186 figures
1998
Hardcover
ISBN 0-471-59520-9

Smirnov, B. M.

Principles of Statistical Physics

Distributions, Structures, Phenomena, Kinetics of Atomic Systems

474 pages with 101 figures and 55 tables
2006
Hardcover
ISBN 3-527-40613-1

Gene F. Mazenko

Nonequilibrium Statistical Mechanics

WILEY-VCH Verlag GmbH & Co. KGaA

The Author

Gene F. Mazenko
The University of Chicago
The James Franck Institute
5640 S. Ellis Avenue
Chicao, IL 60637, USA
gfm@ma.uchicago.edu

Cover
The cover illustration has been created by the author.

■ All books published by Wiley-VCH are carefully produced. Nevertheless, editors, authors and publisher do not warrant the information contained in these books to be free of errors. Readers are advised to keep in mind that statements, data, illustrations, procedural details or other items may inadvertently be inaccurate.

Library of Congress Card No.:
applied for

British Library Cataloguing-in-Publication Data:
A catalogue record for this book is available from the British Library.

Bibliographic information published by the Deutsche Nationalbibliothek
The Deutsche Nationalbibliothek lists this publication in the Deutsche Nationalbibliografie; detailed bibliographic data are available in the Internet at http://dnb.d-nb.de

© 2006 WILEY-VCH Verlag GmbH & Co KGaA, Weinheim

All rights reserved (including those of translation into other languages). No part of this book may be reproduced in any form – by photocopying, microfilm, or any other means – nor transmitted or translated into a machine language without written permission from the publishers. Registered names, trademarks, etc. used in this book, even when not specifically marked as such, are not to be considered unprotected by law.

Printed in the Federal Republic of Germany
Printed on acid-free and chlorine-free paper
Composition: Steingraeber Satztechnik GmbH, Ladenburg
Printing: Strauss GmbH, Mörlenbach
Bookbinding: Litges & Dopf Buchbinderei GmbH, Heppenheim

ISBN-13: 978-3-527-40648-7
ISBN-10: 3-527-40648-4

Contents

1	**Systems Out of Equilibrium** *1*	
1.1	Problems of Interest *1*	
1.2	Brownian Motion *6*	
1.2.1	Fluctuations in Equilibrium *6*	
1.2.2	Response to Applied Forces *13*	
1.3	References and Notes *15*	
1.4	Problems for Chapter 1 *16*	
2	**Time-Dependent Phenomena in Condensed-Matter Systems** *19*	
2.1	Linear Response Theory *19*	
2.1.1	General Comments *19*	
2.1.2	Linear Response Formalism *19*	
2.1.3	Time-Translational Invariance *27*	
2.1.4	Vector Operators *29*	
2.1.5	Example: The Electrical Conductivity *29*	
2.1.6	Example: Magnetic Resonance *32*	
2.1.7	Example: Relaxation From Constrained Equilibrium *37*	
2.1.8	Field Operators *40*	
2.1.9	Identification of Couplings *41*	
2.2	Scattering Experiments *42*	
2.2.1	Inelastic Neutron Scattering from a Fluid *42*	
2.2.2	Electron Scattering *49*	
2.2.3	Neutron Scattering: A More Careful Analysis *50*	
2.2.4	Magnetic Neutron Scattering *52*	
2.2.5	X-Ray and Light Scattering *55*	
2.2.6	Summary of Scattering Experiments *57*	
2.3	References and Notes *58*	
2.4	Problems for Chapter 2 *59*	

Nonequilibrium Statistical Mechanics. Gene F. Mazenko
Copyright © 2006 WILEY-VCH Verlag GmbH & Co. KGaA, Weinheim
ISBN: 3-527-40648-4

3 General Properties of Time-Correlation Functions 63

3.1 Fluctuation-Dissipation Theorem 63
3.2 Symmetry Properties of Correlation Functions 67
3.3 Analytic Properties of Response Functions 70
3.4 Symmetries of the Complex Response Function 73
3.5 The Harmonic Oscillator 75
3.6 The Relaxation Function 77
3.7 Summary of Correlation Functions 81
3.8 The Classical Limit 82
3.9 Example: The Electrical Conductivity 83
3.10 Nyquist Theorem 85
3.11 Dissipation 87
3.12 Static Susceptibility (Again) 89
3.13 Sum Rules 91
3.14 References and Notes 96
3.15 Problems for Chapter 3 96

4 Charged Transport 101

4.1 Introduction 101
4.2 The Equilibrium Situation 101
4.3 The Nonequilibrium Case 104
4.3.1 Setting up the Problem 104
4.3.2 Linear Response 106
4.4 The Macroscopic Maxwell Equations 113
4.5 The Drude Model 116
4.5.1 Basis for Model 116
4.5.2 Conductivity and Dielectric Function 118
4.5.3 The Current Correlation Function 119
4.6 References and Notes 120
4.7 Problems for Chapter 4 121

5 Linearized Langevin and Hydrodynamical Description of Time-Correlation Functions 123

5.1 Introduction 123
5.2 Spin Diffusion in Itinerant Paramagnets 124
5.2.1 Continuity Equation 124
5.2.2 Constitutive Relation 126
5.2.3 Hydrodynamic Form for Correlation Functions 128
5.2.4 Green–Kubo Formula 130
5.3 Langevin Equation Approach to the Theory of Irreversible Processes 134
5.3.1 Choice of Variables 134

5.3.2	Equations of Motion	*134*
5.3.3	Example: Heisenberg Ferromagnet	*136*
5.3.4	Example: Classical Fluid	*138*
5.3.5	Summary	*142*
5.3.6	Generalized Langevin Equation	*142*
5.3.7	Memory-Function Formalism	*144*
5.3.8	Memory-Function Formalism: Summary	*147*
5.3.9	Second Fluctuation-Dissipation Theorem	*147*
5.4	Example: The Harmonic Oscillator	*149*
5.5	Theorem Satisfied by the Static Part of the Memory Function	*154*
5.6	Separation of Time Scales: The Markoff Approximation	*155*
5.7	Example: Brownian Motion	*156*
5.8	The Plateau-Value Problem	*158*
5.9	Example: Hydrodynamic Behavior; Spin-Diffusion Revisited	*161*
5.10	Estimating the Spin-Diffusion Coefficient	*165*
5.11	References and Notes	*170*
5.12	Problems for Chapter 5	*171*

6 Hydrodynamic Spectrum of Normal Fluids *175*

6.1	Introduction	*175*
6.2	Selection of Slow Variables	*175*
6.3	Static Structure Factor	*177*
6.4	Static Part of the Memory Function	*182*
6.5	Spectrum of Fluctuations with No Damping	*188*
6.6	Dynamic Part of the Memory Function	*191*
6.7	Transverse Modes	*192*
6.8	Longitudinal Modes	*194*
6.9	Fluctuation Spectrum Including Damping	*196*
6.10	References and Notes	*202*
6.11	Problems for Chapter 6	*203*

7 Kinetic Theory *205*

7.1	Introduction	*205*
7.2	Boltzmann Equation	*206*
7.2.1	Ideal Gas Law	*206*
7.2.2	Mean-Free Path	*210*
7.2.3	Boltzmann Equation: Kinematics	*213*
7.2.4	Boltzmann Collision Integral	*215*
7.2.5	Collisional Invariants	*219*
7.2.6	Approach to Equilibrium	*221*
7.2.7	Linearized Boltzmann Collision Integral	*223*
7.2.8	Kinetic Models	*225*

7.2.9	Single-Relaxation-Time Approximation	*228*
7.2.10	Steady-State Solutions	*231*
7.3	Traditional Transport Theory	*233*
7.3.1	Steady-State Currents	*233*
7.3.2	Thermal Gradients	*237*
7.3.3	Shear Viscosity	*241*
7.3.4	Hall Effect	*244*
7.4	Modern Kinetic Theory	*246*
7.4.1	Collisionless Theory	*250*
7.4.2	Noninteracting Gas	*252*
7.4.3	Vlasov Approximation	*253*
7.4.4	Dynamic Part of Memory Function	*256*
7.4.5	Approximations	*257*
7.4.6	Transport Coefficients	*260*
7.5	References and Notes	*265*
7.6	Problems for Chapter 7	*266*

8 Critical Phenomena and Broken Symmetry 271

8.1	Dynamic Critical Phenomena	*271*
8.1.1	Order Parameter as a Slow Variable	*271*
8.1.2	Examples of Order Parameters	*273*
8.1.3	Critical Indices and Universality	*277*
8.1.4	The Scaling Hypothesis	*277*
8.1.5	Conventional Approximation	*279*
8.2	More on Slow Variables	*283*
8.3	Spontaneous Symmetry Breaking and Nambu–Goldstone Modes	*285*
8.4	The Isotropic Ferromagnet	*286*
8.5	Isotropic Antiferromagnet	*290*
8.6	Summary	*295*
8.7	References and Notes	*295*
8.8	Problems for Chapter 8	*296*

9 Nonlinear Systems 299

9.1	Historical Background	*299*
9.2	Motivation	*301*
9.3	Coarse-Grained Variables and Effective Hamiltonians	*302*
9.4	Nonlinear Coarse-Grained Equations of Motion	*306*
9.4.1	Generalization of Langevin Equation	*306*
9.4.2	Streaming Velocity	*307*
9.4.3	Damping Matrix	*310*
9.4.4	Generalized Fokker–Planck Equation	*311*

9.4.5	Nonlinear Langevin Equation *312*
9.5	Discussion of the Noise *314*
9.5.1	General Discussion *314*
9.5.2	Gaussian Noise *314*
9.5.3	Second Fluctuation-Dissipation Theorem *315*
9.6	Summary *316*
9.7	Examples of Nonlinear Models *317*
9.7.1	TDGL Models *317*
9.7.2	Isotropic Magnets *320*
9.7.3	Fluids *322*
9.8	Determination of Correlation Functions *326*
9.8.1	Formal Arrangements *326*
9.8.2	Linearized Theory *329*
9.8.3	Mode-Coupling Approximation *329*
9.8.4	Long-Time Tails in Fluids *330*
9.9	Mode Coupling and the Glass Transition *335*
9.10	Mode Coupling and Dynamic Critical Phenomena *336*
9.11	References and Notes *336*
9.12	Problems for Chapter 9 *338*

10 Perturbation Theory and the Dynamic Renormalization Group   *343*

10.1	Perturbation Theory *343*
10.1.1	TDGL Model *343*
10.1.2	Zeroth-Order Theory *344*
10.1.3	Bare Perturbation Theory *345*
10.1.4	Fluctuation-Dissipation Theorem *349*
10.1.5	Static Limit *352*
10.1.6	Temperature Renormalization *356*
10.1.7	Self-Consistent Hartree Approximation *360*
10.1.8	Dynamic Renormalization *361*
10.2	Perturbation Theory for the Isotropic Ferromagnet *369*
10.2.1	Equation of Motion *369*
10.2.2	Graphical Expansion *370*
10.2.3	Second Order in Perturbation Theory *374*
10.3	The Dynamic Renormalization Group *379*
10.3.1	Group Structure *379*
10.3.2	TDGL Case *380*
10.3.3	Scaling Results *388*
10.3.4	Wilson Matching *390*
10.3.5	Isotropic Ferromagnet *391*
10.4	Final Remarks *399*
10.5	References and Notes *399*

Contents

10.6 Problems for Chapter 10 *400*

11 Unstable Growth *403*
11.1 Introduction *403*
11.2 Langevin Equation Description *407*
11.3 Off-Critical Quenches *411*
11.4 Nucleation *413*
11.5 Observables of Interest in Phase-Ordering Systems *416*
11.6 Consequences of Sharp Interfaces *418*
11.7 Interfacial motion *420*
11.8 Scaling *423*
11.9 Theoretical Developments *425*
11.9.1 Linear Theory *425*
11.9.2 Mean-Field Theory *426*
11.9.3 Auxiliary Field Methods *428*
11.9.4 Auxiliary Field Dynamics *432*
11.9.5 The Order Parameter Correlation Function *435*
11.9.6 Extension to n-Vector Model *437*
11.10 Defect Dynamics *439*
11.11 Pattern Forming Systems *445*
11.12 References and Notes *446*
11.13 Problems for Chapter 11 *450*

Appendices

A Time-Reversal Symmetry *455*

B Fluid Poisson Bracket Relations *461*

C Equilibrium Average of the Phase-Space Density *463*

D Magnetic Poisson Bracket Relations *465*

E Noise and the Nonlinear Langevin Equation *467*

Index *471*

Preface

This is the third volume in a series of graduate level texts on statistical mechanics. Volume 1, *Equilibrium Statistical Mechanics* (ESM), is a first year graduate text treating the fundamentals of statistical mechanics. Volume 2, *Fluctuations, Order and Defects* (FOD), treats ordering, phase transitions, broken symmetry, long-range spatial correlations and topological defects. This includes the development of modern renormalization group methods for treating critical phenomena. The mathematical level of both texts is typically at the level of mean-field theory.

In this third volume, *Nonequilibrium Statistical Mechanics* (NESM), I treat nonequilibrium phenomena. The book is divided into three main sections. The first, Chapters 1–4, discusses the connection, via linear response theory, between experiment and theory in systems near equilibrium. Thus I develop the interpretation of scattering and transport experiments in terms of equilibrium-averaged time-correlation functions. The second part of the book, Chapters 5–8, develops the ideas of linear hydrodynamics and the generalized Langevin equation approach. This is also known as the memory-function method. The theory is applied, in detail, to spin diffusion and normal fluids. In these applications the Green–Kubo equations connecting transport coefficients and time integrals over current–current time-correlation functions are established. It is then demonstrated that this memory-function approach is very useful beyond the hydrodynamic regime. It is shown in Chapter 7 how these ideas can be used to develop modern kinetic theory. In Chapter 8 the generalized Langevin equation approach is used to develop the conventional theory of dynamic critical phenomena and linearized hydrodynamics in systems with broken continuous symmetry and traveling Nambu–Goldstone modes.

The third part of the book is devoted to nonlinear processes. In Chapter 9 the generalized Langevin approach is used to derive the generalized Fokker–Planck equation governing the dynamics of the reduced probability distribution for a set of slow variables. These dynamic equations lead to the nonlinear Langevin equations that serve as the basis for the theory of dynamic critical phenomena and the theory of the kinetics of first-order phase transitions.

Analytic methods of treatment of these nonlinear equations are discussed in Chapters 9–11. In particular the methods of Ma and Mazenko for carrying out the dynamic renormalization group are introduced in Chapter 10 in the important cases of the relaxational time-dependent Ginzburg–Landau (TDGL) model and the case of the isotropic ferromagnet. In Chapter 11 we discuss the strongly nonequilibrium behavior associated with phase-ordering systems.

This text is compatible with one of the central themes in FOD. In FOD we developed the idea of coarse-grained effective Hamiltonians governing the long-distance equilibrium correlations for a variety of systems: magnets, superfluids, superconductors, liquid crystals, etc. Here I indicate how one generates a coarse-grained dynamics consistent with these effective Hamiltonians.

The methods of attack on the nonlinear models discussed in Chapters 9 and 10 are very useful for treating systems at the lowest order in perturbation theory. There exist less physical and mathematically more powerful methods for handling higher order calculations. These methods will be discussed in the final volume of this series.

I thank my sister Debbie for crucial help and my wife Judy for her support.

Gene Mazenko
Chicago, July 2006

Abbreviations

COP	Conserved order parameter
ESM	Equilibrium Statistical Mechanics (Volume 1)
FOD	Fluctuations, Order and Defects (Volume 2)
FTMCM	Field Theory Methods in Condensed Matter Physics (Volume 4)
FDT	Fluctuation-dissipation theorem
GCE	Grand canonical ensemble
LGW	Landau–Ginzburg–Wilson
NCOP	Nonconserved order parameter
NG	Nambu–Goldstone
RG	Renormalization group
TDGL	Time-dependent Ginzburg–Landau

1
Systems Out of Equilibrium

1.1
Problems of Interest

The field of nonequilibrium statistical mechanics is wide and far-reaching. Using the broadest interpretation it includes the dynamics of all macroscopic systems. This definition is far too inclusive for our purposes here, and certainly beyond what is generally understood. Rather than discussing matters in abstract generality, let us introduce some examples of nonequilibrium phenomena of interest:

- A very familiar example of time-dependent phenomena is the propagation of sound through air from the speaker's mouth to the listener's ears. If the intensity of the sound is not too great, then the velocity of sound and its attenuation are properties of the medium propagating the sound. This is a very important point since it says that *sound* has significance independent of the mouth and ears generating and receiving it. A question of interest to us is: how can we relate the sound speed and attenuation to the microscopic properties of the air propagating the sound?

- Next, consider a thermally insulated bar of some homogeneous material at a temperature T_0. (See Fig. 1.1). We bring one end of this bar into contact at time t_0 with a heat bath at a temperature $T_1 > T_0$. For times $t >$

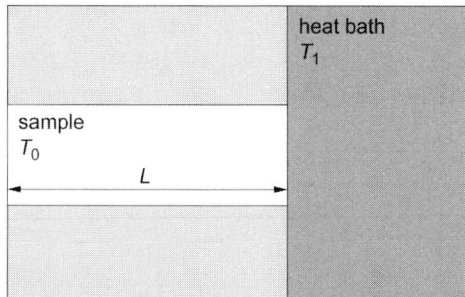

Fig. 1.1 Thermal conductivity experiment. See text for discussion.

Nonequilibrium Statistical Mechanics. Gene F. Mazenko
Copyright © 2006 WILEY-VCH Verlag GmbH & Co. KGaA, Weinheim
ISBN: 3-527-40648-4

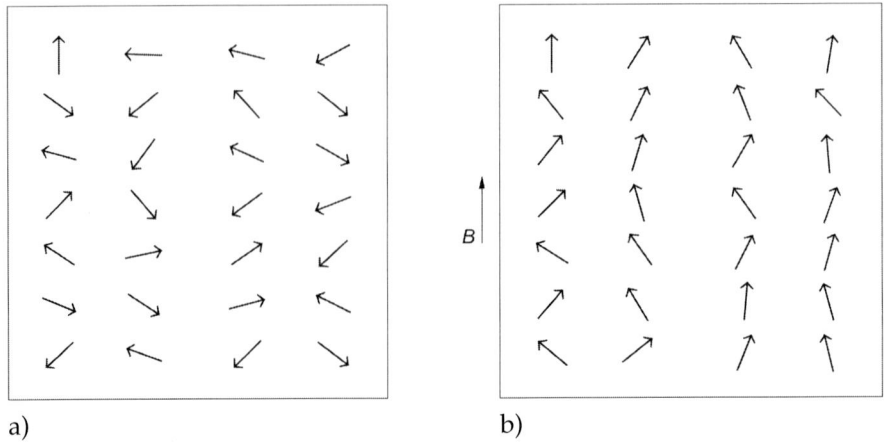

Fig. 1.2 Configurations for a set of paramagnetic spins. **a** Zero external magnetic field. **b** Subject to an external magnetic field along the direction shown.

t_0 heat will flow toward the cold end of the bar and eventually the bar will equilibrate at the new temperature T_1. We know from elementary courses in partial differential equations that this heat flow process is governed by Fourier's law [1], which tells us that the heat current **J** is proportional to the gradient of the temperature:

$$\mathbf{J} = -\lambda \vec{\nabla} T \tag{1}$$

with the thermal conductivity λ being the proportionality constant. Combining this constitutive relation with the continuity equation reflecting conservation of energy leads to a description (see Problem 1.1) in terms of the heat equation. The thermal conductivity is a property of the type of bar used in the experiment. A key question for us is: How does one determine the thermal conductivity for a material? From a theoretical point of view this requires a careful analysis establishing Fourier's law.

- A paramagnet is a magnetic system with no net magnetization in zero applied external field. In Fig. 1.2a we represent the paramagnet as a set of moments (or spins) $\vec{\mu}(\mathbf{R})$ localized on a periodic lattice at sites **R**. At high enough temperatures, in zero externally applied magnetic field, the system is in a disordered state where the average magnetization vanishes,

$$\langle \mathbf{M} \rangle = \left\langle \sum_{\mathbf{R}} \vec{\mu}(\mathbf{R}) \right\rangle = 0 , \tag{2}$$

due to symmetry. Each magnetic moment is equally likely to point in any direction. If one applies an external magnetic field **B** to a paramagnet the magnetic moments, on average, line up along the field:

$$\mathbf{M} \approx \mathbf{B} \ . \tag{3}$$

As shown in Fig. 1.2b, the spins deviate in detail from the *up* orientation along **B** because of thermal fluctuations (for nonzero temperatures). When we turn off **B** at time t_0, the spins relax to the original disordered equilibrium state, where $\langle \mathbf{M} \rangle = 0$, via thermal agitation. How can we quantitatively describe this relaxation process? As we discuss in Chapter 5, this process is analogous to heat diffusion.

- Suppose we fill a bowl with water and put it in a freezer. Clearly, over time, the water freezes. How do we describe the time evolution of this process? What does this process depend upon? In this case we have a dynamic process that connects thermodynamic states across a phase (liquid–solid) boundary.

The common elements in these situations is that we have externally disturbed the system by:

1. Mechanically pushing the air out of ones mouth;
2. Putting heat into a bar;
3. Turning off a magnetic field;
4. Drawing heat out of a system.

For the most part in this text we will focus on situations, such as examples 1–3, which can be understood in terms of the *intrinsic* dynamical properties of the condensed-matter system probed and do not depend in an essential way on *how* the system is probed. Such processes are part of a very important class of experiments that do not strongly disturb the thermodynamic state of the system. Thus, when one talks in a room one does not expect to change the temperature and pressure in the room. We expect the sound velocity and attenuation to depend on the well-defined thermodynamic state of the room.

In cases 1–3 we have applied an external force that has shifted the system from thermal equilibrium. If we remove the applied external force the system will return to the original equilibrium state. These intrinsic properties, which are connected to the return to equilibrium, turn out to be independent of the probe causing the nonequilibrium disturbance. Thus the speed of sound and its attenuation in air, the thermal conductivity of a bar and the paramagnetic relaxation rates are all properties of the underlying many-body systems.

We need to distinguish *weak*, linear or intrinsic response of a system from *strong* or nonlinear response of a system. Linear response, as we shall discuss

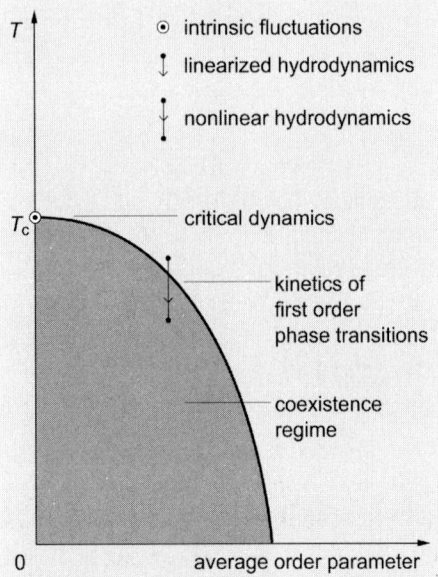

Fig. 1.3 Schematic of the type of dynamic processes studied in terms of movement on a generic phase diagram.

in detail, corresponds to those situations where a system remains near thermal equilibrium during the time-dependent process. In strongly nonlinear processes one applies strong forces to a system that fundamentally change the state of the system. The freezing of the bowl of water falls into this second category. Other nonlinear processes include:

- Nucleation where we rapidly flip the applied magnetic field such that a system becomes metastable. The system wants to follow but has a barrier to climb.

- Spinodal decomposition, where we quench the temperature of a fluid across a phase boundary into an unstable portion of the phase diagram.

- Material deposition where one builds [2] up a film layer by layer.

- Turbulence where we continuously drive a fluid by stirring.

In these examples an understanding of the dynamics depends critically on how, how hard and when we hit a system.

In organizing dynamical processes we can think of two classes of processes. The first set, which will be the primary concern in this text, are processes that connect points on the equilibrium phase diagram. The second set of processes involves driven systems that are sustained in intrinsically nonequilibrium states. The first set of processes can be roughly summarized as shown in Fig. 1.3 where five basic situations are shown:

1. intrinsic fluctuations in equilibrium;

2. linear response (perturbations that change the state of the system infinitesimally;

3. nonlinear hydrodynamics – substantial jumps in the phase diagram within a thermodynamic phase;

4. critical dynamics – dynamic processes near the critical point.

5. kinetics of first-order phase transitions – jumps across phase boundaries.

It is shown in Chapter 2 that processes in categories 1 and 2 are related by the fluctuation-dissipation theorem. Nonlinear hydrodynamics is developed and explored in Chapters 9, 10 and 11. Critical dynamics is treated in Chapters 8, 9, and 10. Finally the kinetics of first-order phase transitions is treated in Chapter 11.

In the second set of processes, like turbulence and interfacial growth [2], systems are maintained in states well out of equilibrium. In a Rayleigh–Benard experiment [3] (Fig. 1.4) where we maintain a temperature gradient across a sample, we can generate states with rolls, defects, chaos and turbulence, which are not associated with any equilibrium state. These more complicated sets of problems, such as driven steady-state nonequilibrium problems [4], will not be treated here.

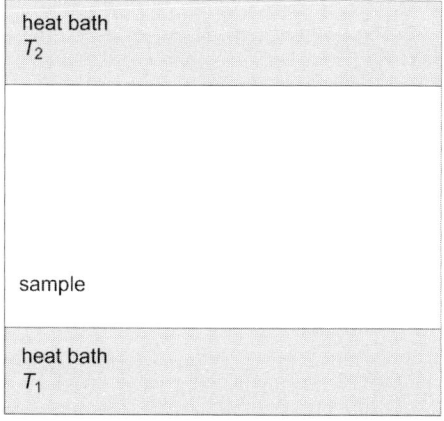

Fig. 1.4 Schematic of the Rayleigh–Benard experiment. A fluid sample is between two plates held at different temperatures $T_2 > T_1$. As the temperature difference increases a sequence of nonequilibrium behaviors occurs, including convection, rolls and turbulence.

1.2
Brownian Motion

1.2.1
Fluctuations in Equilibrium

Before we begin to look at the formal structure of the theory for systems evolving near equilibrium, it is useful to look at the historically important problem of Brownian motion [5]. It will turn out that many intuitive notions about the dynamics of large systems that evolve out of this analysis are supported by the full microscopic development. Indeed this discussion suggests a general approach to such problems.

Consider Fig. 1.5, showing the process of Brownian motion as taken from the work of Jean Perrin [6] near the turn of the previous century. Brownian motion corresponds to the irregular motion of *large* particles suspended in fluids. The general character of this motion was established by Robert Brown [7] in 1828. He showed that a wide variety of organic and inorganic particles showed the same type of behavior. The first quantitative theory of Brownian motion was due to Einstein [8] in 1905. Einstein understood that one needed an underlying atomic bath to provide the necessary fluctuations to account for the erratic motion of the large suspended particle. He realized that many random collisions, which produce no net effect on average, give rise to the observed *random walk* behavior [9].

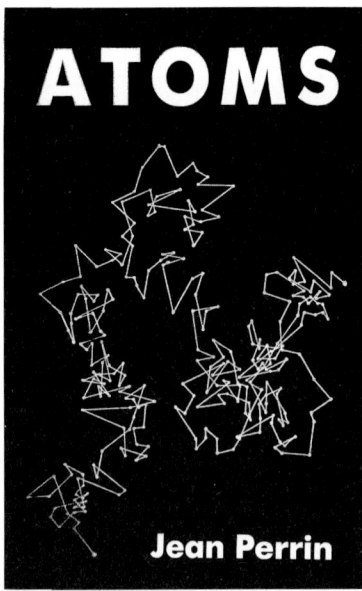

Fig. 1.5 Brownian motion path from cover of Ref. [6].

1.2 Brownian Motion

Let us consider a very large particle, with mass M and velocity $V(t)$ at time t, which is embedded in a fluid of relatively small particles. For simplicity let us work in one dimension. Assume the particle has velocity V_0 at time t_0. We are interested in the velocity of the particle for times $t > t_0$. In the simplest theory the basic assumption is that the force on the large particle can be decomposed into two parts. The first part is a frictional force F_1 opposing the persistent velocity V of the particle and is proportional to the velocity of the large particle:

$$F_1 = -M\gamma V \ , \tag{4}$$

where γ is the friction constant [10]. The second contribution to the force, representing the random buffeting the particle suffers from the small particles, is given by:

$$F_2 = M\eta \ , \tag{5}$$

where η is called the noise. Newton's law then takes the form:

$$M\dot{V} = -M\gamma V + M\eta \ . \tag{6}$$

This is in the form of the simplest *Langevin equation* [11]:

$$\dot{V} = -\gamma V + \eta \ . \tag{7}$$

Next we need to solve this equation. The first step is to write:

$$V(t) = e^{-\gamma t}\phi(t) \ . \tag{8}$$

Taking the time derivative of this equation gives:

$$\dot{V} = -\gamma V + e^{-\gamma t}\dot{\phi} \ . \tag{9}$$

Substituting this result back into the Langevin equation we obtain:

$$e^{-\gamma t}\dot{\phi}(t) = \eta(t) \ . \tag{10}$$

Clearly we can integrate this equation using the initial value for $V(t)$ to obtain:

$$\phi(t) = e^{\gamma t_0} V_0 + \int_{t_0}^{t} d\tau e^{\gamma \tau}\eta(\tau) \tag{11}$$

or in terms of the velocity:

$$V(t) = e^{-\gamma(t-t_0)} V_0 + \int_{t_0}^{t} d\tau e^{-\gamma(t-\tau)}\eta(\tau) \ . \tag{12}$$

The physical interpretation seems clear. The velocity of the particle loses *memory* of the initial value V_0 exponentially with time. $V(t)$ is determined by the sequence of bumps with the noise for $t \gg t_0$.

To go further we must make some simple assumptions about the properties of the noise. We will assume that the noise is a random variable described by its statistical properties. The first assumption is that the noise produces no net force:

$$\langle \eta(t_1) \rangle = 0 \ . \tag{13}$$

Next we need to specify the variance $\langle \eta(t_1)\eta(t_2) \rangle$. Physically we expect that the kicks due to the small particles will be of very short-time duration and noise at different times will be uncorrelated. Thus it is reasonable to assume that we have *white noise*:

$$\langle \eta(t_1)\eta(t_2) \rangle = A\delta(t_1 - t_2) \ , \tag{14}$$

where we will need to consider the proper choice for the value of the constant A. The other important consideration is causality. The velocity of the large particle can not depend on the noise at some later time:

$$\langle \eta(t_1) V(t_2) \rangle = 0 \quad \text{if } t_1 > t_2 \ . \tag{15}$$

We can now investigate the statistical properties of the velocity. The average velocity is given by:

$$\langle V(t) \rangle = e^{-\gamma(t-t_0)} \langle V_0 \rangle + \int_{t_0}^{t} d\tau\, e^{-\gamma(t-\tau)} \langle \eta(\tau) \rangle \ . \tag{16}$$

Since the average of the noise is zero, the average of the velocity is proportional to the average over the initial conditions. If the initial directions of the velocity of the pollen are randomly distributed (as in the case where the system – particle plus fluid – is in thermal equilibrium, then $\langle V_0 \rangle = 0$ and the average velocity is zero:

$$\langle V(t) \rangle = 0 \ . \tag{17}$$

Thus if the system is in equilibrium we expect no net motion for a collection of Brownian particles. If the pollen molecules are introduced with a net average velocity, the system will lose memory of this as time evolves.

We turn next to the velocity autocorrelation function defined by the average:

$$\psi(t, t') = \langle V(t) V(t') \rangle \ . \tag{18}$$

If we multiply the solution for $V(t)$, given by Eq. (12), by that for $V(t')$ and average we see that the cross terms vanish since:

$$\langle \eta(t) V_0 \rangle = 0 \quad \text{for } t > t_0 \tag{19}$$

and we have:

$$\psi(t,t') = \int_{t_0}^{t} d\tau \int_{t_0}^{t'} d\tau' e^{-\gamma(t-\tau)} e^{-\gamma(t'-\tau')} \langle \eta(\tau)\eta(\tau') \rangle$$
$$+ e^{-\gamma(t+t'-2t_0)} \langle V_0^2 \rangle \quad . \tag{20}$$

Using the statistical properties of the noise, Eq. (14), gives:

$$\psi(t,t') = \int_{t_0}^{t} d\tau \int_{t_0}^{t'} d\tau' e^{-\gamma(t-\tau)} e^{-\gamma(t'-\tau')} A\delta(\tau-\tau')$$
$$+ e^{-\gamma(t+t'-2t_0)} \psi(t_0,t_0) \quad . \tag{21}$$

It is left as a problem (Problem 1.2) to show that after performing the τ and τ' integrations one obtains the result:

$$\psi(t,t') = \frac{A}{2\gamma} e^{-\gamma|t-t'|} + \left[\psi(t_0,t_0) - \frac{A}{2\gamma}\right] e^{-\gamma(t+t'-2t_0)} \quad . \tag{22}$$

Notice that the initial condition is properly maintained.

Suppose the system is initially in equilibrium at temperature T_0. This allows us to determine the value of:

$$\psi(t_0,t_0) = \langle V_0^2 \rangle \quad . \tag{23}$$

This is because in equilibrium we can assume that the velocity of the particle satisfies Maxwell–Boltzmann statistics:

$$P[V_0] \approx e^{-\beta_0 \frac{M}{2} V_0^2} \quad , \tag{24}$$

where $\beta_0^{-1} = k_B T_0$, where k_B is the Boltzmann constant. One can then evaluate the average velocity squared as:

$$\langle V_0^2 \rangle = \frac{\int dV_0\, V_0^2 e^{-\beta_0 \frac{M}{2} V_0^2}}{\int dV_0\, e^{-\beta_0 \frac{M}{2} V_0^2}} \quad . \tag{25}$$

It is easy enough to evaluate these Gaussian integrals and obtain:

$$\langle V_0^2 \rangle = \frac{kT_0}{M} \quad , \tag{26}$$

which is just a form of the equipartition theorem:

$$\frac{M}{2} \langle V_0^2 \rangle = \frac{kT_0}{2} \quad . \tag{27}$$

Let us put this back into our expression for the velocity correlation function and concentrate on the case of equal times $t = t'$ where we have:

$$\psi(t,t) = \frac{A}{2\gamma} + \left[\frac{kT_0}{M} - \frac{A}{2\gamma}\right] e^{-\frac{2\gamma}{M}(t-t_0)} \quad . \tag{28}$$

If the system is in equilibrium, then there is nothing special about the time t_0. Unless we disturb the system from equilibrium it is in equilibrium at all times t and we expect $\psi(t,t)$ to be time independent and equal to $\frac{kT_0}{M}$. For this to be true we require:

$$\frac{kT_0}{M} - \frac{A}{2\gamma} = 0 , \tag{29}$$

which allows us to determine:

$$A = 2\gamma \frac{k_B T_0}{M} . \tag{30}$$

This means that we have determined that the autocorrelation for the noise is given by:

$$\langle \eta(t)\eta(t') \rangle = 2\frac{k_B T_0}{M}\gamma\delta(t-t') . \tag{31}$$

Thus the level of the noise increases with temperature as expected. Note also that the noise is related to the friction coefficient. In this particular problem, because we know that the velocity has a Gaussian (Maxwell–Boltzmann) distribution, we can infer (see Problem 1.9) that the noise must also have a Gaussian distribution.

Inserting this result for A into Eq. (22) for the velocity autocorrelation function we now obtain, for an arbitrary initial condition,

$$\psi(t,t') = \frac{kT_0}{M}e^{-\gamma|t-t'|} + \left[\psi(t_0,t_0) - \frac{kT_0}{M}\right]e^{-\gamma(t+t'-2t_0)} . \tag{32}$$

The assumption here is that the background fluid is at some temperature T_0 and we can insert a set of Brownian particles at some time t_0 with a velocity correlation $\psi(t_0, t_0)$ without disturbing the equilibrium of the fluid in any significant way. Then, as time evolves and t and t' become large, the system loses memory of the initial condition and:

$$\psi(t,t') = \frac{kT_0}{M}\chi_V(t-t') , \tag{33}$$

where the normalized equilibrium-averaged velocity autocorrelation function is given by:

$$\chi_V(t-t') = e^{-\gamma|t-t'|} . \tag{34}$$

This is interpreted as the velocity decorrelating with itself exponentially with time. There is little correlation between the velocity at time t and that at t' if the times are well separated. Note that our result is symmetric in $t \leftrightarrow t'$, as we require. Note also that it depends only on the time difference, which reflects the time-translational invariance of the system in equilibrium.

Since the velocity of the large particle is related to its position by:

$$V(t) = \frac{dx(t)}{dt} \, , \tag{35}$$

we can also investigate the root-mean-square displacement of the particle performing Brownian motion. Thus we need to integrate:

$$\frac{d}{dt}\frac{d}{dt'}\langle x(t)x(t')\rangle = \frac{kT_0}{M}e^{-\gamma|t-t'|} \, . \tag{36}$$

These integrations are tedious (see Problem 1.6) and lead to the final result:

$$\langle (x(t)-x(t_0))(x(t')-x(t_0))\rangle = \frac{kT_0}{M}\int_{t_0}^{t}d\tau \int_{t_0}^{t'}d\tau' e^{-\gamma|\tau-\tau'|} \tag{37}$$

$$= \frac{kT_0}{M\gamma}\left[t+t'-|t-t'|-2t_0+\frac{1}{\gamma}\left[e^{-\gamma(t-t_0)}+e^{-\gamma(t'-t_0)}-1-e^{-\gamma|t-t'|}\right]\right]. \tag{38}$$

This is of particular interest for equal times where:

$$\langle [x(t)-x(t_0)]^2\rangle = 2\frac{kT_0}{M\gamma}\left[t-t_0-\frac{1}{\gamma}\left(1-e^{-\gamma(t-t_0)}\right)\right] \, . \tag{39}$$

For long times we see that the averaged squared displacement is linear with time. If we have free particle or ballistic motion (see Problem 1.4), the displacement of the particle is linear in time. The random forcing of the noise causes the average displacement to go as the square root of time.

It is worth stopping to connect up this development to the behavior of density fluctuations for large particles moving in a fluid background. If $n(x,t)$ is the density of the Brownian particles, then because the number of Brownian particles is conserved, we have the continuity equation:

$$\frac{\partial n}{\partial t} = -\frac{\partial J}{\partial x} \, , \tag{40}$$

where J is the particle current. Since the Brownian particles share momentum with the background fluid, the current J is not, as in a simple fluid, itself conserved. Instead, for macroscopic processes, J satisfies Fick's law [12]:

$$J = -D\frac{\partial n}{\partial x} \, , \tag{41}$$

where D is the diffusion coefficient. Clearly Fick's law is similar to Fourier's law, but for particle transport rather than heat transport. We discuss such constitutive relations in detail in Chapter 5. Putting Eq. (41) back into Eq. (40)

one finds that on the longest length and time scales the density, $n(x, t)$, satisfies the diffusion equation:

$$\frac{\partial n}{\partial t} = D \frac{\partial^2 n}{\partial x^2} \ . \tag{42}$$

It is shown in Problem 1.7, for initial conditions where the density fluctuation is well localized in space, near $x(t_0)$, so that we can define:

$$\langle [x(t) - x(t_0)]^2 \rangle = \frac{1}{N} \int dx\, x^2 n(x,t) \tag{43}$$

and:

$$N = \int dx\, n(x,t) \ , \tag{44}$$

then:

$$\langle [x(t) - x(t_0)]^2 \rangle = 2Dt \tag{45}$$

for long times. Comparing with Eq. (39) we find that the diffusion constant is related to the friction coefficient by:

$$D = \frac{k_B T_0}{M\gamma} \ . \tag{46}$$

It was well known at the time of Einstein's work [13], starting from the equations of hydrodynamics, that the drag on a sphere of radius a in a flowing liquid with viscosity ν is given by the Stoke's law result:

$$M\gamma = 6\pi \nu a \ . \tag{47}$$

If we put this back into the equation for the diffusion coefficient we obtain the *Stokes–Einstein* relation [13]:

$$D = \frac{k_B T}{6\pi \nu a} \ . \tag{48}$$

If we know the viscosity and temperature of the liquid and measure the diffusion coefficient D through an observation of the Brownian motion, then we can determine a. If a is known, then this offers a method for determining Avogadro?s number: $N_A = R/k_B$ where R is the gas constant. Solving Eq. (48) for Boltzmann's constant and using Eq. (45) we find:

$$N_A = \frac{R}{k_B} = \frac{t}{\langle (\delta x)^2 \rangle} \frac{RT}{3\pi a \nu} \ . \tag{49}$$

Perrin found, for example for gamboge grains, that $a \approx 0.5\ \mu m$ and $N_A \approx 80 \times 10^{22}$.

1.2.2
Response to Applied Forces

Suppose now that we apply an external force, $F(t)$, to our particle. Clearly our equation of motion, Eq. (6), is then modified to read:

$$M\dot{V}(t) = -M\gamma V(t) + M\eta(t) + F(t) \ . \tag{50}$$

Now the average velocity of the particle is nonzero since the average of the external force is nonzero. Since the average over the noise is assumed to remain zero (suppose the background particles are neutral while the large particles are charged) we have, on averaging the equation of motion:

$$M\langle\dot{V}(t)\rangle = -M\gamma\langle V(t)\rangle + F(t) \ . \tag{51}$$

We assume that the force is weak, such that γ can be assumed to be independent of F. We can solve Eq. (51) again using an integrating factor, to obtain:

$$\langle V(t)\rangle = e^{-\gamma(t-t_0)}\langle V(t_0)\rangle + \frac{1}{M}\int_{t_0}^{t} d\tau\, e^{-\gamma(t-\tau)} F(\tau) \ . \tag{52}$$

For $t \gg t_0$ the average loses memory of the initial condition and:

$$M\langle V(t)\rangle = \int_{t_0}^{t} d\tau\, e^{-\gamma(t-\tau)} F(\tau) \ . \tag{53}$$

Notice that the response to the force can be written as a product of terms:

$$M\langle V(t)\rangle = \int_{t_0}^{t} d\tau\, \chi_V(t-\tau) F(\tau) \ . \tag{54}$$

It can be written as a product of an internal equilibrium response of the system times a term that tells how hard we are forcing the system.

The conclusions we can draw from this simple example have a surprisingly large range of validity.

- Friction coefficients like γ, which are intrinsic properties of the system, govern the evolution of almost all nonequilibrium systems near equilibrium.

- Thermal noise like η is essential to keep the system in thermal equilibrium. Indeed for a given γ we require:

$$\langle \eta(t)\eta(t')\rangle = 2\gamma\frac{k_B T_0}{M}\delta(t-t') \ . \tag{55}$$

So there is a connection between the friction coefficient and the statistics of the noise. Clearly the noise amplitude squared is proportional to the temperature.

- The response of the system to an external force can be written as a product of a part that depends on how the system is driven and a part that depends only on the fluctuations of the system in equilibrium.

One of the major unanswered questions in this formulation is: how do we determine γ? A strategy, which turns out to be general, is to relate the kinetic coefficient back to the velocity correlation function, which is microscopically defined. Notice that we have the integral:

$$\int_0^\infty dt\, \chi_V(t) = \int_0^\infty dt\, e^{-\gamma t} = \frac{1}{\gamma} = D\frac{M}{kT_0}. \quad (56)$$

This can be rewritten in the form:

$$D = \frac{kT_0}{M}\int_0^\infty dt\, \chi_V(t) = \int_0^\infty dt\, \psi(t,0). \quad (57)$$

Thus if we can evaluate the velocity–time correlation function we can determine D.

You may find it odd that we start with a discussion of such an apparently complex situation as pollen performing a random walk in a dense liquid. Historically, nonequilibrium statistical mechanics was built on the *Boltzmann paradigm* [14] where there are N spherical particles in an isolated enclosed box, allowed to evolve in time according to Newton's laws. Out of this dynamical process comes the mixing and irreversible behavior from which we can extract all of the dynamical properties of the system: viscosities, thermal conductivities and speeds of sound. This situation appears cleaner and more appealing to a physicist than the *Langevin paradigm* [15], where the system of interest is embedded in a *bath* of other particles. The appeal of the Boltzmann paradigm is somewhat illusory once one takes it seriously, since it leads to the difficult questions posed by ergodic theory [16] and whether certain isolated systems decay to equilibrium. We will assume that irreversibility is a physical reality. A system will remain [17] out of equilibrium only if we act to keep the system out of equilibrium. While the Langevin paradigm appears less universal, we shall see that this is also something of an illusion. In the Langevin description there is the unknown parameter γ. However, if we can connect this parameter back to the equilibrium fluctuations as in Eq. (56), then we have a complete picture. To do this we must develop a microscopic theory including the background fluid degrees of freedom to determine $\chi_V(t)$ as a function of time. Then we can extract $1/\gamma$ as an integral over a very short time period. The Langevin equation then controls the behavior on the longer time scales of particle diffusion.

The notion of a set of rapid degrees of freedom driving the evolution of slower degrees of freedom is a vital and robust idea. The separation of time scales in the case of Brownian motion comes about because of the larger mass of the pollen compared to the mass of the particles forming the background fluid. More generally, we have a separation of time scales for the fundamental reasons of conservation laws, Nambu–Goldstone modes associated with broken continuous symmetry and slowing down near certain phase transitions.

1.3
References and Notes

1 Jean Baptiste Joseph Fourier began his work on the theory of heat in 1804. He completed his seminal paper *On the Propagation of Heat in Solid Bodies* in 1807. It was read to the Paris Institute on 21 December 1807. For more information, see http://www-history.mcs.st-andrews.ac.uk/Mathematicians/Fourier.html

2 As an extreme example we have the process of diffusion-limited aggregation, which leads to a highly tenuous fractal system. See T. A. Witten and L. S. Sander, Phys. Rev. Lett. **47**, 1400 (1981).

3 See the review by M. C. Cross and P. C. Hohenberg, Rev. Mod. Phys. **65**, 851 (1993), where there is an effort to organize some of these strongly nonequilibrium situations.

4 See, for example, D. Ruelle, Phys. Today, May 2004, *Conversations on Nonequilibrium Physics with an Extraterrestrial*.

5 For a brief discussion of the history, see Chapter 1 in *Brownian Motion: Fluctuations, Dynamics and Applications* by R. M. Mazo, Clarendon, Oxford (2002).

6 Jean Perrin, *Atoms*, Ox Bow, Woodbridge, Conn. (1990). Perrin was awarded the Nobel prize in physics in 1926 *for his work on the discontinuous structure of matter*.

7 R. Brown, Philos. Mag. **4**, 161, (1828); Ann. Phys. Chem. **14**, 294 (1828).

8 A collection of Einstein's work on this topic is to be found in Albert Einstein, *Investigation on the Theory of the Brownian Movement*, Dover, N.Y. (1956).

9 The case of a random walk on a lattice is discussed in some detail in Section 7.2 in FOD.

10 It is more conventional to define the friction coefficient as $\gamma_p = M\gamma$.

11 P. Langevin, C. R. Acad. Sci. Paris **146**, 530 (1908).

12 A. Fick, Ann. Phys.(Leipzig) **94**, 59 (1855).

13 A. Einstein, Ann. Phys.(Leipzig) **17**, 59 (1905). For more background see E. Frey and K. Kloy, Ann. Phys. (Leipzig) **14**, 20 (2005) cond-mat/0502602 (2005).

14 See p. 13 in ESM, Sect. 1.5.1, p. 13.

15 See p. 14 in ESM, Sect. 1.5.1, p. 14.

16 A general overview is given in Appendix C of ESM.

17 This relaxation may take a very long time if there are degrees of freedom that get *trapped* and one has *quenched* impurities. Such behavior is very important in solid systems and leads to phenomena such as localization.

1.4
Problems for Chapter 1

Problem 1.1: If ε is the energy density and $\mathbf{J}$ the heat current, then the local expression of conservation of energy is given by the continuity equation:

$$\frac{\partial \varepsilon}{\partial t} = -\vec{\nabla} \cdot \mathbf{J} \ .$$

The heat current can be related to the local temperature by Fourier's law given by Eq. (1). The energy density density can be related to the local temperature if the system can be taken to be in local equilibrium:

$$\frac{\partial \varepsilon}{\partial t} = \frac{\partial \varepsilon}{\partial T} \frac{\partial T}{\partial t} \ ,$$

where the thermodynamic derivative is equal to the specific heat per unit volume:

$$c_V = \frac{\partial \varepsilon}{\partial T} \ .$$

Show that these steps lead to the heat equation for the local temperature:

$$\frac{\partial T}{\partial t} = D_T \nabla^2 T$$

and identify the thermal diffusivity D_T in terms of the other parameters in the problem.

Consider the problem discussed in the chapter where a plate (half-plane $z > 0$), initially at temperature T_0, is brought into contact with a constant heat source at temperature $T_1 > T_0$ along the plane $z = 0$. Solve the initial boundary value problem for the temperature and show that:

$$T(z) - T_0 = (T_1 - T_0) erfc \left[\frac{z}{2\sqrt{D_T t}}\right] \ .$$

Problem 1.2: Show that the double integral in Eq. (21) leads to the results given in Eq. (22).

Problem 1.3: Suppose, instead of white noise, one has colored noise satisfying the autocorrelation function,

$$\langle \eta(t)\eta(t') \rangle = \frac{2 k_B T}{M} \gamma_0 e^{-|t-t'|/\tau} \ .$$

Determine the equilibrium-averaged velocity autocorrelation function for this case. Focus on the case where $\gamma_0 \tau \ll 1$.

Problem 1.4: Starting with the expression for the displacement given by Eq. (39), investigate the behavior for short-times $t - t_0$. Does the result have a simple physical interpretation?

Problem 1.5: Determine the correlation between the velocity of a tagged particle at time t' and the displacement at time t: $\langle \delta x(t) V(t') \rangle$.

Problem 1.6: Show that carrying out the integrations in Eq. (37) leads to the results given in Eq. (38).

Problem 1.7: Suppose the density of tagged particles, $n(x,t)$, satisfies the diffusion equation:

$$\frac{\partial n}{\partial t} = D \frac{\partial^2 n}{\partial x^2} .$$

Suppose also that the initial conditions are such that the density profile is well localized in space and we can define the displacement squared:

$$\langle [x(t) - x(t_0)]^2 \rangle = \frac{1}{N} \int dx \, x^2 n(x,t) ,$$

where:

$$N = \int dx \, n(x,t) .$$

Show, under these conditions, that for long-times,

$$\langle [x(t) - x(t_0)]^2 \rangle = 2Dt$$

Problem 1.8: Extend Problem 1.7 to d dimensions.

Problem 1.9: We know that, in equilibrium, the velocity is governed by the Maxwell–Boltzmann distribution, which is a Gaussian distribution, and that one of the properties of a Gaussian distribution is that:

$$\langle V^4 \rangle = 3 \langle V^2 \rangle^2 . \tag{58}$$

On the other hand we know that, in equilibrium,

$$V(t) = \int_{t_0}^{t} d\tau \, e^{-\gamma(t-\tau)} \eta(\tau) .$$

Show that Eq. (58) is satisfied if the noise is gaussianly distributed and therefore satisfies:

$$\begin{aligned}
\langle \eta(\tau_1)\eta(\tau_2)\eta(\tau_3)\eta(\tau_4) \rangle &= \langle \eta(\tau_1)\eta(\tau_2) \rangle \langle \eta(\tau_3)\eta(\tau_4) \rangle \\
&+ \langle \eta(\tau_1)\eta(\tau_3) \rangle \langle \eta(\tau_2)\eta(\tau_4) \rangle \\
&+ \langle \eta(\tau_1)\eta(\tau_4) \rangle \langle \eta(\tau_3)\eta(\tau_2) \rangle .
\end{aligned}$$

2
Time-Dependent Phenomena in Condensed-Matter Systems: Relationship Between Theory and Experiment

2.1
Linear Response Theory

2.1.1
General Comments

We have all, from our everyday experience, an intuitive feel for speeds of sound, viscosity and thermal conductivity. Most of us do not have an intuitive sense for the microscopic origins of these quantities. Why does a certain fluid have a high viscosity? What is the temperature dependence of the viscosity? As we discussed in Chapter 1, properties like speeds of sound and transport coefficients are intrinsic to the systems of interest and independent of the weak external probe testing the system. This means that we can investigate such properties by disturbing the system in the most convenient manner. It seems reasonable that the most convenient probe is an infinitesimal external field that couples weakly to the degrees of freedom of an otherwise isolated physical system. In this case we can carry out a complete and theoretically well-defined analysis of the nonequilibrium situation. This development, known as *linear response theory*, has been fundamental in the establishment of a well-defined theoretical approach to *intrinsic* dynamic properties. This work was pioneered in the mid to late 1950s. Substantial credit for the development of linear response theory goes to Kubo [1] and Green [2]. We follow here an approach closer to that of Kadanoff and Martin [3,4] and Forster [5].

2.1.2
Linear Response Formalism

Let us begin with a rather general discussion that highlights the key assumptions in the development. We will return later to a number of specific examples. We consider a quantum-mechanical system, which, in the absence of externally applied fields, is governed by a Hamiltonian H. The equilibrium state for this system is given by the density matrix [6] or probability operator:

$$\rho_{eq} = \frac{e^{-\beta H}}{Z} , \qquad (1)$$

where:

$$Z = Tr\, e^{-\beta H} \qquad (2)$$

is the partition function, Tr is the quantum-mechanical trace over states, and $\beta = 1/k_B T$ where T is temperature and k_B is Boltzmann's constant. We will be interested in the averages of various *observables*. These observables can generally be represented as quantum operators $A(t)$ and the time dependence t is generated by the Heisenberg equations of motion:

$$-i\hbar \frac{\partial A(t)}{\partial t} = [H, A(t)] \qquad (3)$$

where $[A, B] = AB - BA$ is the commutator between the operators A and B. The equilibrium average of these fields is given by:

$$\langle A(t) \rangle_{eq} = Tr\, \rho_{eq} A(t) . \qquad (4)$$

As a specific example we can think of the $A(t)$ as the z-component of the total magnetization in a magnetic system:

$$M_z(t) = \sum_i \mu_z^i(t) \qquad (5)$$

where $\vec{\mu}_i$ is the magnetic moment of the ith molecule or atom in the system.

We next assume that we can apply to our system an external time-dependent field $h(t)$. We assume that the *total* Hamiltonian governing the system can be written in the form [7]:

$$H_T(t) = H(t) - B(t)h(t) . \qquad (6)$$

The external field couples [8] to the quantum operator $B(t)$, which may be related to $A(t)$. We assume that $h(t)$ vanishes for times earlier than some time t_0, and the system is in equilibrium for times prior to t_0 with the equilibrium density matrix ρ_{eq} given in terms of H by Eq. (1). In our magnetic example the external field is the externally applied magnetic field $\mathbf{H}(t)$ that couples to the magnetization via the usual Zeeman term. In a given physical system there is typically a collaborative effort between theory and experiment to identify the nature of the coupling as represented by the operator B and the external field h.

Let us work in the Heisenberg representation [9] where the density matrix is time independent and operators evolve according to the Heisenberg equations of motion:

$$-i\hbar \frac{\partial A(t)}{\partial t} = [H_T(t), A(t)] \ . \tag{7}$$

The operator $B(t)$, which appears in the coupling in Eq. (6), develops in time according to:

$$-i\hbar \frac{\partial B(t)}{\partial t} = [H_T(t), B(t)] \ . \tag{8}$$

Nonequilibrium averages can be written in the form:

$$\langle A(t) \rangle_{ne} = Tr \rho A(t) \ , \tag{9}$$

where the operator A, corresponding to an observable, carries the time dependence and the density matrix, ρ, is time independent. This is a particularly useful approach if we know that the system is initially in equilibrium. Thus if for $t \leq t_0$, $\rho = \rho_{eq}$, then, in the Heisenberg representation, we have for all subsequent times:

$$\rho = \rho_{eq} \tag{10}$$

and nonequilibrium averages are given by:

$$\langle A(t) \rangle_{ne} = Tr \, \rho_{eq} A(t) \ . \tag{11}$$

The result given by Eq. (11) is really quite simple to understand from the point of view of classical physics. Suppose we have a system of N particles in three dimensions. In this case it is clear that if we know the values of the $6N$ phase-space coordinates, $q(t) = \{q_1(t), q_2(t) \ldots q_{6N}(t)\}$, at some time $t = t_0$, $q(t_0) = \{q_1(t_0), q_2(t_0) \ldots q_{6N}(t_0)\}$, then the value can be determined at a later time as a solution of the initial-value problem. Thus we can write [10]:

$$A[q(t)] = A[q(0), t] \ . \tag{12}$$

The average of $A(t)$ requires knowing the probability distribution governing the initial configuration: $P[q(0)]$. Clearly the average of any function of the phase-space coordinates as a function of time t is given by:

$$\langle A(t) \rangle = \int dq(0) P[q(0)] A[q(0), t] \ . \tag{13}$$

But this equation is just the classical version of Eq. (11). For further discussion see Problem 2.1.

It is well known in standard quantum mechanics that one can work either in the Heisenberg representation, as assumed above, where the density matrix is time independent and operators evolve according to the Heisenberg equations of motion, or in the Schroedinger representation where the density matrix evolves in time and operators (observables) are time independent. We now want to discuss how we can go between these two representations, since it will indicate the best strategy for carrying out perturbation theory in the external field $h(t)$. The first step in the analysis is to define the time evolution operator $U(t,t')$ satisfying:

$$i\hbar \frac{\partial}{\partial t} U(t,t') = H_T^s(t) U(t,t') \tag{14}$$

with the boundary condition:

$$U(t,t) = 1 \quad, \tag{15}$$

and $H_T^s(t)$ is given by:

$$H_T^s(t) = H - Bh(t) \tag{16}$$

where the only time dependence is the explicit time dependence carried by $h(t)$.

We also define the inverse operator, U^{-1}, by:

$$U^{-1}(t,t') U(t,t') = U(t,t') U^{-1}(t,t') = 1 \quad. \tag{17}$$

Taking the derivative of Eq. (17) with respect to t we obtain:

$$i\hbar \left[\frac{\partial}{\partial t} U^{-1}(t,t') \right] U(t,t') + U^{-1}(t,t') i\hbar \frac{\partial}{\partial t} U(t,t') = 0 \quad. \tag{18}$$

Using the equation of motion given by Eq. (14) and applying U^{-1} from the right, we obtain:

$$i\hbar \frac{\partial}{\partial t} U^{-1}(t,t') = - U^{-1}(t,t') H_T^s(t) \quad. \tag{19}$$

Taking the hermitian adjoint of Eq. (14), using $(AB)^\dagger = B^\dagger A^\dagger$, and the fact that $H_T^s(t)$ is hermitian we find,

$$-i\hbar \frac{\partial}{\partial t} U^\dagger(t,t') = U^\dagger(t,t') H_T^s(t) \quad. \tag{20}$$

Comparing Eqs. (19) and (20) we see that U is unitary:

$$U^\dagger = U^{-1} \quad. \tag{21}$$

We introduce U since it can be used to formally integrate the Heisenberg equations of motion in the form:

$$A(t) = U^{-1}(t, t_0) A U(t, t_0) \,, \tag{22}$$

where $A = A(t_0)$. To prove this assertion we take the time derivative of Eq. (22):

$$\begin{aligned} i\hbar \frac{\partial A(t)}{\partial t} &= \left[i\hbar \frac{\partial}{\partial t} U^{-1}(t, t_0) \right] A U(t, t_0) + U^{-1}(t, t_0) A i\hbar \frac{\partial}{\partial t} U(t, t_0) \\ &= -U^{-1}(t, t_0) H_T^s(t) A U(t, t_0) + U^{-1}(t, t_0) A H_T^s(t) U(t, t_0) \\ &= -[H_T(t), A(t)] \,, \end{aligned}$$

where we have used Eqs. (14) and (19) and identified:

$$H_T(t) = U^{-1}(t, t_0) H_T^s(t) U(t, t_0) = H(t) - B(t) h(t) \,. \tag{23}$$

Clearly this equation of motion agrees with Eq. (6). Thus Eq. (22) gives a formal solution to Eq. (9) where the dependence on the external field is through $U(t, t_0)$.

It is at this stage in the development where we can understand how one goes back and forth from the Heisenberg to the Schroedinger representation in the case of an external time-dependent field. In the Heisenberg representation we have:

$$\langle A(t) \rangle_{\text{ne}} = \text{Tr} \, \rho_{\text{eq}} U^{-1}(t, t_0) A U(t, t_0) \,. \tag{24}$$

Using the cyclic invariance of the quantum-mechanical trace,

$$\text{Tr} AB = \text{Tr} BA, \tag{25}$$

for any two operators A and B, we can write:

$$\langle A(t) \rangle_{\text{ne}} = \text{Tr} \, U(t, t_0) \rho_{\text{eq}} U^{-1}(t, t_0) A \,. \tag{26}$$

This result corresponds to working in the Schroedinger representation where the density matrix carries the time dependence:

$$\rho(t) = U(t, t_0) \rho_{\text{eq}} U^{-1}(t, t_0) \,. \tag{27}$$

Notice that the density matrix satisfies the equation of motion:

$$i\hbar \frac{\partial}{\partial t} \rho(t) = [H_T(t), \rho(t)] \,, \tag{28}$$

which has the opposite sign from the equation of motion satisfied by operators corresponding to observables. If we work in the Schroedinger representation, then we focus on the time evolution of probability distributions. This is the conventional path to transport descriptions like the Boltzmann equation.

While we have a well-defined problem at this stage, the analysis for arbitrary $h(t)$ is extremely difficult. The problem is simplified considerably if we assume that $h(t)$ is infinitesimally small. The first step in our development is to obtain a more explicit expression for the dependence of $A(t)$ on $h(t)$. This involves organizing the problem in the *interaction representation*. As a step toward this development we need to look at the dynamics for the case where we turn off the external interaction.

When $h = 0$ the time evolution operator $U = U_0$ satisfies:

$$i\hbar \frac{\partial}{\partial t} U_0(t, t_0) = H\, U_0(t, t_0) \; , \tag{29}$$

which has the solution:

$$U_0(t, t_0) = e^{-iH(t-t_0)/\hbar} \; . \tag{30}$$

This is the usual unitary time-evolution operator for a time-independent Hamiltonian, and, using Eq. (22), we see that operators evolve in this case according to:

$$A(t) = e^{iH(t-t_0)/\hbar} A e^{-iH(t-t_0)/\hbar} \; . \tag{31}$$

Equilibrium averages, in the absence of an external driving force, are given by:

$$\langle A(t) \rangle_{h=0} = \frac{1}{Z} Tr\, e^{-\beta H} e^{iH(t-t_0)/\hbar} A e^{-iH(t-t_0)/\hbar} \; . \tag{32}$$

Since H commutes with itself,

$$e^{-\beta H} e^{iH(t-t_0)/\hbar} = e^{iH(t-t_0)/\hbar} e^{-\beta H} \; , \tag{33}$$

and using the cyclic invariance of the trace we have:

$$\langle A(t) \rangle_{h=0} = \frac{1}{Z} Tr\, e^{-iH(t-t_0)/\hbar} e^{iH(t-t_0)/\hbar} e^{-\beta H} A = \langle A \rangle_{eq} \; . \tag{34}$$

These averages in equilibrium are time independent [11].

Let us return to the case of nonzero $h(t)$ and define the operator $\hat{U}$ via:

$$U(t, t_0) \equiv U_0(t, t_0) \hat{U}(t, t_0) \; . \tag{35}$$

Taking the time derivative of Eq. (35) and using Eq. (14) we obtain:

$$i\hbar \frac{\partial}{\partial t} U(t, t_0) = \left[i\hbar \frac{\partial}{\partial t} U_0(t, t_0) \right] \hat{U}(t, t_0) + U_0(t, t_0) i\hbar \frac{\partial}{\partial t} \hat{U}(t, t_0)$$

$$\begin{aligned} &= HU_0(t,t_0)\hat{U}(t,t_0) + U_0(t,t_0)i\hbar\frac{\partial}{\partial t}\hat{U}(t,t_0)\\ &= [H + H_E^s(t)]U(t,t_0) \;, \end{aligned} \qquad (36)$$

where we define the external contribution to the Hamiltonian in the Schroedinger representation as:

$$H_E^s(t) = -Bh(t) \;. \qquad (37)$$

Canceling the $HU(t,t_0)$ terms in Eq. (36), we obtain:

$$\begin{aligned} U_0(t,t_0)i\hbar\frac{\partial}{\partial t}\hat{U}(t,t_0) &= H_E^s(t)U(t,t_0)\\ &= H_E^s(t)U_0(t,t_0)\hat{U}(t,t_0) \;. \end{aligned} \qquad (38)$$

Multiplying from the left by U_0^{-1} we obtain:

$$i\hbar\frac{\partial}{\partial t}\hat{U}(t,t_0) = U_0^{-1}(t,t_0)H_E^s(t)U_0(t,t_0)\hat{U}(t,t_0) \;. \qquad (39)$$

It is convenient to define:

$$H_E^I(t) = U_0^{-1}(t,t_0)H_E^s(t)U_0(t,t_0) = -B^I(t)h(t) \qquad (40)$$

where:

$$B^I(t) = U_0^{-1}(t,t_0)BU_0(t,t_0) \qquad (41)$$

is in the interaction representation. $\hat{U}$ then satisfies the differential equation:

$$i\hbar\frac{\partial}{\partial t}\hat{U}(t,t_0) = H_E^I(t)\hat{U}(t,t_0) \;. \qquad (42)$$

After integrating Eq. (42) from t_0 to t and using the boundary condition $\hat{U}(t_0,t_0) = 1$, we obtain the key integral equation:

$$\hat{U}(t,t_0) = 1 - \frac{i}{\hbar}\int_{t_0}^{t}dt'\, H_E^I(t')\hat{U}(t',t_0) \;. \qquad (43)$$

Then next step is to express the nonequilibrium average in terms of $\hat{U}$. We first note that:

$$U^{-1}(t,t_0) = U^\dagger(t,t_0) = [U_0(t,t_0)\hat{U}(t,t_0)]^\dagger = \hat{U}^\dagger(t,t_0)U_0^\dagger(t,t_0) \;, \qquad (44)$$

then Eq. (24) can be written:

$$\begin{aligned} \langle A(t)\rangle_{ne} &= \text{Tr}\,\rho_{eq}\hat{U}^\dagger(t,t_0)U_0^\dagger(t,t_0)AU_0(t,t_0)\hat{U}(t,t_0)\\ &= \text{Tr}\,\rho_{eq}\hat{U}^\dagger(t,t_0)A^I(t)\hat{U}(t,t_0) \;. \end{aligned} \qquad (45)$$

The entire dependence on H_E is now isolated in the operator $\hat{U}$. In general the best we can do (see Problem 2.2) for arbitrary H_E is to write the formal solution:

$$\hat{U}(t,t_0) = \mathcal{T}\, e^{\left[-\frac{i}{\hbar}\int_{t_0}^{t} dt'\, H_E^I(t')\right]} \tag{46}$$

to the integral equation for $\hat{U}$. In Eq. (46), $\mathcal{T}$ is the time-ordering operator that orders all operators according to time sequence, the latest time standing furthest to the left, for example,

$$\mathcal{T}[A(t_3)B(t_2)C(t_1)] = C(t_1)B(t_2)A(t_3) \tag{47}$$

when $t_1 > t_2 > t_3$.

It is at this point that the assumption of *weak* coupling or small H_E enters. We can iterate the integral equations for $\hat{U}$ to first order in H_E to obtain:

$$\hat{U}(t,t_0) = 1 - \frac{i}{\hbar}\int_{t_0}^{t} dt'\, H_E^I(t') + \ldots \tag{48}$$

$$\hat{U}^+(t,t_0) = 1 + \frac{i}{\hbar}\int_{t_0}^{t} dt'\, H_E^I(t') + \ldots \tag{49}$$

Using these results, we can write the nonequilibrium average as:

$$\begin{aligned}
\langle A(t)\rangle_{ne} &= \mathrm{Tr}\rho_{eq}\left[1 - \frac{1}{i\hbar}\int_{t_0}^{t} dt'\, H_E^I(t') + \cdots\right] \\
&\quad \times A^I(t)\left[1 + \frac{1}{i\hbar}\int_{t_0}^{t} dt'\, H_E^I(t') + \cdots\right] \\
&= \mathrm{Tr}\rho_{eq}[A^I(t) + \frac{1}{i\hbar}\int_{t_0}^{t} dt'[A^I(t)H_E^I(t') - H_E^I(t')A^I(t)] \\
&\quad + \mathcal{O}(H_E^{I\,2}) \\
&= \langle A^I(t)\rangle_{eq} + \frac{1}{i\hbar}\int_{t_0}^{t} dt'\langle [A^I(t), H_E^I(t')]\rangle_{eq} + \mathcal{O}(H_E^{I\,2}) \,.
\end{aligned} \tag{50}$$

Inserting the expression for H_E^I given by Eq. (40) we have

$$\langle A(t)\rangle_{ne} = \langle A_\alpha^I(t)\rangle_{eq} - \frac{1}{i\hbar}\int_{t_0}^{t} dt'\langle [A^I(t), B^I(t')]\rangle_{eq} h(t') + \mathcal{O}(h^2) \,. \tag{51}$$

Note that the averages are now over the equilibrium ensemble with a time dependence governed only by the *intrinsic* Hamiltonian H. We have already shown:

$$\langle A^I(t)\rangle_{eq} = \langle A\rangle_{eq} \,. \tag{52}$$

It will be convenient to drop the label I on the operators A and B in the second average with the understanding that they evolve with U_0.

We now draw out our main result. The deviation from equilibrium due to a very weak external field can be written as a product of two terms. One is the external field and the other is an *equilibrium*-averaged time-correlation function that is independent of the external field:

$$\delta A(t) = \langle A(t) \rangle_{ne} - \langle A(t) \rangle_{eq} = 2i \int_{t_0}^{t} dt' \chi''_{A,B}(t,t') h(t') + \mathcal{O}(h^2) , \qquad (53)$$

where we have defined the *response function*:

$$\chi''_{A,B}(t,t') = \frac{1}{2\hbar} \langle [A(t), B(t')] \rangle_{eq} . \qquad (54)$$

For a simple example see Problem 2.3. We can always choose $h(t')$ such that we can extend the lower limit of the time integration to $-\infty$ and we can define the *dynamic susceptibility*:

$$\chi_{AB}(t,t') = 2i\theta(t-t') \chi''_{AB}(t,t') \qquad (55)$$

so that:

$$\delta A(t) = \int_{-\infty}^{+\infty} dt' \chi_{AB}(t,t') h(t') . \qquad (56)$$

The step function θ occurs naturally in χ since the system is causal. We can not feel the effects of h until after it has acted. This result, first obtained in the early 1950s, is one of the most important in the area of nonequilibrium statistical mechanics. We shall now study this result in some detail in order to appreciate its significance. A key point is that we have reduced our nonequilibrium problem to that of *equilibrium* averaged time-correlation functions. It is of great advantage that we can use the high symmetry of the equilibrium state in analyzing χ'' and χ. As a first example, consider the consequences of time-translational invariance.

2.1.3
Time-Translational Invariance

Consider the equilibrium-averaged time-correlation function (we drop the subscript *eq* on the average when there is no confusion):

$$\langle A(t) B(t') \rangle = \text{Tr } \rho_{eq} A(t) B(t') , \qquad (57)$$

or, writing this out more fully

$$\langle A(t) B(t') \rangle = \text{Tr } \rho_{eq} U_0^+(t,t_0) A U_0(t,t_0) U_0^+(t',t_0) B U_0(t',t_0) . \qquad (58)$$

The factor between operators A and B can be written in a more convenient form:

$$\begin{aligned} U_0(t,t_0) U_0^+(t',t_0) &= e^{-iH(t-t_0)/\hbar} e^{+iH(t'-t_0)/\hbar} \\ &= e^{-iH(t-t')/\hbar} \\ &= U_0(t,t') \ . \end{aligned} \quad (59)$$

Then we can use the cyclic invariance of the trace to move $U_0(t',t_0)$ through ρ_{eq} (since ρ_{eq} commutes with U_0 and U_0^+),

$$U_0(t',t_0)\rho_{eq} = \rho_{eq} U_0(t',t_0) \ , \quad (60)$$

and then use the adjoint of Eq. (59) to obtain:

$$\begin{aligned} \langle A(t) B(t') \rangle &= \mathrm{Tr}\, \rho_{eq} U_0(t',t) A U_0(t,t') B \\ &= \mathrm{Tr}\, \rho_{eq} U_0^+(t,t') A U_0(t,t') B \\ &= \langle A(t-t') B \rangle \end{aligned} \quad (61)$$

Thus we see that time-translational invariance, which is equivalent to the statement that the time evolution operator $U(t,t')$ commutes with the density matrix, demands that equilibrium time-correlation functions depend only on time differences and not on any time origin (e.g., the choice of t_0). This result for $\langle A(t) B(t') \rangle$ immediately implies, after putting back the indices,

$$\chi''_{AB}(t,t') = \chi''_{AB}(t-t') \quad (62)$$

and:

$$\chi_{AB}(t,t') = \chi_{AB}(t-t') \ . \quad (63)$$

This means that our expression for $\delta A(t)$, as in the case of Brownian motion, is in the form of a convolution:

$$\delta A(t) = \int_{-\infty}^{+\infty} dt'\, \chi_{AB}(t-t') h(t') \ . \quad (64)$$

It is therefore natural to introduce Fourier transforms:

$$\delta A(\omega) = \int_{-\infty}^{+\infty} dt\, e^{i\omega t} \delta A(t) \ , \quad (65)$$

and the inverse:

$$h(t) = \int_{-\infty}^{+\infty} \frac{d\omega}{2\pi} e^{-i\omega t} h(\omega) \quad (66)$$

to obtain:

$$\delta A(\omega) = \chi_{AB}(\omega) h(\omega) \ . \quad (67)$$

This is our main result. The nonequilibrium response of a system to a weak external field $h(\omega)$ is proportional to the response function $\chi(\omega)$. Consequently the calculation of the nonequilibrium response of a system to a weak external field reduces to a purely equilibrium calculation. Calculation of quantities like $\chi(\omega)$ are not simple for practical systems, but they are well defined. We expect that information about transport properties (thermal conductivities, viscosities, etc) must be available from χ. Our eventual goal is to see how to extract this information.

2.1.4
Vector Operators

Thus far we have discussed the response of an observable A to an external perturbation that couples to the observable B. We need to generalize these results to a set of observables A_α where the α is a discrete index. The simplest example is where the observable is a vector $A_\alpha \rightarrow \mathbf{A}$. It is natural in this case to consider a vector coupling where $h \rightarrow h_\alpha$, which couples to the vector field B_α. The discrete set may also range over some set of variables like the number of particles, the momentum and the energy, as would occur in fluids. It is easy to see that our linear response results easily generalize to this case in the form:

$$\delta A_\alpha(t) = \sum_\beta \int_{-\infty}^{+\infty} dt' \chi_{A_\alpha B_\beta}(t - t') h_\beta(t') \tag{68}$$

$$\chi_{A_\alpha B_\beta}(t, t') = 2i\theta(t - t') \chi''_{A_\alpha B_\beta}(t, t') \tag{69}$$

$$\chi''_{A_\alpha, B_\beta}(t, t') = \frac{1}{2\hbar} \langle [A_\alpha(t), B_\beta(t')] \rangle_{eq} \ . \tag{70}$$

2.1.5
Example: The Electrical Conductivity

As a first explicit example of our linear response formalism, consider the case of a uniform, external, slowly varying electric field $\mathbf{E}(t)$ applied to a metal. This initial treatment is somewhat naïve; later we will give a more rigorous analysis [12]. The interaction energy for a set of N particles, with charge q_i and located at positions $\mathbf{r}_i$, is given by:

$$H_I = \sum_{i=1}^{N} q_i \phi(\mathbf{r}_i) \tag{71}$$

where $\phi(x)$ is the electrostatic scalar potential at position **x**. We can rewrite this interaction energy as:

$$H_I = \sum_i q_i \int d^d x \delta(\mathbf{x} - \mathbf{r}_i) \phi(\mathbf{r}_i, t) = \int d^d x \sum_i q_i \delta(\mathbf{x} - \mathbf{r}_i) \phi(\mathbf{x}, t)$$

$$= \int d^d x \rho(\mathbf{x}) \phi(\mathbf{x}, t) \tag{72}$$

where:

$$\rho(\mathbf{x}) = \sum_{i=1}^{N} q_i \delta(\mathbf{x} - \mathbf{r}_i) \tag{73}$$

is the charge density and we assume we are working in d spatial dimensions. The specific case of interest here, an applied a uniform electric field, can be written in a simpler form. In general, the electric field is related to the scalar potential by:

$$\mathbf{E}(\mathbf{x}, t) = -\vec{\nabla}_x \phi(\mathbf{x}, t) \ . \tag{74}$$

If **E** is a constant in space, then we can write:

$$\phi(\mathbf{x}, t) = -\mathbf{x} \cdot \mathbf{E}(t) \tag{75}$$

and the interaction energy take the form:

$$H_I = \int d^d x \sum_i q_i \delta(\mathbf{x} - \mathbf{r}_i)(-\mathbf{x} \cdot \mathbf{E}(t))$$

$$= -\sum_i q_i \mathbf{r}_i \cdot \mathbf{E}(t) = -\mathbf{R} \cdot \mathbf{E}(t) \tag{76}$$

where:

$$\mathbf{R} = \sum_i q_i \mathbf{r}_i \tag{77}$$

is the *center of charge* times the number of particles N. In terms of our original notation we identify $h(t) \to \mathbf{E}(t)$ and $B(t) \to \mathbf{R}$.

In this case we are interested in the current flow in the system when we apply the electric field. If $J_\alpha(t)$ is the charge-current density operator, then in the absence of an applied field, there is no net flow of charge:

$$\langle J_\alpha(t) \rangle_{eq} = 0 \ . \tag{78}$$

In the linear response regime where $\mathbf{E}(t)$ is small, we have immediately, using Eq. (68), that:

$$\langle J_\alpha(t) \rangle_{ne} = \sum_\beta \int_{-\infty}^{+\infty} dt' \, \chi_{J_\alpha, R_\beta}(t - t') E_\beta(t') \ . \tag{79}$$

Clearly this allows one to define a nonlocal electrical conductivity:

$$\langle J_\alpha(t)\rangle_{ne} = \sum_\beta \int_{-\infty}^{+\infty} dt'\, \sigma_{\alpha\beta}(t-t') E_\beta(t') \;, \tag{80}$$

where:

$$\sigma_{\alpha\beta}(t-t') = 2i\theta(t-t')\frac{1}{2\hbar}\langle [J_\alpha(t), R_\beta(t')]\rangle_{eq} \;. \tag{81}$$

Thus we obtain an explicit correlation function expression for the electrical conductivity. In terms of Fourier transforms, Eq. (79) reduces to:

$$\langle J_\alpha(\omega)\rangle_{ne} = \sum_\beta \sigma_{\alpha\beta}(\omega) E_\beta(\omega) \;. \tag{82}$$

These results can be used to determine σ explicitly for simple model systems.

If we return to our example of Brownian motion in Chapter 1 where the large particles have a charge q, then the average equation of motion in the presence of an applied electric field is given by:

$$M\frac{d}{dt}\langle V_\alpha(t)\rangle = -M\gamma \langle V_\alpha(t)\rangle + qE_\alpha(t) \tag{83}$$

where we have extended our previous results to d dimensions and included the vector index α on the average velocity and the applied field. If we Fourier transform over time using:

$$\int_{-\infty}^{\infty} dt\, e^{i\omega t}\frac{d}{dt}\langle V_\alpha(t)\rangle = -i\omega \langle V_\alpha(\omega)\rangle \;, \tag{84}$$

then the equation of motion becomes:

$$(-i\omega + \gamma)\langle V_\alpha(\omega)\rangle = \frac{q}{m} E_\alpha(\omega) \;. \tag{85}$$

The average charge-current density is given by:

$$\langle J_\alpha(t)\rangle_{ne} = nq\langle V_\alpha(t)\rangle_{ne} \;, \tag{86}$$

where $n = N/V$ is the charge carrier density in the system. Then, after Fourier transformation of Eq. (86) and using Eq. (85), we obtain the result for the nonequilibrium charge-current density:

$$\langle J_\alpha(\omega)\rangle_{ne} = \frac{nq^2}{(-i\omega + \gamma)M} E_\alpha(\omega) \;. \tag{87}$$

Comparing this result with the linear response result, we can identify the frequency-dependent conductivity:

$$\sigma_{\alpha\beta}(\omega) = \frac{nq^2}{(-i\omega + \gamma)M}\delta_{\alpha\beta} \;. \tag{88}$$

This is the Drude theory [13] result for the conductivity. Note that the conductivity tensor is diagonal. The conductivity is identified in this case with the low-frequency component:

$$\sigma = \lim_{\omega \to 0} \sigma(\omega) = \frac{nq^2}{M\gamma}. \tag{89}$$

Recognizing that the inverse friction constant can be interpreted as a relaxation time $\tau = \gamma^{-1}$, the Drude result can be written in the conventional form:

$$\sigma = \frac{nq^2 \tau}{M}. \tag{90}$$

This result can serve as a definition of the relaxation time τ. From measurements of σ we can determine τ. For example [14], for copper at a temperature of 77 K, $\tau = 21 \times 10^{-14}$ s, at 273 K, $\tau = 2.7 \times 10^{-14}$ s, and at 373 K, $\tau = 1.9 \times 10^{-14}$ s.

Notice that Eq. (89) serves as a microscopic definition of the conductivity. It expresses this transport coefficient in terms of the low-frequency limit, or time-integral, of the equilibrium-averaged time-correlation function given by Eq. (81). We discuss this result in more detail in Chapter 4.

In the case of a dielectric material where we apply an external electric field, we induce a net nonzero polarization field. The coefficient of proportionality between the applied field and the measured response is a frequency-dependent dielectric function. We also discuss this result in some detail in Chapter 4.

2.1.6
Example: Magnetic Resonance

There are a number of magnetic resonance experiments [nuclear magnetic resonance (NMR), electron spin resonance (ESR) etc.] that are interpreted using linear response theory. There are many texts devoted to this topic [15, 16]. As an introduction to this technique and to give some sense as to the physical content contained in the dynamic susceptibility, let us consider the case of a paramagnetic system aligned along a primary external magnetic field:

$$\mathbf{H}_0 = H_0 \hat{z}. \tag{91}$$

This clearly corresponds to a case of constrained equilibrium where the average magnetization aligns along the applied field:

$$\langle \mathbf{M} \rangle = \chi_0 \mathbf{H}_0 \equiv \mathbf{M}_0 \tag{92}$$

where χ_0 is the associated static susceptibility. Now, assume that we perturb this system with a weak, transverse time-dependent external field:

$$\mathbf{h}(t) = h_x(t)\hat{x} + h_y(t)\hat{y}. \tag{93}$$

2.1 Linear Response Theory

What are the responses of the various components of the magnetization to this *transverse* field? If we think of this problem in terms of linear response where H_0 is fixed and the transverse field is small, then the variation of the average magnetization with transverse field is given by:

$$\langle M_\alpha(\omega) \rangle = \sum_\beta \chi_{\alpha\beta}(\omega) h_\beta(\omega) \tag{94}$$

where:

$$\chi_{\alpha\beta}(\omega) = \int_{-\infty}^{\infty} d(t-t')\, e^{i\omega(t-t')} \chi_{\alpha\beta}(t,t') \tag{95}$$

and:

$$\chi_{\alpha\beta}(t,t') = 2i\theta(t-t') \frac{1}{2\hbar} \langle [M_\alpha(t), M_\beta(t')] \rangle_{eq}^{H_0} \tag{96}$$

where the equilibrium state used in the average is in the presence of H_0.

To go further in understanding the information contained in the linear response function in this case we need a model. We adopt the phenomenological model of Bloch [17], which captures the essence of the problem. Later, in Chapter 9, we will discuss a more careful derivation of equations of motion of this type. Here we rely on physical arguments. The Bloch equation of motion, satisfied by the average magnetization, consists of two contributions, one of microscopic origins and one of a more macroscopic source. The first reversible contribution to the equation of motion comes from a term in the microscopic Heisenberg equation of motion:

$$\left[\frac{d}{dt} M_\alpha \right]_{\text{reversible}} = \frac{i}{\hbar} [H_Z, M_\alpha] \ . \tag{97}$$

where H_Z is the Zeeman contribution to the Hamiltonian:

$$H_Z = -\mathbf{M} \cdot \mathbf{H}_T \tag{98}$$

and $\mathbf{H}_T(t) = \mathbf{H}_0 + \mathbf{h}(t)$ is the total external magnetic field. Evaluation of this contribution requires us to be a bit more specific. The total magnetization is given by $\mathbf{M} = \sum_i \vec{\mu}_i$ where $\vec{\mu}_i$ is the magnetic moment of the ith molecule in the system. If $\mathbf{J}_i$ is the total angular momentum for the ith molecule in the system, then it is proportional to the magnetic moment,

$$\vec{\mu}_i = \gamma \mathbf{J}_i \ , \tag{99}$$

where the proportionality constant γ is the gyromagnetic ratio. The total angular momenta satisfy the quantum-mechanical commutation relations:

$$[J_i^\alpha, J_j^\beta] = i\hbar \delta_{ij} \sum_\nu \varepsilon_{\alpha\beta\nu} J_i^\nu \ . \tag{100}$$

If we multiply by γ^2 and sum over i and j we obtain for the components of the total magnetization:

$$[M_\alpha, M_\beta] = i\hbar\gamma \sum_\nu \varepsilon_{\alpha\beta\nu} M_\nu \ . \tag{101}$$

We therefore have the contribution to the equation of motion for the total magnetization:

$$\frac{i}{\hbar}[H_Z, M_\alpha] = -\frac{i}{\hbar} \sum_\beta H_T^\beta [M_\beta, M_\alpha]$$

$$= -\frac{i}{\hbar} \sum_\beta H_T^\beta i\hbar\gamma \sum_\nu \varepsilon_{\beta\alpha\nu} M_\nu$$

$$= -\gamma \sum_{\beta\nu} \varepsilon_{\alpha\beta\nu} H_\beta M_\nu = -\gamma (\mathbf{H}_T \times \mathbf{M})_\alpha \ , \tag{102}$$

which is easily recognized as a term that leads to precession of the magnetic moments.

The second term in the equation of motion is a damping term, like the frictional term in the Brownian motion example, which tends to drive the magnetization to the equilibrium value M_α^0. Our equation of motion then takes the general form:

$$\frac{d}{dt} M_\alpha = -\gamma (\mathbf{H}(t) \times \mathbf{M})_\alpha - \Gamma_\alpha(H_0) \left(M_\alpha - M_\alpha^0\right) \ , \tag{103}$$

where the Γ_α are damping (or friction) coefficients. We should allow for the fact, in the presence of H_0, that damping associated with fluctuations along and perpendicular to the ordering may be different. Following the standard convention we define: $\Gamma_z = \frac{1}{T_1}$ and $\Gamma_x = \Gamma_y = \frac{1}{T_2}$ where T_1 and T_2 are functions of H_0. In the limit $H_0 \to 0$, $T_1 = T_2$. Notice that the external field $\mathbf{H}_0$ breaks any overall rotational symmetry in the problem. One can still, however, have rotational symmetry in the xy plane. The Bloch equation, Eq. (103), then takes the form:

$$\frac{d}{dt} M_z = -\gamma (\mathbf{H}(t) \times \mathbf{M})_z - \frac{1}{T_1}(M_z - M_0) \tag{104}$$

for the longitudinal component, while in the xy plane,

$$\frac{d}{dt} \mathbf{M}_T = -\gamma (\mathbf{H}(t) \times \mathbf{M})_T - \frac{1}{T_2} \mathbf{M}_T \tag{105}$$

where the subscript T means in the transverse plane. The applied field can then be written as the sum:

$$\mathbf{H}(t) = H_0 \hat{z} + \mathbf{h}(t) \ , \tag{106}$$

where the transverse field $\mathbf{h}(t)$ is small in magnitude compared with H_0. It is then easy to estimate that $M_z \approx \mathcal{O}(H_0)$ while the two transverse components of the magnetization are of order h. If we evaluate the cross product terms we obtain:

$$[\mathbf{H}(t) \times \mathbf{M}]_z = H_x M_y - H_y M_x , \tag{107}$$

which is of $\mathcal{O}(h^2)$. Similarly we obtain, to $\mathcal{O}(h)$,

$$[\mathbf{H}(t) \times \mathbf{M}]_x = H_y M_z - H_z M_y = h_y M_z - H_0 M_y \tag{108}$$

and:

$$[\mathbf{H}(t) \times \mathbf{M}]_y = H_z M_x - H_x M_z = H_0 M_x - h_x M_z . \tag{109}$$

The Bloch equations, keeping terms of $\mathcal{O}(h)$, then reduce to:

$$\frac{d}{dt} M_z = -\frac{1}{T_1} (M_z - M_0) \tag{110}$$

$$\left[\frac{d}{dt} + \frac{1}{T_2}\right] M_x = \gamma H_0 M_y - \gamma M_z h_y(t) \tag{111}$$

$$\left[\frac{d}{dt} + \frac{1}{T_2}\right] M_y = -\gamma H_0 M_x + \gamma M_z h_x(t) . \tag{112}$$

If we look at these equations in a steady-state situation where the fields were turned on in the distant past (see Problem 2.4), then we see that the consistent solution for the longitudinal magnetization is that it be a constant:

$$M_z = M_0 . \tag{113}$$

Thus all of the *action* at linear order in $h(t)$ is in the transverse directions. Using Eq. (113) we find that the transverse equations of motion then take the form:

$$\left[\frac{d}{dt} + \frac{1}{T_2}\right] M_x = \gamma H_0 M_y - \gamma M_0 h_y(t) \tag{114}$$

$$\left[\frac{d}{dt} + \frac{1}{T_2}\right] M_y = -\gamma H_0 M_x + \gamma M_0 h_x(t) . \tag{115}$$

Fourier transforming these equations in time and defining the cyclotron frequency $\omega_0 = \gamma H_0$, we obtain the coupled set of linear equations:

$$\left(-i\omega + \frac{1}{T_2}\right) M_x(\omega) - \omega_0 M_y(\omega) = -\gamma M_0 h_y(\omega) \tag{116}$$

$$\omega_0 M_x(\omega) + \left(-i\omega + \frac{1}{T_2}\right) M_y(\omega) = \gamma M_0 h_x(\omega) \ . \tag{117}$$

It is easy enough to invert (see Problem 2.5) this 2×2 matrix equation to obtain:

$$M_\alpha(\omega) = \sum_\beta \chi_{\alpha\beta}(\omega) h_\beta(\omega) \tag{118}$$

and we can easily read off the results:

$$\chi_{xx}(\omega) = \chi_{yy}(\omega) = \frac{\omega_0 \gamma M_0}{\omega_0^2 + (-i\omega + T_2^{-1})^2} \tag{119}$$

and:

$$\chi_{yx}(\omega) = -\chi_{xy}(\omega) = \frac{\gamma M_0(-i\omega + T_2^{-1})}{\omega_0^2 + (-i\omega + T_2^{-1})^2} \ . \tag{120}$$

Let us focus on $\chi_{xx}(\omega)$, which we can write in the form:

$$\chi_{xx}(\omega) = \frac{\omega_0 \gamma M_0}{[\omega_0 + \omega + iT_2^{-1})][\omega_0 - \omega - iT_2^{-1})]} \ . \tag{121}$$

Now suppose our probing transverse field is harmonic:

$$\mathbf{h}(t) = h \cos(\omega_p t) \hat{x} \tag{122}$$

and we probe at frequency ω_p. The Fourier transform of the applied field is:

$$h(\omega) = 2\pi \frac{h}{2} \left[\delta(\omega - \omega_p) + \delta(\omega + \omega_p)\right] \tag{123}$$

and the measured x-component of the magnetization in the time domain is given by:

$$\begin{aligned} M_x(t) &= \int \frac{d\omega}{2\pi} e^{i\omega t} \chi_{xx}(\omega) 2\pi \frac{h}{2} \left[\delta(\omega - \omega_p) + \delta(\omega + \omega_p)\right] \\ &= \frac{h}{2} \left[e^{i\omega_p t} \chi_{xx}(\omega_p) + e^{-i\omega_p t} \chi_{xx}(-\omega_p)\right] \\ &= \frac{h}{2} \Big\{ \cos(\omega_p t) [\chi_{xx}(\omega_p) + \chi_{xx}(-\omega_p)] \\ &\quad + i \sin(\omega_p t) [\chi_{xx}(\omega_p) - \chi_{xx}(-\omega_p)] \Big\} \ . \end{aligned} \tag{124}$$

Since we see from inspection of Eq. (119) that:

$$\chi_{xx}(-\omega) = \chi_{xx}^*(\omega) \ , \tag{125}$$

we have:

$$\chi_{xx}(\omega) + \chi_{xx}(-\omega) = 2\chi'_{xx}(\omega) \tag{126}$$

$$\chi_{xx}(\omega) - \chi_{xx}(-\omega) = 2i\chi''_{xx}(\omega) \ . \tag{127}$$

Inserting these results back into the expression for the x-component of the magnetization gives:

$$M_x(t) = h\left[\cos(\omega_p t)\chi'_{xx}(\omega_p) - \sin(\omega_p t)\chi''_{xx}(\omega_p)\right] \ . \tag{128}$$

The real and imaginary parts of χ_{xx} can be easily extracted if we write Eq. (121) in the form:

$$\chi_{xx}(\omega) = \omega_0 \gamma M_0 \frac{[\omega_0 + \omega - iT_2^{-1}][\omega_0 - \omega + iT_2^{-1}]}{[(\omega_0 + \omega)^2 + T_2^{-2})][(\omega_0 - \omega)^2 + T_2^{-2})]} \tag{129}$$

with the results:

$$\chi'_{xx}(\omega) = \omega_0 \gamma M_0 \frac{[\omega_0^2 - \omega^2 + T_2^{-2}]}{[(\omega_0 + \omega)^2 + T_2^{-2}][(\omega_0 - \omega)^2 + T_2^{-2}]} \tag{130}$$

and:

$$\chi''_{xx}(\omega) = \omega_0 \gamma M_0 \frac{2\omega T_2^{-1}}{[(\omega_0 + \omega)^2 + T_2^{-2}][(\omega_0 - \omega)^2 + T_2^{-2}]} \ . \tag{131}$$

Then, as one sweeps ω, one finds a large response (resonance) as one passes through $\omega = \pm\omega_0$. The smaller the damping, T_2^{-1}, the sharper the resonance. Clearly this method can be used to measure the resonance frequency, ω_0, and the damping T_2^{-1}.

2.1.7
Example: Relaxation From Constrained Equilibrium

An important set of experiments, particularly from a conceptual point of view, involve those where we start with a system in equilibrium at $t = -\infty$, adiabatically turn on an external field, which brings the system into a constrained equilibrium state, and then, at $t = 0$, turn off the field. This is the situation described earlier for a paramagnet. As a prelude to this case let us first consider a slightly simpler situation.

Suppose we adiabatically turn on an external field, h_B, that couples to observable B and then maintain this time-independent field for times greater than some time t_1:

$$h_B(t) = \left[e^{\eta(t-t_1)}\theta(t_1 - t) + \theta(t - t_1)\right]h_B \ . \tag{132}$$

The adiabatic requirement implies that η is very small. We have, then, combining Eq. (53) with Eq. (132), that the linear response to this field is:

$$\delta A(t) = 2i \int_{-\infty}^{t} dt' \chi''_{A,B}(t-t') h_B [e^{\eta(t'-t_1)} \theta(t_1-t') + \theta(t'-t_1)] \quad (133)$$

It is shown in Problem 2.13 that after introducing the Fourier transform of $\chi''_{AB}(t-t')$, the time integral can be carried out and in the adiabatic limit one obtains:

$$\delta A(t) = \chi_{A,B} h_B , \quad (134)$$

where the *static susceptibility* is given by:

$$\chi_{A,B} = \int \frac{d\omega}{\pi} \frac{\chi''_{A,B}(\omega)}{\omega} . \quad (135)$$

Note that $\delta A(t)$ is time independent for $t > t_1$. If we allow $t_1 \to -\infty$ then this gives the response of the system to a time-independent external field.

As a consistency check this result should be the same as one would obtain from the equilibrium calculation of $\langle A \rangle$ where one has the density matrix:

$$\rho_{eq} = e^{-\beta(H+H_E)} / Tr e^{-\beta(H+H_E)} , \quad (136)$$

with $H_E = -Bh_B$, which is treated as a perturbation on H. In Problem 2.6 you are asked to show Eq. (134) holds directly. We will return to this later.

Note, since h is infinitesimal, we have from Eq. (134) the result:

$$\frac{\delta A}{\delta h_B} = \chi_{AB} . \quad (137)$$

This relation helps us make the connection with thermodynamic derivatives.

Let us now return to the dynamic situation where we create a disturbance by turning off an adiabatic field at time $t = t_1 = 0$. Our external field is given by:

$$h_B(t) = e^{\eta t} h_B \theta(-t) . \quad (138)$$

The linear response is given in this case by Eq. (53) with $h(t')$ replaced by $h_B(t')$ given by Eq. (138):

$$\delta A(t) = \int_{-\infty}^{0} dt' \, 2i\theta(t-t') \chi''_{A,B}(t-t') e^{\eta t'} h_B . \quad (139)$$

For $t \geq 0$ the step function is equal to 1 and:

$$\delta A(t) = h_B \int \frac{d\omega}{2\pi} 2i\chi''_{A,B}(\omega) \int_{-\infty}^{0} dt' \, e^{(\eta+i\omega)t'} e^{-i\omega t}$$

$$= h_B \int \frac{d\omega}{\pi} \chi''_{A,B}(\omega) i e^{-i\omega t} \frac{1}{\eta + i\omega}$$
$$= \int \frac{d\omega}{\pi} \frac{\chi''_{A,B}(\omega) e^{-i\omega t}}{\omega - i\eta} h_B . \tag{140}$$

We established previously the static result:

$$\delta A(t=0) = \chi_{A,B} h_B . \tag{141}$$

We can invert this result to obtain:

$$h_B = [\chi_{A,B}]^{-1} \delta A(t=0) , \tag{142}$$

which gives an expression for the *forces* driving the system in terms of a static susceptibility and the initial constrained equilibrium values. We can then write Eq. (140) in the form:

$$\delta A(t) = R_{AB}(t) \delta A(t=0) , \tag{143}$$

where the relaxational function:

$$R_{AB}(t) = \int \frac{d\omega}{\pi} \frac{\chi''_{A,B}(\omega)}{\omega - i\eta} e^{-i\omega t} \chi_{A,B}^{-1} \tag{144}$$

is independent of the applied forces. Note that we have reduced the problem to an *initial-value problem*.

If the observable A and the field h carries a discrete index then we end up with matrix equations for the initial value of the observable:

$$\delta A_\alpha(t=0) = \sum_\beta \chi_{A_\alpha, B_\beta} h_{B_\beta}. \tag{145}$$

Evaluation of h_{B_β} in terms of the initial value requires the construction of the matrix inverse:

$$\sum_\mu [\chi_{A,B}]^{-1}_{\alpha\mu} \chi_{A_\mu, B_\beta} = \delta_{\alpha\beta} . \tag{146}$$

Then we have:

$$h_\alpha = \sum_\mu [\chi_{A,B}]^{-1}_{\alpha\mu} \delta A_\mu(t=0) . \tag{147}$$

Again we can use this result to write the relaxation in terms of a initial value problem:

$$\delta A_\alpha(t) = \sum_\beta R_{A_\alpha B_\beta}(t) \delta A_\beta(t=0) \tag{148}$$

where the relaxational function:

$$R_{A_\alpha B_\beta}(t) = \sum_\gamma \int \frac{d\omega}{\pi} \frac{\chi''_{A_\alpha, B_\gamma}(\omega)}{\omega - i\eta} e^{-i\omega t} [\chi_{A,B}]^{-1}_{\gamma\beta} \quad (149)$$

requires the construction of the matrix inverse $[\chi_{A,B}]^{-1}_{\gamma\beta}$.

2.1.8
Field Operators

Suppose we further extend our analysis to include observable fields $A \to A_\alpha(\mathbf{x})$ and couplings $h \to h_\alpha(\mathbf{x})$, which depend on a position variable $\mathbf{x}$. As a specific example we can think of the $A_\alpha(\mathbf{x}, t)$ as the components of the magnetization density in a magnetic system:

$$\mathbf{M}(\mathbf{x},t) = \sum_i \tfrac{1}{2}[\vec{\mu}_i(t)\delta(\mathbf{x}-\mathbf{r}_i(t)) + \delta(\mathbf{x}-\mathbf{r}_i(t))\vec{\mu}_i(t)] \quad (150)$$

where $\vec{\mu}_i$ is the ith magnetic moment in the system located at position $\mathbf{r}_i$. In this case one can couple, in d dimensions, to the magnetization through the introduction of an inhomogeneous external magnetic field:

$$H^s_E(t) = -\int d^d x \, \mathbf{M}(\mathbf{x}) \cdot \mathbf{H}(\mathbf{x}, t) \quad . \quad (151)$$

In the case of an inhomogeneous scalar potential $\phi(\mathbf{x},t)$ applied to a charged system we found previously:

$$H^s_E(t) = \int d^d x \, \rho(\mathbf{x})\phi(\mathbf{x},t) \quad (152)$$

where $\rho(\mathbf{x})$ is the charge density.

It is then easy to extend our linear response analysis to give the rather general results:

$$\delta A_\alpha(\mathbf{x},t) = \int d^d x' \sum_\beta \int_{-\infty}^{+\infty} dt' \chi_{A_\alpha B_\beta}(\mathbf{x}, \mathbf{x}', t - t') h_\beta(\mathbf{x}', t') \quad , \quad (153)$$

$$\chi_{A_\alpha B_\beta}(\mathbf{x}, \mathbf{x}', t - t') = 2i\theta(t - t')\chi''_{A_\alpha B_\beta}(\mathbf{x}, \mathbf{x}', t - t') \quad (154)$$

and:

$$\chi''_{A_\alpha, B_\beta}(\mathbf{x}, \mathbf{x}'; t - t') = \frac{1}{2\hbar}\langle [A_\alpha(\mathbf{x},t), B_\beta(\mathbf{x}', t')]\rangle_{eq} \quad . \quad (155)$$

If we also have translational invariance in space :

$$\chi_{A_\alpha B_\beta}(\mathbf{x}, \mathbf{x}', t - t') = \chi_{A_\alpha B_\beta}(\mathbf{x} - \mathbf{x}', t - t') \quad , \quad (156)$$

then we can Fourier transform over space,

$$h_B(\mathbf{x}, \omega) = \int \frac{d^d k}{(2\pi)^d} e^{+i\mathbf{k}\cdot\mathbf{x}} h_B(\mathbf{k}, \omega) \ , \tag{157}$$

to obtain the linear response result:

$$\delta A_\alpha(\mathbf{k}, \omega) = \sum_\beta \chi_{A_\alpha B_\beta}(\mathbf{k}, \omega) h_B(\mathbf{k}, \omega) + \mathcal{O}(h^2) \ . \tag{158}$$

2.1.9
Identification of Couplings

There are important situations where the response to a perturbation is not obviously expressible in terms of an interaction Hamiltonian $H_E(t)$. Examples include gradients in temperature and in chemical potential across a sample that induce heat and mass flow. We can gain some feeling [18] for the situation by considering a fluid where we set up a small temperature gradient (or inhomogeneity) in the system holding the system at constant pressure. We will *assume* there exists an adiabatic external driving force $h_B(\mathbf{x})$ that couples linearly to an observable field $B(\mathbf{x})$ to produce a temperature gradient. We assume that the external interaction can be written in the conventional form:

$$H_E(t) = -\int d^d x \, B(\mathbf{x}) h_B(\mathbf{x}) [e^{\eta t} \theta(-t) + \theta(t)] \ . \tag{159}$$

We show in Chapter 6 that if A is a conserved density (it need not be conserved in the classical limit) and we work in the grand canonical ensemble (GCE), then the change in the average of A due to changes in temperature and pressure are given by:

$$\delta \bar{A} = \chi_{n\bar{A}} \frac{\delta p}{\bar{n}} + \chi_{qA} \frac{\delta T}{T} \ , \tag{160}$$

where the χ's are the static susceptibilities and q the heat density defined by:

$$q(\mathbf{x}) = \varepsilon(\mathbf{x}) - \frac{(\bar{\varepsilon} + \bar{p})}{\bar{n}} n(\mathbf{x}) \ , \tag{161}$$

where $\varepsilon(\mathbf{x})$ is the energy density, $n(x)$ the particle density and $\bar{\varepsilon}$ and $\bar{n}$ their equilibrium averages. From Eq. (160) we can read-off that the external coupling generating thermal gradients is given by:

$$h_q(\mathbf{x}) = \frac{\delta T(\mathbf{x})}{T} \tag{162}$$

and it couples to the heat density:

$$B(\mathbf{x}) = q(\mathbf{x}) \ . \tag{163}$$

The *external* coupling generating thermal gradients can then be written as:

$$H_E(t) = -\int d^d x \, q(\mathbf{x}) \frac{\delta T(\mathbf{x})}{T} \left[e^{\eta t} \theta(-t) + \theta(t) \right] . \tag{164}$$

Similarly pressure variations, at fixed temperature, are produced by:

$$H_E(t) = -\int d^d x \, n(\mathbf{x}) \frac{\delta p(\mathbf{x})}{n} \left[e^{\eta t} \theta(-t) + \theta(t) \right] , \tag{165}$$

and one couples directly to the density. In Eq. (161), $\varepsilon(\mathbf{x})$ and $n(\mathbf{x})$ are operators that can be defined without reference to the thermodynamic state of the system. This is not true for the heat density, which depends on the equilibrium-averaged quantities, $(\bar{\varepsilon} + \bar{p})/\bar{n}$.

Let us now turn to a discussion of of scattering experiments in a many-particle system context. This discussion will eventually lead back to $\chi''_{A,B}(\mathbf{k}, \omega)$.

2.2
Scattering Experiments

2.2.1
Inelastic Neutron Scattering from a Fluid

Let us consider a prototype of a wide class of modern experimental techniques: neutron scattering. Neutron scattering from a condensed matter system is only one of a large number of scattering experiments where we probe a system by sending in a small projectile and then look at the end products of the scattering. We can also use electrons, positrons, helium atoms, visible light or x-rays. At present, we shall focus on neutron scattering since the kinematics are the simplest in this case.

Consider the experimental situation where we allow a beam of thermal neutrons to scatter from some macroscopic sample (see Fig. 2.1). We assume

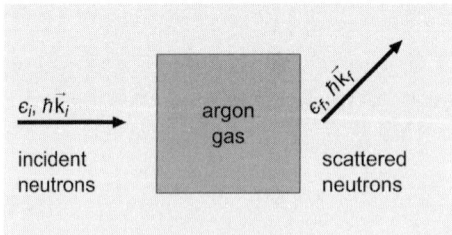

Fig. 2.1 Schematic of the kinematics of neutron scattering. The incident neutron has momentum $\hbar \mathbf{k}_i$ and energy ε_i. The scattered neutron has momentum $\hbar \mathbf{k}_f$ and energy ε_f.

2.2 Scattering Experiments

that the beam of neutrons is incident with energy ε_i and momentum $\hbar k_i$. We mean by thermal neutrons that ε_i is on the order of thermal ($k_B T$) energies. Let us assume for simplicity that the sample is a container of argon gas. Argon, a noble gas, is one of the simplest fluids. The neutrons, which interact with argon nuclei via a coupling between their magnetic moments, will be scattered by the fluid. We then measure, using a bank of neutron detectors, neutrons scattered in a final state labeled by an energy ε_f and a momentum $\hbar k_f$. We assume that initially the argon system can be described by the many-particle quantum state $|i\rangle$. The neutron, through its interaction with the argon, will cause a transition in the argon to some final state $|f\rangle$.

In a slightly simplified version of the real world (which we will make more rigorous later), we can assume that the interaction of a neutron at position **x** and the N argon atoms at positions $\mathbf{r}_j$ can be written as a sum of pair interactions:

$$H_I(\mathbf{x}) = \sum_{j=1}^{N} V(\mathbf{r}_j - \mathbf{x}) = \int d^d y \, n(\mathbf{y}) V(\mathbf{x} - \mathbf{y}) \tag{166}$$

where $V(\mathbf{x})$ is the interaction potential and $n(\mathbf{x}, t)$ is the argon particle density:

$$n(\mathbf{x}, t) = \sum_{j=1}^{N} \delta[\mathbf{x} - \mathbf{r}_j(t)]. \tag{167}$$

Notice that our interaction potential is of the linear response type. In a scattering experiment, when we measure the scattering angle and the change in energy, we measure the inelastic double differential scattering cross section per unit solid angle, $d\Omega$, per unit energy, $d\varepsilon_f$: $\frac{d^2\sigma(i \to f)}{d\Omega d\varepsilon_f}$. The $i \to f$ indicates that the argon is taken from the initial quantum state i to the final state f during the scattering process. We need a theoretical expression for the scattering cross section. Since we are only *tickling* the system (a single neutron will only mildly effect a large sample), the Born approximation can be used to evaluate the cross section in the form referred to as Fermi's golden rule [19]:

$$\frac{d^2\sigma(i \to f)}{d\Omega d\varepsilon_f} = \frac{1}{\mathcal{V}} \frac{k_f}{k_i} \left(\frac{M_N}{2\pi\hbar^2} \right)^2 |\langle \mathbf{k}_f f | H_I | i \mathbf{k}_i \rangle|^2 \, \delta(\hbar\omega - E_f + E_i) \, , \tag{168}$$

where M_N is the mass of the neutron, $\mathcal{V}$ the volume of the sample, $\hbar\omega = \varepsilon_i - \varepsilon_f$ is the energy transfer of the neutron, and E_i and E_f are the initial and final energies of the argon system. The δ-function corresponds to energy conservation:

$$\varepsilon_i + E_i = \varepsilon_f + E_f \, . \tag{169}$$

The initial and final states are product states since the neutron and argon are asymptotically uncoupled:

$$\langle \mathbf{k}_f f| = \langle \mathbf{k}_f| \times \langle f| , \qquad (170)$$

where $\langle \mathbf{k}_f|$ is the final plane wave state for the neutron.

We can evaluate the interaction matrix element more explicitly if we assume the interaction potential between the neutron and argon is Fourier transformable:

$$\begin{aligned}
\langle \mathbf{k}_f f|H_I|i\mathbf{k}_i\rangle &= \left\langle \mathbf{k}_f f \left| \sum_{j=1}^{N} V(\mathbf{r}_j - \mathbf{x}) \right| i\mathbf{k}_i \right\rangle \\
&= \left\langle \mathbf{k}_f f \left| \sum_{j=1}^{N} \int \frac{d^3q}{(2\pi)^3} V(\mathbf{q}) e^{+i\mathbf{q}\cdot(\mathbf{r}_j - \mathbf{x})} \right| i\mathbf{k}_i \right\rangle \\
&= \int \frac{d^3q}{(2\pi)^3} V(\mathbf{q}) \left\langle \mathbf{k}_f | e^{-i\mathbf{q}\cdot\mathbf{x}} | \mathbf{k}_i \right\rangle \left\langle f \left| \sum_{j=1}^{N} e^{+i\mathbf{q}\cdot\mathbf{r}_j} \right| i \right\rangle, \quad (171)
\end{aligned}$$

where we have used the separability of the initial and final states into neutron and argon product states. We can evaluate the *neutron* matrix element by recognizing that $\mathbf{x}$, the position operator for the neutron, is diagonal in the coordinate representation:

$$\mathbf{x}|\mathbf{x}'\rangle = \mathbf{x}'|\mathbf{x}'\rangle . \qquad (172)$$

Then using the completeness of the coordinate representation:

$$1 = \int d^3x' |\mathbf{x}'\rangle\langle \mathbf{x}'| , \qquad (173)$$

we have:

$$\begin{aligned}
\langle \mathbf{k}_f | e^{-i\mathbf{q}\cdot\mathbf{x}} | \mathbf{k}_i \rangle &= \int d^3x' \langle \mathbf{k}_f | e^{-i\mathbf{q}\cdot\mathbf{x}} | \mathbf{x}' \rangle \langle \mathbf{x}' | \mathbf{k}_i \rangle \\
&= \int d^3x' e^{-i\mathbf{q}\cdot\mathbf{x}'} \langle \mathbf{k}_f | \mathbf{x}' \rangle \langle \mathbf{x}' | \mathbf{k}_i \rangle . \quad (174)
\end{aligned}$$

Since the free neutron states are plane waves,

$$\langle \mathbf{x}'|\mathbf{k}_i\rangle = e^{+i\mathbf{k}_i\cdot\mathbf{x}'} , \qquad (175)$$

then:

$$\begin{aligned}
\langle \mathbf{k}_f | e^{-i\mathbf{q}\cdot\mathbf{x}} | \mathbf{k}_i \rangle &= \int d^3x' e^{-i\mathbf{q}\cdot\mathbf{x}'} e^{-i\mathbf{k}_f\cdot\mathbf{x}'} e^{+i\mathbf{k}_i\cdot\mathbf{x}'} \\
&= (2\pi)^3 \delta(\mathbf{k}_i - \mathbf{k}_f - \mathbf{q}) . \quad (176)
\end{aligned}$$

It is convenient to define the neutron momentum exchange in a collision as:

$$\hbar \mathbf{k} = \hbar \mathbf{k}_i - \hbar \mathbf{k}_f \quad , \tag{177}$$

so that the interaction matrix element can be written:

$$\langle \mathbf{k}_f f | H_1 | i \mathbf{k}_i \rangle = \int \frac{d^3 q}{(2\pi)^3} V(\mathbf{q}) (2\pi)^3 \delta(\mathbf{k} - \mathbf{q}) \left\langle f \left| \sum_{j=1}^{N} e^{+i \mathbf{q} \cdot \mathbf{r}_j} \right| i \right\rangle$$

$$= V(\mathbf{k}) \left\langle f \left| \sum_{j=1}^{N} e^{+i \mathbf{k} \cdot \mathbf{r}_j} \right| i \right\rangle \quad . \tag{178}$$

The scattering cross section can then be written as:

$$\frac{d^2 \sigma(i \to f)}{d\Omega d\varepsilon_f} \tag{179}$$

$$= \frac{1}{V} \frac{k_f}{k_i} \left(\frac{M_N}{2\pi \hbar^2} \right)^2 |V(\mathbf{k})|^2 \delta(\hbar \omega - E_f + E_i) \left| \left\langle f \left| \sum_{j=1}^{N} e^{+i \mathbf{k} \cdot \mathbf{r}_j} \right| i \right\rangle \right|^2 .$$

This expression takes us from an arbitrary initial state $|i\rangle$ in the argon fluid to some final state $|f\rangle$. In reality, we will perform measurements over an ensemble of initial states and we sum over all of the possible final states. What we measure then is:

$$\frac{d^2 \sigma}{d\Omega d\varepsilon_f} = \sum_{i,f} \rho_i \frac{d^2 \sigma(i \to f)}{d\Omega d\varepsilon_f} \quad , \tag{180}$$

where ρ_i is the density matrix or probability distribution governing the initial argon system in state i. If we assume the system is in thermal equilibrium initially, then:

$$\rho_i = \frac{\langle i | e^{-\beta H} | i \rangle}{\sum_j \langle j | e^{-\beta H} | j \rangle} = \frac{e^{-\beta E_i}}{Z} \quad , \tag{181}$$

and we then have for the measured cross section:

$$\frac{d^2 \sigma}{d\Omega d\varepsilon_f} = \frac{1}{V} \frac{k_f}{k_i} \left(\frac{M_N}{2\pi \hbar^2} \right)^2 |V(\mathbf{k})|^2$$

$$\times \sum_{i,f} \delta(\hbar \omega - E_f + E_i) \rho_i \left| \left\langle f \left| \sum_{j=1}^{N} e^{+i \mathbf{k} \cdot \mathbf{r}_j} \right| i \right\rangle \right|^2 . \tag{182}$$

Let us now focus on the quantity:

$$\bar{S}_T(\mathbf{k}, \omega) \equiv \frac{\hbar 2\pi}{V} \sum_{i,f} \delta(\hbar \omega - E_f + E_i) \rho_i \left| \left\langle f \left| \sum_{j=1}^{N} e^{+i \mathbf{k} \cdot \mathbf{r}_j} \right| i \right\rangle \right|^2 . \tag{183}$$

If we use the integral representation for the δ-function:

$$2\pi\hbar\delta(\hbar\omega + E_i - E_f) = \int_{-\infty}^{+\infty} dt\, e^{+i(\omega - (E_f - E_i)/\hbar)t} \quad , \tag{184}$$

we can write:

$$\bar{S}_T(\mathbf{k}, \omega) = \sum_{i,f} \rho_i \frac{1}{V} \int_{-\infty}^{+\infty} dt\, e^{+i(\omega - (E_f - E_i)/\hbar)t} \left\langle f \left| \sum_{\ell=1}^{N} e^{+i\mathbf{k}\cdot\mathbf{r}_\ell} \right| i \right\rangle$$

$$\times \left\langle i \left| \sum_{j=1}^{N} e^{-i\mathbf{k}\cdot\mathbf{r}_j} \right| f \right\rangle \quad . \tag{185}$$

We next observe, remembering

$$H|i\rangle = E_i|i\rangle \quad , \tag{186}$$

that:

$$e^{-i(E_f - E_i)t/\hbar} \left\langle i \left| \sum_{j=1}^{N} e^{-i\mathbf{k}\cdot\mathbf{x}_j} \right| f \right\rangle = \left\langle i \left| e^{iE_i t/\hbar} \sum_{j=1}^{N} e^{-i\mathbf{k}\cdot\mathbf{r}_j} e^{-iE_f t/\hbar} \right| f \right\rangle$$

$$= \left\langle i \left| e^{iHt/\hbar} \sum_{j=1}^{N} e^{-i\mathbf{k}\cdot\mathbf{r}_j} e^{-iHt/\hbar} \right| f \right\rangle$$

$$= \left\langle i \left| \sum_{j=1}^{N} e^{-i\mathbf{k}\cdot\mathbf{r}_j(t)} \right| f \right\rangle , \tag{187}$$

where

$$\mathbf{r}_j(t) = e^{iHt/\hbar} \mathbf{r}_j e^{-iHt/\hbar} \tag{188}$$

is the position operator for the jth argon atom at time t. Using this result in Eq. (185) we obtain:

$$\bar{S}_T(\mathbf{k}, \omega) = \int_{-\infty}^{+\infty} \frac{dt}{V} e^{+i\omega t} \sum_{i,f} \rho_i \left\langle f \left| \sum_{l=1}^{N} e^{i\mathbf{k}\cdot\mathbf{r}_l} \right| i \right\rangle \left\langle i \left| \sum_{j=1}^{N} e^{-i\mathbf{k}\cdot\mathbf{r}_j(t)} \right| f \right\rangle \quad . \tag{189}$$

It is now convenient to rewrite Eq. (189) in terms of the density operator for the argon system. The density at point $\mathbf{x}$ and time t is:

$$n(\mathbf{x}, t) = \sum_{j=1}^{N} \delta[\mathbf{x} - \mathbf{r}_j(t)] \quad . \tag{190}$$

The Fourier-transform of the density can be defined:

$$n_{\mathbf{k}}(t) = \frac{1}{\sqrt{V}} \int d^3x \, e^{-i\mathbf{k}\cdot\mathbf{x}} n(\mathbf{x},t)$$

$$= \frac{1}{\sqrt{V}} \sum_{j=1}^{N} e^{-i\mathbf{k}\cdot\mathbf{r}_j(t)} \, . \tag{191}$$

We can then write:

$$\bar{S}_T(\mathbf{k},\omega) = \int_{-\infty}^{+\infty} dt \, e^{+i\omega t} \sum_{i,f} \rho_i \langle f|n_{-\mathbf{k}}|i\rangle \langle i|n_{\mathbf{k}}(t)|f\rangle \, . \tag{192}$$

Since the set of final states is complete $\sum_f |f\rangle\langle f| = 1$, we have:

$$\bar{S}_T(\mathbf{k},\omega) = \int_{-\infty}^{+\infty} dt \, e^{+i\omega t} \sum_i \rho_i \langle i|n_{\mathbf{k}}(t)n_{-\mathbf{k}}(0)|i\rangle$$

$$= \int_{-\infty}^{+\infty} dt \, e^{+i\omega t} \sum_i \frac{e^{-\beta E_i}}{Z} \langle i|n_{\mathbf{k}}(t)n_{-\mathbf{k}}(0)|i\rangle$$

$$= \int_{-\infty}^{+\infty} dt \, e^{+i\omega t} \langle n_{\mathbf{k}}(t)n_{-\mathbf{k}}(0)\rangle, \tag{193}$$

and $\bar{S}_T(\mathbf{k},\omega)$ is just the Fourier transform of a density–density time-correlation function. Returning to Eqs. (182) and (183) we have our final result, that the cross section can be written:

$$\frac{d^2\sigma}{d\Omega d\varepsilon_f} = \frac{1}{2\pi\hbar} \frac{k_f}{k_i} \left(\frac{M_N}{2\pi\hbar^2}\right)^2 |V(k)|^2 \, \bar{S}_T(\mathbf{k},\omega) \, . \tag{194}$$

We note that the factors multiplying $\bar{S}_T(\mathbf{k},\omega)$ depend on the interaction of the neutron with the argon. They depend on how one probes the system. The quantity $\bar{S}_T(\mathbf{k},\omega)$ however depends only on the properties of the argon system. It is completely independent of the nature of the probe. As in the linear response case, the dynamics intrinsic to the system can be written in terms of an equilibrium-averaged time-correlation function. Equation (193) can be written as:

$$\bar{S}_T(\mathbf{k},\omega) = \int_{-\infty}^{+\infty} dt \int \frac{d^3x d^3x'}{V} e^{+i\omega t} e^{-i\mathbf{k}\cdot(\mathbf{x}-\mathbf{x}')} \bar{S}_T(\mathbf{x},\mathbf{x}',t) \, , \tag{195}$$

where:

$$\bar{S}_T(\mathbf{x},\mathbf{x}',t) = \langle n(\mathbf{x},t)n(\mathbf{x}',0)\rangle \, . \tag{196}$$

The result expressing the double differential scattering cross section in terms of the dynamic structure factor was first obtained by van Hove [20].

A key question one should ask: is the correlation function $\langle n(\mathbf{x},t)n(\mathbf{x}',0)\rangle$ related to the response function $\langle[n(\mathbf{x},t),n(\mathbf{x}',0)]\rangle$? We shall see that they are closely related, but it will take some development to make the connection.

Before going on there is one technical point concerning the scattering case we should straighten out. Since there should be no correlation between density elements separated by large distances in a system with full translational invariance, we expect that the correlation function will factorize:

$$\lim_{|\mathbf{x}-\mathbf{x}'|\to\infty} \langle n(\mathbf{x},t)n(\mathbf{x}',0)\rangle \to \langle n(\mathbf{x},t)\rangle\langle n(\mathbf{x}',0)\rangle \ . \tag{197}$$

It is therefore mathematically desirable to write:

$$\langle n(\mathbf{x},t)n(\mathbf{x}',0)\rangle = \bar{S}(\mathbf{x}-\mathbf{x}',t) + \langle n(\mathbf{x},t)\rangle\langle n(\mathbf{x}',0)\rangle \tag{198}$$

where:

$$\begin{aligned}\bar{S}(\mathbf{x}-\mathbf{x}',t) &= \langle n(\mathbf{x},t)n(\mathbf{x}',0)\rangle - \langle n(\mathbf{x},t)\rangle\langle n(\mathbf{x}',0)\rangle \\ &= \langle \delta n(\mathbf{x},t)\delta n(\mathbf{x}',0)\rangle \end{aligned} \tag{199}$$

and:

$$\delta n(\mathbf{x},t) = n(\mathbf{x},t) - \langle n(\mathbf{x},t)\rangle \tag{200}$$

is the fluctuation in the density about its average value. Note then that:

$$\lim_{|\mathbf{x}-\mathbf{x}'|\to\infty} \bar{S}(\mathbf{x}-\mathbf{x}',t) \to 0 \ . \tag{201}$$

We have then that Eq. (195) can be written:

$$\begin{aligned}\bar{S}_T(\mathbf{k},\omega) &= \int_{-\infty}^{+\infty} dt\, e^{+i\omega t} \int \frac{d^3x}{\sqrt{V}} \int \frac{d^3x'}{\sqrt{V}} e^{-i\mathbf{k}\cdot(\mathbf{x}-\mathbf{x}')} \\ &\quad \times [\bar{S}(\mathbf{x}-\mathbf{x}',t) + \langle n(\mathbf{x}',0)\rangle\langle n(\mathbf{x},t)\rangle] \\ &= \bar{S}(\mathbf{k},\omega) + S_{el}(\mathbf{k},\omega) \ , \end{aligned} \tag{202}$$

where $\bar{S}(\mathbf{k},\omega)$ is the dynamic structure factor and the elastic or *forward* part of the scattering is given by:

$$S_{el}(\mathbf{k},\omega) = n^2(2\pi)^4\delta(\omega)\delta(\mathbf{k}) \tag{203}$$

in the case where the system has full translational invariance and $\langle n(\mathbf{x},t)\rangle$ is independent of $\mathbf{x}$ and t:

$$\langle n(\mathbf{x},t)\rangle = n \ . \tag{204}$$

Note we have chosen our normalizations such that:

$$\int \frac{d^3x}{\sqrt{V}} \int \frac{d^3x'}{\sqrt{V}} e^{+i\mathbf{k}\cdot(\mathbf{x}-\mathbf{x}')} = \int \frac{d^3x}{V} \int d^3r\, e^{+i\mathbf{k}\cdot\vec{r}} = \frac{V}{V}(2\pi)^3\delta(\mathbf{k}) \ . \tag{205}$$

In the case of periodic solids, where one has long-range order and the breakdown of translational invariance:

$$\langle n(\mathbf{x}, t)\rangle = \langle n(\mathbf{x} + \mathbf{R}, t)\rangle \tag{206}$$

where $\mathbf{R}$ is a lattice vector. In Problem 2.7 we discuss the dynamic structure factor for a solid in the harmonic approximation.

2.2.2
Electron Scattering

In the neutron-scattering case, the interaction potential could be written in the form:

$$H_I(\mathbf{x}) = \int d^3y \, V(\mathbf{x} - \mathbf{y}) n(\mathbf{y}) \tag{207}$$

where $n(\mathbf{x})$ is the particle density. We see that the neutron couples directly to the density fluctuations of the argon system. In the case of electron scattering, the interaction between the incident electron and a system of bound charged particles is given by the Coulomb interaction:

$$H_I = \sum_{j=1}^{N} \frac{e q_j}{|\mathbf{x} - \mathbf{r}_j|}, \tag{208}$$

where e is the electron charge and q_j is the charge of particle j located at position $\mathbf{x}_j$. We can rewrite Eq. (208) in the form:

$$H_I = \int d^3y \, \frac{e}{|\mathbf{x} - \mathbf{y}|} \rho(\mathbf{y}), \tag{209}$$

where:

$$\rho(\mathbf{x}) = \sum_{j=1}^{N} q_j \, \delta(\mathbf{x} - \mathbf{r}_j) \tag{210}$$

is the charge-density operator. We can then obtain the scattering cross section for electron scattering from the neutron-scattering result if we make the replacements: $M_N \to m_e =$ mass of electron, $n \to \rho$ and:

$$V(\mathbf{k}) = \int d^3x \, e^{-i\mathbf{k}\cdot\mathbf{x}} \frac{e}{|\mathbf{x}|} = \frac{4\pi e}{k^2}. \tag{211}$$

The scattering cross section is given then by:

$$\frac{d^2\sigma}{d\Omega d\varepsilon_f} = \frac{1}{2\pi\hbar} \frac{k_f}{k_i} \left(\frac{m_e}{2\pi\hbar^2}\right)^2 \left(\frac{4\pi e}{k^2}\right)^2 \bar{S}_{\rho\rho}(\mathbf{k}, \omega) \tag{212}$$

where $\tilde{S}_{\rho\rho}(\mathbf{k},\omega)$ is the charge-density dynamic structure factor. Thus electrons measure charge fluctuations.

2.2.3
Neutron Scattering: A More Careful Analysis

A main difficulty in connecting theory and scattering experiments is finding a tractable form for the interaction Hamiltonian H_I. The case of electron scattering is very straightforward. Generally things are not so simple and theorists and experimentalists have to work together to determine the appropriate coupling. A realistic treatment of neutron scattering is slightly more complicated than electron scattering and light scattering is quite complicated. Here we will first concentrate on neutron scattering.

In the case of thermal neutron scattering, momentum transfers $\hbar\mathbf{k}$ are sufficiently small that we can assume $kL \ll 1$ where L is the *range* of the interaction between the neutron and the argon nucleus. Thus we can replace:

$$V(\mathbf{k}) \to V(0) = \left(\frac{2\pi\hbar^2}{M_N}\right) a \ , \tag{213}$$

where a is the Fermi scattering length or pseudo-potential weight. The interaction between a neutron and a nucleus typically possesses a spin-interaction,

$$a \to a_1 + a_2 \mathbf{s}_i \cdot \vec{\sigma}_N \tag{214}$$

where $\mathbf{s}_i$ is the spin-operator for the ith nucleus and $\vec{\sigma}_N$ is the Pauli spin-matrix for the neutron. The scattering lengths a_1 and a_2 can be determined experimentally and are known (see Table 2.1) for many materials.

Table 2.1 Coherent ($\sigma_c = 4\pi|a_1|^2$) and incoherent ($\sigma_i = 4\pi|a_2|^2$) neutron-scattering cross sections in barns= 10^{-22}cm^2. Source: [25], except ^{1}H, ^{2}H, ^{3}He, C, and V, which are taken from [26]

Element	σ_c	σ_i	Element	σ_c	σ_i
^{1}H	1.8	80.2	Rb	6.3	0.003
^{2}H	5.6	2	Na	1.6	1.8
^{3}He	4.9	1.2	Sn	4.7	0.2
^{4}He	1.13	0.0	Bi	9.3	0.1
C	5.6	0.0	Pb	11.1	0.3
Ne	2.7	0.2	Al	1.5	0.0
A	76.6	0.54	Cu	7.3	1.2
Ga	6.5	1.0	V	0.02	5.0

The interaction between a neutron and argon atoms can then be written as:

$$H_I(\mathbf{x}) = \left(\frac{2\pi\hbar^2}{M_N}\right) [a_1 n(\mathbf{x}) + a_2 \vec{\sigma}_N \cdot \mathbf{S}(\mathbf{x})] \ , \tag{215}$$

where:

$$S(\mathbf{x}) = \sum_{i=1}^{N} \frac{1}{2} [\mathbf{s}_i \delta(\mathbf{x} - \mathbf{r}_i) + \delta(\mathbf{x} - \mathbf{r}_i) \mathbf{s}_i] \tag{216}$$

is the nuclear spin density for the argon system. The correlation function entering the cross section is then, remembering that the Pauli matrices are hermitian,

$$\begin{aligned} |a|^2 \langle n_\mathbf{k}(t) n_{-\mathbf{k}} \rangle \rightarrow &|a_1|^2 \langle n_\mathbf{k}(t) n_{-\mathbf{k}} \rangle + a_2 a_1^* \langle \mathbf{S}_\mathbf{k}(t) n_{-\mathbf{k}} \rangle \cdot \vec{\sigma}_N \\ &+ a_1 a_2^* \langle n_\mathbf{k}(t) \mathbf{S}_{-\mathbf{k}} \rangle \cdot \vec{\sigma}_N \\ &+ |a_2|^2 \sum_{\alpha\beta} \langle S_\mathbf{k}^\alpha(t) S_{-\mathbf{k}}^\beta \rangle \sigma_N^\alpha \sigma_N^\beta \quad . \end{aligned} \tag{217}$$

Normally, unless one does spin-polarized neutron scattering, one averages over the nuclear spins:

$$\langle \vec{\sigma}_N \rangle = 0 \tag{218}$$

$$\langle \sigma_N^\alpha \sigma_N^\beta \rangle = \delta_{\alpha\beta} \langle \sigma^\alpha \sigma^\alpha \rangle = \delta_{\alpha\beta} \tag{219}$$

and

$$\frac{d^2\sigma}{d\Omega d\varepsilon_f} = \frac{1}{2\pi\hbar} \frac{k_f}{k_i} \left[|a_1|^2 \bar{S}_{nn}(\mathbf{k}, \omega) + |a_2|^2 \bar{S}_{ss}(\mathbf{k}, \omega) \right] , \tag{220}$$

where $\bar{S}_{nn}(\mathbf{k}, \omega)$ is the density correlation function discussed earlier and

$$\begin{aligned} \bar{S}_{ss}(\mathbf{k}, \omega) = &\int_{-\infty}^{+\infty} dt \int \frac{d^3x d^3x'}{V} e^{+i\omega(t-t')} e^{-i\mathbf{k}\cdot(\mathbf{x}-\mathbf{x}')} \\ &\times \langle \mathbf{S}(\mathbf{x}, t) \cdot \mathbf{S}(\mathbf{x}', t') \rangle \end{aligned} \tag{221}$$

is the spin-density correlation function. It is a matter of practical importance that one measures a sum of $\bar{S}_{nn}$ and $\bar{S}_{ss}$ in inelastic neutron scattering. There are several ways of separating $\bar{S}_{nn}$ and $\bar{S}_{ss}$. The first is to choose a system where $|a_1|^2$ or $|a_2|^2$ is very small. Alternatively one can vary a_1 and a_2 by changing isotopes (which presumably does not change $\bar{S}_{nn}$ and $\bar{S}_{ss}$). This then allows one to measure $\bar{S}_{nn}$ and $\bar{S}_{ss}$ separately. Such a program has been carried out on argon, for example [21]. In this case measurements were made on naturally occurring argon, ^{36}A, and separate measurements on a mixture of ^{36}A and ^{40}A. Values for $|a_1|^2$ and $|a_2|^2$ are given for a selected set of elements in Table 2.1.

There are certain situations where the nuclear spins are uncorrelated. This is true of *classical* fluids like argon. This is not true of quantum fluids like

^{3}He at low temperatures. Consider the spin-density correlation function in the *classical* limit:

$$\langle \mathbf{S}(\mathbf{x},t) \cdot \mathbf{S}(\mathbf{x}',t') \rangle = \left\langle \sum_{i=1}^{N} \mathbf{S}_i \cdot \delta[\mathbf{x} - \mathbf{r}_i(t)] \sum_{j=1}^{N} \mathbf{S}_j \delta[\mathbf{x}' - \mathbf{r}_j(t')] \right\rangle . \quad (222)$$

If the spins of different particles are uncorrelated, then:

$$\left\langle \sum_{i \neq j=1}^{N} \mathbf{S}_i \cdot \mathbf{S}_j \delta[\mathbf{x} - \mathbf{r}_i(t)] \delta[\mathbf{x}' - \mathbf{r}_j(t')] \right\rangle = 0 \quad (223)$$

and:

$$\langle \mathbf{S}(\mathbf{x},t) \cdot \mathbf{S}(\mathbf{x}',t') \rangle = \left\langle \sum_{i=1}^{N} S_i^2 \delta[\mathbf{x} - \mathbf{r}_i(t)] \delta[\mathbf{x}' - \mathbf{r}_j(t')] \right\rangle . \quad (224)$$

Assuming that the nuclear state is in the well-behaved state of total spin $S_i^2 = \hbar^2 S(S+1)$, we have:

$$\langle \mathbf{S}(\mathbf{x},t) \cdot \mathbf{S}(\mathbf{x}',t') \rangle = \hbar^2 S(S+1) \left\langle \sum_{i=1}^{N} \delta[\mathbf{x} - \mathbf{x}_i(t)] \delta[\mathbf{x}' - \mathbf{x}_i(t')] \right\rangle$$
$$= \hbar^2 S(S+1) S_s(\mathbf{x} - \mathbf{x}', t - t') \quad (225)$$

and S_s is known as the van Hove self-correlation function [20], which measures correlations of a particle with itself as a function of position and time. A measurement of S_s gives one a method for tagging a particle in a fluid and measuring single particle properties. We write, finally,

$$\frac{d^2\sigma}{d\Omega d\varepsilon_f} = \frac{1}{2\pi\hbar} \frac{k_f}{k_i} \left[a_{coh}^2 \bar{S}_{nn}(\mathbf{k},\omega) + a_{inc}^2 \hbar^2 S(S+1) \bar{S}_s(\mathbf{k},\omega) \right] \quad (226)$$

where $a_{coh}^2 = |a_1|^2$ is associated with the *coherent* part of the scattering and $a_{inc}^2 = |a_2|^2$ is associated with the *incoherent* part of the scattering.

In Fig. 2.2 we show the data for liquid argon for both coherent $\bar{S}_{nn}(\mathbf{k},\omega)$ and incoherent $\bar{S}_s(\mathbf{k},\omega)$ contribution to the neutron-scattering cross sections.

2.2.4
Magnetic Neutron Scattering

A different type of neutron scattering occurs when one scatters from magnetic atoms or molecules. If we think of spins or total magnetic moments localized on lattice sites $\mathbf{r}_i$, then the interaction is of the form:

$$H_I = \sum_{i=1}^{N} 2\mu^N \sigma_N^\alpha V_{\alpha\beta}(\mathbf{x} - \mathbf{r}_i) s_i^\beta , \quad (227)$$

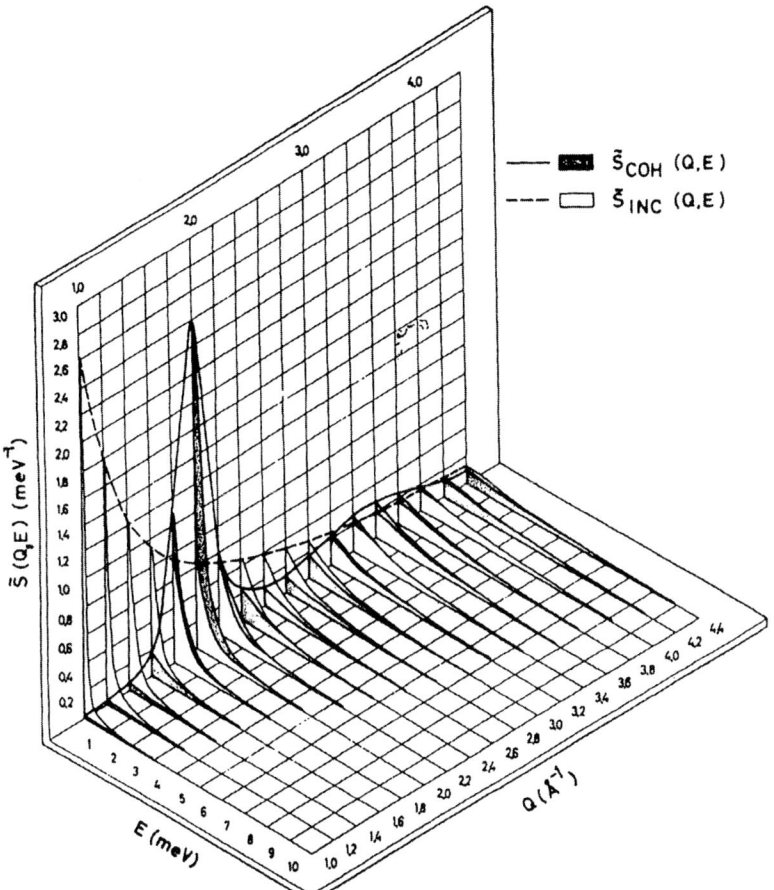

Fig. 2.2 Example of smoothed data for the coherent and incoherent dynamic structure factors for liquid argon at 85.2 K. Source: K. Sköld, J. M. Rowe, G. Ostrowski and P. D. Randolph, Ref. [21].

where μ^N is the neutron magnetic moment

$$\mu^N = 1.91 \, (e\hbar/M_N c) \,, \tag{228}$$

σ_N^α is the α-component of a Pauli spin matrix for the neutron, s_i^β is the βth component of the spin at lattice site i. The dipole interaction potential between the spins is given by:

$$V_{\alpha\beta}(\mathbf{r}) = \frac{1}{4\pi} \left[\delta_{\alpha\beta} \nabla^2 - \nabla_\alpha \nabla_\beta \right] \frac{1}{r} \,. \tag{229}$$

We see then that the interaction given by Eq. (227) can be rewritten as:

$$H_I = \int d^3 x' \, 2\mu^N \sigma_N^\alpha V_{\alpha\beta}(\mathbf{x} - \mathbf{x}') M_\beta(\mathbf{x}') \tag{230}$$

where:

$$M_\beta(\mathbf{x}') = \sum_{j=1}^{N} \frac{1}{2}\left[s_j^\beta \delta(\mathbf{x}' - \mathbf{x}_j) + \delta(\mathbf{x}' - \mathbf{x}_j)s_j^\beta\right] \qquad (231)$$

is the magnetization density. It is then clear that we can obtain the scattering cross section for this case from the neutron-scattering example if we replace $n \to M_\beta$ and:

$$V(\mathbf{k}) \to 2\mu^N \sigma_N^\alpha V_{\alpha\beta}(\mathbf{k}) \quad . \qquad (232)$$

Then:

$$\frac{d^2\sigma}{d\Omega d\varepsilon_f} = \frac{1}{2\pi\hbar} \frac{k_f}{k_i} \left(\frac{M_N}{2\pi\hbar^2}\right)^2 \sum_{\alpha,\beta,\alpha',\beta'} (2\mu^N)^2$$
$$\times \sigma_N^\alpha V_{\alpha\beta}(\mathbf{k})(\sigma_N^{\alpha'} V_{\alpha'\beta'}(\mathbf{k}))^\dagger S_{M_\beta,M_{\beta'}}(\mathbf{k},\omega) \quad . \qquad (233)$$

If we assume we are performing an unpolarized scattering experiment, we average over the neutron spin polarizations. Using Eq. (219) we obtain:

$$\frac{d^2\sigma}{d\Omega d\varepsilon_f} = \frac{1}{2\pi\hbar} \frac{k_f}{k_i} \left(\frac{M_N}{2\pi\hbar^2}\right)^2 (2\mu^N)^2$$
$$\times \sum_{\alpha,\beta,\beta'} V_{\alpha\beta}(\mathbf{k}) V_{\alpha\beta'}(\mathbf{k}) S_{M_\beta,M_{\beta'}}(\mathbf{k},\omega) \quad . \qquad (234)$$

We can easily compute the Fourier transform of the potential:

$$V_{\alpha\beta}(\mathbf{k}) = \int d^3 x\, e^{+i\mathbf{k}\cdot\mathbf{x}} \left(\frac{1}{4\pi}\right) \left[\nabla^2 \delta_{\alpha\beta} - \nabla_\alpha \nabla_\beta\right] \frac{1}{|\mathbf{x}|}$$
$$= \frac{1}{4\pi}\left(-k^2 \delta_{\alpha\beta} - (ik_\alpha)(ik_\beta)\right) \int d^3 x\, e^{+i\mathbf{k}\cdot\mathbf{x}} \frac{1}{|\mathbf{x}|}$$
$$= \frac{-1}{4\pi}\left(k^2 \delta_{\alpha\beta} - k_\alpha k_\beta\right) \frac{4\pi}{k^2} = \hat{k}_\alpha \hat{k}_\beta - \delta_{\alpha\beta} \quad . \qquad (235)$$

Putting this result back into Eq. (234) gives the final result:

$$\frac{d^2\sigma}{d\Omega d\varepsilon_f} = \frac{1}{2\pi\hbar} \frac{k_f}{k_i} \left(\frac{M_N}{2\pi\hbar^2}\right)^2 (2\mu^N)^2$$
$$\times \sum_{\alpha\beta} \left(\delta_{\alpha\beta} - \hat{k}_\alpha \hat{k}_\beta\right) \bar{S}_{M_\alpha M_\beta}(\mathbf{k},\omega) \quad . \qquad (236)$$

In magnetic neutron scattering we couple to the fluctuations in the magnetization density.

2.2.5
X-Ray and Light Scattering

When we treat the scattering of electromagnetic (EM) radiation from a condensed matter system it is more difficult [22] to identify the appropriate coupling Hamiltonian H_I. We know that the scattering by an atomic system occurs as a result of interaction between the EM wave and the atomic electrons. If we have a complicated molecular system, the interaction would be considerably complicated by the coupling of the radiation to the internal degrees of freedom of the molecules. If we restrict ourselves to a simple atomic system we can start with the microscopic expression for the total Hamiltonian:

$$H_T = \sum_{i=1}^{N} \frac{1}{2m_i} \left(\mathbf{p}_i - \frac{q_i}{c} \mathbf{A}(\mathbf{r}_i) \right)^2 + \frac{1}{2} \sum_{i \neq j=1}^{N} V(\mathbf{r}_i - \mathbf{r}_j) + \sum_{i=1}^{N} q_i \phi(\mathbf{r}_i) \quad (237)$$

where $\mathbf{p}_i$, $\mathbf{r}_i$, q_i and m_i are, respectively, the momentum, position, charge and mass of the ith particle in the sample, and c is the speed of light. $\mathbf{A}(\mathbf{x})$ is the vector potential representing the external electromagnetic field and $\phi(\mathbf{r}_i)$ is the associated scalar potential. We can then write:

$$H_T = H + H_I \quad (238)$$

where H is the Hamiltonian for a system in the absence of the external field and the coupling is:

$$\begin{aligned} H_I &= -\sum_{i=1}^{N} \frac{q_i}{2m_i c} (\mathbf{p}_i \cdot \mathbf{A}(\mathbf{r}_i) + \mathbf{A}(\mathbf{r}_i) \cdot \mathbf{p}_i) \\ &+ \sum_{i=1}^{N} \frac{q_i^2}{2m_i c^2} \mathbf{A}^2(\mathbf{r}_i) + \sum_{i=1}^{N} q_i \phi(\mathbf{r}_i) \\ &= \int d^3x \left[-\mathbf{J}_q(\mathbf{x}) \cdot \frac{\mathbf{A}(\mathbf{x})}{c} + \rho(\mathbf{x}) \phi(\mathbf{x}) + \frac{1}{2c^2} \rho_{q^2}(\mathbf{x}) A^2(\mathbf{x}) \right] \end{aligned} \quad (239)$$

where:

$$\mathbf{J}_q(\mathbf{x}) = \sum_{i=1}^{N} q_i \frac{1}{2} \left[\frac{\mathbf{p}_i}{m_i} \delta(\mathbf{x} - \mathbf{r}_i) + \delta(\mathbf{x} - \mathbf{r}_i) \frac{\mathbf{p}_i}{m_i} \right] \quad (240)$$

is the charge-current density,

$$\rho(\mathbf{x}) = \sum_{i=1}^{N} q_i \delta(\mathbf{x} - \mathbf{r}_i) \quad (241)$$

is the charge density and:

$$\rho_{q^2}(\mathbf{x}) = \sum_{i=1}^{N} \frac{q_i^2}{m_i} \delta(\mathbf{x} - \mathbf{r}_i) \quad . \quad (242)$$

If all the charges have the same magnitude $q_i^2 = e^2$ (say electrons and protons) then:

$$\rho_{q^2}(\mathbf{x}) = \frac{e^2}{m_e} n(\mathbf{x}) + \frac{e^2}{m_p} n_p(\mathbf{x}) \tag{243}$$

where m_e is the mass of the electrons, m_p is the mass of the protons, $n(\mathbf{x})$ the density of electrons, and $n_p(\mathbf{x})$ the density of protons. Since $m_p \gg m_e$, to a first approximation, we can neglect the proton contribution.

In the case of electromagnetic radiation we can choose a gauge where $\phi = 0$ and we can write the quantized vector potential:

$$\mathbf{A}(\mathbf{k}) = \left[\frac{hc^2}{\omega(\mathbf{k})} \right]^{1/2} [\mathbf{a}_k + \mathbf{a}_k^\dagger] \tag{244}$$

where $\mathbf{a}_k$ is a photon annihilation operator, $\mathbf{k} \cdot \mathbf{a}_k = 0$ (photons have two transverse polarizations), and $\omega = ck$ where c is the speed of light.

If we use the Born approximation to evaluate the scattering then we see that only the term in H_I proportional to $\mathbf{A}^2$ contributes. This is because the matrix element $\langle \mathbf{k}_i | \mathbf{A} | \mathbf{k}_f \rangle$, where $|\mathbf{k}_i\rangle$ and $|\mathbf{k}_f\rangle$ are incident and final one-photon states, is zero. The analysis of the scattering cross section can then be worked out in detail [23] with the final result:

$$\frac{d^2\sigma}{d\Omega d\varepsilon_f} = \left(\frac{e^2}{m_e c^2} \right)^2 \left(\frac{1 + (\hat{k}_i \cdot \hat{k}_f)^2}{2} \right) S_{nn}(\mathbf{k}, \omega) \ . \tag{245}$$

One measures the density–density correlation function in x-ray scattering.

The difference between x-ray and light scattering is that their wavelengths are very different when compared to typical atomic distances. For x-rays the electromagnetic wavelength is comparable to the atomic size and therefore one must treat the scattering from individual electrons. In the case of light scattering, the electromagnetic wavelength is so much larger than interatomic distances that the atoms see a nearly uniform electric field. In this case, it is meaningful to talk about the coupling of the radiation to a fluctuating dielectric constant $\varepsilon(\mathbf{x}, t)$ for the medium. One then assumes that the fluctuations in density are responsible for fluctuations in dielectric *constant* and:

$$\varepsilon(\mathbf{x}, t) = \left(\frac{\partial \varepsilon}{\partial n} \right)_T n(\mathbf{x}, t) \ . \tag{246}$$

One then obtains after a number of semiclassical arguments that:

$$\frac{d^2\sigma}{d\Omega d\varepsilon_f} = \frac{\sqrt{\varepsilon}}{4} \left(\frac{\partial \varepsilon}{\partial n} \right)_T^2 \left(\frac{\omega_i}{c} \right)^4 \frac{(\hat{k}_i \times \hat{k}_f)^2}{(2\pi)^3} S_{nn}(\mathbf{k}, \omega) \ . \tag{247}$$

For a discussion of various restrictions on the validity of this expression and further discussion of its derivation see Appendix A in Ref. [24]. In the end it is useful to remember that for simple atomic fluids one can think of light as coupling directly to the density.

2.2.6
Summary of Scattering Experiments

We have discussed a number of scattering experiments. In each case, the results can be interpreted in terms of an equilibrium-averaged time-correlation function: We list our results in Table 2.2 below. It is important to emphasize that the characteristic momenta and energies associated with these probes are quite different. In Table 2.3 we give characteristic values for the incident wavenumbers and energies for the various probes.

Table 2.2 Summary of couplings in scattering experiments

Scattering experiment	Couples to	Correlation function measured
Neutron scattering	Density	$S_{nn}(\mathbf{k}, \omega)$
	Single particle motion	$S_s(\mathbf{k}, \omega)$
Magnetic neutron scattering	Magnetization density	$S_{M_\alpha, M_\beta}(\mathbf{k}, \omega)$
Electron scattering	Charge density	$S_{\rho\rho}(\mathbf{k}, \omega)$
X-ray scattering	Mass density	$S_{nn}(\mathbf{k}, \omega)$
Light scattering	Mass density	$S_{nn}(\mathbf{k}, \omega)$

Table 2.3 Characteristic incident wavenumbers and energies in scattering experiments

	k_i [Å^{-1}]	E_i	$\lambda_i = 2\pi/k_i$ [Å]
Neutrons	2	10^{-3} to 1 eV	1–10
Light	10^{-3}	10 eV	6000
X-rays	1	50 keV	0.5–2
Electrons	0.1	50 keV	0.1

We first note that x-rays and electrons are similar in the range of k and ω they probe. They can lead to rather large momentum transfers and therefore can sample microscopic regions in space. However, the initial and final energies are so large compared to the desired energy exchanges (500,000 to 1) that one cannot, with present techniques, energy analyze x-ray and electron scattering. One therefore integrates over all ω and effectively measures the structure factor:

$$\bar{S}(\mathbf{k}) = \int \frac{d\omega}{2\pi} \bar{S}(\mathbf{k},\omega) , \qquad (248)$$

in these experiments.

Neutron and light scattering are complementary in that neutrons measure short-distance, short-time behavior, while light scattering is preferable for longer wavelengths and time phenomena. Light scattering is particularly useful in looking at hydrodynamical phenomena.

In this chapter we have discussed both linear response and scattering experiments and both measure, under appropriate circumstances, equilibrium-averaged time-correlation functions. The rapidly expanding direct visualization method, the modern version of Perrin's experiment, are also conveniently analyzed in terms of time-correlation functions.

2.3
References and Notes

1 R. Kubo, R., in *Lectures in Theoretical Physics, Vol. I*, Interscience, N.Y. (1959), Chapter 4; J. Phys. Soc. Jpn. **12**, 570 (1957). In recognition of his contributions Professor Kubo received the second Boltzmann medal for achievement in statistical physics. For an overview see R. Kubo, Science **233**, 330 (1986).

2 M. S. Green, J. Chem. Phys. **22**, 398 (1954); Phys. Rev. **119**, 829 (1960).

3 L. P. Kadanoff, P. C. Martin, Ann. Phys. **24**, 419 (1963).

4 P. C. Martin, in *Many-Body Physics*, C. DeWitt, R. Balian (eds), Gordon and Breach, N.Y. (1968), p. 39.

5 D. Forster, *Hydrodynamic Fluctuations, Broken Symmetry, and Correlation Functions*, W. A. Benjamin, Reading, Mass. (1975).

6 In the GCE, we replace $H \to H - \mu N$ where μ is the chemical potential and N the number operator.

7 The time dependence of the intrinsic Hamiltonian comes from the time evolution of the operators appearing in the Hamiltonian. When one turns off the external field the intrinsic Hamiltonian is time independent.

8 Thermodynamically B and h are conjugate variables.

9 For a discussion of the different representations see A. Messiah, *Quantum Mechanics, Vol. I*, Wiley, N.Y. (1966), pp. 312, 314.

10 See the discussion in Section 1.3.2 in ESM.

11 For a discussion of the interplay of symmetry and invariance properties see the discussions in Section 1.13 in ESM and Section 1.3.3 in FOD.

12 The worry is that one is dealing with long-range interactions in charged systems and this requires considerable care.

13 P. Drude, Ann. Phys. **1**, 566 (1900); Ann. Phys. **3**, 369 (1900).

14 N. Ashcroft, N. D. Mermin, *Solid State Physics*, Holt, Rinehart and Winston, N.Y. (1976). See Table 7.1.

15 G. E. Pake, *Paramagnetic Resonance*, W. A. Benjamin, N.Y. (1962).

16 C. P. Slichter, *Principles of Magnetic Resonance*, Harper and Row, N.Y. (1963).

17 F. Bloch, Phys. Rev. **70**, 460 (1946).

18 See the discussion on p. 44 of Kadanoff and Martin [3].

19 L. Schiff, *Quantum Mechanics*, McGraw-Hill, N.Y., (1968).

20 L. van Hove, Phys. Rev. **95**, 249 (1954).

21 K. Sköld, J. M. Rowe, G. Ostrowski, P. D. Randolph, Phys. Rev. A **6**, 1107 (1972).

22 B. J. Berne, R. Pecora, *Dynamic Light Scattering*, Wiley, N.Y. (1976).
23 G. Baym, *Lectures on Quantum Mechanics*, W. A. Benjamin, N.Y., (1969), p. 289.
24 D. Forster, *Hydrodynamic Fluctuations, Broken Symmetry and Correlation Functions*, W. A. Benjamin, Reading (1975).
25 J. R. D. Copley, S. W. Lovesey, Rep. Prog. Phys. **38**, 461 (1975).
26 G. L. Squires, *Thermal Neutron Scattering*, Cambridge University Press (1978).

2.4 Problems for Chapter 2

Problem 2.1: Consider a classical set of N harmonic oscillators governed by the Hamiltonian:

$$H = \sum_{i=1}^{N}\left[\frac{\mathbf{p}_i^2}{2m} + \frac{k}{2}\mathbf{r}_i^2\right],$$

where the oscillators have mass m and spring constant k. Assume that the initial phase-space coordinates are governed by a probability distribution:

$$P[\mathbf{r}_i(0), \mathbf{p}_i(0)] = \mathcal{D}\exp\left(-\sum_{i=1}^{N}\left[a\mathbf{p}_i^2 + b\mathbf{r}_i^2\right]\right),$$

where $\mathcal{D}$ is a normalization constant and a and b are constants. Compute the average kinetic energy for this system as a function of time.

Problem 2.2: Show that the formal solution to Eq. (43), given by Eq. (46), is correct by comparing with the direct iteration of Eq. (43) to second order in H_E^I.

Problem 2.3: Consider the one-dimensional harmonic oscillator governed by the Hamiltonian operator:

$$\hat{H} = \frac{\hat{p}^2}{2m} + \frac{k}{2}\hat{x}^2 .$$

Determine the associated linear response function $\chi''_{x,p}(t,t')$ defined by Eq. (54).

Problem 2.4: Consider the equation of motion for the longitudinal component of the magnetization:

$$\frac{d}{dt}M_z = -\frac{1}{T_1}(M_z - M_0) .$$

Solve this equation as an initial-value problem and determine the long-time value of $M_z(t)$.

Problem 2.5: Solve the coupled set of equations Eq. (116) and Eq. (117) to obtain $M_\alpha(\omega)$. From this determine:

$$\chi_{\alpha\beta}(\omega) = \frac{\partial M_\alpha(\omega)}{\partial h_\beta(\omega)} .$$

2 Time-Dependent Phenomena in Condensed-Matter Systems

Problem 2.6: Starting with the constrained equilibrium density matrix:

$$\rho_{eq} = e^{-\beta(H+H_E)} / Tr e^{-\beta(H+H_E)} \ ,$$

where we treat:

$$H_E = -Bh_B$$

as a perturbation, show that the linear response for the average of some observable A is given by:

$$\frac{\delta A}{\delta h_B} = \chi_{AB} \ ,$$

where χ_{AB} is given by Eq. (135).

Problem 2.7: In a solid the density of particles can be written in the form:

$$n(\mathbf{x}) = \sum_{\mathbf{R}} \delta[\mathbf{x} - \mathbf{R} - \mathbf{u}(\mathbf{R})] \ ,$$

where $\mathbf{R}$ labels the lattice vectors and $\mathbf{u}(\mathbf{R})$ is the displacement of the atom at site $\mathbf{R}$ from its equilibrium position. In the harmonic approximation the Hamiltonian governing the lattice displacement degrees of freedom is given by:

$$H = E_0 + \sum_{\mathbf{R}} \frac{\mathbf{p}^2(\mathbf{R})}{2m} + \frac{1}{2} \sum_{\mathbf{R},\mathbf{R}'} \sum_{\alpha\gamma} u_\alpha(\mathbf{R}) u_\gamma(\mathbf{R}') D_{\alpha\gamma}(\mathbf{R} - \mathbf{R}') \ ,$$

where E_0 is the classical ground state energy, m is the atomic mass, $\mathbf{p}(\mathbf{R})$ is the momentum of the atom at site $\mathbf{R}$, and $D_{\alpha\gamma}(\mathbf{R} - \mathbf{R}')$ is the dynamical matrix governing the interaction between atomic displacements. Assume you are given the complete and orthonormal set of eigenfunctions $\varepsilon_\alpha^p(\mathbf{k})$ and the eigenvalues $\lambda_p(\mathbf{k})$ associated with the Fourier transform of the dynamical matrix:

$$\sum_\gamma D_{\alpha\gamma}(\mathbf{k}) \varepsilon_\gamma^p(\mathbf{k}) = \lambda_p(\mathbf{k}) \varepsilon_\alpha^p(\mathbf{k}) \ ,$$

where the index p labels the eigenvalues. In the classical case, express the Fourier transform of the atomic displacement time-correlation function:

$$G_{\alpha\gamma}(\mathbf{k}, t - t') = \frac{1}{V} \sum_{\mathbf{R},\mathbf{R}'} e^{-i\mathbf{k}\cdot((\mathbf{R}-\mathbf{R}'))} \langle u_\alpha(\mathbf{R}, t) u_\gamma(\mathbf{R}', t') \rangle \ ,$$

in terms of the eigenvalues and eigenfunctions.

Problem 2.8: In this problem, building on the results found in Problem 2.7, we explore the form of the dynamical structure factor measured in an inelastic scattering experiments on simple solids. If we scatter from the particle density, then we can write the dynamical structure factor in the form:

$$\bar{S}(\mathbf{k},\omega) = \int_{-\infty}^{\infty} dt e^{i\omega(t-t')} \int \frac{d^3x\, d^3y}{V} e^{-i\mathbf{k}\cdot(\mathbf{x}-\mathbf{y})} \bar{S}_{nn}(\mathbf{x},\mathbf{y},t-t') ,$$

where the density can be written in the form:

$$n(\mathbf{x}) = \sum_{\mathbf{R}} \delta[\mathbf{x} - \mathbf{R} - \mathbf{u}(\mathbf{R})] ,$$

where $\mathbf{R}$ labels the lattice vectors and $\mathbf{u}(\mathbf{R})$ is the displacement of the atom at site $\mathbf{R}$ from its equilibrium position.

(i) We are given the result that in the harmonic approximation the averages needed over the displacement variables in evaluating the dynamical structure factor are of the form:

$$\left\langle e^{\sum_\mathbf{R} \int d\bar{t}\, \mathbf{A}(\mathbf{R},\bar{t})\cdot \mathbf{u}(\mathbf{R},\bar{t})} \right\rangle$$

$$= \exp\left[\frac{1}{2}\sum_{\mathbf{R},\mathbf{R}'}\sum_{\alpha\gamma} \int d\bar{t}_1 \int d\bar{t}_2 A_\alpha(\mathbf{R},\bar{t}_1) A_\gamma(\mathbf{R}',\bar{t}_2) G_{\alpha\gamma}(\mathbf{R},\mathbf{R}',\bar{t}_1 - \bar{t}_2)\right] ,$$

where $\mathbf{A}(\mathbf{R},t)$ is an arbitrary vector that depends on lattice position and time and:

$$G_{\alpha\gamma}(\mathbf{R},\mathbf{R}',t-t') = \langle u_\alpha(\mathbf{R},t) u_\gamma(\mathbf{R}',t') \rangle$$

is the displacement–displacement equilibrium-averaged time-correlation function discussed in the previous problem. Express the dynamic structure factor in terms of integrals and sums over $G_{\alpha\gamma}(\mathbf{R},\mathbf{R}',t-t')$.

(ii) Assuming that the displacement correlation is small at low Temperatures, expand the structure factor in powers of the displacement correlation function. As a function of frequency, where would we expect to see peaks in the dynamic structure factor?

Problem 2.9: Show that one can extract the velocity auto correlation function from the van Hove self-correlation function $S_s(k,t-t')$.

Problem 2.10: Show that the scattering length a, defined by Eq. (213), has dimensions of length.

Problem 2.11: Starting with the Bloch equation for the longitudinal component of the magnetization:

$$\left(\frac{d}{dt} + \frac{1}{T_1}\right) M_z = \frac{M_0}{T_1} - \gamma(h_x M_y - h_y M_x) ,$$

find the second order in $\mathbf{h}$ component of M_z that shows the resonance seen at first order in M_x and M_y.

Assume that $\mathbf{h} = \hat{x}\, h\cos(\omega_p t)$.

Problem 2.12: Assume that our Langevin equation of motion is extended to include the Lorentz force in response to an applied static magnetic field **H**:

$$M\frac{d}{dt}\langle \mathbf{V}(t)\rangle = -M\gamma\langle \mathbf{V}(t)\rangle + q\left(\mathbf{E}(t) + \langle \mathbf{V}\rangle \times \mathbf{H}\right) \ .$$

Assuming $\mathbf{E} = E\hat{z}$ and $\mathbf{H} = H\hat{x}$ determine the conductivity tensor $\sigma_{\alpha\beta}(\omega)$ defined by:

$$\langle J_\alpha\rangle = nq\langle V_\alpha\rangle = \sum_\beta \sigma_{\alpha\beta}(\omega) E_\beta(\omega) \ .$$

Problem 2.13: Consider the expression Eq. (133) for the linear response to an adiabatic field $h_B(t)$ given by Eq. (132). Show, by using the Fourier representation of $\chi''_{A,B}(t-t')$, that one can do the t' integration and obtain, in the adiabatic ($\eta \to 0$) limit,

$$\delta A(t) = \chi_{A,B} h_B$$

for $t \geq t_1$. Also find $\delta A(t)$ in the regime $t \leq t_1$.

3
General Properties of Time-Correlation Functions

Presumably we are all now convinced of the central role of equilibrium-averaged time-correlation functions in treating the kinetics of systems experiencing small deviations from equilibrium. Let us now focus on some of the general properties of these time-correlation functions.

3.1
Fluctuation-Dissipation Theorem

Suppose for a given system we can carry out both a linear response experiment and a scattering experiment—what is the relationship between the information obtained in the two experiments? As a concrete example, look at the response to an external field that couples to the density in a fluid:

$$H_{ext}(t) = -\int d^3x\, U(\mathbf{x},t) n(\mathbf{x}) \tag{1}$$

and also carry out the neutron-scattering experiment discussed earlier. The Fourier transform of the linear response of the density to this external field is given by:

$$\delta n(\mathbf{k},\omega) = \chi_{nn}(\mathbf{k},\omega) U(\mathbf{k},\omega) + \mathcal{O}(U^2) \tag{2}$$

where the dynamic susceptibility is given by:

$$\chi_{nn}(\mathbf{k},\omega) = 2i \int_{-\infty}^{+\infty} dt \int d^3x\, e^{+i\omega(t-t')} e^{-i\mathbf{k}\cdot(\mathbf{x}-\mathbf{x}')}$$
$$\times \Theta(t-t') \chi''_{nn}(\mathbf{x}-\mathbf{x}',t-t') \tag{3}$$

and the linear response function is defined by:

$$\chi''_{nn}(\mathbf{x}-\mathbf{x}',t-t') = \frac{1}{2\hbar} \langle [n(\mathbf{x},t), n(\mathbf{x}',t')] \rangle \,. \tag{4}$$

In the scattering experiment we measure the dynamic structure factor:

$$\bar{S}_{nn}(\mathbf{k},\omega) = \int_{-\infty}^{+\infty} dt \int d^3x\, e^{+i\omega(t-t')} e^{-i\mathbf{k}\cdot(\mathbf{x}-\mathbf{x}')} \bar{S}_{nn}(\mathbf{x}-\mathbf{x}',t-t') \tag{5}$$

Nonequilibrium Statistical Mechanics. Gene F. Mazenko
Copyright © 2006 WILEY-VCH Verlag GmbH & Co. KGaA, Weinheim
ISBN: 3-527-40648-4

where:

$$\bar{S}_{nn}(\mathbf{x}-\mathbf{x}',t-t') = \langle \delta n(\mathbf{x},t)\delta n(\mathbf{x}',t')\rangle \ . \tag{6}$$

Consider the following questions: Do the linear response and scattering experiments give independent information? Put another way: are $\chi(\mathbf{k},\omega)$ and $\bar{S}(\mathbf{k},\omega)$ related? They are related and establishing their precise relationship will expose an important underlying physical principle.

In making the connection between $\bar{S}$ and χ, consider the more general case of the correlation functions:

$$\bar{S}_{AB}(t-t') = \langle A(t)B(t')\rangle \tag{7}$$

and:

$$\chi''_{AB}(t-t') = \frac{1}{2\hbar}\langle [A(t),B(t')]\rangle \ , \tag{8}$$

where A and B are any two operators representing observables. We can always choose A and B to have zero average, $\langle A\rangle = 0$, by choosing $\delta A = A - \langle A\rangle$. For clarity we suppress the δ below. Consider first $\bar{S}_{AB}$. We can write, working in the canonical ensemble,

$$\begin{aligned}\bar{S}_{AB}(t-t') &= \frac{1}{Z}Tr\, e^{-\beta H}A(t)B(t')\\ &= \frac{1}{Z}Tr\, e^{-\beta H}A(t)e^{\beta H}e^{-\beta H}B(t') \end{aligned} \tag{9}$$

since $e^{\beta H}e^{-\beta H} = 1$. Remember that the time evolution of an operator is given by:

$$A(t) = e^{itH/\hbar}Ae^{-itH/\hbar} \ . \tag{10}$$

Extending our definition of time onto the complex plane, we can write:

$$\begin{aligned}e^{-\beta H}A(t)e^{+\beta H} &= e^{-\beta H}e^{+itH/\hbar}Ae^{-itH/\hbar}e^{+\beta H}\\ &= e^{+i(t+i\beta\hbar)H/\hbar}Ae^{-i(t+i\beta\hbar)H/\hbar} = A(t+i\beta\hbar) \end{aligned} \tag{11}$$

and Eq. (9) can be written as:

$$\bar{S}_{AB}(t-t') = \frac{1}{Z} Tr\, A(t+i\beta\hbar)e^{-\beta H}B(t') \ . \tag{12}$$

Using the cyclic invariance of the trace, this can be written as:

$$\bar{S}_{AB}(t-t') = \frac{1}{Z} Tr\, e^{-\beta H}B(t')A(t+i\beta\hbar) = \bar{S}_{BA}(t'-t-i\beta\hbar) \ . \tag{13}$$

In terms of Fourier transforms:

$$\bar{S}_{AB}(\omega) = \int d(t-t')e^{+i\omega(t-t')} \int \frac{d\bar{\omega}}{2\pi} \bar{S}_{BA}(\bar{\omega})e^{-i\bar{\omega}(t'-t-i\beta\hbar)}$$

$$= \int \frac{d\bar{\omega}}{2\pi} \bar{S}_{BA}(\bar{\omega})e^{-\beta\hbar\bar{\omega}} 2\pi\delta(\omega+\bar{\omega})$$

$$\bar{S}_{AB}(\omega) = \bar{S}_{BA}(-\omega)e^{+\beta\hbar\omega} \quad . \tag{14}$$

If we work in the grand canonical ensemble then the density matrix is of the form:

$$\rho = \frac{e^{-\beta(H-\mu N)}}{Z} \tag{15}$$

where N is the number operator. If the operators A and B commute with N (they are defined for fixed particle number) then everything commutes with:

$$e^{+\beta\mu N} \tag{16}$$

and all of our arguments can be carried through. The main point is that $A(t+i\beta\hbar)$ commutes with $e^{\beta\mu N}$. If, however, A or B changes the number of particles, then we must refine our analysis. This would be necessary, for example, if A or B are creation and annihilation operators (see Problem 3.1).

The result given by Eq. (14) can be *understood* using the following arguments. Since, in our neutron-scattering experiment, $\hbar\omega = \varepsilon_i - \varepsilon_f$, we see that $\omega > 0$ means the neutron lost energy in its collision with the target system. This means the neutron created an excitation with energy $\hbar\omega$ in the argon. If $\omega < 0$, then the neutron picked up energy from the target. In this case the neutron destroyed an excitation with energy $\hbar\omega$. We see that it is consistent to interpret $\bar{S}_{nn}(\omega)$ as proportional to the probability of creating an excitation with energy $\hbar\omega$ in the argon and $\bar{S}_{nn}(-\omega)$ as proportional to the probability of destroying an excitation with energy $\hbar\omega$ in the argon. There are usually no restrictions on creating excitations, but in order to destroy one the excitation must exist. The density of excitations with energy $\hbar\omega$ is proportional to the Boltzmann factor $e^{-\beta\hbar\omega}$. These arguments lead back to the result:

$$\bar{S}_{BA}(-\omega)/\bar{S}_{AB}(\omega) = e^{-\beta\hbar\omega} \quad . \tag{17}$$

We can now use this result in treating the linear response function:

$$\chi''_{AB}(t-t') = \frac{1}{2\hbar}\langle[A(t),B(t')]\rangle$$

$$= \frac{1}{2\hbar}[\bar{S}_{AB}(t-t') - \bar{S}_{BA}(t'-t)] \quad . \tag{18}$$

Then, in terms of Fourier transforms,

$$\chi''_{AB}(\omega) = \frac{1}{2\hbar}[\bar{S}_{AB}(\omega) - \bar{S}_{BA}(-\omega)]$$

and, using Eq. (14),

$$\chi''_{AB}(\omega) = \frac{1}{2\hbar}[\bar{S}_{AB}(\omega) - \bar{S}_{AB}(\omega)\,e^{-\beta\hbar\omega}]$$
$$= \frac{(1-e^{-\beta\hbar\omega})}{2\hbar}\bar{S}_{AB}(\omega) \ . \tag{19}$$

We see, at this point, that there is a direct relationship between the response function and the scattering function. Before commenting on this relationship it will be convenient, and it is conventional, to introduce the so-called fluctuation function:

$$S_{AB}(t-t') = \tfrac{1}{2}\langle[A(t),B(t')]_+\rangle \tag{20}$$
$$= \tfrac{1}{2}\langle[A(t)B(t') + B(t')A(t)]\rangle \ . \tag{21}$$

We see that S_{AB} is a symmetrized version of $\bar{S}_{AB}$ and $[,]_+$ indicates an anticommutator. We see that we can write:

$$S_{AB}(t-t') = \tfrac{1}{2}[\bar{S}_{AB}(t-t') + \bar{S}_{BA}(t'-t)] \ , \tag{22}$$

and:

$$S_{AB}(\omega) = \tfrac{1}{2}\left(\bar{S}_{AB}(\omega) + \bar{S}_{BA}(-\omega)\right) \ . \tag{23}$$

Using Eq. (14) again, we have:

$$S_{AB}(\omega) = \tfrac{1}{2}\bar{S}_{AB}(\omega)(1+e^{-\beta\hbar\omega}) \ . \tag{24}$$

Eliminating $\bar{S}_{AB}(\omega)$ in favor of $S_{AB}(\omega)$ in Eq. (19) leads to:

$$\chi''_{AB}(\omega) = \frac{(1-e^{-\beta\hbar\omega})2}{2\hbar(1+e^{-\beta\hbar\omega})}S_{AB}(\omega)$$
$$= \frac{1}{\hbar}\left(\frac{e^{+\beta\hbar\omega/2}-e^{-\beta\hbar\omega/2}}{e^{\beta\hbar\omega/2}+e^{-\beta\hbar\omega/2}}\right)S_{AB}(\omega)$$
$$= \frac{1}{\hbar}\tanh\left(\frac{\beta\hbar\omega}{2}\right)S_{AB}(\omega) \tag{25}$$

or:

$$S_{AB}(\omega) = \hbar\coth\left(\frac{\beta\hbar\omega}{2}\right)\chi''_{AB}(\omega) \ . \tag{26}$$

The relationship connecting the fluctuation function and the response function is known as the *fluctuation-dissipation theorem* [1]. The physical content of this theorem is quite important and was pointed out by Onsager [2] many years ago. $S_{AB}(\omega)$ is the autocorrelation between fluctuations of the operators A

and B. χ_{AB} represents the change one observes in the average of A due to a small external change in B. Because of the symmetry of the situation it also gives one the change in the average of B due to a small external change in A. The fluctuation-dissipation theorem relates spontaneous fluctuations to small external perturbations. This theorem tells us that a system can not tell whether a small fluctuation is spontaneous or is caused by a weak external force.

If we now return to our example of the fluid, we easily obtain, choosing $A = \hat{n}(x)$ and $B = \hat{n}(\mathbf{x}')$,

$$\bar{S}_{nn}(\mathbf{k}, \omega) = \frac{2\hbar}{(1 - e^{-\beta\hbar\omega})} \chi''_{nn}(\mathbf{k}, \omega) \ . \tag{27}$$

Thus if we map out $\bar{S}_{nn}(\mathbf{k}, \omega)$ using neutrons, then we calculate $\chi''_{nn}(\mathbf{k}, \omega)$ using Eq. (27). Alternatively, if we also perform a resonance experiment, this is a fundamental consistency check on the scattering experiments.

3.2
Symmetry Properties of Correlation Functions

There are a number of general symmetries of correlation functions that we can work out. It is convenient to focus first on the response function χ''_{AB} and then use the fluctuation-dissipation theorem to obtain the properties of the fluctuation function S_{AB}.

Let us begin with a quick overview. The basic properties we have at our disposal are that χ'' is defined in terms of a commutator, the operators A and B are hermitian and we presume to know how A and B transform under time reversal. From this set of ingredients we can determine the basic symmetry properties of $\chi''_{AB}(t - t')$ and $\chi''_{AB}(\omega)$.

Starting with the basic definition:

$$\chi''_{AB}(t - t') = \frac{1}{2\hbar} \langle [A(t), B(t')] \rangle \ , \tag{28}$$

we use the fact that χ''_{AB} is a commutator and is therefore antisymmetric under interchange of $A(t)$ and $B(t')$ to obtain:

$$\chi''_{AB}(t - t') = -\chi''_{BA}(t' - t) \tag{29}$$

or, for the Fourier transform,

$$\chi''_{AB}(\omega) = -\chi''_{BA}(-\omega) \ . \tag{30}$$

Next we consider the complex conjugate of χ''. Since χ''_{AB} is the commutator of hermitian operators we have:

$$\chi_{AB}^{\prime\prime*}(t-t') = \frac{1}{2\hbar}\langle[A(t),B(t')]\rangle^*$$

$$= \frac{1}{2\hbar}\sum_{i,j}\frac{e^{-\beta(E_i-\mu N_i)}}{Z}[\langle i|A(t)|j\rangle^*\langle j|B(t')|i\rangle^*$$

$$-\langle i|B(t')|j\rangle^*\langle j|A(t)|i\rangle^*] \ . \tag{31}$$

The complex conjugation gives:

$$\langle i|A(t)|j\rangle^* = \langle j|A^\dagger(t)|i\rangle = \langle j|A(t)|i\rangle \tag{32}$$

since A is hermitian. Then Eq. (31) reduces to:

$$\chi_{AB}^{\prime\prime*}(t-t') = \frac{1}{2\hbar}\sum_{i,j}\frac{e^{-\beta(E_i-\mu N_i)}}{Z}[\langle j|A(t)|i\rangle\langle i|B(t')|j\rangle$$

$$-\langle j|B(t')|i\rangle\langle i|A(t)|j\rangle]$$

$$= \frac{1}{2\hbar}\langle[B(t'),A(t)]\rangle = \chi_{BA}^{\prime\prime}(t'-t) \ . \tag{33}$$

Then, using the result Eq. (29), we find:

$$\chi_{AB}^{\prime\prime*}(t-t') = -\chi_{AB}^{\prime\prime}(t-t') \ , \tag{34}$$

and we find that $\chi_{AB}^{\prime\prime}(t-t')$ is imaginary. Taking the Fourier transform over time and using Eq. (34),

$$\chi_{AB}^{\prime\prime*}(\omega) = \left[\int_{-\infty}^{+\infty} dt\, e^{+i\omega(t-t')}\chi_{AB}^{\prime\prime}(t-t')\right]^*$$

$$= \int_{-\infty}^{+\infty} dt\, e^{-i\omega(t-t')}\chi_{AB}^{\prime\prime*}(t-t')$$

$$= \int_{-\infty}^{+\infty} dt\, e^{-i\omega(t-t')}[-\chi_{AB}^{\prime\prime}(t-t')]$$

$$\chi_{AB}^{\prime\prime}(\omega)^* = -\chi_{AB}^{\prime\prime}(-\omega) \tag{35}$$

In general $\chi_{AB}^{\prime\prime}(\omega)$ need not [3] be real. The reality properties of $\chi_{AB}^{\prime\prime}(\omega)$ are connected with time-reversal symmetry. In Appendix A the properties of the time-reversal operation are discussed. The main point for our present discussion is that if T is the time-reversal operator, and the density matrix is time-reversal invariant, $T\hat{\rho}T^{-1} = \hat{\rho}$, then we have the symmetry principle,

$$\langle \hat{A}\rangle = \left\langle\left(T\hat{A}T^{-1}\right)^\dagger\right\rangle \ . \tag{36}$$

Observables corresponding to operators $\hat{A}(t)$ have definite signatures under time reversal:

$$A'(t) = TA(t)T^{-1} = \varepsilon_A A(-t) \ . \tag{37}$$

For example, $\varepsilon_A = +1$ for positions and electric fields, -1 for momenta, angular momenta and magnetic fields. For an observable field with a definite signature under time reversal, we have the result:

$$\langle \hat{A} \rangle = \varepsilon_A \langle \hat{A}^\dagger \rangle = \varepsilon_A \langle \hat{A} \rangle \tag{38}$$

and the average vanishes if $\varepsilon_A = -1$.

Turning to the time-correlation functions we have:

$$\chi''_{AB}(t-t') = \frac{1}{2\hbar} \langle (T[A(t), B(t')]T^{-1})^\dagger \rangle . \tag{39}$$

Then we have the operator expression:

$$\begin{aligned}
\left[T[A(t), B(t')]T^{-1}\right]^\dagger &= \left[TA(t)T^{-1}TB(t')T^{-1} - TB(t')T^{-1}TA(t)T^{-1}\right]^\dagger \\
&= \varepsilon_A \varepsilon_B \left[A(-t)B(-t') - B(-t')A(-t)\right]^\dagger \\
&= \varepsilon_A \varepsilon_B \left[B(-t')A(-t) - A(-t')B(-t)\right] \\
&= -\varepsilon_A \varepsilon_B [A(-t), B(-t')] \tag{40}
\end{aligned}$$

where we have used the fact that A and B are hermitian. This leads to the result:

$$\chi''_{AB}(t-t') = -\varepsilon_A \varepsilon_B \chi''_{AB}(t'-t) . \tag{41}$$

or, for the Fourier transform,

$$\chi''_{AB}(\omega) = -\varepsilon_A \varepsilon_B \chi''_{AB}(-\omega) . \tag{42}$$

We can now collect the three basic symmetry properties:

$$\chi''_{AB}(\omega) = -\chi''_{BA}(-\omega) \tag{43}$$

$$\chi''^*_{AB}(\omega) = -\chi''_{AB}(-\omega) \tag{44}$$

and:

$$\chi''_{AB}(\omega) = -\varepsilon_A \varepsilon_B \chi''_{AB}(-\omega) . \tag{45}$$

We can rewrite Eq. (45) in the form:

$$\chi''_{AB}(-\omega) = -\varepsilon_A \varepsilon_B \chi''_{AB}(\omega) . \tag{46}$$

Using this on the right-hand sides of Eqs. (43) and (44), we can then collect these various results in their most useful forms:

$$\chi''_{AB}(\omega) = \varepsilon_A \varepsilon_B \chi''_{BA}(\omega) \tag{47}$$

$$\chi''_{AB}(\omega)^* = \varepsilon_A \varepsilon_B \chi''_{AB}(\omega) \tag{48}$$

$$\chi''_{AB}(\omega) = -\varepsilon_A \varepsilon_B \chi''_{AB}(-\omega) \ . \tag{49}$$

If ε_A and ε_B are the same, then $\chi''_{AB}(\omega)$ is even under the exchange of the labels A and B, and real and odd under $\omega \to -\omega$. If the signatures of A and B are different, then $\chi''_{AB}(\omega)$ is odd under exchange of A and B, and imaginary and even under $\omega \to -\omega$.

If we have an autocorrelation function $\chi''_A(\omega) \equiv \chi''_{AA}(\omega)$, then:

$$\chi''^*_A(\omega) = \chi''_A(\omega) \ . \tag{50}$$

$$\chi''_A(\omega) = -\chi''_A(-\omega) \ . \tag{51}$$

If the system is isotropic, then it is invariant under spatial translations, rotations, and reflections (parity). These lead to further symmetry relations. As we shall see these symmetries are rather obvious in physical situations [4].

If we use the fluctuation-dissipation theorem:

$$S_{AB}(\omega) = \hbar \coth(\beta \hbar \omega / 2) \chi''_{AB}(\omega) \ , \tag{52}$$

we can easily write down the symmetry properties satisfied by the fluctuation function $S_{AB}(\omega)$:

$$S_{AB}(\omega) = \varepsilon_A \varepsilon_B S_{BA}(\omega) \tag{53}$$

$$S^*_{AB}(\omega) = \varepsilon_A \varepsilon_B S_{AB}(\omega) \tag{54}$$

$$S_{AB}(\omega) = \varepsilon_A \varepsilon_B S_{AB}(-\omega) \ . \tag{55}$$

3.3
Analytic Properties of Response Functions

Let us now turn to a discussion of the analytic properties of response functions. Remember that it is not $\chi''_{AB}(\omega)$ that we measure directly in a response experiment, but:

$$\chi_{AB}(\omega) = 2i \int_{-\infty}^{+\infty} dt \, e^{+i\omega(t-t')} \theta(t-t') \chi''_{AB}(t-t') \ . \tag{56}$$

3.3 Analytic Properties of Response Functions

We can relate $\chi_{AB}(\omega)$ and $\chi''_{AB}(\omega)$ if we use the integral representation (see Problem 3.4) for the step function:

$$\theta(x) = \int_{-\infty}^{+\infty} \frac{d\omega}{2\pi i} \frac{e^{+ix\omega}}{(\omega - i\eta)}, \tag{57}$$

where η is infinitesimally small and positive. We then have:

$$\begin{aligned}
\chi_{AB}(\omega) &= 2i \int_{-\infty}^{+\infty} dt\, e^{+i\omega(t-t')} \int_{-\infty}^{+\infty} \frac{d\omega'}{2\pi i} \frac{e^{+i(t-t')\omega'}}{(\omega' - i\eta)} \chi''_{AB}(t-t') \\
&= \frac{2i}{2\pi i} \int_{-\infty}^{+\infty} \frac{d\omega'}{\omega' - i\eta} \int_{-\infty}^{+\infty} \frac{d\omega''}{2\pi} \chi''_{AB}(\omega'') \\
&\quad \times \int_{-\infty}^{+\infty} dt\, e^{+i\omega(t-t')} e^{+i\omega'(t-t')} e^{-i\omega''(t-t')} \\
&= \int_{-\infty}^{+\infty} \frac{d\omega'}{\pi} \frac{1}{\omega' - i\eta} \int \frac{d\omega''}{2\pi} \chi''_{AB}(\omega'') 2\pi\, \delta(\omega + \omega' - \omega'') \\
&= \int \frac{d\omega''}{\pi} \frac{\chi''_{AB}(\omega'')}{(\omega'' - \omega - i\eta)}.
\end{aligned} \tag{58}$$

If we use the rather general Plemelj relations [5,6]:

$$\lim_{\eta \to 0^+} \frac{1}{\omega + i\eta} = P\frac{1}{\omega} \pm i\pi\delta(\omega), \tag{59}$$

where P denotes the *principle-value* part, we obtain:

$$\begin{aligned}
\chi_{AB}(\omega) &= \int \frac{d\omega''}{\pi} \chi''_{AB}(\omega'') \left[P\frac{1}{\omega'' - \omega} + i\pi\delta(\omega'' - \omega) \right] \\
&= P \int \frac{d\omega''}{\pi} \frac{\chi''_{AB}(\omega'')}{\omega'' - \omega} + i\chi''_{AB}(\omega).
\end{aligned} \tag{60}$$

In those typical cases where A and B have the same signature under time reversal and $\chi''_{AB}(\omega)$ is real, then we can finally identify $\chi''_{AB}(\omega)$ as the imaginary part of $\chi_{AB}(\omega)$ and:

$$\chi'_{AB}(\omega) = P \int \frac{d\omega'}{\pi} \frac{\chi''_{AB}(\omega')}{\omega' - \omega} \tag{61}$$

as the real part of χ_{AB}. We can now see that the double prime (χ''_{AB}) is associated with it being the imaginary part of $\chi_{AB}(\omega)$. Note also that the real part of χ_{AB} is related to χ''_{AB} via an integral relationship. We can find an inverse relationship if we use the *Poincare-Bertrand* identity [7]:

$$P \int \frac{d\bar{\omega}}{\pi} \frac{1}{(\omega - \bar{\omega})(\omega' - \bar{\omega})} = \int \frac{d\bar{\omega}}{\pi} \left[\frac{1}{\omega - \bar{\omega} - i\eta} - i\pi\delta(\omega - \bar{\omega}) \right]$$

$$\times \left[\frac{1}{\omega' - \bar{\omega} - i\eta} - i\pi\delta(\omega' - \bar{\omega}) \right]$$

$$= -\frac{i\pi}{\pi} \left(\frac{1}{\omega' - \omega - i\eta} + \frac{1}{\omega - \omega' - i\eta} \right) - \frac{\pi^2}{\pi} \delta(\omega - \omega')$$

$$= -i(2\pi i)\delta(\omega - \omega') - \pi\delta(\omega - \omega')$$

$$= \pi\delta(\omega - \omega') \,, \tag{62}$$

where we have used the result:

$$\int \frac{d\bar{\omega}}{\pi} \frac{1}{(\omega - \bar{\omega} - i\eta)} \frac{1}{(\omega' - \bar{\omega} - i\eta)} = 0 \,, \tag{63}$$

which follows if we close the contour integral in the upper half-plane, where there are no singularities. Letting $\omega \to \omega'$ in Eq. (61), multiplying by $(\omega' - \omega)^{-1}$, integrating over ω', dividing by π and taking the principal value we obtain:

$$P\int \frac{d\omega'}{\pi} \frac{\chi'_{AB}(\omega')}{\omega' - \omega} = P\int \frac{d\omega'}{\pi} \frac{1}{\omega' - \omega} \int \frac{d\omega''}{\pi} \frac{\chi''_{AB}(\omega'')}{\omega'' - \omega'}$$

$$= -\int \frac{d\omega''}{\pi} \chi''_{AB}(\omega'') P\int \frac{d\omega'}{\pi} \frac{1}{\omega' - \omega} \frac{1}{\omega' - \omega''}$$

$$= -\chi''_{AB}(\omega) \,. \tag{64}$$

The set of relations:

$$\chi'_{AB}(\omega) = P\int \frac{d\omega'}{\pi} \frac{\chi''_{AB}(\omega')}{\omega' - \omega} \tag{65}$$

$$\chi''_{AB}(\omega) = P\int \frac{d\omega'}{\pi} \frac{\chi'_{AB}(\omega')}{\omega - \omega'} \tag{66}$$

are known as Kramers–Kronig relations [8] and mean we need only χ'_{AB} or χ''_{AB} to construct all of the correlation functions, including χ_{AB}.

The complex response function or susceptibility can be interpreted as the boundary value as z approaches ω on the real axis from above, of the analytic function of z,

$$\chi_{AB}(z) = \int \frac{d\omega'}{\pi} \frac{\chi''_{AB}(\omega')}{\omega' - z} \,. \tag{67}$$

Then, the *causal* part of this response function is given by:

$$\chi_{AB}(\omega) = \chi_{AB}(\omega + i\eta) \,. \tag{68}$$

We introduce $\chi_{AB}(z)$ because, if z is the upper half-plane, $\chi_{AB}(z)$ is just the Laplace transform,

$$\chi_{AB}(z) = 2i \int_0^{+\infty} dt \, e^{+izt} \chi''_{AB}(t) \tag{69}$$

with the imaginary part of z, positive, by Im $z > 0$. If z is the lower half-plane, $\chi(z)$ is determined from negative times,

$$\chi_{AB}(z) = -2i \int_{-\infty}^{0} dt e^{+izt} \chi''_{AB}(t) \tag{70}$$

for Im $z < 0$.

3.4
Symmetries of the Complex Response Function

It is useful to stop and look at the properties satisfied by the response function $\chi_{AB}(\omega)$ when we combine the symmetry properties of $\chi''_{AB}(\omega)$ with the analytic properties of the last section. At the end of this section we will see that these rather formal properties lead to very profound physical results.

Let us begin with the general relation connecting the response function to the dissipation function:

$$\chi_{AB}(\omega) = \int \frac{d\omega''}{\pi} \frac{\chi''_{AB}(\omega'')}{\omega'' - \omega - i\eta} \quad . \tag{71}$$

If we then use the symmetry relation given by Eq. (47) for the interchange of A and B for the dissipation function we obtain:

$$\chi_{AB}(\omega) = \varepsilon_A \varepsilon_B \chi_{BA}(\omega) \quad . \tag{72}$$

Taking the complex conjugate of Eq. (71) and using the symmetry Eq. (48) we obtain:

$$\chi^*_{AB}(\omega) = \varepsilon_A \varepsilon_B \int \frac{d\omega''}{\pi} \frac{\chi''_{AB}(\omega'')}{\omega'' - \omega + i\eta} \quad . \tag{73}$$

Using the Plemelj and Kramers–Kronig relations we easily find:

$$\chi^*_{AB}(\omega) = \varepsilon_A \varepsilon_B \left[\chi'_{AB}(\omega) - i\chi''_{AB}(\omega)\right] \quad . \tag{74}$$

We showed previously that

$$\chi_{AB}(\omega) = \left[\chi'_{AB}(\omega) + i\chi''_{AB}(\omega)\right] \quad . \tag{75}$$

Comparing these results we see that if A and B have the same sign under time reversal then:

$$\chi'_{AB}(\omega) = \text{Re } \chi_{AB}(\omega) \tag{76}$$

$$\chi''_{AB}(\omega) = \text{Im } \chi_{AB}(\mathbf{k}, \omega) \quad . \tag{77}$$

If A and B have the opposite sign under time reversal then:

$$\chi'_{AB}(\omega) = \text{Im}\, \chi_{AB}(\omega) \tag{78}$$

$$\chi''_{AB}(\omega) = \text{Re}\, \chi_{AB}(\omega) \,. \tag{79}$$

Using Eq. (49) to flip the frequency of the dissipation function in Eq. (44) we obtain:

$$\chi_{AB}(\omega) = \int \frac{d\omega'}{\pi} (-\varepsilon_A \varepsilon_B) \frac{\chi''_{AB}(-\omega')}{(\omega' - \omega - i\eta)} \tag{80}$$

$$= (-\varepsilon_A \varepsilon_B) \int \frac{d\omega'}{\pi} \frac{\chi''_{AB}(\omega')}{(-\omega' - \omega - i\eta)} \tag{81}$$

$$= \varepsilon_A \varepsilon_B \int \frac{d\omega'}{\pi} \frac{\chi''_{AB}(\omega')}{(\omega' + \omega + i\eta)} \,. \tag{82}$$

Comparing this result with Eq. (73) we see that:

$$\chi_{AB}(\omega) = \chi^*_{AB}(-\omega) \,. \tag{83}$$

The symmetry reflected by Eq. (72):

$$\chi_{AB}(\omega) = \varepsilon_A \varepsilon_B \chi_{BA}(\omega) \tag{84}$$

has an important physical interpretation: the response of a system as manifested by the nonequilibrium behavior of a variable A as induced by a force coupling to a variable B is, up to a sign, equivalent to the response of variable B to a force coupling to variable A. In the case of magnetic resonance, the application of a field in the y-direction induces a magnetization response in the x-direction that is equivalent to the application of the same force in the x-direction driving a response in the y-direction.

Let us consider a less obvious example of a fluid where temperature or pressure variations couple linearly to the heat density and particle density respectively. In general the external force is given by the contribution to the total Hamiltonian:

$$H_E(t) = -\int d^d x \left[q(\mathbf{x}) \frac{\delta T(\mathbf{x},t)}{T} + n(\mathbf{x}) \frac{\delta p(\mathbf{x},t)}{n} \right] \tag{85}$$

where the heat density q is defined by Eq. (2.178), and δT and δp are imposed variations in temperature and pressure. We can then conclude that the response of the density to the temperature is equal to the heat density variation with the pressure variation in the sense:

$$\frac{\langle q(\mathbf{k},\omega) \rangle_{ne}}{\delta p(\mathbf{k},\omega)/n} = \chi_{qn}(\mathbf{k},\omega) \tag{86}$$

$$\frac{\langle n(\mathbf{k},\omega)\rangle_{ne}}{\delta T(\mathbf{k},\omega)/T} = \chi_{nq}(\mathbf{k},\omega) \tag{87}$$

and because n and q have the same signature (+1) under time reversal [2]:

$$\chi_{qn}(\mathbf{k},\omega) = \chi_{nq}(\mathbf{k},\omega) \; . \tag{88}$$

3.5 The Harmonic Oscillator

It is appropriate at this point to introduce the simplest of examples of a quantum nonequilibrium system: the driven harmonic oscillator. In this case we can work everything out. The isolated system Hamiltonian in this case can be taken to have the form:

$$H = \frac{\hat{p}^2}{2m} + \frac{k}{2}\hat{x}^2 \; , \tag{89}$$

the external coupling:

$$H_E(t) = -\hat{x}h(t) \; , \tag{90}$$

and the total Hamiltonian is given by:

$$H_T(t) = H + H_E(t) \; . \tag{91}$$

In the absence of the applied field we assume the system is in thermal equilibrium at temperature T. With no further work we know that the linear response of the average displacement to the external force is given by:

$$\langle \hat{x}(\omega)\rangle_{ne} = \chi_{xx}(\omega)h(\omega) \; , \tag{92}$$

where $\chi_{xx}(\omega)$ is the response function, which is related by Eq. (58) to the Fourier transform of the dissipation function:

$$\chi''_{xx}(t-t') = \frac{1}{2\hbar}\langle [\hat{x}(t),\hat{x}(t')]\rangle_{eq} \; . \tag{93}$$

Now the simplifying aspect of this example is that we can easily work out the time dependence of the operators $\hat{x}(t)$ and $\hat{p}(t)$ for the isolated system using the equal-time canonical commutation relations:

$$[\hat{x}(t),\hat{p}(t)] = i\hbar \; . \tag{94}$$

The solution to the Heisenberg equations of motion (see Problem 3.5) is given by the usual classical results, but with initial conditions that are operators:

$$\hat{x}(t) = \hat{x}(0)\cos(\omega_0 t) + \frac{\hat{p}(0)}{m\omega_0}\sin(\omega_0 t) \tag{95}$$

$$\hat{p}(t) = -m\omega_0 \hat{x}(0)\sin(\omega_0 t) + \hat{p}(0)\cos(\omega_0 t) \ , \tag{96}$$

where the oscillator frequency is given by:

$$\omega_0^2 = \frac{k}{m} \ . \tag{97}$$

It is then straightforward to determine the commutator for the displacement operator at different times:

$$\begin{aligned}
[\hat{x}(t), \hat{x}(t')] &= \left[\left(\hat{x}(0)\cos(\omega_0 t) + \frac{\hat{p}(0)}{m\omega_0}\sin(\omega_0 t) \right) , \right. \\
&\qquad \left. \left(\hat{x}(0)\cos(\omega_0 t') + \frac{\hat{p}(0)}{m\omega_0}\sin(\omega_0 t') \right) \right] \\
&= \cos(\omega_0 t)\sin(\omega_0 t')\frac{i\hbar}{m\omega_0} + \sin(\omega_0 t)\cos(\omega_0 t')\frac{-i\hbar}{m\omega_0} \\
&= -\frac{i\hbar}{m\omega_0}\sin(\omega_0(t-t')) \ .
\end{aligned} \tag{98}$$

As a check on this result we see that it vanishes when $t = t'$, as it should since an operator commutes with itself at equal times. Note that the commutator, unlike a typical case, is a *c-number* and the average giving the dissipation function is trivial:

$$\chi_{xx}''(t-t') = \frac{(-i)}{2m\omega_0}\sin[\omega_0(t-t')] \ . \tag{99}$$

Notice, as expected, that this autocorrelation function is pure imaginary and odd in $t - t'$. Taking the Fourier transform we easily find:

$$\begin{aligned}
\chi_{xx}''(\omega) &= -\frac{i}{2m\omega_0}\int_{-\infty}^{\infty} dt\, e^{i\omega(t-t')}\left[\frac{e^{i\omega_0(t-t')} - e^{-i\omega_0(t-t')}}{2i} \right] \\
&= \frac{\pi}{2m\omega_0}[\delta(\omega - \omega_0) - \delta(\omega + \omega_0)] \ ,
\end{aligned} \tag{100}$$

which is real and odd under $\omega \to -\omega$. The fluctuation function is given then by the fluctuation-dissipation theorem as:

$$\begin{aligned}
S_{xx}(\omega) &= \hbar\coth\left(\frac{\beta\hbar\omega}{2}\right)\chi_{xx}''(\omega) \\
&= \frac{\hbar\pi}{2m\omega_0}\coth\left(\frac{\beta\hbar\omega_0}{2}\right)[\delta(\omega + \omega_0) + \delta(\omega - \omega_0)] \ ,
\end{aligned} \tag{101}$$

which is real and even in ω.

A very interesting consequence of this result is that we can determine the equilibrium-averaged squared displacement as:

$$\langle x^2 \rangle = \int \frac{d\omega}{2\pi}S_{xx}(\omega) = \frac{\hbar}{2m\omega_0}\coth\left(\frac{\beta\hbar\omega_0}{2}\right) \ . \tag{102}$$

In the classical limit this reduces to:

$$\langle x^2 \rangle = \frac{k_B T}{m\omega_0^2} , \qquad (103)$$

which is equivalent to the equipartition theorem result:

$$\frac{k}{2}\langle x^2 \rangle = \tfrac{1}{2}k_B T . \qquad (104)$$

For zero temperature, $\hbar\omega_0/kT \to \infty$ and $\tanh\left(\frac{\beta\hbar\omega_0}{2}\right) \to 1$ and:

$$\langle x^2 \rangle = \frac{\hbar}{2m\omega_0} , \qquad (105)$$

which is due to *zero-point motion*.

We seem to have evaluated the equilibrium average without appearing to do so! How did this come about? The key is the fluctuation-dissipation theorem, which carries the information about the equilibrium state.

If we look at the cross correlations we find:

$$\chi''_{xp}(t - t') = m\frac{d}{dt'}\chi''_{xx}(t - t') = \frac{i}{2}\cos\left[\omega_0(t - t')\right] . \qquad (106)$$

The Fourier transform is given by:

$$\chi''_{xp}(\omega) = \frac{i\pi}{2}\left[\delta(\omega + \omega_0) + \delta(\omega - \omega_0)\right] \qquad (107)$$

and is even in ω and imaginary as we expect for operators with different signatures under time reversal.

Our explicit results here should be contrasted with those corresponding to the solution for the response function found earlier for the Bloch equations. The significant differences are:

- We included phenomenological damping terms in the Bloch equations.
- The treatment of the Bloch equations was essentially classical.
- The fundamental dynamics of the two systems is different. In the oscillator case the fundamental variables satisfy the commutation relations, $[x, p] = i\hbar$, while for the spin system we have $[s_\alpha, s_\beta] = i\hbar \sum_\gamma \varepsilon_{\alpha\beta\gamma} s_\gamma$ (see Problem 3.6).

3.6
The Relaxation Function

We showed in Chapter 2 that the nonequilibrium dynamics of a variable A relaxing from constrained equilibrium at $t = 0$ can be expressed in the form:

$$\delta A(t) = R_{AB}(t)\chi_{AB}h_B , \qquad (108)$$

where $R_{AB}(t)$ is the relaxation function:

$$R_{AB}(t) = \chi_{AB}^{-1} \int \frac{d\omega}{\pi} \frac{\chi''_{AB}(\omega) e^{+i\omega t}}{\omega - i\eta} \ . \tag{109}$$

Assuming $\chi''_{AB}(\omega = 0) = 0$, which is true if A and B have the same signature under time reversal, we can let $\eta \to 0$ in the denominator.

The complex relaxation function is the Laplace transform of the relaxation function:

$$R_{AB}(z) = \int_0^{+\infty} dt \, e^{+izt} R_{AB}(t) \ , \tag{110}$$

where $z = \omega + i\eta$, with $\eta > 0$. Then, inserting Eq. (109), we find:

$$\begin{aligned} R_{AB}(z) &= \chi_{AB}^{-1} \int \frac{d\omega}{\pi} \frac{\chi''_{AB}(\omega)}{\omega} \int_0^{+\infty} e^{+i(z-\omega)t} \\ &= \chi_{AB}^{-1} \int \frac{d\omega}{i\pi} \frac{\chi''_{AB}(\omega)}{\omega(\omega - z)} \ . \end{aligned} \tag{111}$$

This complex relaxation function can be related back to the complex susceptibility,

$$\chi_{AB}(z) = \int \frac{d\omega}{\pi} \frac{\chi''_{AB}(\omega)}{\omega - z} \ , \tag{112}$$

if we use the identity:

$$\frac{1}{\omega - z} = -\frac{1}{z} + \frac{\omega}{z} \frac{1}{\omega - z} \ . \tag{113}$$

We then have:

$$\begin{aligned} R_{AB}(z) &= \chi_{AB}^{-1} \int \frac{d\omega}{\pi i} \frac{\chi''_{AB}(\omega)}{\omega} \left[-\frac{1}{z} + \frac{\omega}{z} \frac{1}{\omega - z} \right] \\ &= \frac{\chi_{AB}^{-1}}{iz} \left[-\int \frac{d\omega}{\pi} \frac{\chi''_{AB}(\omega)}{\omega} + \chi_{AB}(z) \right] \\ &= \frac{\chi_{AB}^{-1}}{iz} [\chi_{AB}(z) - \chi_{AB}] \ . \end{aligned} \tag{114}$$

So the complex relaxation function can be expressed in terms of the complex susceptibility, which, in turn, can be expressed in terms of $\chi''_{AB}(\omega)$.

We could have approached this relaxation problem from an alternative point of view. Suppose for times $t < t_0$ the system is in constrained thermal equilibrium with the density matrix:

$$\rho_C = \frac{e^{-\beta(H + H_E)}}{Z_C} \tag{115}$$

where:
$$Z_C = Tr\, e^{-\beta(H+H_E)} \tag{116}$$

and:
$$H_E = -Bh_B\ . \tag{117}$$

At time t_0 we switch H_E off and the system propagates forward in time governed by the Hamiltonian H. In the Heisenberg representation the density matrix is time independent, so for $t \geq t_0$ we have:
$$\langle A(t)\rangle_{neq} = Tr\, \rho_C\, A(t)\ , \tag{118}$$

where:
$$A(t) = e^{iH(t-t_0)/\hbar} A e^{-iH(t-t_0)/\hbar}\ . \tag{119}$$

If we expand in powers of h_B this should be equivalent to our previously derived result for constrained equilibrium. This is the approach taken in one of the earliest contributions to linear response theory. Kubo developed perturbation theory for the density matrix rather than for the Heisenberg operators representing observables.

If one focuses on the quantity:
$$\rho_C Z_C = U(\beta) \equiv e^{-\beta(H+H_E)}\ , \tag{120}$$

then one can follow the treatment of the time evolution operator $U(t,t')$ given in Chapter 2. $U(\beta)$ satisfies the *equation of motion*:
$$\frac{\partial U(\beta)}{\partial \beta} = -(H+H_E)U(\beta)\ , \tag{121}$$

and following the development used in treating $U(t,t')$ (see Problem 3.7), we find that the density matrix, to first order in H_E, is given by:
$$\rho_C = \rho_0\left(1 - \int_0^\beta d\lambda\, \delta H_E^I(\lambda) + \cdots\right)\ , \tag{122}$$

where:
$$\delta H_E^I(\lambda) = e^{\lambda H} H_E e^{-\lambda H} - Tr\rho_0 H_E \tag{123}$$

and the unconstrained equilibrium density matrix is given by:
$$\rho_0 = e^{-\beta H}/Tr\, e^{-\beta H}\ . \tag{124}$$

Notice in Eq. (122), to first order in h_B, that ρ_C is properly normalized. The nonequilibrium average is given then to first order in h_B by:

$$\langle A(t)\rangle_{ne} = Tr\rho_0\left[1 - \int_0^\beta d\lambda \delta H_E^I(\lambda)\right] A(t)$$

$$= \langle A(t)\rangle_{eq} - \int_0^\beta d\lambda \langle \delta H_E^I(\lambda) A(t)\rangle \quad (125)$$

or:

$$\delta A(t) = h_B \beta C_{AB}(t) \quad (126)$$

where the *Kubo response* function is defined:

$$C_{AB}(t-t') = \beta^{-1} \int_0^\beta d\lambda \langle e^{\lambda H}\delta B(t') e^{-\lambda H}\delta A(t)\rangle \quad (127)$$

It is shown in Problem 3.12 that the Kubo response function satisfies the relation:

$$\frac{\partial}{\partial t}C_{AB}(t-t') = -\frac{2i}{\beta}\chi''_{AB}(t-t') \quad (128)$$

In terms of Fourier transforms:

$$\chi''_{AB}(\omega) = \int_{-\infty}^{+\infty} dt\, e^{i\omega t}\, \frac{\beta i}{2}\frac{\partial}{\partial t}C_{AB}(t)$$

$$= \left(\frac{\beta\omega}{2}\right) C_{AB}(\omega) \quad . \quad (129)$$

and $C_{AB}(\omega)$ is simply related to $\chi''_{AB}(\omega)$.

The equal-time Kubo response function is related to the static susceptibility by inserting Eq. (129) into:

$$C_{AB}(t=0) = \int \frac{d\omega}{2\pi} C_{AB}(\omega) = \int \frac{d\omega}{2\pi}\frac{2}{\beta\omega}\chi''_{AB}(\omega)$$

$$= \beta^{-1}\chi_{AB} \quad . \quad (130)$$

For times $t > 0$ it is convenient to introduce the complex Kubo function via the Laplace transform:

$$C_{AB}(z) = -i\int_0^{+\infty} dt\, e^{+izt} C_{AB}(t)$$

$$= -i\int_0^{+\infty} dt\, e^{+izt} \int_{-\infty}^{+\infty} \frac{d\omega}{2\pi} e^{-i\omega t} C_{AB}(\omega)$$

$$C_{AB}(z) = \int_{-\infty}^{+\infty} \frac{d\omega}{2\pi} \frac{C_{AB}(\omega)}{z - \omega} . \tag{131}$$

In terms of the Laplace transform, we have for the Kubo approach to the relaxation problem:

$$\begin{aligned} \delta A(z) &= \int_0^{+\infty} dt\, e^{+izt} \delta A(t) \\ &= \int_0^{+\infty} dt\, e^{+izt} \beta C_{AB}(t) h_B \\ &= i\beta C_{AB}(z) h_B . \end{aligned} \tag{132}$$

We can compare this with the result where we adiabatically turned on h_B. Taking the Laplace transform of Eq. (108), remembering Eq. (110), we obtain:

$$\delta A(z) = R_{AB}(z) \chi_{AB} h_B \tag{133}$$

and for the two approaches to be equivalent we must have:

$$i\beta C_{AB}(z) = R_{AB}(z) \chi_{AB} \tag{134}$$

It is left to Problem 3.12 to show that these are equivalent.

3.7
Summary of Correlation Functions

Let us summarize the various definitions of correlation functions we have introduced:

$$\chi''_{AB}(t - t') = \frac{1}{2\hbar} \langle [A(t), B(t')] \rangle \tag{135}$$

$$\chi_{AB}(t - t') = 2i\theta(t - t') \chi''_{AB}(t - t') \tag{136}$$

$$\bar{S}_{AB}(t - t') = \langle \delta A(t) \delta B(t') \rangle \tag{137}$$

$$S_{AB}(t - t') = \tfrac{1}{2} \langle [\delta A(t) \delta B(t') + \delta B(t') \delta A(t)] \rangle \tag{138}$$

$$C_{AB}(t - t') = \beta^{-1} \int_0^\beta d\lambda \langle e^{\lambda H} \delta B(t') e^{-\lambda H} \delta A(t) \rangle . \tag{139}$$

The Fourier transforms of all these quantities are interrelated. We can, for example, express them all in terms of $\chi''(\omega)$:

$$\chi_{AB}(\omega) = \int \frac{d\omega'}{\pi} \frac{\chi''_{AB}(\omega')}{(\omega' - \omega - i\eta)} \tag{140}$$

$$\bar{S}_{AB}(\omega) = 2\hbar(1 - e^{-\beta\hbar\omega})^{-1}\chi''_{AB}(\omega) \tag{141}$$

$$S_{AB}(\omega) = \hbar \coth(\beta\hbar\omega/2)\chi''_{AB}(\omega) \tag{142}$$

$$C_{AB}(\omega) = \frac{2}{\beta\omega}\chi''_{AB}(\omega) \ . \tag{143}$$

We also define the complex susceptibility $\chi_{AB}(z)$, relaxation function $R_{AB}(z)$ and Kubo function $C_{AB}(z)$. These are all linearly related since:

$$\chi_{AB}(z) = \int \frac{d\omega}{\pi} \frac{\chi''_{AB}(\omega)}{\omega - z} \tag{144}$$

$$R_{AB}(z) = \frac{i}{z} \chi_{AB}^{-1}[\chi_{AB} - \chi_{AB}(z)] \ , \tag{145}$$

where:

$$\chi_{AB} = \chi_{AB}(z=0) = \int \frac{d\omega}{\pi} \frac{\chi''_{AB}(\omega)}{\omega} \tag{146}$$

is the static susceptibility, and:

$$C_{AB}(z) = \int \frac{d\omega}{2\pi} \frac{C_{AB}(\omega)}{z - \omega} = -i\beta^{-1}\chi_{AB}R_{AB}(z) \ . \tag{147}$$

The complex fluctuation function,

$$S_{AB}(z) = \int \frac{d\omega}{2\pi} \frac{S_{AB}(\omega)}{z - \omega} \ , \tag{148}$$

can not be expressed directly in terms of $\chi_{AB}(z)$.

If we eliminate $R_{AB}(z)$ between Eqs. (145) and (147) we find:

$$\chi_{AB}(z) = \chi_{AB} - \beta z C_{AB}(z) \ . \tag{149}$$

We will need this result later.

3.8
The Classical Limit

We will, in many cases, be interested in classical systems. In this limit a number of our general relationships, summarized above, simplify. First we note that in the classical limit, where $\hbar \to 0$, the fluctuation-dissipation theorem becomes:

$$S_{AB}(\omega) = \frac{2}{\beta\omega} \chi''_{AB}(\omega) \ . \tag{150}$$

We see then, since Eq. (129) gives:

$$C_{AB}(\omega) = \frac{2}{\beta\omega}\chi''_{AB}(\omega) \; , \tag{151}$$

that:

$$S_{AB}(\omega) = C_{AB}(\omega) \; . \tag{152}$$

Classically the fluctuation and Kubo functions are equal. We can easily understand this equality. Classically, operators commute so:

$$S_{AB}(t-t') = \bar{S}_{AB}(t-t') = \langle \delta A(t)\delta B(t')\rangle \tag{153}$$

and, starting with Eq. (139),

$$C_{AB}(t-t') = \beta^{-1}\int_0^\beta d\lambda \langle \delta B(t')\delta A(t)\rangle = \langle \delta A(t)\delta B(t')\rangle \; . \tag{154}$$

We see that the static susceptibility in the classical limit can be written as:

$$\chi_{AB} = \int \frac{d\omega}{\pi}\frac{\chi''_{AB}(\omega)}{\omega} = \beta \int \frac{d\omega}{2\pi} S_{AB}(\omega) \tag{155}$$
$$= \beta\langle \delta A \delta B\rangle \tag{156}$$

and the susceptibility is simply related [9] to the equal-time correlation function.

3.9
Example: The Electrical Conductivity

We want to show how these various properties can be used to analyze quantities like the electrical conductivity. We have from Eqs. (2.80) and (2.81) that:

$$\sigma_{\alpha\beta}(t-t') = \chi_{J_\alpha,R_\beta}(t-t') \tag{157}$$

where **J** is the charge-current density and $\mathbf{R} = \sum_{i=1}^N q_i \mathbf{r}_i$. In term of Fourier transforms:

$$\sigma_{\alpha\beta}(\omega) = \chi_{J_\alpha,R_\beta}(\omega) \; . \tag{158}$$

Using Eq. (71) we can express the frequency-dependent conductivity in the form:

$$\sigma_{\alpha\beta}(\omega) = \int \frac{d\omega'}{\pi}\frac{\chi''_{J_\alpha,R_\beta}(\omega')}{\omega' - \omega - i\eta} \; . \tag{159}$$

We can rewrite Eq. (159) as:

$$\sigma_{\alpha\beta}(\omega) = \int \frac{d\omega'}{\pi} \frac{1}{(\omega'-\omega-i\eta)\omega'} \omega' \chi''_{J_\alpha,R_\beta}(\omega') \ . \qquad (160)$$

and look at:

$$\begin{aligned}
\omega' \chi''_{J_\alpha,R_\beta}(\omega') &= \int dt\, e^{+i\omega'(t-t')} \omega' \chi''_{J_\alpha,R_\beta}(t-t') \\
&= \int dt\, (-i\frac{\partial}{\partial t} e^{+i\omega'(t-t')}) \chi''_{J_\alpha,R_\beta}(t-t') \\
&= i \int dt\, e^{+i\omega'(t-t')} \frac{d}{\partial t}\chi''_{J_\alpha,R_\beta}(t-t') \\
&= -i \int dt\, e^{+i\omega'(t-t')} \frac{d}{\partial t'}\chi''_{J_\alpha,R_\beta}(t-t') \\
&= -i \int dt\, e^{+i\omega'(t-t')} \chi''_{J_\alpha,\dot R_\beta}(t-t') \\
&= -i\chi''_{J_\alpha,J_\beta^T}(\omega') \\
&= -\frac{i}{V}\chi''_{J_\alpha^T,J_\beta^T}(\omega') \ , \qquad (161)
\end{aligned}$$

where we have used time-translational invariance and noted that the total current can be written as:

$$J_\beta^T(t') = \frac{d}{dt'} R^\beta(t') = \sum_i^N q_i v_i^\beta(t') = V J_\beta(t') \ . \qquad (162)$$

Using Eq. (160) we have:

$$\sigma_{\alpha\beta}(\omega) = -\frac{i}{V} \int \frac{d\omega'}{\pi(\omega'-\omega-i\eta)\omega'} \chi''_{J_\alpha^T,J_\beta^T}(\omega') \ . \qquad (163)$$

However, since the Kubo correlation function is related to χ'' by Eq. (129):

$$C_{J_\alpha^T,J_\beta^T}(\omega) = \frac{2}{\beta\omega} \chi''_{J_\alpha^T,J_\beta^T}(\omega) \ , \qquad (164)$$

we have:

$$\sigma_{\alpha\beta}(\omega) = -\frac{i\beta}{V} \int \frac{d\omega'}{2\pi} \frac{C_{J_\alpha^T,J_\beta^T}(\omega')}{\omega'-\omega-i\eta} \ . \qquad (165)$$

The dc-conductivity is then given by:

$$\sigma_{\alpha\beta}(0) = -\frac{i\beta}{V} \int \frac{d\omega'}{2\pi} \frac{C_{J_\alpha^T,J_\beta^T}(\omega')}{\omega'-i\eta} \ . \qquad (166)$$

If we look at the longitudinal conductivity:

$$\sigma_L = \frac{1}{3}\sum_\alpha \sigma_{\alpha\alpha}(0)$$
$$= \frac{-i\beta}{3V}\sum_\alpha \int \frac{d\omega'}{2\pi} C_{J_\alpha^T,J_\alpha^T}(\omega')\left[P\frac{1}{\omega'} + i\pi\delta(\omega')\right] \quad . \tag{167}$$

Since a Kubo autocorrelation function is even under $\omega \to -\omega$,

$$C_{J_\alpha^T,J_\alpha^T}(\omega) = C_{J_\alpha^T,J_\alpha^T}(-\omega) \quad , \tag{168}$$

and we obtain:

$$\sigma_L = \frac{\beta}{6V}\sum_\alpha C_{J_\alpha^T,J_\alpha^T}(0) \quad . \tag{169}$$

This can be rewritten as a time integral:

$$\sigma_L = \frac{\beta}{6V}\int_{-\infty}^{+\infty} dt\, C_{J^T\cdot J^T}(t) = \frac{\beta}{3V}\int_0^{+\infty} dt\, C_{J^T\cdot J^T}(t) \quad , \tag{170}$$

over a current–current Kubo function. Relationships of this type, giving a transport coefficient in terms of an integral over an equilibrium-averaged time-correlation function, played a central role in the development of linear response theory. They are called Kubo or Green–Kubo formulae. Such relationships give one precise definitions of transport coefficients.

One of the most important ramifications of a Kubo formula is easily seen in the classical limit where:

$$\sigma_L = \frac{\beta}{3V}\int_0^{+\infty} dt\, \langle \mathbf{J}^T(t)\cdot \mathbf{J}^T(0)\rangle \quad . \tag{171}$$

In this limit one has an integral over a simple current–current time fluctuation function. This provides a direct and clean method for determining σ_L using molecular dynamics: direct numerical simulation of a fluctuating equilibrium system.

3.10 Nyquist Theorem

Let us see how the results developed in the last section lead to the famous Nyquist theorem. We start with Eq. (165):

$$\sigma_{xx}(\omega) = -\frac{i\beta}{V}\int \frac{d\omega'}{2\pi}\frac{C_{J_x^T,J_x^T}(\omega')}{\omega'-\omega-i\eta} \quad . \tag{172}$$

If we take the real part of the conductivity we have:

$$\sigma'_{xx}(\omega) = \text{Re}\left[-\frac{i\beta}{V}\int\frac{d\omega'}{2\pi}C_{J_x^T,J_x^T}(\omega')\left[P\frac{1}{\omega'-\omega}+i\pi\delta(\omega'-\omega)\right]\right]$$

$$= \frac{\beta}{2V}C_{J_x^T,J_x^T}(\omega)$$

$$= \frac{1}{2k_BTV}\int_{-\infty}^{\infty}dt e^{i\omega(t-t')}\langle J_x^T(t)J_x^T(t')\rangle \quad . \tag{173}$$

Following Kubo [10] we take the inverse Fourier transform to obtain:

$$\langle J_x^T(t)J_x^T(t')\rangle = 2k_BTV\int_{-\infty}^{\infty}\frac{d\omega}{2\pi}\sigma'_{xx}(\omega)e^{-i\omega(t-t')} \quad . \tag{174}$$

If, over the frequency range we probe current fluctuations, the conductivity can be taken to be a constant, $\sigma'_{xx}(\omega)\approx\sigma$, then:

$$\langle J_x^T(t)J_x^T(t')\rangle = 2k_BT\sigma V\delta(t-t') \quad . \tag{175}$$

The fluctuations in the current in this regime are δ-correlated. It is then consistent to assume that the current density fluctuations can be related to the voltage fluctuations via:

$$V(t) = LJ_x(t)/\sigma \quad . \tag{176}$$

The total current can then be written:

$$J_x^T(t) = VJ_x = \frac{V\sigma}{L}V(t) = S\sigma V(t) \tag{177}$$

$$= \frac{L}{R}V(t) \tag{178}$$

where $S = V/L$ is the cross-sectional area and the resistance for the sample is given by $R = L/\sigma S$. The voltage fluctuations are given then by:

$$\langle V(t)V(t')\rangle = 2k_BT\left(\frac{R}{L}\right)^2 V\frac{L}{RS}\delta(t-t') \quad . \tag{179}$$

$$= 2k_BTR\delta(t-t') \quad . \tag{180}$$

In terms of fluctuations in the frequency domain:

$$\langle V(\omega)V(\omega')\rangle = 2k_BTR2\pi\delta(\omega+\omega') \quad . \tag{181}$$

Looking at the fluctuations in a frequency window $\Delta\omega$:

$$V_\Delta = \int_{\omega_0}^{\omega_0+\Delta\omega}d\omega_1 V(\omega_1) \tag{182}$$

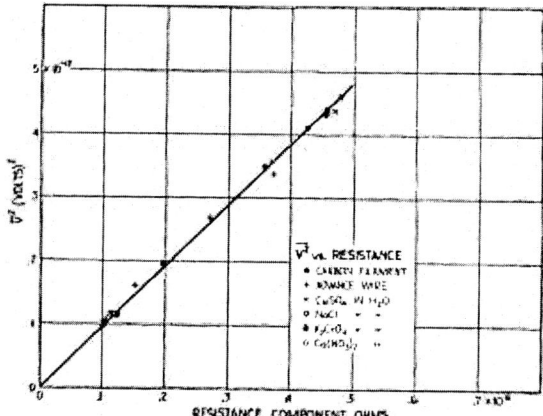

Fig. 3.1 The classic demonstration that Eq. (183) holds is due to J. B. Johnson, Ref. [1] where $\langle V^2 \rangle$ is plotted versus R.

then the fluctuation in the voltage is proportional to the resistance:

$$\langle V_\Delta^2 \rangle = 2k_B T R 2\pi \Delta\omega \ . \tag{183}$$

For several different derivations of this result see the discussion in Kittel [11]. This result is verified in Fig. 3.1.

3.11
Dissipation

Let us consider the work done on our condensed-matter system by the externally applied time-dependent field h. We have, working in the Schroedinger representation, that the work done on the system is given by:

$$W = \int_{-\infty}^{\infty} dt \, Tr \, \rho_{ne}(t) \frac{\partial H_T}{\partial t} \ , \tag{184}$$

where the derivative is with respect to the explicit time dependence of the Hamiltonian (which is evident if we work in the Schroedinger representation). Assuming that the Hamiltonian governing the system is of the form given by Eq. (2.6), Eq. (184) can be written in the form:

$$W = -\int_{-\infty}^{+\infty} dt \, Tr\rho_{ne}(t) \sum_i A_i \frac{\partial h_i(t)}{\partial t}$$

$$= -\int_{-\infty}^{\infty} dt \sum_i \langle A_i(t) \rangle_{ne} \frac{\partial h_i(t)}{\partial t} \ . \tag{185}$$

If we integrate by parts and assume the applied field vanishes at distant times,

$$\lim_{|t|\to\infty} h_i(t) \langle A_i(t) \rangle_{ne} = 0 \ , \tag{186}$$

then:
$$W = \int_{-\infty}^{+\infty} dt \sum_i h_i(t) \frac{\partial}{\partial t}\langle A_i(t)\rangle_{ne} \ . \tag{187}$$

If we work to lowest order in h, then we can use the linear response result:
$$\langle A_i(t)\rangle_{ne} = \langle A_i(t)\rangle_{eq} + \sum_j \int_{-\infty}^{+\infty} dt' \chi_{ij}(t-t')h_j(t') \ , \tag{188}$$

and, since $\langle A_i(t)\rangle_{eq}$ is time independent,
$$W = \int_{-\infty}^{+\infty} dt \int_{-\infty}^{+\infty} dt' \sum_{i,j} h_i(t)\frac{\partial}{\partial t}\chi_{ij}(t-t')h_j(t') + \mathcal{O}(h^3) \ . \tag{189}$$

Replacing $\chi_{ij}(t-t')$ and the applied fields with their Fourier transforms and doing the time integrations:
$$W = \int_{-\infty}^{+\infty} dt \int_{-\infty}^{+\infty} dt' \sum_{i,j} h_i(t) \int \frac{d\omega}{2\pi} e^{-i\omega(t-t')}(-i\omega)\chi_{ij}(\omega)h_j(t')$$
$$= \sum_{i,j} \int \frac{d\omega}{2\pi} h_i^*(\omega)(-i\omega)\chi_{ij}(\omega)h_j(\omega) \ . \tag{190}$$

If we let $\omega \to -\omega$ [note that $h_i(t)$ is real, so $h_i^*(\omega) = h_i(-\omega)$] and let $i \leftrightarrow j$, we obtain:
$$W = \sum_{i,j} \int \frac{d\omega}{2\pi} h_i^*(\omega)(i\omega)\chi_{ji}(-\omega)h_j(\omega) \ . \tag{191}$$

Adding the two expressions for W, Eqs. (190) and (191), and dividing by two gives:
$$W = \sum_{i,j} \int \frac{d\omega}{2\pi} h_i^*(\omega)(-i\omega)h_j(\omega)\tfrac{1}{2}\left[\chi_{ij}(\omega) - \chi_{ji}(-\omega)\right] \ . \tag{192}$$

Using the symmetries we found previously,
$$\chi_{ji}(\omega) = \varepsilon_i\varepsilon_j\chi_{ij}(\omega) \tag{193}$$

and:
$$\chi_{ij}(\omega) = \chi_{ij}^*(-\omega) \ , \tag{194}$$

we can write the factor:
$$\chi_{ij}(\omega) - \chi_{ji}(-\omega) = \chi_{ij}(\omega) - \varepsilon_i\varepsilon_j\chi_{ij}^*(\omega) \tag{195}$$

and obtain:

$$W = \sum_{i,j} \int \frac{d\omega}{2\pi} h_i^*(\omega)(-i\omega)h_j(\omega)\tfrac{1}{2}\left[\chi_{ij}(\omega) - \varepsilon_i\varepsilon_j\chi_{ij}^*(\omega)\right] . \tag{196}$$

In a stable system (one that is decaying to equilibrium), we require that:

$$W > 0 . \tag{197}$$

We must put energy into the system to excite it. Since the fields $h_i^*(\omega)h_j(\omega)$ are arbitrary, we could in principle excite a single mode k, $h_i = h\delta_{i,k}$ with a single Fourier component to obtain:

$$\begin{aligned}W &= \int \frac{d\omega}{2\pi} |h_k(\omega)|^2 (-i\omega)\tfrac{1}{2}[\chi_{kk}(\omega) - \chi_{kk}^*(\omega)] \\ &= \int \frac{d\omega}{2\pi} |h_k(\omega)|^2 \omega\chi_{kk}''(\omega) > 0 ,\end{aligned} \tag{198}$$

which implies:

$$\omega\chi_{kk}''(\omega) \geq 0 \tag{199}$$

for any ω. We also see that we can identify $\chi_{ii}''(\omega)$ with the dissipation in the system.

3.12
Static Susceptibility (Again)

We showed earlier that the linear response to a static external field is given by:

$$\delta A = \chi_{AB} h_B + \mathcal{O}(h_B^2) , \tag{200}$$

where:

$$\chi_{AB} = \int \frac{d\omega}{\pi} \frac{\chi_{AB}''(\omega)}{\omega} \tag{201}$$

is the static susceptibility. If $A(\mathbf{x})$ is some observable density and $h_B(\mathbf{x})$ couples to the density $B(\mathbf{x})$, then:

$$\delta A(\mathbf{x}) = \int d^d y\, \chi_{AB}(\mathbf{x}-\mathbf{y})h_B(\mathbf{y}) . \tag{202}$$

In the case of a uniform external field:

$$\begin{aligned}\delta A(\mathbf{x}) &= h_B \int d^d y\, \chi_{AB}(\mathbf{x}-\mathbf{y}) \\ &= h_B \chi_{AB}(k=0) \end{aligned} \tag{203}$$

and $\delta A(\mathbf{x})$ is uniform. In the limit, as h_B goes to zero, we have the result that the thermodynamic derivative is related to the small k-limit of a susceptibility:

$$\frac{\partial A}{\partial h_B} = \lim_{\mathbf{k}\to 0} \chi_{AB}(\mathbf{k}) \quad . \tag{204}$$

Notice that we can write:

$$\chi_{AB}(k=0) = \int \frac{d^d x d^d y}{V} \chi_{AB}(\mathbf{x}-\mathbf{y}) \tag{205}$$

$$= \frac{\chi_{A_T B_T}}{V} \tag{206}$$

where:

$$A_T = \int d^d x A(\mathbf{x}) \tag{207}$$

is the *total* amount of A. Then:

$$\frac{\partial A}{\partial h_B} = \frac{\chi_{A_T B_T}}{V} \tag{208}$$

or:

$$\frac{\partial A_T}{\partial h_B} = \chi_{A_T B_T} \quad . \tag{209}$$

Let us consider the static susceptibility:

$$\chi_{A_T B_T} = \int \frac{d\omega}{\pi} \frac{\chi''_{A_T B_T}(\omega)}{\omega} \quad . \tag{210}$$

Using Eq. (129) we can write:

$$\chi_{A_T B_T} = \beta \int \frac{d\omega}{2\pi} C_{A_T B_T}(\omega)$$

$$= \beta C_{A_T B_T}(tt) \quad . \tag{211}$$

We can then use Eq. (127) to write:

$$\chi_{A_T B_T} = \int_0^\beta d\lambda \langle e^{\lambda H} \delta B_T(t) e^{-\lambda H} \delta A_T(t) \rangle \quad . \tag{212}$$

If B is conserved it commutes with H and:

$$\chi_{A_T B_T} = \beta \langle \delta B_T(t) \delta A_T(t) \rangle \quad . \tag{213}$$

This also holds if A is conserved. Equations (209) and (213) allow us to write:

$$\frac{\partial A_T}{\partial h_B} = \beta \langle \delta B_T \delta A_T \rangle \quad . \tag{214}$$

Let us consider the example of the magnetic susceptibility in the case where the total magnetization and the applied field are both in the z-direction. The magnetic susceptibility per unit volume is given in zero external field by:

$$\chi_M = \frac{1}{V}\frac{\partial \langle M_z^T \rangle}{\partial B} = \frac{\partial \langle M_z \rangle}{\partial B} \tag{215}$$

$$= \frac{\beta}{V}\langle (\delta M_z^T)^2 \rangle \tag{216}$$

where M_z^T is the total magnetization in the z-direction.

3.13
Sum Rules

There is one final class of formal properties of time-correlation functions that can be of some practical use. These properties are known as sum rules and allow one to calculate quantities like the integral:

$$\int \frac{d\omega}{\pi}\, \omega^n\, \chi''_{AB}(\omega) = \omega_{AB}(n) \tag{217}$$

explicitly for certain small integers n. $\omega_{AB}(n)$ gives the area under $\omega^n \chi''_{AB}(\omega)$. To see how we can compute $\omega_{AB}(n)$, consider first the identity:

$$\left[\left(i\frac{\partial}{\partial t}\right)^n \chi''_{AB}(t-t')\right]_{t=t'} = \frac{1}{2\hbar}\left\langle \left[\left(\frac{i\partial}{\partial t}\right)^n A(t), B(t')\right]\right\rangle_{t=t'}. \tag{218}$$

If we replace $\chi''_{AB}(t-t')$ with its Fourier transform, we easily obtain:

$$\frac{\omega_{AB}(n)}{2} = \int \frac{d\omega}{2\pi}\, \omega^n \chi''_{AB}(\omega) = \frac{1}{2\hbar}\left\langle \left[\left(\frac{i\partial}{\partial t}\right)^n A(t), B(t')\right]\right\rangle_{t=t'}, \tag{219}$$

or:

$$\omega_{AB}(n) = \frac{1}{\hbar}\left\langle \left[\left(\frac{i\partial}{\partial t}\right)^n A(t), B(t')\right]\right\rangle_{t=t'}. \tag{220}$$

If we further note, on repeated use of the Heisenberg equation of motion, that:

$$\left(\frac{i\partial}{\partial t}\right)^n A(t) = \frac{1}{\hbar^n}[\ldots[A(t), H], \ldots, H] , \tag{221}$$

and:

$$\omega_{AB}(n) = \frac{1}{\hbar^{n+1}}\langle [[[\ldots[A(t), H], \ldots, H] \ldots, H], B(t)]\rangle . \tag{222}$$

The sum rules are given then by commutators at *equal* times. Looking at $n = 0$, we have:

$$\omega_{AB}(0) = \frac{1}{\hbar} \langle [A(t), B(t)] \rangle = \int \frac{d\omega}{\pi} \chi''_{AB}(\omega) \tag{223}$$

We know from Eq. (45) that:

$$\chi''_{AB}(\omega) = -\varepsilon_A \varepsilon_B \chi''_{AB}(-\omega) \tag{224}$$

so:

$$\omega_{AB}(0) = \frac{1}{2}(1 - \varepsilon_A \varepsilon_B) \int \frac{d\omega}{\pi} \chi''_{AB}(\omega) \, , \tag{225}$$

and for operators with the same time-reversal signature:

$$\omega_{AB}(0) = 0 \, , \tag{226}$$

while it is in general nonzero for those with the opposite time reversal properties. We note here that the moments $\omega_{AB}(n)$ are associated with the large-frequency behavior of the complex susceptibility.:

$$\chi_{AB}(z) = \int \frac{d\omega}{\pi} \frac{\chi''_{AB}(\omega)}{\omega - z} \, . \tag{227}$$

We can use the identity:

$$\frac{1}{\omega - z} = -\frac{1}{z}\left(\frac{1}{1 - \omega/z}\right) = -\frac{1}{z}\sum_{n=0}^{\infty}\left(\frac{\omega}{z}\right)^n \tag{228}$$

to write:

$$\chi_{AB}(z) = -\frac{1}{z}\sum_{n=0}^{\infty}\frac{\omega_{AB}(n)}{z^n} \, . \tag{229}$$

Clearly this large z expansion can only be asymptotically correct. That is, it is only valid when $|z|$ is large compared to *all frequencies in the system*, which means all frequencies ω for which $\chi''_{AB}(\omega)$ is not effectively zero.

Let us now consider two examples where we can evaluate the sum rules explicitly. First let us consider the case of the density–density response function. We see immediately that:

$$\omega_n(\mathbf{k}, 0) = \int \frac{d\omega}{\pi} \chi''_{nn}(\mathbf{k}, \omega)$$

$$= \int d^3x e^{+i\mathbf{k}\cdot(\mathbf{x}-\mathbf{y})} \frac{1}{2\hbar} \langle [\hat{n}(\mathbf{x}), \hat{n}(\mathbf{y})] \rangle = 0 \, , \tag{230}$$

since the density commutes with itself at different space points. We will be more interested in the next order sum rule:

$$\omega_n(\mathbf{k}, 1) = \int \frac{d\omega}{\pi} \omega \chi''_{nn}(\mathbf{k}, \omega)$$

$$= \int d^3x e^{+i\mathbf{k}\cdot(\mathbf{x}-\mathbf{y})} \omega_n(\mathbf{x}-\mathbf{y}, 1)$$

$$\omega_n(\mathbf{x}-\mathbf{y}, 1) = \frac{1}{\hbar^2} \langle [\hat{n}(\mathbf{x}), H], \hat{n}(\mathbf{y})] \rangle \ . \tag{231}$$

In a simple fluid, the Hamiltonian is given by:

$$H = \sum_{i=1}^{N} \frac{\mathbf{p}_i^2}{2m} + \frac{1}{2} \sum_{i \neq j}^{N} V(\mathbf{r}_i - \mathbf{r}_j) \tag{232}$$

and:

$$\mathbf{p}_i = -i\hbar \vec{\nabla}_i \tag{233}$$

in the coordinate representation.

To determine the value of the sum rule given by Eq. (231), we need to evaluate two commutators. The first commutator is between the density and Hamiltonian. It is important to recall that this commutator is related to the time evolution of the density via:

$$[n, H] = i\hbar \frac{\partial n}{\partial t} \ . \tag{234}$$

We need to evaluate the commutator:

$$[n(\mathbf{x}), H] = \sum_{i,j} \left[\delta(\mathbf{x}-\mathbf{r}_i), \frac{\mathbf{p}_j^2}{2m} \right] \ , \tag{235}$$

where we have used the fact that the density n and the potential energy commute. The rest of the evaluation is carried out in Problem 3.8 with the result:

$$[n(\mathbf{x}), H] = -\frac{i\hbar}{m} \nabla_x \cdot \mathbf{g}(\mathbf{x}, t) \ , \tag{236}$$

where $\mathbf{g}(\mathbf{x}, t)$ is the momentum density,

$$\mathbf{g}(\mathbf{x}, t) = \frac{1}{2} \sum_{i=1}^{N} [\mathbf{p}_i \delta(\mathbf{x}-\mathbf{r}_i) + \delta(\mathbf{x}-\mathbf{r}_i)\mathbf{p}_i] \tag{237}$$

and one has the *continuity equation*:

$$\frac{\partial n(\mathbf{x}, t)}{\partial t} = -\nabla \cdot \mathbf{g}(\mathbf{x}, t)/m \ , \tag{238}$$

which expresses conservation of particle number.

The second commutator to be evaluated in working out $\omega_{nn}(\mathbf{k}, 1)$ is between the particle density and its current $\mathbf{g}$. As we shall see this commutator plays an important role throughout our discussion. We have then:

$$[n(\mathbf{x}), g_\alpha(\mathbf{y})] = \frac{1}{2} \sum_{i=1}^{N} \left[\delta(\mathbf{x} - \mathbf{r}_i), \sum_{k=1}^{N} (p_k^\alpha \delta(\mathbf{y} - \mathbf{r}_k) + \delta(\mathbf{y} - \mathbf{r}_k) p_k^\alpha) \right]$$

$$= \frac{1}{2} \sum_{i,k=1}^{N} \left(\delta(\mathbf{x} - \mathbf{r}_i)(-i\hbar \nabla_k^\alpha) \delta(\mathbf{x}' - \mathbf{r}_k) - (-i\hbar \nabla_k^\alpha) \delta(\mathbf{y} - \mathbf{r}_i) \delta(\mathbf{x} - \mathbf{r}_k) \right.$$

$$\left. + \delta(\mathbf{x} - \mathbf{r}_i) \delta(\mathbf{y} - \mathbf{x}_k)(-i\hbar \nabla_k^\alpha) - \delta(\mathbf{y} - \mathbf{r}_k)(-i\hbar \nabla_k^\alpha) \delta(\mathbf{x} - \mathbf{r}_i) \right)$$

$$= \sum_{i=1}^{N} \delta(\mathbf{y} - \mathbf{r}_i)(i\hbar \nabla_i^\alpha) \delta(\mathbf{x} - \mathbf{r}_i)$$

$$= -i\hbar \nabla_x^\alpha \sum_{i=1}^{N} \delta(\mathbf{y} - \mathbf{r}_i) \delta(\mathbf{x} - \mathbf{r}_i)$$

$$[n(\mathbf{x}), g_\alpha(\mathbf{y})] = -i\hbar \nabla_x^\alpha \left[\delta(\mathbf{x} - \mathbf{y}) n(\mathbf{x}) \right] \quad . \tag{239}$$

We then have for the moment $n = 1$, using Eqs. (236) and (239) in (231):

$$\omega_{nn}(\mathbf{x} - \mathbf{y}, 1) = -\frac{i}{m\hbar} \nabla_x \cdot \langle [\mathbf{g}(\mathbf{x}), n(\mathbf{y})] \rangle$$

$$= -\frac{i}{m\hbar} \nabla_x \cdot \langle i\hbar \nabla_y \left[\delta(\mathbf{x} - \mathbf{y}) n(\mathbf{y}) \right] \rangle$$

$$= \frac{1}{m} \nabla_x \cdot \nabla_y \delta(\mathbf{x} - \mathbf{y}) \bar{n} \tag{240}$$

where $\bar{n}$ is the average density. The Fourier transform is particularly simple:

$$\omega_{nn}(\mathbf{k}, 1) = \frac{k^2 \bar{n}}{m} = \int \frac{d\omega}{\pi} \omega \chi_{nn}''(\mathbf{k}, \omega) \quad . \tag{241}$$

Clearly this serves as a convenient normalization check on, for example, the results of a scattering experiment that gives $\chi_{nn}''(\mathbf{k}, \omega)$. The result given by Eq. (241) holds, of course, in the classical limit where we have:

$$\chi_{nn}''(\mathbf{k}, \omega) = \frac{\beta \omega}{2} S_{nn}(\mathbf{k}, \omega) \tag{242}$$

and:

$$\int \frac{d\omega}{2\pi} \omega^2 S_{nn}(\mathbf{k}, \omega) = \frac{k^2 \bar{n} kT}{m} \quad . \tag{243}$$

In the quantum case, the moments of $\chi_{AB}''(\omega)$ are related to those of the Kubo function via:

$$\int \frac{d\omega}{\pi} \omega^n \chi_{AB}''(\omega) = \beta \int \frac{d\omega}{2\pi} \omega^{n+1} C_{AB}(\omega) \quad . \tag{244}$$

As a second example, consider the electrical conductivity. In that case we have the complex relation given by Eq. (159):

$$\sigma_{\alpha\beta}(\omega) = \int \frac{d\omega'}{\pi} \frac{\chi''_{J_\alpha, R_\beta}(\omega')}{\omega' - \omega - i\eta} . \tag{245}$$

We can determine the large-frequency behavior of $\sigma_{\alpha\beta}(\omega)$ by noting that:

$$\lim_{|\omega|\to\infty} \omega \sigma_{\alpha\beta}(\omega) = -\int \frac{d\omega'}{\pi} \chi''_{J_\alpha, R_\beta}(\omega')$$

$$= -\frac{2}{2\hbar} \langle [J_\alpha, R_\beta] \rangle \tag{246}$$

where the charge-current density is given by:

$$J_\alpha = \frac{1}{V} \sum_{i=1}^{N} \frac{q_i p_i^\alpha}{m} \tag{247}$$

and the electric dipole moment by:

$$R_\beta = \sum_{j=1}^{N} q_j r_j^\beta . \tag{248}$$

The commutator is then given by:

$$[J_\alpha, R_\beta] = \frac{1}{Vm} \sum_{i,j} q_i q_j [p_i^\alpha, r_j^\beta]$$

$$= -\frac{i\hbar}{Vm} \sum_{i=1}^{N} q_i^2 \delta_{\alpha,\beta} = -\frac{i\hbar q^2}{m} \delta_{\alpha\beta} \frac{N}{V} \tag{249}$$

and:

$$\lim_{|\omega|\to\infty} \omega \sigma_{\alpha\beta}(\omega) = \frac{ie^2}{m} \delta_{\alpha\beta} n = i \frac{\omega_p^2}{4\pi} , \tag{250}$$

where ω_p is the plasma frequency. This relation gives a strong constraint on the frequency-dependent conductivity at large frequencies. Here we have assumed that all of the charged particles have the same $q_i^2 = q^2$. For an in-depth discussion of sum rules for charged systems see Ref. [12]. For a discussion of application to neutral fluids see Ref. [13].

3.14
References and Notes

1. The original development of the fluctuation-dissipation theorem in the context of electrical voltage fluctuations is due to H. Nyquist, Phys. Rev. **32**, 110 (1928). The experimental verification of these results discussed in Section 3.10, is given by J. B. Johnson, Phys. Rev. **32**, 97 (1928). The generalization of Nyquist's results is given in H. B. Callen, T. R. Welton, Phys. Rev. **83**, 34 (1951).
2. L. Onsager, Phys. Rev. **37**, 405 (1931); Phys. Rev. **38**, 2265 (1931).
3. In the *typical* case the double prime indicates the imaginary part as discussed below.
4. For a formal discussion of the consequences of translational, rotational, and reflection (parity) symmetry see D. Forster, *Hydrodynamic Fluctuations, Broken Symmetry, and Correlation Functions*, W. A. Benjamin, Reading (1975), p. 50.
5. G. Carrier, M. Krook and C. Pearson, *Functions of a Complex Variable*, McGraw-Hill, N.Y., (1966), p. 414.
6. N. I. Muskhelishvili, *Singular Integral Equations*, Erven P. Nordhoff, NV, Gronigen, Netherlands (1953).
7. Ref. [5], p. 417.
8. G. Arfken, *Mathematical Methods for Physicists*, Academic, Orlando, Fla. (1985), p. 421.
9. The static linear response formalism is discussed in Chapter 2 of FOD.
10. R. Kubo, *Statistical Mechanics*, North-Holland, Amsterdam (1971), p. 415.
11. C. Kittel, *Elementary Statistical Physics*, Wiley, N.Y. (1958).
12. P. C. Martin, Phys. Rev. **161**, 143 (1967).
13. D. Forster, P. C. Martin, S. Yip, Phys. Rev. **170**, 155, 160 (1968).

3.15
Problems for Chapter 3

Problem 3.1: Find the relation analogous to Eq. (3.13) satisfied by:

$$G^<_{ij}(tt') = \langle \psi^\dagger_i(t)\psi_j(t') \rangle$$

and:

$$G^>_{ij}(tt') = \langle \psi_i(t)\psi^\dagger_j(t') \rangle \; ,$$

where the average is over the GCE density matrix and $\psi^\dagger_i$ and $\psi_i(t)$ are the creation and annihilation operators for the system. These operators satisfy the commutation (bosons) and anticommutation (fermions) relations:

$$[\psi_i, \psi^\dagger_j]_\pm = \delta_{ij} \; .$$

Problem 3.2: Assuming that the operators A and B are fields that depend on position, show that $\chi''_{AB}(\mathbf{k}, \omega)$ satisfies Eqs. (47), (48), and (49).

Problem 3.3: If operator A is the magnetization density $\mathbf{M}(\mathbf{x}, t)$ and B is the energy density $\varepsilon(\mathbf{x}, t)$ in a magnetic system, what are the symmetry properties of $\chi''_{M_i, \varepsilon}(\mathbf{k}, \omega)$?

Problem 3.4: Verify that the integral given by Eq. (57) indeed represents a step function.

Problem 3.5: Solve the Heisenberg equations of motion explicitly for a harmonic oscillator and verify Eqs. (95) and (96).

Problem 3.6: Consider a set of N spins $\mathbf{S}(\mathbf{R})$ on a lattice labeled by site index $\mathbf{R}$. These spins satisfy the conventional angular momentum commutation relations:

$$[S_\alpha(\mathbf{R}), S_\beta(\mathbf{R}')] = i\hbar \sum_\gamma \varepsilon_{\alpha\beta\gamma} S_\gamma(\mathbf{R}) \delta_{\mathbf{R},\mathbf{R}'} \quad . \tag{251}$$

Suppose the dynamics of these spins is governed by the Hamiltonian:

$$H = -\sum_{\mathbf{R}} \mu h_z S_z(\mathbf{R}) \tag{252}$$

where h_z is an external magnetic field pointing in the z-direction. Find the Fourier transform $\chi_{\alpha\beta}(\mathbf{k}, \omega)$ of the dissipation function:

$$\chi_{\alpha\beta}(\mathbf{R}, \mathbf{R}', tt') = \frac{1}{2\hbar} \langle [S_\alpha(\mathbf{R}, t), S_\beta(\mathbf{R}', t')] \rangle \quad .$$

Problem 3.7: Show that the iterated solution of Eq. (121) leads to the expression for the equilibrium density matrix given by Eq. (122).

Problem 3.8: Evaluate the commutator defined by Eq. (235) and verify that:

$$[n(\mathbf{x}), H] = -\frac{i\hbar}{m} \nabla_x \cdot \mathbf{g}(\mathbf{x}, t) \quad ,$$

where $\mathbf{g}(\mathbf{x}, t)$ is the momentum density,

$$\mathbf{g}(\mathbf{x}, t) = \frac{1}{2} \sum_{i=1}^{N} [\mathbf{p}_i \delta(\mathbf{x} - \mathbf{r}_i) + \delta(\mathbf{x} - \mathbf{r}_i) \mathbf{p}_i]$$

Problem 3.9: Consider the one-dimensional anharmonic oscillator governed by the Hamiltonian:

$$H = \frac{p^2}{2m} + \frac{k}{2} x^2 + \frac{u}{4} x^4 \quad .$$

Express the $n = 1$ sum rule satisfied by the momentum–momentum correlation function:

$$\omega_{PP}(n) = \int \frac{d\omega}{\pi} \omega^n \chi''_{pp}(\omega)$$

in terms of equilibrium averages of phase-space coordinates.

Problem 3.10: In nonrelativistic systems, an established invariance principle is Galilean invariance. For a system of N particles with position operators $\mathbf{r}_1, \mathbf{r}_2, \ldots, \mathbf{r}_N$ and momentum operators $\mathbf{p}_1, \mathbf{p}_2, \ldots, \mathbf{p}_N$, find the operator $\mathbf{G}$ generating the transformation:

$$A'(\mathbf{v}) = U_G(\mathbf{v}) A U_G^\dagger(\mathbf{v})$$

such that:

$$\mathbf{r}_i'(\mathbf{v}) = \mathbf{r}_i - \mathbf{v}t$$

and:

$$\mathbf{p}_i'(\mathbf{v}) = \mathbf{p}_i - m\mathbf{v} \ ,$$

where $\mathbf{v}$ is a constant vector.

Hint: $U_G(\mathbf{v}) = e^{i\mathbf{v}\cdot\mathbf{G}/\hbar}$

How does the particle density operator $\hat{n}(\mathbf{x}, t)$ transform under a Galilean transformation?

How does the Hamiltonian transform under a Galilean transformation?

Is the conventional grand canonical density matrix invariant under Galilean transformations?

Extra credit: write the generator $\mathbf{G}$ in second-quantized form in terms of creation and annihilation operators.

How does a second-quantized bosonic annihilation operator $\psi(\mathbf{x})$ transform under a Galilean transformation?

Problem 3.11: If we look at the equilibrium fluctuations of a solid at temperature T we can consider the simple model:

$$\frac{dx_\alpha}{dt} = \frac{p_\alpha}{M}$$

$$\frac{dp_\alpha}{dt} = -\kappa x_\alpha - \frac{\Gamma}{M} p_\alpha + \eta_\alpha \ ,$$

where x_α is the displacement and p_α the associated momentum and the noise satisfies:

$$\langle \eta_\alpha(t) \eta_\beta(t') \rangle = 2 k_B T \Gamma \delta_{\alpha\beta} \delta(t - t') \ ,$$

where the kinetic coefficient $\Gamma > 0$.

Compute the momentum correlation function:

$$S_{\alpha\beta}^p(\omega) = \int_{-\infty}^{\infty} dt\, e^{i\omega(t-t')} \langle p_\alpha(t) p_\beta(t') \rangle \ .$$

Find the equal-time quantity $\langle p_\alpha(t) p_\beta(t) \rangle$. Does your result make sense?

Problem 3.12: Starting with the definition of the Kubo response function, defined by Eq. (127), show that:

$$\frac{\partial}{\partial t} C_{AB}(t - t') = -\frac{2i}{\beta} \chi''_{AB}(t - t') \ .$$

Using this result show that:

$$\chi_{AB} = \beta C_{AB}(tt) \ .$$

Finally show that the complex Kubo function, defined by Eq. (127) is related to the complex relaxation function by:

$$C_{AB}(z) = \frac{1}{i\beta} R_{AB}(z) \chi_{AB} \ .$$

Problem 3.13: Verify that the symmetry properties:

$$\chi_{AB}(\omega) = \chi^*_{AB}(-\omega)$$

$$\chi_{AB}(\omega) = \varepsilon_A \varepsilon_B \chi_{BA}(\omega) \ ,$$

are satisfied by solutions of the Bloch equations describing paramagnetic resonance in Chapter 2.

Problem 3.14: For the case where variables A and B have different signatures under time reversal, evaluate the relaxation function:

$$\frac{\delta A(t)}{h_B} = \int \frac{d\omega}{\pi} \frac{\chi''_{AB}}{\omega - i\eta} e^{i\omega t} \ .$$

Show that $\delta A(t)$ is real and look at the short-time regime.

Problem 3.15: Assuming a dispersion relation representation for the frequency-dependent conductivity:

$$\sigma(\omega) = \int \frac{d\omega'}{\pi i} \frac{\sigma'(\omega')}{\omega' - \omega - i\eta} \ ,$$

find the consequences of Eq. (250).

Problem 3.16: Show that the sum rule given by Eq. (250) is satisfied by $\sigma_{\alpha\beta}(\omega)$ given by Eq. (165) in the classical limit.

4
Charged Transport

4.1
Introduction

One of the most important set of properties of a material is its response to external electric fields. Interesting electrical properties result typically from *free* conduction band electrons in metals or polarization effects in dielectrics. In both cases we are interested in the influence of an electric field on a condensed-matter system comprised of charged particles. We discuss here the linear response approach to this problem, with care to treat the long-range interactions between the particles.

We will assume that we have a classical system with full translational symmetry [1]. A quantum-mechanical treatment for a system defined relative to a lattice proceeds along very similar lines. In the first sections of this chapter we follow the development due to Martin [2,3].

4.2
The Equilibrium Situation

Let us consider a classical condensed-matter system of volume $\mathcal{V}$ comprised of a set of N charged particles. Let us assume that the ith particle, with mass m_i and charge q_i, is located at position $\mathbf{r}_i$ and has a velocity $\mathbf{v}_i$.

The electromagnetic fields in the sample generated by these charges are governed by Maxwell's equations in the form:

$$\nabla \cdot \mathbf{E}(\mathbf{x}, t) = 4\pi \rho(\mathbf{x}, t) \qquad (1)$$

$$\nabla \cdot \mathbf{B}(\mathbf{x}, t) = 0 \qquad (2)$$

$$\nabla \times \mathbf{E}(\mathbf{x}, t) = -\frac{1}{c}\frac{\partial \mathbf{B}(\mathbf{x}, t)}{\partial t} \qquad (3)$$

$$\nabla \times \mathbf{B} = \frac{4\pi \mathbf{J}}{c} + \frac{1}{c}\frac{\partial \mathbf{E}(\mathbf{x}, t)}{\partial t} \qquad (4)$$

Nonequilibrium Statistical Mechanics. Gene F. Mazenko
Copyright © 2006 WILEY-VCH Verlag GmbH & Co. KGaA, Weinheim
ISBN: 3-527-40648-4

where the charge density is given by:

$$\rho(\mathbf{x},t) = \sum_{i=1}^{N} q_i \delta[\mathbf{x} - \mathbf{r}_i(t)] \ , \tag{5}$$

the charge-current density is given by:

$$\mathbf{J}(\mathbf{x},t) = \sum_{i=1}^{N} \mathbf{v}_i(t) q_i \delta[\mathbf{x} - \mathbf{r}_i(t)] \tag{6}$$

and c is the speed of light. Since electric charge is conserved, the charge density and current are related by the continuity equation:

$$\frac{\partial}{\partial t}\rho(\mathbf{x},t) + \nabla \cdot \mathbf{J}(\mathbf{x},t) = 0 \ . \tag{7}$$

These equations must be supplemented by the dynamical equations of motion satisfied by the particles. The Hamiltonian for this set of particles is given by:

$$\begin{aligned} H_0 &= \sum_{i=1}^{N} \frac{1}{2m_i} \left(\mathbf{p}_i - \frac{q_i}{c}\mathbf{A}(\mathbf{r}_i)\right)^2 + \sum_{i=1}^{N} q_i \phi(x_i) \\ &\quad + \frac{1}{2} \sum_{i,j} V_s(\mathbf{r}_i - \mathbf{r}_j) \end{aligned} \tag{8}$$

where $\mathbf{p}_i$ is the canonical momentum, $\mathbf{A}(\mathbf{x})$ is the vector potential, $\phi(\mathbf{x})$ the scalar potential and V_s describes any short-range two-body interactions between the charged particles. The electric and magnetic fields are related to the potentials by:

$$\mathbf{E} = -\nabla \phi - \frac{1}{c}\frac{\partial \mathbf{A}}{\partial t} \tag{9}$$

$$\mathbf{B} = \nabla \times \mathbf{A} \ . \tag{10}$$

The equations of motion satisfied by the particles are given by Hamilton's equations:

$$\frac{d}{dt}\mathbf{r}_i = \nabla_{p_i} H_0 \tag{11}$$

$$\frac{d}{dt}\mathbf{p}_i = -\nabla_{r_i} H_0 \ . \tag{12}$$

We then have an impressively complicated set of equations to solve. When we are in thermal equilibrium, the probability distribution (for simplicity we work in the canonical ensemble) is given by:

$$P_{eq} = \frac{e^{-\beta H_0}}{Z} \tag{13}$$

with the partition function:

$$Z = \int d^3r_1...d^3r_N \frac{d^3p_1}{(2\pi\hbar)^3}...\frac{d^3p_N}{(2\pi\hbar)^3} e^{-\beta H_0} \quad . \tag{14}$$

The average of any function A of the phase-space coordinates $(\mathbf{r}_i, \mathbf{p}_i)$ is given by:

$$\langle A \rangle = \int d^3r_1...d^3r_N \frac{d^3p_1}{(2\pi\hbar)^3}...\frac{d^3p_N}{(2\pi\hbar)^3} P_{eq} \, A(\mathbf{r}_1, \mathbf{r}_2, .., \mathbf{p}_1, \mathbf{p}_2, ..) \quad . \tag{15}$$

Taking the average of the microscopic Maxwell's equations gives:

$$\nabla \cdot \langle \mathbf{E} \rangle = 4\pi \langle \rho \rangle \tag{16}$$

$$\nabla \cdot \langle \mathbf{B} \rangle = 0 \tag{17}$$

$$\nabla \times \langle \mathbf{E} \rangle = -\frac{1}{c}\frac{\partial \langle \mathbf{B} \rangle}{\partial t} \tag{18}$$

$$\nabla \times \langle \mathbf{B} \rangle = \frac{4\pi}{c}\langle \mathbf{J} \rangle + \frac{1}{c}\frac{\partial}{\partial t}\langle \mathbf{E} \rangle \tag{19}$$

and the problem reduces in this case to knowing the average charge density and the average current density. It is when we begin to look at the problem from this perspective that we begin to obtain some degree of simplicity. For a system in the absence of an externally applied electromagnetic field with no net charge (electrically neutral), the average charge density and charge-current density are zero: $\langle \rho \rangle = 0$ and $\langle \mathbf{J} \rangle = 0$. This says that all of the local fluctuations in charge and current balance out. The average Maxwell equations are given then by:

$$\nabla \cdot \langle \mathbf{E} \rangle = 0 \tag{20}$$

$$\nabla \cdot \langle \mathbf{B} \rangle = 0 \tag{21}$$

$$\nabla \times \langle \mathbf{E} \rangle = -\frac{1}{c}\frac{\partial}{\partial t}\langle \mathbf{B} \rangle \tag{22}$$

$$\nabla \times \langle \mathbf{B} \rangle = -\frac{1}{c}\frac{\partial}{\partial t}\langle \mathbf{E} \rangle \quad . \tag{23}$$

There are several possible solutions to this set of equations:

1. These equations support wave solutions. This implies either that the system is radiating or that electromagnetic waves are bombarding our sample. In the first case our system would be losing energy and therefore would not be in equilibrium. We have thus far been assuming there are no external fields acting. Consequently we can rule out these *radiation* solutions.

2. There are symmetry-broken solutions [4]:

$$\langle \mathbf{E} \rangle = \mathbf{E}_0 \tag{24}$$

$$\langle \mathbf{B} \rangle = \mathbf{B}_0 \tag{25}$$

where $\mathbf{E}_0$ and $\mathbf{B}_0$ are independent of time. For example $\mathbf{E}_0 = 0$, $\mathbf{B}_0 \neq 0$ corresponds to a ferromagnet. Alternatively $\mathbf{E}_0 \neq 0$ and $\mathbf{B}_0 = 0$ corresponds to a ferroelectric.

3. In the third, and most typical situation,

$$\langle \mathbf{E} \rangle = \langle \mathbf{B} \rangle = 0 \ . \tag{26}$$

Whether case 2 or case 3 is appropriate depends on the material in question and the thermodynamic state of the system. Thus, for example, PbTiO$_3$ has $\langle \mathbf{E} \rangle = 0$ for temperatures greater than 763°K and $\langle \mathbf{E} \rangle \neq 0$ for temperatures below 763° K. Similarly Fe has $\langle \mathbf{B} \rangle = 0$ for temperatures greater than 1043° K and is ferromagnetic for temperatures less than 1043° K. The temperatures where $\mathbf{E}_0$ or $\mathbf{B}_0$ go to zero are called the Curie temperatures and at these temperatures the system is undergoing a phase transition from a high-temperature paramagnetic or paraelectric to ferromagnetic or ferroelectric phase.

The situation where we have no net charge and no externally applied field is relatively simple. The system will be in equilibrium and the average electric and magnetic fields will be uniform in space and unchanging in time.

4.3
The Nonequilibrium Case

4.3.1
Setting up the Problem

Thus far we have assumed that there are no external fields acting on our system. Let us turn now to the case where we couple electrically to our system. The situation we discuss is where there is an external charge density $\rho_{\text{ext}}(\mathbf{x}, t)$ and an external charge-current density $\mathbf{J}_{\text{ext}}(\mathbf{x}, t)$ [5]. These external charges satisfy the continuity equation:

$$\frac{\partial \rho_{\text{ext}}(\mathbf{x}, t)}{\partial t} + \nabla \cdot \mathbf{J}_{\text{ext}}(\mathbf{x}, t) = 0 \ . \tag{27}$$

These charges will set up external electromagnetic fields satisfying:

$$\nabla \cdot \mathbf{E}_{ext}(\mathbf{x}, t) = 4\pi \rho_{ext}(\mathbf{x}, t) \tag{28}$$

$$\nabla \cdot \mathbf{B}_{ext}(\mathbf{x}, t) = 0 \tag{29}$$

$$\nabla \times \mathbf{B}_{ext}(\mathbf{x}, t) = \frac{4\pi}{c} \mathbf{J}_{ext}(\mathbf{x}, t) + \frac{1}{c} \frac{\partial \mathbf{E}_{ext}(\mathbf{x}, t)}{\partial t} \tag{30}$$

$$\nabla \times \mathbf{E}_{ext}(\mathbf{x}, t) = -\frac{1}{c} \frac{\partial \mathbf{B}_{ext}(\mathbf{x}, t)}{\partial t} \, . \tag{31}$$

In the presence of these external fields that are varying in space and time, the Hamiltonian describing our N-particle system is given by:

$$H(t) = \sum_{i=1}^{N} \frac{1}{2m_i} \left[\mathbf{p}_i - \frac{q_i}{c} \mathbf{A}_T(\mathbf{x}_i, t) \right]^2 + V_s \, , \tag{32}$$

where the total vector potential is:

$$\mathbf{A}_T(\mathbf{x}, t) = \mathbf{A}(\mathbf{x}) + \mathbf{A}_{ext}(\mathbf{x}, t) \, , \tag{33}$$

where $\mathbf{A}_{ext}$ is the external vector potential. We will work in a gauge [6] where the total scalar potential is zero. V_s is the short-ranged part of the interaction between particles. The Maxwell equations satisfied by the total electric and magnetic fields are:

$$\nabla \cdot \mathbf{E}_T(\mathbf{x}, t) = 4\pi \rho_T(\mathbf{x}, t) \tag{34}$$

$$\nabla \cdot \mathbf{B}_T(\mathbf{x}, t) = 0 \tag{35}$$

$$\nabla \times \mathbf{E}_T(\mathbf{x}, t) = -\frac{1}{c} \frac{\partial \mathbf{B}_T(\mathbf{x}, t)}{\partial t} \tag{36}$$

$$\nabla \times \mathbf{B}_T(\mathbf{x}, t) = \frac{4\pi}{c} \mathbf{J}_T(\mathbf{x}, t) + \frac{1}{c} \frac{\partial \mathbf{E}_T(\mathbf{x}, t)}{\partial t} \tag{37}$$

where:

$$\mathbf{E}_T(\mathbf{x}, t) = -\frac{1}{c} \frac{\partial}{\partial t} \mathbf{A}_T(\mathbf{x}, t) \tag{38}$$

$$\mathbf{B}_T(\mathbf{x}, t) = \nabla \times \mathbf{A}_T(\mathbf{x}, t) \, . \tag{39}$$

and:

$$\rho_T(\mathbf{x},t) = \rho(\mathbf{x},t) + \rho_{\text{ext}}(\mathbf{x},t) \qquad (40)$$

$$\mathbf{J}_T(\mathbf{x},t) = \mathbf{J}_s(\mathbf{x},t) + \mathbf{J}_{\text{ext}}(\mathbf{x},t) , \qquad (41)$$

where, as before, the system contribution to the charge density is given by:

$$\rho(\mathbf{x},t) = \sum_{i=1}^{N} q_i \delta[\mathbf{x} - \mathbf{r}_i(t)] \qquad (42)$$

and the system part of the charge-current density is:

$$\mathbf{J}_s(\mathbf{x},t) = \sum_{i=1}^{N} \frac{q_i}{m_i} \left(\mathbf{p}_i - \frac{q_i}{c} \mathbf{A}_T(\mathbf{r}_i,t) \right) \delta[\mathbf{x} - \mathbf{r}_i(t)] , \qquad (43)$$

since the velocity is related to the canonical momentum by:

$$\mathbf{v}_i = \frac{1}{m_i} \left(\mathbf{p}_i - \frac{q_i}{c} \mathbf{A}_T(\mathbf{r}_i,t) \right) . \qquad (44)$$

4.3.2
Linear Response

A general attack on this problem is totally impractical. However, for many purposes, it is sufficient to analyze the case where the external field is *weak* and the system is only slightly removed from equilibrium. Things simplify greatly in the *linear* regime. Our basic approximation will be to keep only the linear term coupling the canonical momentum of the particles to the external field. We will neglect terms of order A_{ext}^2. We can then write the Hamiltonian in the form:

$$\begin{aligned} H &= \sum_{i=1}^{N} \frac{1}{2m_i} \left[\mathbf{p}_i - \frac{q_i}{c} (\mathbf{A}(\mathbf{r}_i) + \mathbf{A}_{\text{ext}}(\mathbf{r}_i,t)) \right]^2 + V_s \\ &= H_0 - \sum_{i=1}^{N} \frac{q_i}{cm_i} \left(\mathbf{p}_i - \frac{q_i}{c} \mathbf{A}(\mathbf{r}_i) \right) \cdot \mathbf{A}_{\text{ext}}(\mathbf{r}_i,t) + \mathcal{O}(A_{\text{ext}}^2) \\ &= H_0 - \frac{1}{c} \int d^3x \, \mathbf{J}(\mathbf{x}) \cdot \mathbf{A}_{\text{ext}}(\mathbf{x},t) + \ldots \end{aligned} \qquad (45)$$

where H_0 is the Hamiltonian for the system in the absence of external fields discussed earlier and $\mathbf{J}(\mathbf{x})$ is given by:

$$\mathbf{J}(\mathbf{x},t) = \sum_{i=1}^{N} \frac{q_i}{m_i} \left(\mathbf{p}_i - \frac{q_i}{c} \mathbf{A}(\mathbf{x},t) \right) \delta[\mathbf{x} - \mathbf{r}_i(t)] \qquad (46)$$

The problem is now clearly in the form where our linear response approach is valid. Thus the external field is $\mathbf{A}_{\text{ext}}/c$, which couples to the charge-current density. The linear response or *induced* charge density is given by our result from Chapter 2, Eq. (2.56), which in this case reads:

$$\langle \rho(\mathbf{k}, \omega) \rangle_I = \sum_\alpha \chi_{\rho J_\alpha}(\mathbf{k}, \omega) \frac{1}{c} A^\alpha_{\text{ext}}(\mathbf{k}, \omega) \; . \tag{47}$$

It is useful to rewrite this result in terms of the associated correlation function. We showed previously, Eq. (3.149), for general observables A and B the dynamic susceptibility can be expressed as:

$$\chi_{AB}(\mathbf{k}, \omega) = \chi_{AB}(\mathbf{k}) - \beta \omega C_{AB}(\mathbf{k}, \omega) \; , \tag{48}$$

where $\chi_{AB}(\mathbf{k})$ is the static susceptibility and $C_{AB}(\mathbf{k}, \omega)$ is the complex (Kubo) correlation function. Since the complex correlation function is defined:

$$C_{AB}(\mathbf{k}, \omega) = -i \int_0^\infty dt \, e^{i\omega t} C_{AB}(\mathbf{k}, t) \; , \tag{49}$$

we can use integration by parts and write:

$$\begin{aligned} \omega C_{AB}(\mathbf{k}, \omega) &= \int_0^\infty dt \left(-\frac{d}{dt} e^{i\omega t} \right) C_{AB}(\mathbf{k}, t) \\ &= C_{AB}(\mathbf{k}, t=0) + i C_{\dot{A}B}(\mathbf{k}, \omega) \; . \end{aligned} \tag{50}$$

We know from Eq. (3.130) that:

$$\beta C_{AB}(\mathbf{k}, t=0) = \chi_{AB}(\mathbf{k}) \tag{51}$$

Using Eqs. (50) and (51) back in Eq. (48) we obtain the useful result:

$$\chi_{AB}(\mathbf{k}, \omega) = -i\beta C_{\dot{A}B}(\mathbf{k}, \omega) \; . \tag{52}$$

In the very important special case where A is conserved,

$$\frac{\partial A}{\partial t} = -\nabla \cdot \mathbf{J}_A \; , \tag{53}$$

where $\mathbf{J}_A$ is the current associated with A, we have:

$$\frac{\partial A(\mathbf{k}, t)}{\partial t} = i\mathbf{k} \cdot \mathbf{J}_A(\mathbf{k}, t) \; . \tag{54}$$

Using this result in Eq. (52) we find:

$$\chi_{AB}(\mathbf{k}, \omega) = \beta \sum_\mu k_\mu C_{J_A^\mu B}(\mathbf{k}, \omega) \; . \tag{55}$$

Using this result in Eq. (47) we have:

$$\langle \rho(\mathbf{k},\omega)\rangle_I = \sum_{\alpha,\mu} k_\mu \frac{\beta}{c} C_{J_\mu J_\alpha}(\mathbf{k},\omega) A^\alpha_{\text{ext}}(\mathbf{k},\omega) \ . \tag{56}$$

If we remember that in the $\phi = 0$ gauge:

$$\mathbf{E}_{\text{ext}}(\mathbf{x},t) = -\frac{1}{c}\frac{\partial}{\partial t}\mathbf{A}_{\text{ext}}(\mathbf{x},t) \ , \tag{57}$$

then:

$$\mathbf{E}_{\text{ext}}(\mathbf{k},\omega) = \frac{i\omega}{c}\mathbf{A}_{\text{ext}}(\mathbf{k},\omega) \ . \tag{58}$$

The induced charge density can be written as:

$$\langle \rho(\mathbf{k},\omega)\rangle_I = \sum_{\alpha,\mu} k_\mu \beta C_{J_\mu J_\alpha}(\mathbf{k},\omega) \frac{E^\alpha_{\text{ext}}(\mathbf{k},\omega)}{(i\omega)} \ . \tag{59}$$

Let us turn now to the calculation of the average charge-current density. The total charge-current density is given by Eq. (41):

$$\mathbf{J}_T(\mathbf{x},t) = \mathbf{J}_{\text{ext}}(\mathbf{x},t) + \mathbf{J}_s(\mathbf{x},t) \tag{60}$$

where $\mathbf{J}_s$ is the current given by Eq. (43):

$$\mathbf{J}_s(\mathbf{x},t) = \sum_{i=1}^{N} \frac{q_i}{m_i}\left(\mathbf{p}_i - \frac{q_i}{c}\mathbf{A}_T(\mathbf{x})\right)\delta[\mathbf{x}-\mathbf{r}_i(t)] \tag{61}$$

where $\mathbf{A}_T$ is the total vector potential. Working to lowest order in the external field we have for the total charge-current density:

$$\mathbf{J}_T(\mathbf{x},t) = \mathbf{J}_{\text{ext}}(\mathbf{x},t) + \mathbf{J}(\mathbf{x},t) - \sum_{i=1}^{N}\frac{q_i^2}{mc}\delta[\mathbf{x}-\mathbf{r}_i(t)]\mathbf{A}_{\text{ext}}(\mathbf{x},t) \tag{62}$$

where $\mathbf{J}(\mathbf{x},t)$ is given by Eq. (46). The average total current density is given then by:

$$\langle \mathbf{J}_T(\mathbf{x},t)\rangle_{neq} = \mathbf{J}_{\text{ext}}(\mathbf{x},t) - \frac{\omega_p^2}{4\pi c}\mathbf{A}_{\text{ext}}(\mathbf{x},t) + \langle \mathbf{J}(\mathbf{x},t)\rangle_{neq} \ , \tag{63}$$

where, to linear order in $\mathbf{A}_{\text{ext}}$, the average of the term multiplying the explicit factor of $\mathbf{A}_{\text{ext}}$ defines the *plasma frequency* ω_p,

$$\frac{\omega_p^2}{4\pi} = \left\langle \sum_{i}^{N}\frac{q_i^2}{m_i}\delta(\mathbf{x}-\mathbf{r}_i)\right\rangle \ . \tag{64}$$

4.3 The Nonequilibrium Case

In the case where there are electrons and very heavy ions, we can write to a good approximation:

$$\frac{\omega_p^2}{4\pi} \approx \frac{q^2 n}{m} \tag{65}$$

where q is the electron charge, m its mass and n the electron particle density.

It should be clear that the nonequilibrium average of the current is given in linear response by:

$$\langle J^\alpha(\mathbf{k},\omega)\rangle_{neq} = \sum_\mu \chi_{J_\alpha J_\mu}(\mathbf{k},\omega) \frac{1}{c} A^\mu_{ext}(\mathbf{k},\omega) \ . \tag{66}$$

Using Eq. (51) in Eq. (48), where A and B are components of the currents:

$$\chi_{J_\alpha J_\mu}(\mathbf{k},\omega) = \beta\left[C_{J_\alpha J_\mu}(\mathbf{k},t=0) - \omega C_{J_\alpha J_\mu}(\mathbf{k},\omega)\right] \ . \tag{67}$$

Using Eq. (58), expressing $\mathbf{A}_{ext}$ in terms of $\mathbf{E}_{ext}$, we obtain:

$$\langle J^\alpha(\mathbf{k},\omega)\rangle_{neq} = \beta \sum_\mu \left[iC_{J_\alpha J_\mu}(\mathbf{k},\omega) + \frac{C_{J_\alpha J_\mu}(\mathbf{k},t=0)}{i\omega}\right] E^\mu_{ext}(\mathbf{k},\omega) \ . \tag{68}$$

The quantity $C_{J_\alpha J_\mu}(\mathbf{k},t=0)$ is evaluated explicitly in Problem 4.1. We find:

$$C_{J_\alpha J_\beta}(\mathbf{k},t=0) = \delta_{\alpha,\beta} \beta^{-1} \frac{\omega_p^2}{4\pi} \ . \tag{69}$$

Putting this back into Eq. (68) gives:

$$\langle J^\alpha(\mathbf{k},\omega)\rangle_{neq} = \sum_\beta \left[i\beta C_{J_\alpha J_\beta}(\mathbf{k},\omega) + \delta_{\alpha,\beta} \frac{\omega_p^2}{4\pi i\omega}\right] E^\beta_{ext}(\mathbf{k},\omega) \ . \tag{70}$$

The total average charge-current density follows from Eq. (63) and is given by:

$$\begin{aligned}\langle J_T^\alpha(\mathbf{k},\omega)\rangle_{neq} &= J^\alpha_{ext}(\mathbf{k},\omega) - \frac{\omega_p^2}{4\pi c} A^\alpha_{ext}(\mathbf{k},\omega) \\ &+ \sum_\beta \left[i\beta C_{J_\alpha J_\beta}(\mathbf{k},\omega) + \frac{\delta_{\alpha,\beta}\omega_p^2}{4\pi i\omega}\right] E^\beta_{ext}(\mathbf{k},\omega) \\ &= J^\alpha_{ext}(\mathbf{k},\omega) + \sum_\beta (i\beta) C_{J_\alpha J_\beta}(\mathbf{k},\omega) E^\beta_{ext}(\mathbf{k},\omega) \ ,\end{aligned} \tag{71}$$

where the terms proportional to ω_p^2 cancel. The induced current density is given then by:

$$\langle J^\alpha(\mathbf{k},\omega)\rangle_I = \langle J^\alpha_T(\mathbf{k},\omega)\rangle - \langle J^\alpha_{\text{ext}}(\mathbf{k},\omega)\rangle$$
$$= \sum_\beta i\beta C_{J_\alpha J_\beta}(\mathbf{k},\omega) E^\beta_{\text{ext}}(\mathbf{k},\omega) \quad . \tag{72}$$

We can write this in the form:

$$\langle J^\alpha(\mathbf{k},\omega)\rangle_I = \sum_\beta \sigma_0^{\alpha\beta}(\mathbf{k},\omega) E^\beta_{\text{ext}}(\mathbf{k},\omega) \quad , \tag{73}$$

where we define:

$$\sigma_0^{\alpha\beta}(\mathbf{k},\omega) = i\beta C_{J_\alpha J_\beta}(\mathbf{k},\omega) \quad . \tag{74}$$

We do not identify σ_0 as the conductivity, since the conductivity gives the response to the total electric field not the external field.

Comparing the expressions for the induced density, Eq. (56), expressed in terms of the external electric field,

$$\langle \rho(\mathbf{k},\omega)\rangle_I = -\sum_{\alpha,\beta} \frac{k^\alpha}{\omega}(\sigma_0^{\alpha\beta}(\mathbf{k},\omega) E^\beta_{\text{ext}}(\mathbf{k},\omega) \quad , \tag{75}$$

and the induced current density, we see that the continuity equation is satisfied:

$$-i\omega\langle\rho(\mathbf{k},\omega)\rangle_I = i\mathbf{k}\cdot\langle \mathbf{J}(\mathbf{k},\omega)\rangle_I \quad . \tag{76}$$

The next step in the development is to realize that it is conventional to express the induced current in terms of the *total* electric field. As shown in Problem 4.7, in the $\phi = 0$ gauge the Maxwell's equations for the external vector potential are given by:

$$-\frac{1}{c}\frac{\partial}{\partial t}(\nabla\cdot\mathbf{A}_{\text{ext}}) = 4\pi\rho_{\text{ext}}(\mathbf{x},t) \tag{77}$$

and:

$$\left(-\nabla^2 + \frac{1}{c^2}\frac{\partial^2}{\partial t^2}\right)\mathbf{A}_{\text{ext}} + \nabla(\nabla\cdot\mathbf{A}_{\text{ext}}) = \frac{4\pi}{c}\mathbf{J}_{\text{ext}}(\mathbf{x},t) \quad . \tag{78}$$

Since:

$$\mathbf{E}_{\text{ext}}(\mathbf{x},t) = -\frac{1}{c}\frac{\partial \mathbf{A}_{\text{ext}}}{\partial t} \quad , \tag{79}$$

we can convert Eq. (78) into an equation for $\mathbf{E}_{\text{ext}}$:

$$\left(-\nabla^2 + \frac{1}{c^2}\frac{\partial^2}{\partial t^2}\right)\mathbf{E}_{\text{ext}}(\mathbf{x},t) + \nabla[\nabla\cdot\mathbf{E}_{\text{ext}}(\mathbf{x},t)] = -\frac{4\pi}{c^2}\frac{\partial}{\partial t}\mathbf{J}_{\text{ext}}(\mathbf{x},t) \quad . \tag{80}$$

Similarly, we have for the total system:

$$\left(-\nabla^2 + \frac{1}{c^2}\frac{\partial^2}{\partial t^2}\right)\mathbf{E}_T(\mathbf{x},t) + \nabla[\nabla \cdot \mathbf{E}_T(\mathbf{x},t)] = -\frac{4\pi}{c^2}\frac{\partial}{\partial t}\langle \mathbf{J}_T(\mathbf{x},t)\rangle \ . \tag{81}$$

Fourier transforming the last two equations we obtain:

$$(k^2 - \omega^2/c^2)\mathbf{E}_{ext}(\mathbf{k},\omega) - \mathbf{k}[\mathbf{k}\cdot\mathbf{E}_{ext}(\mathbf{k},\omega)] = \frac{4\pi}{c^2}(i\omega)\mathbf{J}_{ext}(\mathbf{k},\omega) \tag{82}$$

and:

$$(k^2 - \omega^2/c^2)\mathbf{E}_T(\mathbf{k},\omega) - \mathbf{k}[\mathbf{k}\cdot\mathbf{E}_T(\mathbf{k},\omega)] = \frac{4\pi}{c^2}(i\omega)\langle \mathbf{J}_T(\mathbf{k},\omega)\rangle \ . \tag{83}$$

Subtracting Eq. (82) from Eq. (83) we find:

$$\begin{aligned}(k^2 - \omega^2/c^2)E_T^\alpha(\mathbf{k},\omega) &- k^\alpha[\mathbf{k}\cdot\mathbf{E}_T(\mathbf{k},\omega)] \\ = (k^2 - \omega^2/c^2)E_{ext}^\alpha(\mathbf{k},\omega) &- k^\alpha[\mathbf{k}\cdot\mathbf{E}_{ext}(\mathbf{k},\omega)] \\ &+ \frac{4\pi}{c^2}(i\omega)\sum_\beta \sigma_0^{\alpha\beta}(\mathbf{k},\omega)E_{ext}^\beta(\mathbf{k},\omega) \ , \end{aligned} \tag{84}$$

where we have used Eqs. (72) and (75). To simplify matters let us assume the system is isotropic in space. Then:

$$\sigma_0^{\alpha\beta}(\mathbf{k},\omega) = \hat{k}_\alpha\hat{k}_\beta \sigma_0^L(\mathbf{k},\omega) + \left(\delta_{\alpha\beta} - \hat{k}_\alpha\hat{k}_\beta\right)\sigma_0^T(\mathbf{k},\omega) \tag{85}$$

and the electric field can be decomposed into longitudinal and transverse parts:

$$\mathbf{E}_T(\mathbf{k},\omega) = \hat{k}E_T^L(\mathbf{k},\omega) + \mathbf{E}_T^T(\mathbf{k},\omega) \ , \tag{86}$$

where $\mathbf{k}\cdot\mathbf{E}_T^T(\mathbf{k},\omega) = 0$. Taking the transverse and longitudinal components of Eq. (84) gives:

$$\begin{aligned}\left(k^2 - \frac{\omega^2}{c^2}\right)\mathbf{E}^T(\mathbf{k},\omega) &= \left(k^2 - \frac{\omega^2}{c^2}\right)\mathbf{E}_{ext}^T(\mathbf{k},\omega) \\ &+ \frac{4\pi i\omega}{c^2}\sigma_0^T(\mathbf{k},\omega)\mathbf{E}_{ext}^T(\mathbf{k},\omega)\end{aligned} \tag{87}$$

and:

$$\begin{aligned}\left(-\frac{\omega^2}{c^2}\right)E_T^L(\mathbf{k},\omega) &= \left(-\frac{\omega^2}{c^2}\right)E_{ext}^L(\mathbf{k},\omega) \\ &+ \frac{4\pi i\omega}{c^2}\sigma_0^L(\mathbf{k},\omega)E_{ext}^L(\mathbf{k},\omega) \ . \end{aligned} \tag{88}$$

We can then express the external fields in terms of the total fields:

$$\mathbf{E}_{ext}^T(\mathbf{k},\omega) = \frac{k^2 - \frac{\omega^2}{c^2}}{k^2 - \frac{\omega^2}{c^2} + \frac{4\pi i\omega}{c^2}\sigma_0^T(\mathbf{k},\omega)} \mathbf{E}_T^T(\mathbf{k},\omega) \tag{89}$$

and:

$$E_{ext}^L(\mathbf{k},\omega) = \frac{\omega^2}{\omega^2 - 4\pi i\omega \sigma_0^L(\mathbf{k},\omega)} E_T^L(\mathbf{k},\omega) \ . \tag{90}$$

In terms of components:

$$\langle \rho(\mathbf{k},\omega) \rangle_I = -\frac{k}{\omega} \sigma_0^L(\mathbf{k},\omega) E_{ext}^L(\mathbf{k},\omega)$$

$$\langle J^L(\mathbf{k},\omega) \rangle_I = \sigma_0^L(\mathbf{k},\omega) E_{ext}^L(\mathbf{k},\omega)$$

$$\langle \mathbf{J}^T(\mathbf{k},\omega) \rangle_I = \sigma_0^T(\mathbf{k},\omega) \mathbf{E}_{ext}^T(\mathbf{k},\omega) \ . \tag{91}$$

Using Eqs. (89) and (90) to eliminate $\mathbf{E}_{ext}$ in terms of $\mathbf{E}_T$ we find:

$$\langle \rho(\mathbf{k},\omega) \rangle_I = -\frac{\sigma_0^L(\mathbf{k},\omega)}{\omega} \frac{\omega^2 [k E_T^L(\mathbf{k},\omega)]}{[\omega^2 - 4\pi i\omega \sigma_0^L(\mathbf{k},\omega)]} \tag{92}$$

and:

$$\langle \mathbf{J}^T(\mathbf{k},\omega) \rangle_I = \frac{\sigma_0^T(\mathbf{k},\omega)\left(\omega^2 - c^2 k^2\right)}{[\omega^2 - c^2 k^2 - 4\pi i\omega \sigma_0^T(\mathbf{k},\omega)]} \mathbf{E}_T^T(\mathbf{k},\omega) \tag{93}$$

and:

$$\langle J^L(\mathbf{k},\omega) \rangle_I = \frac{\sigma_0^L(\mathbf{k},\omega)\omega^2}{[\omega^2 - 4\pi i\omega \sigma_0^L(\mathbf{k},\omega)]} E_T^L(\mathbf{k},\omega) \ . \tag{94}$$

It is conventional to define the physical conductivity tensor via:

$$\langle J^\alpha(\mathbf{k},\omega) \rangle_I = \sum_\beta \sigma_{\alpha\beta}(\mathbf{k},\omega) E_T^\beta(\mathbf{k},\omega) \ , \tag{95}$$

which has the components:

$$\sigma^T(\mathbf{k},\omega) = \frac{\sigma_0^T(\mathbf{k},\omega)\left(\omega^2 - c^2 k^2\right)}{\omega^2 - c^2 k^2 - 4\pi i\omega \sigma_0^T(\mathbf{k},\omega)} \tag{96}$$

$$\sigma^L(\mathbf{k},\omega) = \frac{\sigma_0^L(\mathbf{k},\omega)\omega^2}{\omega^2 - 4\pi i\omega \sigma_0^L(\mathbf{k},\omega)} \ . \tag{97}$$

4.4
The Macroscopic Maxwell Equations

Let us see how these results can be used to rewrite Maxwell's equations. The induced charge density can be written as:

$$\langle \rho(\mathbf{k}, \omega) \rangle_I = -\frac{k}{\omega} \sigma^L(\mathbf{k}, \omega) E_T^L(\mathbf{k}, \omega) . \tag{98}$$

The induced polarization, **P**, is defined in this case by:

$$\langle \rho(\mathbf{x}, t) \rangle_I = -\nabla \cdot \mathbf{P}(\mathbf{x}, t) \tag{99}$$

or, in terms of Fourier transforms,

$$\langle \rho(\mathbf{k}, \omega) \rangle_I = i\mathbf{k} \cdot \mathbf{P}(\mathbf{k}, \omega) . \tag{100}$$

Comparing Eqs. (98) and (100) we can identify:

$$\mathbf{P}^{L,T}(\mathbf{k}, \omega) = \left[\frac{\sigma^{L,T}(\mathbf{k}, \omega)}{-i\omega} \right] \mathbf{E}_T^{L,T}(\mathbf{k}, \omega) . \tag{101}$$

If we put Eq. (99) into Gauss's law,

$$\begin{aligned}
\nabla \cdot \mathbf{E}_T(\mathbf{x}, t) &= 4\pi \langle \rho_T(\mathbf{x}, t) \rangle \\
&= 4\pi \rho_{\text{ext}}(\mathbf{x}, t) + 4\pi \langle \rho(\mathbf{x}, t) \rangle_I \\
&= 4\pi \rho_{\text{ext}}(\mathbf{x}, t) - 4\pi \nabla \cdot \mathbf{P}(\mathbf{x}, t) ,
\end{aligned} \tag{102}$$

we see that it is convenient to introduce the displacement field:

$$\mathbf{D}(\mathbf{x}, t) = \mathbf{E}_T(\mathbf{x}, t) + 4\pi \mathbf{P}(\mathbf{x}, t) \tag{103}$$

so that Gauss's law takes the form:

$$\nabla \cdot \mathbf{D}(\mathbf{x}, t) = 4\pi \rho_{\text{ext}}(\mathbf{x}, t) . \tag{104}$$

Using our result relating the polarization to the total field, putting Eq. (101) back into Eq. (103), we obtain:

$$\begin{aligned}
\mathbf{D}^{L,T}(\mathbf{k}, \omega) &= \left[1 - \frac{4\pi \sigma^{L,T}(\mathbf{k}, \omega)}{i\omega} \right] \mathbf{E}_T^{L,T}(\mathbf{k}, \omega) \\
&= \varepsilon^{L,T}(\mathbf{k}, \omega) \mathbf{E}^{L,T}(\mathbf{k}, \omega) ,
\end{aligned} \tag{105}$$

which defines the generalized dielectric function:

$$\varepsilon^{L,T}(\mathbf{k}, \omega) = 1 - \frac{4\pi \sigma^{L,T}(\mathbf{k}, \omega)}{i\omega} . \tag{106}$$

Turning to Ampere's law, we have on Fourier transformation:

$$-i\mathbf{k} \times \mathbf{B}(\mathbf{k},\omega) = \frac{4\pi}{c} \langle \mathbf{J}_T(\mathbf{k},\omega) \rangle - \frac{i\omega}{c} \mathbf{E}_T(\mathbf{k},\omega) \ . \tag{107}$$

We can then write for the current:

$$\langle \mathbf{J}_T(\mathbf{k},\omega) \rangle = \mathbf{J}_{\text{ext}}(\mathbf{k},\omega) + \langle \mathbf{J}(\mathbf{k},\omega) \rangle_I \tag{108}$$

where the induced current,

$$\begin{aligned}\langle \mathbf{J}^{L,T}(\mathbf{k},\omega) \rangle_I &= \sigma^{L,T}(\mathbf{k},\omega) \mathbf{E}_T^{L,T}(\mathbf{k},\omega) \\ &= [\varepsilon^{L,T}(\mathbf{k},\omega) - 1]\frac{(-i\omega)}{4\pi} \mathbf{E}_T^{L,T}(\mathbf{k},\omega) \ , \end{aligned} \tag{109}$$

can be written as:

$$\langle \mathbf{J}(\mathbf{k},\omega) \rangle_I = -\frac{i\omega}{4\pi}[\mathbf{D}(\mathbf{k},\omega) - \mathbf{E}_T(\mathbf{k},\omega)] \ . \tag{110}$$

Putting this back into Eq. (107) we have:

$$\begin{aligned}-i\mathbf{k} \times \mathbf{B}(\mathbf{k},\omega) =\ &\frac{4\pi}{c} \mathbf{J}_{\text{ext}}(\mathbf{k},\omega) \\ &- \frac{i\omega}{c}[\mathbf{D}(\mathbf{k},\omega) - \mathbf{E}_T(\mathbf{k},\omega)] - \frac{i\omega}{c} \mathbf{E}_T(\mathbf{k},\omega)\end{aligned} \tag{111}$$

and, canceling the terms depending on $\mathbf{E}_T(\mathbf{k},\omega)$, gives:

$$-i\mathbf{k} \times \mathbf{B}(\mathbf{k},\omega) = \frac{4\pi}{c} \mathbf{J}_{\text{ext}}(q,\omega) - \frac{i\omega}{c} \mathbf{D}(q,\omega) \tag{112}$$

and we obtain, taking the inverse Fourier transforms, the macroscopic Maxwell equations for dielectric or conducting materials:

$$\nabla \cdot \mathbf{D}(\mathbf{x},t) = 4\pi \rho_{\text{ext}}(\mathbf{x},t) \tag{113}$$

$$\nabla \times \mathbf{B}(\mathbf{x},t) = \frac{4\pi}{c} \mathbf{J}_{\text{ext}}(\mathbf{x},t) + \frac{1}{c}\frac{\partial}{\partial t}\mathbf{D}(\mathbf{x}) \ . \tag{114}$$

These equations are supplemented by Eq. (105) giving the displacement field in terms of the electric field where the dielectric function is defined by Eq. (106) in terms of the conductivity. The conductivity, in turn, is defined by Eqs. (96) and (97) in terms of a bare conductivity $\sigma_0(\mathbf{k},\omega)$, which is microscopically determined in terms of the current–current correlation function given by Eq. (74). Furthermore, the induced charge and polarization are given by Eqs. (98) and (101) respectively. If we can determine $C_{JJ}(\mathbf{k},\omega)$, then all of the other quantities follow. In the next section we see how all of this holds together for particular physical situations.

These results for the conductivity and dielectric constant can be written in somewhat different forms. The first step is to use the continuity equation to show (see Problem 4.2) that the density and longitudinal current response function are related by:

$$\omega^2 \chi_{\rho\rho}(\mathbf{k}, \omega) = q^2 \left[\chi_{JJ}^L(\mathbf{k}, \omega) - \frac{\omega_P^2}{4\pi} \right] . \quad (115)$$

We also have from the classical fluctuation-dissipation theorem, Eqs. (67) and (69),

$$\chi_{JJ}^L(\mathbf{k}, \omega) = \frac{\omega_P^2}{4\pi} - \beta\omega C_{JJ}^L(\mathbf{k}, \omega) . \quad (116)$$

Starting with Eq. (74) we have then, using in turn Eq. (116) and then Eq. (115):

$$\begin{aligned}
\sigma_0(\mathbf{k}, \omega) &= i\beta C_{JJ}^L(\mathbf{k}, \omega) \\
&= \frac{i}{\omega} \left(\frac{\omega_P^2}{4\pi} - \chi_{JJ}^L(\mathbf{k}, \omega) \right) \\
&= -i \frac{\omega}{k^2} \chi_{\rho\rho}(\mathbf{k}, \omega) .
\end{aligned} \quad (117)$$

Putting this result back into Eq. (97) we have for the longitudinal conductivity:

$$\begin{aligned}
\sigma^L(\mathbf{k}, \omega) &= -i \frac{\omega}{k^2} \frac{\omega^2 \chi_{\rho\rho}(\mathbf{k}, \omega)}{\left[\omega^2 - 4\pi i \omega \left(-\frac{i}{\omega} \frac{\omega^2}{k^2} \chi_{\rho\rho}(\mathbf{k}, \omega) \right) \right]} \\
&= -i \frac{\omega}{k^2} \frac{\chi_{\rho\rho}(\mathbf{k}, \omega)}{[1 - \frac{4\pi}{k^2} \chi_{\rho\rho}(\mathbf{k}, \omega)]} .
\end{aligned} \quad (118)$$

The dielectric function can then be written:

$$\begin{aligned}
\varepsilon^L(\mathbf{k}, \omega) &= 1 + \frac{4\pi i}{\omega} \sigma^L(\mathbf{k}, \omega) \\
&= 1 + \frac{4\pi}{k^2} \frac{\chi_{\rho\rho}(\mathbf{k}, \omega)}{1 - \frac{4\pi}{k^2} \chi_{\rho\rho}(\mathbf{k}, \omega)} \\
&= \frac{1}{1 - \frac{4\pi}{k^2} \chi_{\rho\rho}(\mathbf{k}, \omega)} .
\end{aligned} \quad (119)$$

It is conventional to introduced the *screened* response function:

$$\chi_{\rho\rho}^{(s)}(\mathbf{k}, \omega) = \frac{\chi_{\rho\rho}(\mathbf{k}, \omega)}{1 - \frac{4\pi}{k^2} \chi_{\rho\rho}(\mathbf{k}, \omega)} . \quad (120)$$

then the conductivity can be written as:

$$i\frac{\sigma^L(\mathbf{k},\omega)}{\omega} = \frac{\chi_{\rho\rho}^{(s)}(\mathbf{k},\omega)}{k^2} . \quad (121)$$

The Kubo formula for the conductivity is given by the double limit:

$$\sigma^L = \lim_{\omega\to 0}\lim_{k\to 0} -i\frac{\omega}{k^2}\chi_{\rho\rho}^{(s)}(\mathbf{k},\omega) . \quad (122)$$

In this case we can not directly express this result into the form of a time integral over a current–current correlation function.

In our approach here we have first treated the induced current assuming the applied external field is small. Then we solved Maxwell's equation to find the total electric field in terms of the applied field as given by Eqs. (89) and (90). In an alternative approach one can treat the total vector potential contribution to the kinetic energy as small and expand in terms of the total vector potential, which, in turn, can be expressed in terms of the total electric field. This is discussed in some detail in Section 3.7 in Ref. [7]. This approach looks superficially simpler, but leads to some difficult questions of consistency. In particular one must deal with the fact that the total field is a fluctuating quantity. See the critique in Ref. [3].

4.5
The Drude Model

4.5.1
Basis for Model

We can gain an appreciation for the formal development in the previous section by again using the Drude model [8] introduced in Section 2.1.5. While there are important wavenumber-dependent phenomena in these charged systems, we will be satisfied with looking at the long wavelength, $k = 0$, behavior. Under these circumstances we expect σ^L and σ^T to be equal and we write:

$$\sigma^L(0,\omega) = \sigma^T(0,\omega) = \sigma(0,\omega) \quad (123)$$
$$\varepsilon^L(0,\omega) = \varepsilon^L(0,\omega) = \varepsilon(0,\omega) . \quad (124)$$

The Drude model was developed to treat charge flow in a solid. In this model the charged particles are assumed to be acted upon by four types of average or *effective* forces.

- The first force we take into account is the force due to electric field that acts on the particle i with charge q: $\mathbf{F}_1 = q\mathbf{E}(t)$ where $\mathbf{E}$ is the total electric field representing the sum of the externally applied field and

4.5 The Drude Model

the average electric field due to the other charged particles. We assume that **E** is uniform.

- The second force acting is a *viscous* or drag force. A particle moving through a fluid experiences an irreversible frictional force opposing the motion: $\mathbf{F}_2 = -m\gamma\mathbf{V}$, where m is the effective mass, $\mathbf{V}$ is the velocity of the particle and γ is the coefficient of friction. The microscopic physics behind γ is associated with collisions of the charged particles with ions and impurities. A key physical assumption is that the long-ranged Coulomb interaction is *screened* and only the screened electron–electron interactions contribute to the effective collisions.

- In a solid there is a restoring force [9] in the system attracting the charged particle back to its equilibrium lattice position, $\mathbf{F}_3 = -k\mathbf{r}$ where k is the spring constant. To simplify matters somewhat, we assume here that the equilibrium position of the particle of interest is at the origin. Different particles will have different equilibrium positions. It is convenient to introduce an oscillator frequency:

$$k = m\omega_0^2 \; . \tag{125}$$

- There is, as in Chapter 1, a noise $\vec{\eta}(t)$, acting on the particle of interest. The noise has zero average $\langle \eta_i(t) \rangle = 0$ and variance:

$$\langle \eta_i(t)\eta_j(t') \rangle = 2k_B T \Gamma \delta(t-t')\delta_{ij} \; . \tag{126}$$

Thus $\eta_i(t)$ is assumed to be *white* noise (see Problem 4.6 for the determination of Γ in terms of γ). The equation of motion satisfied by our selected particle is given by:

$$m\frac{d^2}{dt^2}\mathbf{r}(t) = q\mathbf{E}(t) - m\gamma\frac{d}{dt}\mathbf{r}(t) - m\omega_0^2\mathbf{r}(t) + m\vec{\eta}(t) \; . \tag{127}$$

Taking the average of Eq. (127) we obtain for the average displacement:

$$m\left[\frac{d^2}{dt^2}\langle\mathbf{r}(t)\rangle + \frac{d}{dt}\gamma\langle\mathbf{r}(t)\rangle + \omega_0^2\langle\mathbf{r}(t)\rangle\right] = q\mathbf{E}(t) \; . \tag{128}$$

In this model γ and ω_0^2 are phenomenological parameters. We can easily solve our equation of motion via the Fourier transformation:

$$\langle\mathbf{r}(\omega)\rangle = \int dt\, e^{-i\omega t}\langle\mathbf{r}(t)\rangle \tag{129}$$

to obtain:

$$m[-\omega^2 - i\omega\gamma + \omega_0^2]\langle\mathbf{r}(\omega)\rangle = q\mathbf{E}(\omega) \tag{130}$$

and:

$$\langle \mathbf{r}(\omega) \rangle = \frac{q\mathbf{E}(\omega)}{m[-\omega^2 - i\omega\gamma + \omega_0^2]} \quad . \tag{131}$$

The average velocity of our selected particle is given by:

$$\langle \mathbf{V}(\omega) \rangle = -i\omega \langle \mathbf{r}(\omega) \rangle \tag{132}$$

$$= \frac{i\omega q \mathbf{E}(\omega)}{m[\omega^2 - \omega_0^2 + i\omega\gamma]} \quad . \tag{133}$$

4.5.2
Conductivity and Dielectric Function

The spatial Fourier-transform of the induced current in our driven system is given by:

$$\langle \mathbf{J}(\mathbf{k},t) \rangle_I = \frac{1}{V} \langle \sum_i q_i \mathbf{V}_i e^{+i\mathbf{k}\cdot\mathbf{r}_i(t)} \rangle \quad . \tag{134}$$

For uniform motion, $\mathbf{k} \to 0$ and, assuming the charges of the mobile particles are equal, we have:

$$\langle \mathbf{J}(0,t) \rangle_I = \frac{1}{V} \left\langle \sum_{i=1}^{N} q_i \mathbf{V}_i(t) \right\rangle = \frac{Nq}{V} \langle \mathbf{V}(t) \rangle \quad . \tag{135}$$

as in Chapter 2. Fourier transforming over time:

$$\langle \mathbf{J}(0,\omega) \rangle_I = nq \langle \mathbf{V}(\omega) \rangle \quad ,$$

and using Eq. (135) we find for the induced current:

$$\langle \mathbf{J}(0,\omega) \rangle_I = \frac{i\omega q^2}{m[\omega^2 - \omega_0^2 + i\omega\gamma]} n \mathbf{E}(\omega)$$

$$= \frac{1}{4\pi} \frac{i\omega \omega_p^2}{[\omega^2 - \omega_0^2 + i\omega\gamma]} \mathbf{E}(\omega) \quad , \tag{136}$$

where in the last step we used $\omega_p^2 = 4\pi n q^2/m$. Since $\langle \mathbf{J} \rangle_I = \sigma \mathbf{E}$, we have for the conductivity:

$$\sigma(\omega) = \frac{i\omega \omega_p^2}{4\pi[\omega^2 - \omega_0^2 + i\omega\gamma]} \quad . \tag{137}$$

The dielectric function is given then by Eq. (106):

$$\varepsilon(\omega) = 1 - \frac{4\pi \sigma(\omega)}{i\omega} = 1 + \frac{\omega_p^2}{-\omega^2 - i\omega\gamma + \omega_0^2} \quad . \tag{138}$$

We can gain some insight into the meaning of the parameters in the model by going to the dc limit $\omega \to 0$. We have then that for small ω that the conductivity vanishes:

$$\sigma(\omega) = -\frac{i\omega\omega_p^2}{4\pi\omega_0^2} \tag{139}$$

and the dielectric constant is given by:

$$\varepsilon(0) = 1 + \frac{\omega_p^2}{\omega_0^2} \ . \tag{140}$$

As things stand we see that the conductivity goes to zero as $\omega \to 0$. This model does not represent a metal. Clearly it represents an insulator or dielectric material. This is because the charged particles are *bound*. This binding is represented by the oscillator frequency ω_0. We note that ω_0^2 can be empirically determined from the dielectric function [10]. Clearly ω_0 is essentially the Einstein or optical frequency in a dielectric. For nonzero ω_0 we have an insulator for *those* degrees of freedom. If, for some degrees of freedom $\omega_0 = 0$, then we return to the Drude model of Chapter 2 where the frequency-dependent conductivity is given by:

$$\sigma(\omega) = \frac{\sigma(0)}{1 - i\omega\tau} \ , \tag{141}$$

where the dc conductivity is given then by:

$$\sigma(0) = \frac{nq^2\tau}{m} \ , \tag{142}$$

with the relaxation time $\tau = 1/\gamma$.

Note that in a metal the dielectric constant, Eq. (138) with $\omega_0 = 0$ is not smooth at low frequencies. Indeed, it is a complex quantity and we know that this is the reason that light does not propagate in a metal.

4.5.3
The Current Correlation Function

It is very interesting to calculate the density–density correlation function that would be measured in a scattering experiment. Remember that:

$$\sigma_0(0, \omega) = i\beta C_{JJ}(0, \omega) \tag{143}$$

and the physical conductivity is given by:

$$\sigma = \frac{\sigma_0 \omega^2}{\omega^2 - 4\pi i \omega \sigma_0} \ . \tag{144}$$

We can invert this equation to obtain σ_0:

$$\sigma_0(0,\omega) = \frac{\omega^2\,\sigma(0,\omega)}{\omega^2 + 4\pi i \omega \sigma(0,\omega)} \quad . \tag{145}$$

Assuming that $\sigma(0,\omega)$ is given by Eq. (137), the *bare* conductivity is given by:

$$\sigma_0(0,\omega) = \frac{i\omega(\omega_p^2/4\pi)}{\omega^2 + i\omega\gamma - \omega_0^2 - \omega_p^2} \quad . \tag{146}$$

The current correlation function in a solid is given by:

$$C_{JJ}(\mathbf{0},\omega) = \beta^{-1}\,\frac{i\omega(\omega_p^2/4\pi)}{\omega^2 + i\omega\gamma - \omega_0^2 - \omega_p^2} \quad . \tag{147}$$

C_{JJ} in the time regime is explored in Problem 4.5. In contrast to a simple single-component fluid, the current for a charged neutral system is not conserved. Momentum is transferred to the lattice in a solid or between types of charge in a fluid. There is damping even at $k = 0$. Looking at the denominator of C_{JJ} we find poles at:

$$\omega = -i\frac{\gamma}{2} \pm \sqrt{\omega_0^2 + \omega_p^2 - \gamma^2/4} \quad . \tag{148}$$

Thus we have damped oscillations for $\omega_0^2 + \omega_p^2 > \gamma^2/4$. We expect oscillations in a solid, which are interpreted as optical phonons. In a metal, where $\omega_0 = 0$ we have plasma oscillations [11]. We discuss in Chapter 7 how kinetic theory can be used to develop a more microscopic approach to charged transport.

4.6
References and Notes

1 The quantum-mechanical formulation is set up in Chapter 5 in ESM.
2 P. C. Martin, in *Many-Body Physics*, C. De Witt, R. Balian (eds), Gordon and Breach, N.Y. (1968), p. 39.
3 P. C. Martin, Phys. Rev. **161**, 143 (1967).
4 See the discussion in Chapter 5 in ESM and Chapter 6 in FOD.
5 The case of a static set of charges is discussed in Chapter 6 of FOD for the case of a dielectric material.
6 For a discussion of gauge invariance in this context see Ref. [2], p. 104.
7 G. D. Mahan, *Many-Particle Physics*, Plenum, N.Y. (1981).
8 P. Drude, Ann. Phys. **1**, 566 (1900); Ann. Phys. **3**, 369 (1900).
9 A realistic model treats the normal modes in the system.
10 See the discussion in N. Ashcroft, N. D. Mermin, *Solid State Physics*, Holt, Rinehart and Winston, N.Y. (1976), p. 543.
11 C. Kittel, *Introduction to Solid State Physics*, 7th edn., Wiley, N.Y. (1996), p. 278.

4.7 Problems for Chapter 4

Problem 4.1: Evaluate the classical static charge current–current correlation function:

$$C_{J_\alpha J_\beta}(\mathbf{k}, 0) = \int d^3x\, e^{+i\mathbf{k}\cdot(\mathbf{x}-\mathbf{x}')} \langle J_\alpha(\mathbf{x}) J_\beta(\mathbf{x}') \rangle \ .$$

Problem 4.2: Use the continuity equation to show that the density and current response functions are related by:

$$\omega^2 \chi_{\rho\rho}(\mathbf{k}, \omega) = k^2 \left[\chi_{JJ}^L(\mathbf{k}, \omega) - \frac{\omega_p^2}{4\pi} \right] \ .$$

Problem 4.3: Rederive the expression giving the conductivity tensor in terms of response function in the fully quantum-mechanical case. Show that Eq. (118) holds in this case where the response functions are the quantum-mechanical quantities.

Problem 4.4: Show that Eqs. (96) and (97) are consistent with Eqs. (123) and Eqs. (124).

Problem 4.5: Consider the Langevin equation, Eq. (127), giving the displacement in terms of the noise. Solve for the velocity autocorrelation function in the absence of a driving external field:

$$\psi(\omega) = \langle |\mathbf{v}(\omega)|^2 \rangle \ .$$

Use the equipartition theorem to determine the noise strength Γ in terms of the damping γ.

Problem 4.6: Determine the time-dependent current–current correlation function $C_{JJ}(0.t)$ by taking the inverse Fourier transform Eq. (147).

Problem 4.7: Show that the $\phi = 0$ Maxwell equations lead to Eqs. (77) and (78).

5
Linearized Langevin and Hydrodynamical Description of Time-Correlation Functions

5.1
Introduction

We are now interested in determining the time-correlation functions for the variety of physical systems of interest. From a fundamental point of view we could set out to develop completely microscopic, first-principle calculations. After all the correlation functions are well-defined mathematical objects and we should simply *evaluate* them. This is the basic point of view, as we shall see, in kinetic theory. In practice, such a first-principles analysis of time-correlation functions is very difficult. In many cases they can be evaluated numerically using molecular dynamics or Monte Carlo simulations [1]. It turns out, however, that there are a number of situations where we can gain a feeling for the physics without solving the entire problem. The most important such situation is that involving the long-distance and time behavior of a system with *slow variables*. In many cases of interest there is a separation of time scales between these slow variables and the huge number of fast microscopic degrees of freedom. In these cases, there is much we can say about the dynamics of these slow variables. We can identify these slow degrees of freedom for reasons fundamental to physics. One of the main sources of slow variables are conservation laws [2]. With every symmetry principle [3] one can identify an observable that generates the symmetry. Since, by definition, if there is a symmetry principle, this observable commutes with the equilibrium probability distribution or density matrix. This means it commutes with the Hamiltonian. Since this observable commutes with the Hamiltonian it is conserved; does not change with time. As we show in detail below, the dynamics of the local density associated with the conserved generator is a slow variable. Other slow variables are associated with the breaking of a continuous symmetry as one passes into the ordered state. These slow variables are called Nambu–Goldstone modes [4].

It is fundamental to realize that it is precisely the same reasoning we develop here that is at work in arguing for the existence of thermodynamics. The slow variables are just the extensive thermodynamic state variables. For simple fluid systems with no broken continuous symmetries, the conservation laws give the state variables: the number of particles N, the energy E, the total momentum $\mathbf{P}$ and the total angular momentum $\mathbf{L}$, and we have the constraint of a fixed volume $\mathcal{V}$. If the system is not translating or rotating we can take the total momentum $\mathbf{P}$ and total angular momentum $\mathbf{L}$ to be zero and thermodynamics can be organized in terms of N, E, and $\mathcal{V}$. When we break a continuous symmetry, as in the case of a superfluid or a solid, we generate slow Nambu–Goldstone variables. These variables must be added to the list of thermodynamic variables. The classic examples are the elastic displacement degrees of freedom in a solid. We will return to this issue later.

A third example of a slow variable is the order parameter near a second-order phase transition where the system is experiencing critical slowing down. We discuss this in detail in Chapters 8 and 10. Slow variables can occur for other special reasons: The motion of a very large particle moving in a fluid of much smaller particles is slow. This was the case in our treatment of a Brownian particle in earlier chapters.

Let us first discuss from a semiphenomenological point of view the case of a conserved variable in the very simple case of spin diffusion in a paramagnet. We will then come back and develop a more general and rigorous approach to this and other problems.

5.2
Spin Diffusion in Itinerant Paramagnets

5.2.1
Continuity Equation

Let us consider a fluid of N particles of mass m interacting via the usual Hamiltonian:

$$H = \sum_{i=1}^{N} \frac{\mathbf{p}_i^2}{2m} + \frac{1}{2} \sum_{i \neq j=1}^{N} V(\mathbf{r}_i - \mathbf{r}_j) \tag{1}$$

where $\mathbf{p}_i$ and $\mathbf{r}_i$ are the momentum and position of the ith particle. We assume that these particles carry a spin S_i and we assume there is no interaction between spins on different particles. To keep things as simple as possible we assume a scalar spin as for an Ising model. We are then interested in the magnetization density:

$$M(\mathbf{x}, t) = \sum_{i=1}^{N} S_i \delta[\mathbf{x} - \mathbf{r}_i(t)] \ . \tag{2}$$

Since the Hamiltonian is independent of the spin, S_i is time independent. The spins are independent of the phase-space coordinates; the average of a spin at

a site is zero:
$$\langle S_i \rangle = 0 , \tag{3}$$
and averages over products of spins at different sites is zero:
$$\langle S_i S_j \rangle = \delta_{ij} \langle S_i^2 \rangle , \tag{4}$$
where $S_i^2 = S^2$ is independent of the particular site.

This is a simplified example of liquid ^{3}He in the normal phase. This model is restricted to the paramagnetic regime:
$$\langle M(\mathbf{x}, t) \rangle = 0 .$$

We choose this example because it has the simplest nontrivial structure of a conserved variable that does not couple to any other degrees of freedom in the system. Thus, because of the symmetry of the magnetization density, for example, the static correlation function between the magnetization and particle density vanishes:
$$\langle M(\mathbf{x}) n(\mathbf{y}) \rangle = 0 . \tag{5}$$

Let us first investigate the equation of motion satisfied by $M(\mathbf{x}, t)$. We have the Heisenberg equation of motion:
$$\begin{aligned} \frac{\partial M(\mathbf{x}, t)}{\partial t} &= \frac{i}{\hbar} [H, M(\mathbf{x}, t)] \\ &= \frac{i}{\hbar} \sum_{i=1}^{N} [\frac{p_i^2(t)}{2m}, M(\mathbf{x}, t)] \end{aligned} \tag{6}$$

since $M(\mathbf{x}, t)$ commutes with the potential energy. We can then work out the commutation relation between the magnetization density and the kinetic energy:
$$\begin{aligned} \frac{\partial M(\mathbf{x}, t)}{\partial t} &= \frac{i}{2m\hbar} \sum_{i=1}^{N} (\mathbf{p}_i(t) \cdot [\mathbf{p}_i(t), M(\mathbf{x}, t)] \\ &\quad + [\mathbf{p}_i(t), M(\mathbf{x}, t)] \cdot \mathbf{p}_i(t)) , \end{aligned} \tag{7}$$

where:
$$\begin{aligned} [\mathbf{p}_i(t), M(\mathbf{x}, t)] &= \sum_{j=1}^{N} (-i\hbar \nabla_{r_i} \delta[\mathbf{x} - \mathbf{r}_j(t)]) S_j \\ &= i\hbar \nabla_x \delta[\mathbf{x} - \mathbf{r}_i(t)] S_i . \end{aligned} \tag{8}$$

Putting Eq. (8) back into Eq. (6), we arrive at the continuity equation:
$$\frac{\partial M(\mathbf{x}, t)}{\partial t} = -\nabla_x \cdot \sum_{i=1}^{N} S_i \frac{1}{2} \left(\frac{\mathbf{p}_i(t)}{m} \delta[\mathbf{x} - \mathbf{r}_i(t)] + \delta[\mathbf{x} - \mathbf{r}_i(t)] \frac{\mathbf{p}_i(t)}{m} \right)$$

$$= -\nabla_x \cdot \mathbf{J}(\mathbf{x}, t) \ . \tag{9}$$

J is the magnetization-density current given by:

$$\mathbf{J}(\mathbf{x}, t) = \sum_{i=1}^{N} S_i \frac{1}{2} \left[\frac{\mathbf{p}_i(t)}{m}, \delta[\mathbf{x} - \mathbf{r}_i(t)] \right]_+ \tag{10}$$

and $[A, B]_+ = AB + BA$ is the anticommutator. In the classical limit, where $\mathbf{p}_i$ and $\mathbf{r}_i$ commute, this reduces to the expected magnetization-density current:

$$\mathbf{J}(\mathbf{x}, t) = \sum_{i=1}^{N} S_i \frac{\mathbf{p}_i(t)}{m} \delta[\mathbf{x} - \mathbf{r}_i(t)] \ . \tag{11}$$

We easily derive from the continuity equation that the total magnetization:

$$M_T(t) = \int d^3x M(\mathbf{x}, t) = \sum_{i=1}^{N} S_i \tag{12}$$

is conserved:

$$\frac{d}{dt} M_T(t) = \frac{d}{dt} \int d^3x M(\mathbf{x}, t) = 0 \ . \tag{13}$$

This is consistent with the result:

$$[H, M_T] = 0 \ . \tag{14}$$

The continuity equation is an operator equation, valid independent of the state of the system.

5.2.2
Constitutive Relation

Suppose now we take the nonequilibrium average of the continuity equation:

$$\frac{\partial}{\partial t} \langle M(\mathbf{x}, t) \rangle_{ne} + \nabla \cdot \langle \mathbf{J}(\mathbf{x}, t) \rangle_{ne} = 0 \ . \tag{15}$$

This gives us one equation and two unknowns. We can obtain a second equation by using the fact that $\langle M(\mathbf{x}, t) \rangle_{ne}$ is conserved and in equilibrium the average of $M(\mathbf{x}, t)$ is uniform. Thus if there is a local excess of spins, since the spin is conserved, it has to be spread over time throughout the rest of the system not spontaneously destroyed. This means that the magnetization-density current, in an effort to equilibrate toward a uniform state, will flow from regions of positive $M(\mathbf{x}, t)$ to regions of negative $M(\mathbf{x}, t)$. Consider the following simple situation as shown in Fig. 5.1. Suppose we can determine the spin density at two points, $\mathbf{x} - a\hat{z}$ and $\mathbf{x} + a\hat{z}$, in our fluid. Also suppose

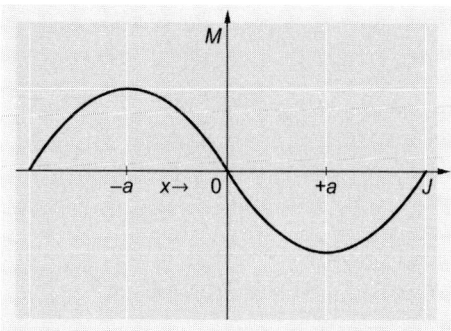

Fig. 5.1 Schematic of the flow of spin-density current from regions of positive magnetization density toward regions of negative magnetization density.

we measure the magnetization-density current flowing in the region between $\mathbf{x} - a\hat{z}$ and $\mathbf{x} + a\hat{z}$. If a is not too large then $\mathbf{J}$ will be approximately uniform over the region of interest. We expect, from the discussion above, that the current is proportional to the difference in the magnetization densities at the points $\mathbf{x} - a\hat{z}$ and $\mathbf{x} + a\hat{z}$:

$$J_z(\mathbf{x}, t) \approx [M(\mathbf{x} - a\hat{z}, t) - M(\mathbf{x} + a\hat{z}, t)] \quad . \tag{16}$$

We write the constant of proportionality as:

$$J_z(\mathbf{x}, t) = \frac{D}{2a}[M(\mathbf{x} - a\hat{z}, t) - M(\mathbf{x} + a\hat{z}, t)] \quad . \tag{17}$$

If $M(\mathbf{x} - a\hat{z}) > M(\mathbf{x} + a\hat{z})$ we expect a positive current flow, so D should be a positive number. We divide by $2a$ since we expect that J will be finite as a becomes small. In particular, in the limit $a \to 0$,

$$\begin{aligned} J_z(\mathbf{x}, t) &= \lim_{a \to 0} \frac{D}{2a}[M(\mathbf{x}, t) - M(\mathbf{x}, t) - 2a\hat{z} \cdot \nabla_x M(\mathbf{x}, t) + \mathcal{O}(a^2)] \\ &= -D \frac{d}{dz} M(\mathbf{x}, t) \quad . \end{aligned} \tag{18}$$

We can write more generally that:

$$\langle \mathbf{J}(\mathbf{x}, t) \rangle_{ne} = -D \vec{\nabla} \langle M(\mathbf{x}, t) \rangle_{ne} \tag{19}$$

when we realize that the current we measure is some nonequilibrium-induced *average* current. This relation connecting the average current to the gradient of the average of a conserved field is called a **constitutive relation**. Constitutive relations are valid on a time and distance scale that is large compared to microscopic processes.

The proportionality constant in our constitutive equation, D, is a transport coefficient known [5] as the spin-diffusion coefficient. Constitutive relations, relating currents to the gradient of a field, show up in many areas of many-body physics: Ohm's law [6], Fourier's law [7], Fick's law [8], etc.

5.2.3
Hydrodynamic Form for Correlation Functions

If we now insert our constitutive equation into the continuity equation we obtain the closed equation:

$$\frac{\partial}{\partial t}\langle M(\mathbf{x},t)\rangle_{ne} - D\nabla^2\langle M(\mathbf{x},t)\rangle_{ne} = 0 \; . \tag{20}$$

This is, of course, in the form of a heat or diffusion equation and can be solved by introducing the Fourier–Laplace transform:

$$\delta M(\mathbf{k},z) = \int_0^{+\infty} dt \, e^{izt} \int d^3x \, e^{i\mathbf{k}\cdot\mathbf{x}} \langle \delta M(\mathbf{x},t)\rangle_{ne} \tag{21}$$

where $\delta M(\mathbf{x},t) = M(\mathbf{x},t) - \langle M(\mathbf{x},t)\rangle_{eq}$ and the equilibrium average of $M(\mathbf{x},t)$ trivially satisfies the diffusion equation. We find immediately that Eq. (20) leads to the solution:

$$\delta M(\mathbf{k},z) = \frac{i}{z + iDk^2} \langle \delta M(\mathbf{k}, t=0)\rangle \; . \tag{22}$$

Notice that $\delta M(\mathbf{k},z)$ possesses a *hydrodynamic* pole on the negative imaginary axis:

$$z_{pole} = -iDk^2 \; . \tag{23}$$

If we invert the Laplace transform, we see that the main contribution to the time dependence is due to this pole and we have in the time domain:

$$\delta M(\mathbf{k},t) = e^{-Dk^2 t} \delta M(\mathbf{k},0) \; . \tag{24}$$

Consequently $\delta M(\mathbf{k},t)$ is exponentially damped with a lifetime $\tau(k) = (Dk^2)^{-1}$. It is characteristic of hydrodynamic modes to have a lifetime inversely proportional to k^2. Consequently, for small k and long wavelengths hydrodynamic modes decay very slowly. In other words, a long wavelength excitation of a conserved variable (spread over a large distance) takes a long time to dissipate or spread throughout the system as a whole. Note that we require $D > 0$ for a stable, decaying system.

We now want to connect our hydrodynamic result with our previous discussion of linear response. We showed, taking the Laplace transform of Eq. (2.143), that:

$$\delta M(\mathbf{k},z) = R_M(\mathbf{k},z)\delta M(\mathbf{k}, t=0) \; , \tag{25}$$

where $R_M(\mathbf{k},z)$ is the relaxation function. If we compare with our hydrodynamic result given by Eq. (22), we easily identify the relaxation function as:

$$R_M(\mathbf{k},z) = \frac{i}{z + iDk^2} \quad . \tag{26}$$

We then recall Eq. (3.145), that the relaxation function is related to the complex response function $\chi_M(\mathbf{k},z)$ by:

$$R_M(\mathbf{k},z) = \frac{1}{iz}[\chi_M(\mathbf{k},z)/\chi_M(\mathbf{k}) - 1] \quad , \tag{27}$$

where $\chi_M(\mathbf{k})$ is the static susceptibility. Comparing Eqs. (26) and (27), we easily identify the dynamic susceptibility:

$$\chi_M(\mathbf{k},z) = \frac{iDk^2 \chi_M(\mathbf{k})}{z + iDk^2} \quad . \tag{28}$$

$\chi_M(\mathbf{k},z)$ is related to the dissipation function $\chi_M''(\mathbf{k},\omega)$ via:

$$\chi_M(\mathbf{k},z) = \int \frac{d\omega'}{\pi} \frac{\chi_M''(\mathbf{k},\omega')}{\omega' - z} \quad . \tag{29}$$

From this equation, we see:

$$\chi_M''(\mathbf{k},\omega) = Im\, \chi_M(\mathbf{k},\omega + i\eta) \tag{30}$$

and we can then determine:

$$\chi_M''(\mathbf{k},\omega) = \chi_M(\mathbf{k}) \frac{\omega Dk^2}{\omega^2 + (Dk^2)^2} \quad . \tag{31}$$

This is the hydrodynamical form for the dissipation or linear response function. We expect it to be valid in the limit of small frequencies and wavenumbers or long times and distances.

Let us pause to check that χ_M'' has the expected properties. We first note that we have an autocorrelation function since we couple only to the magnetization density. Thus, as expected, we find that χ_M'' is real and odd as $\omega \to -\omega$. We also see that:

$$\omega \chi_M''(\mathbf{k},\omega) = \chi_M(\mathbf{k}) \frac{\omega^2 Dk^2}{\omega^2 + (Dk^2)^2} > 0 \tag{32}$$

if $\chi_M(\mathbf{k}) > 0$. Recall that in the long wavelength limit $\chi_M(\mathbf{k})$ reduces to a thermodynamic derivative and for thermodynamic stability we require that the static susceptibility be positive!

Since we know $\chi_M''(\mathbf{k},\omega)$, we can calculate the hydrodynamical forms for all the other correlation functions. The Kubo function measured using incoherent neutron scattering from liquid ^{3}He is related to the dissipation function by Eq. (3.143):

$$C_M(\mathbf{k}, \omega) = \frac{2}{\beta\omega} \chi_M''(\mathbf{k}, \omega) \ . \tag{33}$$

In the hydrodynamic regime:

$$C_M(\mathbf{k}, \omega) = S_M(\mathbf{k}) \frac{2Dk^2}{\omega^2 + (Dk^2)^2} \ . \tag{34}$$

where the static structure factor $S_M(\mathbf{k})$, defined by:

$$S_M(\mathbf{k}) = \langle |M(\mathbf{k})|^2 \rangle = \int \frac{d\omega}{2\pi} C_M(\mathbf{k}, \omega) \ , \tag{35}$$

is related to the wavenumber-dependent susceptibility by:

$$S_M(\mathbf{k}) = k_B T \chi_M(\mathbf{k}) \ . \tag{36}$$

We find in our neutron-scattering experiment that in the hydrodynamical region the spectrum is Lorentzian with a width Dk^2. As we shall see, typical hydrodynamic spectra are Lorentzians and have widths proportional to a transport coefficient times k^2.

5.2.4
Green–Kubo Formula

An appropriate question at this point is: what is D? We introduced it through the constitutive relation as a phenomenological constant. We could simply say that we measure D. Carry out a scattering experiment in the small k and ω limit and identify the coefficient of k^2 as the width D. This is and continues to be an important experimental question to be investigated for magnetic systems. For our purposes the interesting theoretical question is: how do we calculate D? This question could only be answered in a very roundabout fashion before the pioneering work by Green [9] and Kubo [10] in the 1950s. The calculation of transport coefficients prior to that time consisted of solving some approximate nonequilibrium (like the Boltzmann equation) theory for $\langle M(\mathbf{k}, t) \rangle_{ne}$ directly, take the limit of $\mathbf{k} \to 0, t \to \infty$ and verify that the solution is of the decaying form. From this result one could read off the transport coefficient. This is a very cumbersome procedure and it was very difficult to understand the nature of the transport coefficient.

With the development of linear response theory and the introduction of time-correlation functions it became increasingly clear that these transport coefficients are properties of the equilibrium system and, as long as the deviation from equilibrium is small, they are essentially independent of how

5.2 Spin Diffusion in Itinerant Paramagnets

the nonequilibrium situation was set up. It seems reasonable, in analogy with our earlier treatments of particle diffusion and the electrical conductivity, that we can express transport coefficients, like D for a spin system, or the shear viscosity in a fluid, in terms of time-correlation functions.

In the spin problem we have been discussing we can see this rather directly. We start with the hydrodynamical form for the correlation function in the classical limit (for simplicity),

$$C_M(\mathbf{k},\omega) = 2\frac{S_M(\mathbf{k})Dk^2}{\omega^2 + (Dk^2)^2} \ . \tag{37}$$

Then if we multiply $C_M(\mathbf{k},\omega)$ by k^{-2} and take the limit $\mathbf{k} \to 0$, we obtain:

$$\lim_{\mathbf{k}\to 0} \frac{1}{k^2} C_M(\mathbf{k},\omega) = 2\frac{S_M D}{\omega^2} \tag{38}$$

where $S_M = S_M(0)$. We can then solve [11] for DS_M as,

$$DS_M = \frac{1}{2}\lim_{\omega \to 0}\lim_{\mathbf{k}\to 0} \frac{\omega^2}{k^2} C_M(\mathbf{k},\omega) \ . \tag{39}$$

This gives D in terms of S_M and C_M, which are equilibrium correlation functions. While this form for D is useful in some contexts, it is conventional to write this equation in a slightly different way. We note that we can write:

$$\lim_{\omega \to 0}\lim_{\mathbf{k}\to 0} \frac{\omega^2}{k^2} C_M(\mathbf{k},\omega) = \lim_{\omega \to 0}\lim_{\mathbf{k}\to 0} \int_{-\infty}^{+\infty} d(t-t')e^{+i\omega(t-t')}$$

$$\times \frac{1}{k^2}\frac{\partial^2}{\partial t \partial t'} C_M(\mathbf{k}, t - t') \tag{40}$$

where we have integrated by parts twice.

At this stage we need to take a brief detour to discuss the relationship of the spatial Fourier $C_M(\mathbf{k},t,t')$ and the Fourier transform of the field $M(\mathbf{x},t)$. Using our previous conventions we have:

$$C_M(\mathbf{k},t,t') = \int d^d x \, e^{i\mathbf{k}\cdot(\mathbf{x}-\mathbf{y})} C_M(\mathbf{x}-\mathbf{y},t,t') \ . \tag{41}$$

Because of translational invariance this expression is independent of $\mathbf{y}$, and therefore we can write:

$$\begin{aligned} C_M(\mathbf{k},t,t') &= \int \frac{d^d y}{V} \int d^d x \, e^{i\mathbf{k}\cdot(\mathbf{x}-\mathbf{y})} C_M(\mathbf{x}-\mathbf{y},t,t') \\ &= \int \frac{d^d x d^d y}{V} e^{i\mathbf{k}\cdot(\mathbf{x}-\mathbf{y})} \langle M(\mathbf{x},t)M(\mathbf{y},t')\rangle \\ &= \langle M(\mathbf{k},t)M(-\mathbf{k},t')\rangle \end{aligned} \tag{42}$$

and we have defined the Fourier transform of the microscopic fields in the symmetric fashion:

$$M(\mathbf{k}, t) = \int \frac{d^d x}{\sqrt{V}} e^{i\mathbf{k}\cdot\mathbf{x}} M(\mathbf{x}, t) \ . \tag{43}$$

With this set of conventions, we can Fourier transform the continuity equation to obtain:

$$\frac{\partial}{\partial t} M(\mathbf{k}, t) + i\mathbf{k} \cdot \mathbf{J}(\mathbf{k}, t) = 0 \ , \tag{44}$$

so:

$$\frac{\partial}{\partial t}\frac{\partial}{\partial t'} C(\mathbf{k}, t - t') = \left\langle \frac{\partial M(\mathbf{k}, t)}{\partial t} \frac{\partial M(-\mathbf{k}, t')}{\partial t'} \right\rangle$$

$$= \sum_{i,j} k_i k_j \langle J_i(\mathbf{k}, t) J_j(-\mathbf{k}, t') \rangle \ . \tag{45}$$

Then, for small k,

$$\lim_{\omega \to 0} \lim_{\mathbf{k} \to 0} \frac{\omega^2}{k^2} C_M(\mathbf{k}, \omega) = \lim_{\omega \to 0} \int_{-\infty}^{+\infty} d(t - t') e^{+i\omega(t - t')} \lim_{\mathbf{k} \to 0}$$

$$\times \sum_{i,j} \frac{k_i k_j}{k^2} \frac{1}{V} \langle J_i^T(t) J_j^T(t') \rangle \tag{46}$$

where:

$$J_i^T(t) = \int d^3x J_i(\mathbf{x}, t) \ . \tag{47}$$

For an isotropic system in three dimensions:

$$\langle J_i^T(t) J_j^T(t') \rangle = \frac{1}{3} \delta_{ij} \langle \mathbf{J}^T(t) \cdot \mathbf{J}^T(t') \rangle \ . \tag{48}$$

So, putting this together, the diffusion coefficient is given by:

$$DS_M = \frac{1}{2V} \lim_{\mathbf{k} \to 0} \lim_{\omega \to 0} \int_{-\infty}^{+\infty} d(t - t') e^{+i\omega(t - t')} \frac{k^2}{k^2} \frac{1}{3} \langle \mathbf{J}^T(t) \cdot \mathbf{J}^T(t') \rangle$$

$$= \frac{1}{3V} \lim_{\varepsilon \to 0} \int_0^{+\infty} dt \, e^{-\varepsilon t} \langle \mathbf{J}^T(t) \cdot \mathbf{J}^T(0) \rangle \ , \tag{49}$$

where we have used the fact that the autocorrelation function is even under letting $t \to -t$. This is the conventional Green–Kubo form [12] expressing a transport coefficient in terms of a time integral over a current–current correlation function.

This representation is very convenient because it is a useful starting point for computer molecular dynamics experiments. One can compute $\langle \mathbf{J}^T(t) \cdot \mathbf{J}^T(0) \rangle$

directly from Newton's equations and from this determine DS_M. Since the total current $\mathbf{J}^T(t)$ is not conserved, one expects the current–current correlation function to decay to zero on a microscopic time scale, which is convenient for simulations. This has been a particularly useful approach in developing the dynamical theory of liquids.

Note that it is the product form DS_M that occurs naturally in the Green–Kubo equation. It is typical that an associated equilibrium susceptibility like S_M enters the analysis and we need to determine it from a separate calculation. We will look at this from a more general perspective in the next section.

We show here that we have also treated in this analysis the problem of self-diffusion. It is shown in Problem 5.2, because the spins at different sites are uncorrelated, that:

$$C_M(\mathbf{x}-\mathbf{y}, t-t') = S^2 S_S(\mathbf{x}-\mathbf{y}, t-t') \ , \tag{50}$$

where S_S is the van Hove self-correlation function. We discuss here how we can simplify the Green–Kubo equation given by Eq. (49). We first treat the static correlation function. Working in the classical limit:

$$\begin{aligned}
S(\mathbf{x}-\mathbf{y}) &= \langle \delta M(\mathbf{x}) \delta M(\mathbf{y}) \rangle \\
&= \sum_{i=1}^{N} \sum_{j=1}^{N} \langle S_i S_j \delta(\mathbf{x}-\mathbf{R}_i) \delta(\mathbf{y}-\mathbf{R}_j) \rangle \\
&= \sum_{i=1}^{N} \sum_{j=1}^{N} \delta_{ij} \langle S^2 \rangle \langle \delta(\mathbf{x}-\mathbf{R}_i) \delta(\mathbf{y}-\mathbf{R}_j) \rangle \\
&= \langle S^2 \rangle \langle n \rangle \delta(\mathbf{x}-\mathbf{y}) = \bar{S} \delta(\mathbf{x}-\mathbf{y})
\end{aligned} \tag{51}$$

where $\bar{S} = S^2 \langle n \rangle$ is a constant and the static structure factor has a constant Fourier transform:

$$S(\mathbf{q}) = \bar{S} = nS^2 \ . \tag{52}$$

Turning to the right-hand side of Eq. (49), we need the spin current given by Eq. (11) and the total current:

$$\mathbf{J}_T(t) = \int d^d x\, \mathbf{J}(\mathbf{x}, t) = \sum_{i=1}^{N} S_i \mathbf{v}_i(t) \ . \tag{53}$$

Inserting this result in Eq. (49) we are led to the result:

$$\begin{aligned}
DnS^2 &= \frac{1}{3V} \int_0^\infty dt \left\langle \sum_{i=1}^{N} S_i \mathbf{v}_i(t) \cdot \sum_{j=1}^{N} S_j \mathbf{v}_j(0) \right\rangle \\
&= \frac{1}{3V} \int_0^\infty dt\, S^2 \langle \sum_{i=1}^{N} \mathbf{v}_i(t) \cdot \mathbf{v}_i(0) \rangle
\end{aligned}$$

$$= \frac{1}{3\mathcal{V}} \int_0^\infty dt\, S^2 N \langle \mathbf{v}_i(t) \cdot \mathbf{v}_i(0) \rangle$$

$$= \frac{nS^2}{3} \int_0^\infty dt\, \langle \mathbf{v}_i(t) \cdot \mathbf{v}_i(0) \rangle \ . \tag{54}$$

Canceling common terms gives finally:

$$D = \frac{1}{3} \int_0^\infty dt\, \langle \mathbf{v}_i(t) \cdot \mathbf{v}_i(0) \rangle \ . \tag{55}$$

This is the Green–Kubo equation found in Chapter 1.

5.3
Langevin Equation Approach to the Theory of Irreversible Processes

5.3.1
Choice of Variables

Let us investigate a theoretical approach to the determination of time-correlation functions, which is a generalization of the approximate Langevin equation approach discussed in Chapters 1, 2 and 4. We learned in the preceding section that under certain circumstances particular variables, out of the fantastically large number of degrees of freedom in the system, have relatively simple long-time and -distance behavior. In general we are only interested in a few variables ψ_i in a system.

It will be convenient to allow the subscript i to index the type of variable as well as any vector or coordinate label. We will, for now, choose ψ_i such that it has a zero equilibrium average [13]:

$$\langle \psi_i \rangle = 0 \ . \tag{56}$$

For a ferromagnet we might choose ψ_i to be the magnetization density $\mathbf{M}(\mathbf{x})$. In a fluid we might include in ψ_i the particle density $n(\mathbf{x})$, the momentum density $\mathbf{g}(\mathbf{x})$ and the energy density $\varepsilon(\mathbf{x})$.

5.3.2
Equations of Motion

In general the variables ψ_i satisfy the Heisenberg equation of motion written in the form:

$$\frac{\partial \psi_i(t)}{\partial t} = iL\psi_i(t) \ , \tag{57}$$

where L is the Liouville operator. Quantum mechanically:

$$L\psi_i = \frac{1}{\hbar}[H, \psi_i] \tag{58}$$

is just the commutator of the dynamical variable with the Hamiltonian.

It is instructive to look at the form of the equation of motion given by Eq. (57) in the classical limit. If the variable ψ_i depends on the phase-space coordinates $\{r_1, r_2, \ldots, r_N; p_1, p_2, \ldots, p_N\}$ for a system of N particles then it satisfies the equation of motion,

$$\frac{\partial \psi_i(t)}{\partial t} = \sum_{j,\alpha} \left[\frac{\partial \psi_i(t)}{\partial r_j^\alpha(t)} \dot{r}_j^\alpha(t) + \frac{\partial \psi_i(t)}{\partial p_j^\alpha(t)} \dot{p}_j^\alpha(t) \right] \tag{59}$$

where we have used the chain-rule for differentiation. After using Hamilton's equations:

$$\dot{r}_i^\alpha(t) = \frac{\partial H}{\partial p_i^\alpha(t)} \tag{60}$$

$$\dot{p}_i^\alpha(t) = -\frac{\partial H}{\partial r_i^\alpha(t)}, \tag{61}$$

the equation of motion for $\psi_i(t)$ can be written as:

$$\frac{\partial \psi_i(t)}{\partial t} = \sum_{j,\alpha} \left[\frac{\partial \psi_i(t)}{\partial r_j^\alpha(t)} \frac{\partial H}{\partial p_j^\alpha(t)} - \frac{\partial \psi_i(t)}{\partial p_j^\alpha(t)} \frac{\partial H}{\partial r_j^\alpha(t)} \right]$$
$$= \{\psi_i, H\} \tag{62}$$

where we have introduced the Poisson bracket for two variables A and B:

$$\{A, B\} = \sum_{i,\alpha} \left[\frac{\partial A}{\partial r_i^\alpha(t)} \frac{\partial B}{\partial p_i^\alpha(t)} - \frac{\partial A}{\partial p_i^\alpha(t)} \frac{\partial B}{\partial r_i^\alpha(t)} \right]. \tag{63}$$

Comparing the classical and quantum equations of motion,

$$\frac{\partial \psi_i(t)}{\partial t} = \begin{cases} \{\psi_i, H\} & \text{classical} \\ \frac{1}{i\hbar} [\psi_i, H] & \text{quantum mechanical} \end{cases},$$

we find the quantum-classical mapping, first suggested by Dirac [14],

$$\frac{1}{i\hbar}[A, B] \to \{A, B\}. \tag{64}$$

The classical form for the Liouville operator is given by:

$$L\psi_i = \frac{1}{\hbar}[H, \psi_i] \to i\{H, \psi_i\} \tag{65}$$

and we have for the equation of motion:

$$\frac{\partial \psi_i(t)}{\partial t} = iL\psi_i(t) = -\{H, \psi_i(t)\} \ . \tag{66}$$

We can formally integrate the equation of motion to obtain:

$$\psi_i(t) = e^{iLt}\psi_i \equiv U(t)\psi_i \ , \tag{67}$$

where $\psi_i = \psi_i(t=0)$ and $U(t) = e^{iLt}$ is known as Koopman's operator [15].

We will find it very convenient to deal with the time Laplace transform of $\psi_i(t)$, defined:

$$\psi_i(z) = -i \int_0^\infty dt\, e^{izt} \psi_i(t) \ , \tag{68}$$

where z has a small positive imaginary piece. If we put Eq. (67) into Eq. (68) we easily obtain:

$$\psi_i(z) = R(z)\psi_i \ , \tag{69}$$

where the resolvant operator is defined:

$$R(z) = (z+L)^{-1} \ . \tag{70}$$

The equation of motion for $\psi(z)$ can now be obtained by using the operator identity:

$$zR(z) = 1 - LR(z) \ . \tag{71}$$

We find:

$$z\psi_i(z) = \psi_i - L\psi_i(z) \ . \tag{72}$$

In the next two sections we work out examples of nontrivial equations of motion for magnetic and fluid systems.

5.3.3
Example: Heisenberg Ferromagnet

In order to appreciate the content of Eq. (72) and to make things more concrete, let us consider two examples. First let us consider a quantum-mechanical Heisenberg ferromagnet [16] described by the Hamiltonian [17]:

$$H = -\frac{1}{2}\sum_{\mathbf{R}\neq\mathbf{R}'} J(\mathbf{R}-\mathbf{R}')\mathbf{M}(\mathbf{R})\cdot\mathbf{M}(\mathbf{R}') - \sum_{\mathbf{R}} \mathbf{M}(\mathbf{R})\cdot\mathbf{B} \ , \tag{73}$$

where $J(\mathbf{R}-\mathbf{R}')$ is the exchange interaction between spins at lattice sites $\mathbf{R}$ and $\mathbf{R}'$ and $\mathbf{B}$ is an applied external magnetic field. We can then easily calculate the effect of L on $\mathbf{M}(\mathbf{R})$ using the usual spin-commutation relations (we adsorb the factors of μ_0 relating the magnetic moment and the spin in the definition of the exchange coupling):

$$L\mathbf{M}(\mathbf{R}) = \frac{1}{\hbar}[H, \mathbf{M}(\mathbf{R})] \ .$$

The commutator is worked out in Problem 5.3 with the result:

$$L\mathbf{M}(\mathbf{R}) = -i\sum_{\mathbf{R}'} J(\mathbf{R} - \mathbf{R}')\mathbf{M}(\mathbf{R}) \times \mathbf{M}(\mathbf{R}') - i\mathbf{B} \times \mathbf{M}(\mathbf{R}) \ . \qquad (74)$$

The equation of motion is given for this system by:

$$\frac{\partial \mathbf{M}(\mathbf{R})}{\partial t} = iL\mathbf{M}(\mathbf{R}) = \sum_{\mathbf{R}'} J(\mathbf{R} - \mathbf{R}')\mathbf{M}(\mathbf{R}) \times \mathbf{M}(\mathbf{R}') + \mathbf{B} \times \mathbf{M}(\mathbf{R}) \ . \qquad (75)$$

Note that Eq. (75) is explicitly a nonlinear equation for $\mathbf{M}$.

Let us restrict ourselves to the case of zero applied external field, $\mathbf{B} = 0$. It is easy to show, for this isotropic system, that the total magnetization,

$$\mathbf{M}_T = \sum_{\mathbf{R}} \mathbf{M}(\mathbf{R}) \ , \qquad (76)$$

is conserved. Summing Eq. (75) over all $\mathbf{R}$ we obtain:

$$\frac{d}{dt}\mathbf{M}_T = \sum_{\mathbf{R},\mathbf{R}'} J(\mathbf{R} - \mathbf{R}')\mathbf{M}(\mathbf{R}) \times \mathbf{M}(\mathbf{R}') \ . \qquad (77)$$

If $J(\mathbf{R} - \mathbf{R}') = J(\mathbf{R}' - \mathbf{R})$ (we have translational and rotational invariance), the sums on the right vanish due to the odd symmetry for $\mathbf{R} \leftrightarrow \mathbf{R}'$, and $\mathbf{M}_T$ is conserved:

$$\frac{d\mathbf{M}_T}{dt} = 0 \qquad (78)$$

$$L\mathbf{M}_T = 0 \ . \qquad (79)$$

Since we have a conservation law we should be able to write a continuity equation and extract an expression for the spin-density current. Because the spins are interacting here we expect this process to be a bit more complicated when compared to the analysis in the first section of this chapter. The first step is to introduce the spin or magnetization density:

$$\mathbf{M}(\mathbf{x}, t) = \sum_{\mathbf{R}} \mathbf{M}(\mathbf{R}, t)\delta(\mathbf{x} - \mathbf{R}) \ . \qquad (80)$$

After multiplying Eq. (75) by $\delta(\mathbf{x} - \mathbf{R})$ and summing over $\mathbf{R}$, we obtain:

$$\frac{\partial \mathbf{M}(\mathbf{x})}{\partial t} = \sum_{\mathbf{R},\mathbf{R}'} \delta(\mathbf{x} - \mathbf{R}) J(\mathbf{R} - \mathbf{R}')\mathbf{M}(\mathbf{R}) \times \mathbf{M}(\mathbf{R}')$$

$$= \frac{1}{2} \sum_{R,R'} (\delta(x-R) - \delta(x-R'))$$
$$\times J(R-R')M(R) \times M(R') \ . \tag{81}$$

We can identify the spin-current if we use a trick [18]. Consider the identity,

$$\frac{\partial}{\partial s}\delta(x-sR) = -R \cdot \nabla_x \delta(x-sR) \ , \tag{82}$$

where s is a real continuous variable. Integrate this expression over s from 0 to 1:

$$\delta(x-R) - \delta(x) = -\int_0^1 ds \, R \cdot \nabla_x \delta(x-sR) \ . \tag{83}$$

Using this result we can write:

$$\delta(x-R) - \delta(x-R') = -\int_0^1 ds \, \nabla_x \cdot \left[R\delta(x-sR) - R'\delta(x-sR')\right] \ . \tag{84}$$

Inserting Eq. (84) in Eq. (81), we find:

$$\frac{\partial M^i(x,t)}{\partial t} = -\sum_\beta \nabla_x^\beta J_\beta^i(x,t) \ , \tag{85}$$

where:

$$J_\beta^i(x,t) = \frac{1}{2} \sum_{R,R'} \sum_{jk} \varepsilon_{ijk} \int_0^1 ds \left[R_\beta \delta(x-sR) - R'_\beta \delta(x-sR')\right]$$
$$\times J(R-R')M^j(R,t)M^k(R',t) \ . \tag{86}$$

Integrating over all x we obtain the total current:

$$J_{T,\beta}^i(t) = \frac{1}{2} \sum_{R,R'} \sum_{jk} \varepsilon_{ijk}[R-R']_\beta J(R-R') M_R^j(t) M_{R'}^k(t) \ . \tag{87}$$

This expression for the total current can be used in the associated Green–Kubo expression.

5.3.4
Example: Classical Fluid

As a second example of an equation of motion, we consider a classical fluid of N point particles with mass m whose dynamics involve the phase-space coordinates r_i and p_i. The Hamiltonian describing these particles is assumed to be of the form:

$$H = \sum_{i=1}^N \frac{p_i^2}{2m} + \frac{1}{2} \sum_{i \neq j=1}^N V(r_i - r_j) \ , \tag{88}$$

where $V(\mathbf{r})$ is the pair potential acting between particles. An interesting dynamical variable in this case (a good choice for ψ_i) is the phase-space density:

$$\hat{f}(\mathbf{x}, \mathbf{p}) = \sum_{i=1}^{N} \delta(\mathbf{x} - \mathbf{r}_i)\delta(\mathbf{p} - \mathbf{p}_i) \ . \tag{89}$$

We can construct the particle density, momentum density and kinetic energy density from $\hat{f}(\mathbf{x}, \mathbf{p})$ by multiplying by 1, $\mathbf{p}$ and $\mathbf{p}^2/2m$, respectively, and integrating over $\mathbf{p}$:

$$n(\mathbf{x}) = \int d^3p \, \hat{f}(\mathbf{x}, \mathbf{p}) = \sum_{i=1}^{N} \delta(\mathbf{x} - \mathbf{r}_i) \tag{90}$$

$$\mathbf{g}(\mathbf{x}) = \int d^3p \, \mathbf{p} \, \hat{f}(\mathbf{x}, \mathbf{p}) = \sum_{i=1}^{N} \mathbf{p}_i \delta(\mathbf{x} - \mathbf{r}_i) \tag{91}$$

$$K(\mathbf{x}) = \int d^3p \, \frac{\mathbf{p}^2}{2m} \hat{f}(\mathbf{x}, \mathbf{p}) = \sum_{i=1}^{N} \frac{\mathbf{p}_i^2}{2m} \delta(\mathbf{x} - \mathbf{r}_i) \ . \tag{92}$$

The Liouville operator in this case can be written:

$$L = i\{H, \}$$
$$= i \sum_{i=1}^{N} (\nabla_{\mathbf{r}_i} H \cdot \nabla_{\mathbf{p}_i} - \nabla_{\mathbf{p}_i} H \cdot \nabla_{\mathbf{r}_i})$$
$$= -i \sum_{i=1}^{N} \frac{\mathbf{p}_i}{m} \cdot \nabla_{\mathbf{r}_i} + i \sum_{i \neq j=1}^{N} \nabla_{\mathbf{r}_i} V(\mathbf{r}_i - \mathbf{r}_j) \cdot \nabla_{\mathbf{p}_i}$$
$$= L_0 + L_I \ , \tag{93}$$

where the *free-streaming* noninteracting part is given by:

$$L_0 = -i \sum_{i=1}^{N} \frac{\mathbf{p}_i}{m} \cdot \nabla_{\mathbf{r}_i} \tag{94}$$

and the interacting part by:

$$L_I = i \sum_{i \neq j=1}^{N} \nabla_{\mathbf{r}_i} V(\mathbf{r}_i - \mathbf{r}_j) \cdot \nabla_{\mathbf{p}_i} \ . \tag{95}$$

It is useful to look at the effect of allowing the Liouville operator to act on the phase-space density:

$$L\hat{f}(\mathbf{x}, \mathbf{p}) = L_0 \hat{f}(\mathbf{x}, \mathbf{p}) + L_I \hat{f}(\mathbf{x}, \mathbf{p}) \ . \tag{96}$$

Let us look first at the free-streaming contribution:

$$\begin{aligned}
L_0 \hat{f}(\mathbf{x}, \mathbf{p}) &= -i \sum_{i=1}^{N} \frac{\mathbf{p}_i}{m} \cdot \nabla_{\mathbf{r}_i} \sum_{j=1}^{N} \delta(\mathbf{x} - \mathbf{r}_j) \delta(\mathbf{p} - \mathbf{p}_j) \\
&= -i \sum_{i=1}^{N} \frac{\mathbf{p}_i}{m} \cdot \nabla_{\mathbf{r}_i} \delta(\mathbf{x} - \mathbf{r}_i) \delta(\mathbf{p} - \mathbf{p}_i) \\
&= i \nabla_x \cdot \sum_{i=1}^{N} \frac{\mathbf{p}_i}{m} \delta(\mathbf{x} - \mathbf{r}_i) \delta(\mathbf{p} - \mathbf{p}_i) \\
&= i \frac{\mathbf{p} \cdot \nabla_x}{m} \hat{f}(\mathbf{x}, \mathbf{p}) \\
&\equiv -L_0(\mathbf{x}, \mathbf{p}) \hat{f}(\mathbf{x}, \mathbf{p}) \ .
\end{aligned} \qquad (97)$$

Notice that the operator L_0, given by Eq. (94), acts on phase-space coordinates, while $L_0(\mathbf{x}, \mathbf{p})$, defined by Eq. (97), acts on the external arguments labeling the phase-space densities. Turning to the interacting part of the Liouville operator:

$$\begin{aligned}
L_I \hat{f}(\mathbf{x}, \mathbf{p}) &= i \sum_{i \neq j=1}^{N} \nabla_{\mathbf{r}_i} V(\mathbf{r}_i - \mathbf{r}_j) \cdot \nabla_{\mathbf{p}_i} \sum_{k=1}^{N} \delta(\mathbf{x} - \mathbf{r}_k) \delta(\mathbf{p} - \mathbf{p}_k) \\
&= i \sum_{i \neq j=1}^{N} \nabla_{\mathbf{r}_i} V(\mathbf{r}_i - \mathbf{r}_j) \cdot \nabla_{\mathbf{p}_i} \delta(\mathbf{x} - \mathbf{r}_i) \delta(\mathbf{p} - \mathbf{p}_i) \\
&= -i \nabla_{\mathbf{p}} \cdot \sum_{i \neq j=1}^{N} \nabla_{\mathbf{r}_i} V(\mathbf{r}_i - \mathbf{r}_j) \delta(\mathbf{x} - \mathbf{r}_i) \delta(\mathbf{p} - \mathbf{p}_i) \ .
\end{aligned} \qquad (98)$$

In order to write $L_I \hat{f}(\mathbf{x}, \mathbf{p})$ as an operator like $L_0(\mathbf{x}, \mathbf{p})$ acting on a product of phase-space densities, we insert the identity to obtain:

$$\begin{aligned}
L_I \hat{f}(\mathbf{x}, \mathbf{p}) &= -i \nabla_{\mathbf{p}} \cdot \sum_{i \neq j=1}^{N} \nabla_{\mathbf{r}_i} V(\mathbf{r}_i - \mathbf{r}_j) \int d^3 x_1 d^3 p_1 \delta(\mathbf{x}_1 - \mathbf{r}_j) \delta(\mathbf{p}_1 - \mathbf{p}_j) \\
&\quad \times \delta(\mathbf{x} - \mathbf{r}_i) \delta(\mathbf{p} - \mathbf{p}_i) \\
&= -i \nabla_{\mathbf{p}} \cdot \int d^3 x_1 d^3 p_1 \nabla_x V(\mathbf{x} - \mathbf{x}_1) \hat{f}(\mathbf{x}_1, \mathbf{p}_1) \hat{f}(\mathbf{x}, \mathbf{p}) \\
&= -\int d^3 x_1 d^3 p_1 L_I(\mathbf{x} - \mathbf{x}_1, \mathbf{p}, \mathbf{p}_1) \hat{f}(\mathbf{x}_1, \mathbf{p}_1) \hat{f}(\mathbf{x}, \mathbf{p}) \ ,
\end{aligned} \qquad (99)$$

where we define the two-particle interaction part of the Liouville operator:

$$L_I(\mathbf{x} - \mathbf{x}_1, \mathbf{p}, \mathbf{p}_1) = i \nabla_x V(\mathbf{x} - \mathbf{x}_1) \cdot (\nabla_{\mathbf{p}} - \nabla_{\mathbf{p}_1}) \ , \qquad (100)$$

and in the next to the last step in obtaining Eq. (99) we have dropped a self-interacting term, $i = j$, which can be eliminated through a careful choice for the potential. We can then write the effect of the Liouville operator acting on the phase-space density in the form:

$$L\hat{f}(\mathbf{x},\mathbf{p}) = -L_0(\mathbf{x},\mathbf{p})\hat{f}(\mathbf{x},\mathbf{p})$$
$$- \int d^3x_1 d^3p_1 L_I(\mathbf{x}-\mathbf{x}_1,\mathbf{p},\mathbf{p}_1)\hat{f}(\mathbf{x}_1,\mathbf{p}_1)\hat{f}(\mathbf{x},\mathbf{p}) \quad . \tag{101}$$

Note there is a piece of $L\hat{f}$ that is linear in $\hat{f}$ (the kinetic energy term) that plays a formal role similar to the $\mathbf{B} \times \mathbf{M}$ term in the magnetic example. There is also a nonlinear interaction term $\approx \hat{f}^2$, as in the spin case.

The equation of motion is given then by:

$$\frac{\partial}{\partial t}\hat{f}(\mathbf{x},\mathbf{p}) = -iL_0(\mathbf{x},\mathbf{p})\hat{f}(\mathbf{x},\mathbf{p})$$
$$-i \int d^3x_1 d^3p_1 L_I(\mathbf{x}-\mathbf{x}_1,\mathbf{p},\mathbf{p}_1)\hat{f}(\mathbf{x},\mathbf{p})\hat{f}(\mathbf{x}_1,\mathbf{p}_1) \quad . \tag{102}$$

We see that we have *projected* the equation of motion onto the space spanned by the labels of the phase-space densities. The price we pay is that the equations of motion are now nonlinear. As discussed in Chapter 7, Eq. (102) serves as the basis for developing kinetic theory. We get started in such an approach by assuming we have a low-density system and particles spend most of their time free streaming, where the nonlinear part of the equation of motion can be treated as a small perturbation.

Notice that one can use this equation of motion, Eq. (102), to look at the conservation laws. If we simply integrate over $\mathbf{p}$ we obtain the continuity equation:

$$\frac{\partial n(\mathbf{x},t)}{\partial t} = -\nabla \cdot \frac{\mathbf{g}(\mathbf{x},t)}{m} \quad . \tag{103}$$

If we multiply Eq. (102) by $\mathbf{p}$ and integrate over $\mathbf{p}$, we can, after some manipulations discussed in Appendix F of ESM, show that conservation of momentum is expressed locally as:

$$\frac{\partial g_\alpha(\mathbf{x})}{\partial t} = -\sum_\beta \nabla_x^\beta \sigma_{\alpha\beta}(\mathbf{x}) \quad , \tag{104}$$

where the microscopic *stress tensor* is given by

$$\sigma_{\alpha\beta}(\mathbf{x}) = \int d^3p \frac{p_\alpha p_\beta}{m}\hat{f}(\mathbf{x},\mathbf{p}) - \frac{1}{4}\int_{-1}^{+1} ds \int d^3r \frac{r_\alpha r_\beta}{r}\frac{\partial V(r)}{\partial r}$$
$$\times n(\mathbf{x}+(1/2)(s+1)\mathbf{r},t) n(\mathbf{x}+(1/2)(s-1)\mathbf{r},t) \quad . \tag{105}$$

Similarly the local statement of conservation of energy is given by:

$$\frac{\partial \varepsilon(\mathbf{x},t)}{\partial t} = -\nabla \cdot \mathbf{J}_\varepsilon(\mathbf{x},t) \quad , \tag{106}$$

where the energy current is given by:

$$J_\varepsilon^\alpha(\mathbf{x},t) = \int d^3p \left(\frac{p^2}{2m} + \int d^3r \frac{1}{2}V(\mathbf{x}-\mathbf{r})n(\mathbf{r})\right) p_\alpha \hat{f}(\mathbf{x},\mathbf{p},t)$$

$$-\frac{1}{4}\int_{-1}^{+1} ds \int d^3r \int d^3 p d^3 p' \sum_\beta \frac{r_\alpha r_\beta}{r} \frac{\partial V(r)}{\partial r} (p_\beta + p'_\beta)$$

$$\times \hat{f}[\mathbf{x} - (1/2)(s+1)\mathbf{r}, \mathbf{p}, t]\hat{f}[\mathbf{x} - (1/2)(s-1)\mathbf{r}, \mathbf{p}', t] \quad . \quad (107)$$

5.3.5
Summary

In summary we expect that the microscopic equation of motion for the variables ψ_i can be written in the generic form:

$$\dot{\psi}_i = \sum_j i\varepsilon_{ij}\psi_j + iN_i \, , \tag{108}$$

where N_i represents the nonlinear terms in the equation of motion:

$$N_i = \sum_{jk} V_{ijk}\psi_j\psi_k + \cdots \quad . \tag{109}$$

5.3.6
Generalized Langevin Equation

The equation of motion for $\psi_i(t)$ is, in general, a complicated nonlinear functional of $\psi_i(t)$. Thus a direct microscopic attack on the problem appears very difficult. One should also keep in mind that the examples of equations of motion discussed above are for the simplest of systems. If we want to look at more complex systems like liquid crystals and polymers we must develop a more flexible approach that is less tied to the microscopic details that may not be important in describing long-distance and long-time motions. Toward this end, and despite the fact that $L\psi_i(z)$ is in general a complicated nonlinear functional of $\psi_i(t)$, it is useful to assume that $L\psi_i(z)$ can be written as the sum of two pieces, one of which is linear in $\psi_i(z)$ and something that is left over:

$$-L\psi_i(z) = \sum_j K_{ij}(z)\psi_j(z) + if_i(z) \quad . \tag{110}$$

In Eq. (110) $K_{ij}(z)$ is a function and $f_i(z)$ is a dynamical variable in the same sense as $\psi_i(z)$. The first piece on the right-hand side of Eq. (110) can be thought of as the *projection* of $-L\psi_i(z)$ back along $\psi_i(z)$, and $f_i(z)$ is what is left over. The physics, which we will draw out as we go along, is that the linear piece, proportional to $\sum_j K_{ij}(z)\psi_j$, corresponds to the friction terms in the Brownian motion problem, while f_i corresponds to the noise.

We first note, taking the equilibrium average of Eq. (110), that:

$$\langle f_i(z) \rangle = 0 \tag{111}$$

since $\langle \psi_i(z) \rangle = 0$ [from Eq. (57)] and:

$$\langle L\psi_i(z) \rangle = -i \int_0^\infty dt\, e^{izt} (-i) \frac{d}{dt} \langle \psi_i(t) \rangle = 0 \,, \tag{112}$$

which follows from time-translational invariance of the equilibrium ensemble.

It should be clear that Eq. (110) is still not unique, since we have one equation and two unknowns. A second equation comes from requiring that the noise $f_i(z)$ have no projection onto the initial value of $\psi_i(t)$:

$$\langle \psi_j f_i(z) \rangle = 0 \,. \tag{113}$$

As we shall see, this is a powerful requirement [19]. To keep things simple, we assume we have a classical theory where the ψ's commute. This is not a necessary [20] requirement, but its elimination would necessitate additional, but essentially unilluminating, further formal development.

As we shall see Eqs. (110) and (113) uniquely determine K and f. If we put Eq. (110) back into Eq. (72) we obtain the form for the equation of motion:

$$\sum_j [z\delta_{ij} - K_{ij}(z)] \psi_j(z) = \psi_i + i f_i(z) \,. \tag{114}$$

If we take the inverse Laplace transform (see Problem 5.19), we find:

$$\frac{\partial \psi_i}{\partial t}(t) + \sum_j \int_0^t d\bar{t}\, K_{ij}(t - \bar{t}) \psi_j(\bar{t}) = f_i(t) \,. \tag{115}$$

This equation is known as the *generalized Langevin equation*. Indeed this is a generalization of the Langevin equation we studied in our discussion of Brownian motion. The integral kernel $K_{ij}(t - t')$ is known as the memory function. The classic works on the generalized Langevin equation are due to Zwanzig [21] and Mori [22] who used projection operator techniques. We will follow a somewhat different procedure [23].

In the Langevin equation formulation one thinks in terms of two sets of dynamical variables ψ_i and f_i, which are initially independent. As time proceeds however the ψ_i and f_i are mixed by the nonlinearities in the equation of motion and $\psi_i(z)$ has components along both ψ_i and $f_i(z)$. In the next section we show how $K_{ij}(t)$ can be determined.

5.3.7
Memory-Function Formalism

We will be interested in calculating the equilibrium-averaged time-correlation functions:

$$C_{ij}(t) = \langle \psi_j \psi_i(t) \rangle \tag{116}$$

for the set of slow variables. It is convenient to introduce the Fourier transform:

$$C_{ij}''(\omega) = \int_{-\infty}^{\infty} dt\, e^{i\omega t} C_{ij}(t) \tag{117}$$

and the Laplace transform:

$$C_{ij}(z) = -i \int_0^{\infty} dt\, e^{izt} C_{ij}(t) \tag{118}$$

$$= \int_{-\infty}^{\infty} \frac{d\omega}{2\pi} \frac{C_{ij}''(\omega)}{(z-\omega)} \tag{119}$$

and obtain the usual relation between the Laplace and Fourier transforms. Remembering Eq. (68), it is clear that we can identify:

$$C_{ij}(z) = \langle \psi_j \psi_i(z) \rangle = \langle \psi_j R(z) \psi_i \rangle \;. \tag{120}$$

If we multiply Eq. (114) by ψ_i, take the equilibrium average and use Eq. (113), we obtain:

$$z C_{ij}(z) = S_{ij} + \sum_l K_{il}(z) C_{lj}(z) \;, \tag{121}$$

where:

$$S_{ij} = \langle \psi_j \psi_i \rangle \tag{122}$$

is the static or equal-time correlation function among the slow variables. We assume we know the S_{ij} from equilibrium statistical mechanics. If we know $K_{ij}(z)$ then we can solve for $C_{ij}(z)$. However, to some extent, Eq. (121) simply defines the new function $K_{ij}(z)$ in terms of the correlation function. The introduction of $K_{ij}(z)$ is useful only if it is easier to calculate or approximate than $C_{ij}(z)$ itself. Obviously this will turn out to be the case or we would not introduce it.

We now want to show how one can express the memory function $K_{ij}(z)$ in terms of time-correlation functions. If we multiply Eq. (72) by ψ_j and take the equilibrium average we obtain:

$$z C_{ij}(z) = S_{ij} - \langle \psi_j R(z) L \psi_i \rangle \;. \tag{123}$$

If we compare this equation with Eq. (121) we can identify:

$$\sum_l K_{il}(z) C_{lj}(z) = -\langle \psi_j R(z) L \psi_i \rangle \quad . \tag{124}$$

If we multiply Eq. (124) by z, use Eq. (123) on the left-hand side in $C_{lj}(z)$ and Eq. (71) on the right-hand side, we obtain:

$$\sum_l K_{il}(z)[S_{lj} - \langle \psi_j L R(z) \psi_l \rangle] = -\langle \psi_j L \psi_i \rangle + \langle \psi_j L R(z) L \psi_i \rangle \quad . \tag{125}$$

After using the property (see Problem 5.8) of the Liouville operator:

$$\langle ALB \rangle = -\langle (LA)B \rangle \tag{126}$$

(which follows from the time-translational invariance of the equilibrium distribution function) we can write Eq. (125) in the form:

$$K_{ij}(z) = K_{ij}^{(s)} + K_{ij}^{(d)}(z) \quad , \tag{127}$$

where:

$$\Gamma_{ij}^{(s)} = \sum_l K_{il}^{(s)} S_{lj} = -\langle \psi_j L \psi_i \rangle \quad , \tag{128}$$

and:

$$\Gamma_{ij}^{(d)}(z) = \sum_l K_{il}^{(d)}(z) S_{lj} = -\langle (L\psi_j) R(z) (L\psi_i) \rangle$$
$$- \sum_l K_{il}(z) \langle (L\psi_j) R(z) \psi_l \rangle \quad . \tag{129}$$

If we introduce the matrix inverse:

$$\sum_l C_{il}(z) C_{lj}^{-1}(z) = \delta_{ij} \quad , \tag{130}$$

then matrix multiply Eq. (124) from the right by $C^{-1}(z)$, we obtain:

$$K_{il}(z) = -\sum_k \langle \psi_k R(z) L \psi_i \rangle C_{kl}^{-1}(z) \quad . \tag{131}$$

Using his result in Eq. (129) gives:

$$\Gamma_{ij}^{(d)}(z) = -\langle (L\psi_j) R(z) (L\psi_i) \rangle$$
$$+ \sum_{k,l} \langle \psi_k R(z) (L\psi_i) \rangle C_{kl}^{-1}(z) \langle (L\psi_j) R(z) \psi_l \rangle \quad . \tag{132}$$

Let me summarize these results for the memory function:

$$\Gamma_{ij}^{(d)}(z) = \sum_k K_{ik}^{(s)} S_{kj} + \sum_k K_{ik}^{(d)}(z) S_{kj} \quad , \tag{133}$$

where the *static* or z-independent part of the memory function is given by:

$$\Gamma_{ij}^{(s)} = -\langle \psi_j L \psi_i \rangle \tag{134}$$

and the *dynamical* part by:

$$\Gamma_{ij}^{(d)}(z) = -\langle (L\psi_j) R(z)(L\psi_i) \rangle + \sum_{k,l} \langle \psi_k R(z)(L\psi_i) \rangle C_{kl}^{-1}(z) \langle (L\psi_j) R(z)\psi_l \rangle \ . \tag{135}$$

We note that $K_{ij}^{(s)}$ is independent of z. Taking the inverse Laplace transform we see that in the time domain:

$$K_{ij}^{(s)}(t) = 2i K_{ij}^{(s)} \delta(t) \ . \tag{136}$$

The generalized Langevin equation then takes the form:

$$\frac{\partial \psi_i(t)}{\partial t} + i \sum_j K_{ij}^{(s)} \psi_j(t) + \sum_j \int_0^t d\bar{t}\, K_{ij}^{(d)}(t-\bar{t}) \psi_j(\bar{t}) = f_i(t) \ . \tag{137}$$

The static part of the memory function, $K^{(s)}$ can be expressed in terms of static correlation functions. The evaluation of the dynamic part, $K^{(d)}$, however, involves a direct confrontation with the many-body dynamics. In the case of fluids where $\psi_i \to \hat{f}(\mathbf{x}, \mathbf{p})$, there has been considerable work [23–25] carried out to evaluate $K^{(d)}$ and Eq. (135) is a convenient starting point for detailed microscopic calculations. Similarly, for spin systems a rather straightforward analysis of Eq. (135) leads to the mode-coupling approximation for $K^{(d)}$ discussed by Resibois and De Leener [26] and by Kawasaki [27]. We discuss this further below.

The expression for $K^{(d)}$ given by Eq. (135) looks very complicated. There is one crucial and simplifying property satisfied by $K^{(d)}$. Suppose we can decompose $L\psi_i$ into a piece linear in ψ and the rest:

$$L\psi_i = \sum_m \varepsilon_{im} \psi_m + N_i \ , \tag{138}$$

where N_i includes nonlinear products of the ψ's and contributions that cannot be expressed in terms of the ψ's. Let us substitute Eq. (138) into Eq. (135) for $K^{(d)}$ and focus on the terms proportional to ε_{im}. We have the contribution:

$$-\sum_m \varepsilon_{im} [\langle (L\psi_j) R(z) \psi_m \rangle - \sum_{k,l=1} \langle \psi_k R(z) \psi_m \rangle C_{kl}^{-1}(z) \times \langle (L\psi_j) R(z) \psi_l \rangle] \ , \tag{139}$$

but in the second term, we have the combination:

$$\sum_k \langle \psi_k R(z) \psi_m \rangle C_{kl}^{-1}(z) = \delta_{lm} \ . \tag{140}$$

Using this result in Eq. (139) we see that the two terms cancel. Similarly we can make the same replacement with respect to $L\psi_j$ and show:

$$\Gamma_{ij}^{(d)}(z) = -\langle N_j R(z) N_i \rangle + \sum_{k,l} \langle \psi_k R(z) N_i \rangle C_{kl}^{-1}(z) \langle N_j R(z) \psi_l \rangle. \quad (141)$$

One can therefore conclude that any linear part of $L\psi$ does not contribute to $K^{(d)}$. It is carefully treated by the static part of the memory function. It is in this sense that $K^{(d)}$ is a *one-particle* or *one-body* irreducible quantity. $K^{(d)}$ is determined by the interactions between variables (nonlinearities) or by fluctuations that are not included in ψ. If there is a nonlinear coupling u such that $N_i \approx u$, then $K^{(d)} \approx \mathcal{O}(u^2)$ and the multiplying correlation functions to lowest order in u can be evaluated for $u = 0$. This is discussed in Problems 5.4 and 5.5.

5.3.8
Memory-Function Formalism: Summary

Here we summarize the structure of the memory-function formalism. We consider an equilibrium-averaged time correlation:

$$C_{ij}(t) = \langle \psi_j \psi_i(t) \rangle \quad (142)$$

for a set of dynamic variables ψ_i. The Laplace transform $C_{ij}(z)$, Eq. (118), satisfies a kinetic equation:

$$\sum_\ell [z\delta_{i\ell} - K_{i\ell}(z)] C_{\ell j}(z) = S_{ij} \quad (143)$$

where $S_{ij} = \langle \psi_j \psi_i \rangle$ is the static correlation function and the memory function $K_{ij}(z)$ is the sum of static and dynamic parts;:

$$K_{ij}(z) = K_{ij}^{(s)} + K_{ij}^{(d)}(z) \quad , \quad (144)$$

where:

$$\sum_\ell K_{i\ell}^{(s)} S_{\ell j} = -\langle \psi_j L \psi_i \rangle \quad (145)$$

with L the Liouville operator for the system, and the dynamic part of the memory function is given by Eq. (135). If $L\psi_i$ can be separated into a component linear in ψ_i and nonlinear contributions, Eq. (138), then the dynamic part of the memory function can be written in the form given by Eq. (141).

5.3.9
Second Fluctuation-Dissipation Theorem

Now that we have a mechanism for determining the memory function K_{ij}, we can view Eq. (138) as a defining equation for the *noise* $f_i(z)$. It is the part of $-L\psi_i(z)$ that is not *along* the vector $\psi_i(z)$. It is important to note that the autocorrelation of the noise with itself is given by the simple result:

$$\langle f_j(t') f_i(t) \rangle = \sum_l K_{il}^{(d)}(t - t') S_{lj} \quad . \tag{146}$$

The first step in the proof of Eq. (146) is, as shown in Problem 5.6, establishment of the identity:

$$\psi_i(t) = \sum_{k,\ell} C_{ik}(t) S_{k\ell}^{-1} \psi_\ell + \sum_{k\ell} \int_0^t ds \, C_{ik}(t - s) S_{k\ell}^{-1} f_\ell(s) \quad . \tag{147}$$

Next we define the quantity:

$$F_i(t) = \frac{\partial \psi_i(t)}{\partial t} + i \sum_j K_{ij}^{(s)} \psi_j(t) \quad . \tag{148}$$

Using the generalized Langevin equation, Eq. (137), we can write:

$$F_i(t) + \sum_\ell \int_0^t ds \, i K_{i\ell}^{(d)}(t - s) \psi_\ell(s) = f_i(t) \quad . \tag{149}$$

Multiply this equation by ψ_j and take the equilibrium average:

$$\langle F_i(t) \psi_j \rangle + \sum_\ell \int_0^t ds \, K_{i\ell}^{(d)}(t - s) C_{\ell j}(s) = 0 \quad , \tag{150}$$

where we have remembered that $f_i(t)$ and ψ_j are orthogonal. Next we can write:

$$\langle F_i(t) \psi_j \rangle = \langle F_i \psi_j(-t) \rangle \quad . \tag{151}$$

We have from Eq. (149) at $t = 0$ that:

$$F_i = f_i \quad , \tag{152}$$

so:

$$\langle F_i(t) \psi_j \rangle = \langle f_i \psi_j(-t) \rangle \quad . \tag{153}$$

Next we use Eq. (147) for $\psi_j(-t)$ in Eq. (153):

$$\langle F_i(t) \psi_j \rangle = \left\langle f_i \left[\sum_{k,\ell} C_{jk}(-t) S_{k\ell}^{-1} \psi_\ell \right.\right.$$

$$\left.\left. + \sum_{k\ell} \int_0^{-t} ds \, C_{jk}(-t - s) S_{k\ell}^{-1} f_\ell(s) \right] \right\rangle \quad . \tag{154}$$

Since $\langle f_i \psi_\ell \rangle = 0$, this reduces to:

$$\langle F_i(t)\psi_j \rangle = \sum_{k\ell} \int_0^{-t} ds\, C_{jk}(-t-s)S_{k\ell}^{-1}\langle f_i f_\ell(s) \rangle . \tag{155}$$

Using Eq. (150) to substitute for the left-hand side and letting $s \to -s$ on the right we find:

$$\sum_\ell \int_0^t ds\, K_{i\ell}^{(d)}(t-s)C_{\ell j}(s) = \sum_{k\ell} \int_0^t ds\, C_{jk}(s-t)S_{k\ell}^{-1}\langle f_i f_\ell(-s) \rangle . \tag{156}$$

Let $s = t - y$ on the right, then:

$$\sum_\ell \int_0^t ds\, K_{i\ell}^{(d)}(t-s)C_{\ell j}(s) = \sum_{k\ell} \int_0^t dy\, C_{jk}(-y)S_{k\ell}^{-1}\langle f_i f_\ell(y-t) \rangle$$

$$= \sum_{k\ell} \int_0^t dy\, C_{kj}(y)S_{k\ell}^{-1}\langle f_i(t-y) f_\ell \rangle , \tag{157}$$

where we have used time-translational invariance in the equilibrium averages. Stripping off the common factor of $C_{\ell j}(s)$ from the right gives:

$$K_{ij}^{(d)}(t) = \sum_\ell \langle f_i(t) f_\ell \rangle S_{\ell j}^{-1} \tag{158}$$

and we arrive at the desired result:

$$\langle f_i(t) f_j \rangle = \sum_\ell K_{i\ell}^{(d)}(t) S_{\ell j} = \Gamma_{ij}^{(d)}(t) . \tag{159}$$

The result for the autocorrelation of the noise indicates that there is a fundamental relationship between the noise and the dynamic part of the memory function. This is known as the second fluctuation-dissipation theorem [28].

5.4
Example: The Harmonic Oscillator

In order to see how all of this formal structure hangs together let us consider the simple example of a set of uncoupled harmonic oscillators. This example is not very realistic since there is no irreversible behavior, but it does indicate how some of the manipulations in the theory are carried out.

Consider a set of N one-dimensional uncoupled classical harmonic oscillators described by a Hamiltonian:

$$H = \sum_{i=1}^N \left[\frac{P_i^2}{2M} + \frac{\kappa}{2} R_i^2 \right] , \tag{160}$$

where P_i is the momentum of the ith oscillator and R_i the position. The Liouville operator in this case is given by Eqs. (93), (94) and (95), which take the form here:

$$L = -i\sum_{i=1}^{N}\left(\frac{\partial}{\partial P_i}H\frac{\partial}{\partial R_i} - \frac{\partial}{\partial R_i}H\frac{\partial}{\partial P_i}\right)$$

$$= -i\sum_{i=1}^{N}\left[\frac{P_i}{M}\frac{\partial}{\partial R_i} - \kappa R_i\frac{\partial}{\partial P_i}\right] \ . \tag{161}$$

We have then, for example:

$$\frac{\partial R_i}{\partial t} = iLR_i = \frac{P_i}{M} \tag{162}$$

and:

$$\frac{\partial P_i}{\partial t} = iLP_i = -\kappa R_i \ . \tag{163}$$

From these familiar equations we immediately obtain Newton's law:

$$\frac{\partial^2 R_i}{\partial t^2} = \frac{1}{M}\frac{\partial P_i}{\partial t} = -\frac{\kappa}{M}R_i \ . \tag{164}$$

Defining:

$$\omega_0^2 = \frac{\kappa}{M} \ , \tag{165}$$

and incorporating the initial conditions we have the usual solutions:

$$R_i(t) = R_i(0)\cos\omega_0 t + \frac{P_i(0)}{M\omega_0}\sin\omega_0 t \tag{166}$$

$$P_i(t) = -R_i(0)M\omega_0\sin\omega_0 t + P_i(0)\cos\omega_0 t \ . \tag{167}$$

The formal structure of the theory requires us to work out the equal-time correlation functions. Noting that $\langle R_i \rangle = \langle P_i \rangle = 0$, we have explicitly:

$$\langle R_i(0)R_j(0) \rangle = \frac{1}{Z}\int dR_1 \ldots dR_N dP_1 \ldots dP_N \, e^{-\beta H} R_i R_j$$

$$= \delta_{ij}\frac{\int dR_i e^{-\beta\kappa R_i^2} R_i^2}{\int dR_i e^{-\beta\kappa}}$$

$$= \delta_{ij}\frac{k_B T}{\kappa} \ , \tag{168}$$

$$\langle R_i(0)P_j(0)\rangle = 0 \ , \tag{169}$$

and:
$$\langle P_i(0)P_j(0)\rangle = Mk_BT\delta_{ij} \;, \tag{170}$$

which are just statements of the equipartition theorem. Using these results and Eqs. (166) and (167) in the time-correlation functions we obtain:

$$\langle R_i(t)R_j(0)\rangle = \delta_{ij}\frac{k_BT}{\kappa}\cos\omega_0 t \tag{171}$$

$$\langle R_i(t)P_j(0)\rangle = \delta_{ij}Mk_BT\frac{\sin\omega_0 t}{M\omega_0} \tag{172}$$

$$\langle P_i(t)R_j(0)\rangle = -\delta_{ij}\frac{k_BT}{\kappa}M\omega_0\sin\omega_0 t \tag{173}$$

$$\langle P_i(t)P_j(0)\rangle = \delta_{ij}Mk_BT\cos\omega_0 t \;. \tag{174}$$

Note that:
$$\langle R_i(t)P_j(0)\rangle = \delta_{ij}\frac{k_BT}{\omega_0}\sin\omega_0 t = \langle R_i(0)P_j(-t)\rangle, \tag{175}$$

which follows from time-translational invariance.

Let us turn now to our memory-function formalism. Let us choose as our variables:
$$\psi_\alpha = \delta_{\alpha,R}R + \delta_{\alpha,P}P \;, \tag{176}$$

where, since lattice sites are uncoupled, we drop the lattice index. One lesson derived from this example is the virtue of a good choice of notation. We then need the equal-time matrix:

$$\begin{aligned}S_{\alpha\beta} &= \langle\psi_\alpha\psi_\beta\rangle \\ &= \delta_{\alpha\beta}\langle\psi_\alpha^2\rangle \\ &= \delta_{\alpha\beta}k_BTM_\alpha \;, \end{aligned} \tag{177}$$

where we have defined:
$$M_\alpha = \kappa^{-1}\delta_{\alpha,R} + M\delta_{\alpha,P} \;. \tag{178}$$

Next we need to work out the action of the Liouville operator on the basic field. Remembering that:

$$LR = -i\frac{P}{M} \tag{179}$$

$$LP = i\kappa R \, , \qquad (180)$$

we can write the quantity $L\psi_\alpha$ in a convenient form. We have:

$$\begin{aligned} L\psi_\alpha &= \delta_{\alpha,R}\left(-i\frac{P}{M}\right) + \delta_{\alpha,P}(i\kappa R) \\ &= \delta_{\alpha,R}\left(-i\frac{\psi_P}{M_P}\right) + \delta_{\alpha,P}\left(\frac{i\psi_R}{M_R}\right) \\ &= -i\sum_\gamma \varepsilon_{\alpha\gamma}\frac{\psi_\gamma}{M_\gamma} = -\sum_\gamma K_{\alpha\gamma}\psi_\gamma \, , \end{aligned} \qquad (181)$$

where we have introduced the antisymmetric tensor:

$$\varepsilon_{RP} = 1 = -\varepsilon_{PR} \; ; \; \varepsilon_{RR} = 0 = \varepsilon_{PP} \, , \qquad (182)$$

and, in the last line, we have defined the matrix:

$$K_{\alpha\beta} = i\frac{\varepsilon_{\alpha,\beta}}{M_\beta} \, . \qquad (183)$$

Next we consider the static part of the memory function:

$$\sum_\gamma K^{(s)}_{\alpha\gamma} S_{\gamma\beta} = -\langle \psi_\beta L\psi_\alpha \rangle \qquad (184)$$

and, using Eq. (177) on the left-hand side and Eq. (181) on the right-hand side, we obtain:

$$\begin{aligned} K^{(s)}_{\alpha\beta} M_\beta k_B T &= \sum_\gamma K_{\alpha\gamma} S_{\gamma\beta} \\ &= K_{\alpha\beta} M_\beta k_B T \, , \end{aligned} \qquad (185)$$

so:

$$K^{(s)}_{\alpha\beta} = K_{\alpha\beta} \qquad (186)$$

Clearly, since $L\psi_\alpha$ is linear in ψ, the dynamic part of the memory function vanishes and the full memory function is given by $K_{\alpha\beta}$, which is listed above. Our general correlation function expression, from Eq. (114) is given by:

$$zC_{\alpha\beta}(z) - \sum_\gamma K_{\alpha\gamma} C_{\gamma\beta}(z) = S_{\alpha\beta} \qquad (187)$$

or, inserting the explicit results from Eqs. (177) and (183),

$$zC_{\alpha\beta}(z) + \sum_\gamma \frac{i\varepsilon_{\alpha\gamma}}{M_{-\gamma}} C_{\gamma\beta}(z) = \delta_{\alpha\beta} M_\alpha k_B T \, . \qquad (188)$$

The equation is solved in Problem 5.7 with the result:

$$C_{\alpha\beta}(z) = \frac{k_B T}{z^2 - \omega_0^2} [z\delta_{\alpha\beta} M_\alpha + i\varepsilon_{\alpha\beta}] \ . \tag{189}$$

Suppose we make a poor choice of variables. How does this reflect itself in the memory-function formalism? Let us make the choice:

$$\psi = R \ . \tag{190}$$

Then:

$$L\psi = LR = -i\frac{P}{M} \tag{191}$$

and $L\psi$ cannot be expressed in terms of ψ. Then we find that the static part of the memory function is zero,

$$K^{(s)} S_{RR} = -\langle RLR \rangle = 0 \ , \tag{192}$$

while the dynamic part, given by Eq. (132), takes the form:

$$\begin{aligned}
K^{(d)}(z) S_{RR} &= -\left\langle \left(-i\frac{P}{M}\right) R(z) \left(-i\frac{P}{M}\right) \right\rangle \\
&\quad + \left\langle RR(z) \left(-i\frac{P}{M}\right) \right\rangle C_{RR}^{-1}(z) \left\langle \left(-i\frac{P}{M}\right) R(z) R \right\rangle \\
&= \frac{1}{M^2} C_{PP}(z) - \frac{1}{M^2} C_{PR}(z) C_{RR}^{-1}(z) C_{RP}(z) \ ,
\end{aligned} \tag{193}$$

or:

$$K^{(d)}(z) \frac{M^2 k_B T}{\kappa} = C_{PP}(z) - C_{PR}(z) C_{RR}^{-1}(z) C_{RP}(z) \ . \tag{194}$$

The kinetic equation for the correlation function Eq. (121) is trivial to solve in this case:

$$C_{RR}(z) = \frac{k_B T}{\kappa} \frac{1}{[z - K^{(d)}(z)]} \ . \tag{195}$$

Let us now evaluate $K^{(d)}$ using the known expressions, Eq. (189) for $C_{\alpha\beta}(z)$:

$$\begin{aligned}
K^{(d)}(z) &= \frac{\kappa}{M^2 k_B T} [C_{PP}(z) - C_{PR}(z) C_{RR}^{-1}(z) C_{RP}(z)] \\
&= \frac{\kappa}{M^2(z^2 - \omega_0^2)} [zM - \frac{\kappa}{z}] \\
&= \frac{\omega_0^2}{z} \ .
\end{aligned} \tag{196}$$

Inserting Eq. (196) back into Eq. (195) we recover:

$$C_{RR}(z) = \frac{k_B T}{\kappa} \frac{z}{z^2 - \omega_0^2} . \tag{197}$$

Note that $K^{(d)}(z)$ does *not* have a simple long-time, small-frequency behavior. This is the indicator that one has chosen in this case the wrong set of variables. The extension of the theory to the case of a nonlinear oscillator is discussed in Problems 5.4 and 5.5.

5.5
Theorem Satisfied by the Static Part of the Memory Function

In the case of classical systems, the static part of the memory function satisfies a very nice theorem that will be useful to us in several contexts. We start with the general expression:

$$\Gamma^{(s)}_{\alpha\beta} \equiv \sum_\gamma K^{(s)}_{\alpha\gamma} S_{\gamma\beta} = -\langle \psi_\beta L \psi_\alpha \rangle . \tag{198}$$

The slow variables, in this system, are functions of the phase-space coordinates $\mathbf{r}_i$ and $\mathbf{p}_i$ and Poisson brackets are defined as:

$$\{A, B\} = \sum_i \left(\frac{\partial A}{\partial \mathbf{r}_i} \cdot \frac{\partial B}{\partial \mathbf{p}_i} - \frac{\partial A}{\partial \mathbf{p}_i} \cdot \frac{\partial B}{\partial \mathbf{r}_i} \right) . \tag{199}$$

We can then use the definition of the Liouville operator in terms of the Poisson bracket to obtain:

$$\begin{aligned}\Gamma^{(s)}_{\alpha\beta} &= -i\langle \psi_\beta \{H, \psi_\alpha\} \rangle \\ &= -i\frac{1}{Z} \operatorname{Tr} e^{-\beta H} \psi_\beta \{H, \psi_\alpha\} .\end{aligned} \tag{200}$$

This is written for convenience in the canonical ensemble, but can easily be worked out in the GCE. Because of the first-order derivatives in the Poisson bracket, we can write:

$$e^{-\beta H}\{H, \psi_\alpha\} = -\beta^{-1}\{e^{-\beta H}, \psi_\alpha\} , \tag{201}$$

so that:

$$\Gamma^{(s)}_{\alpha\beta} = i\frac{\beta^{-1}}{Z} \operatorname{Tr} \psi_\beta \{e^{-\beta H}, \psi_\alpha\} . \tag{202}$$

Consider, then, the quantity:

$$\operatorname{Tr} \psi_\beta \{e^{-\beta H}, \psi_\alpha\}$$

$$= Tr \sum_i \psi_\beta \left[\left(\frac{\partial}{\partial \mathbf{r}_i} e^{-\beta H}\right) \cdot \frac{\partial}{\partial \mathbf{p}_i} \psi_i - \left(\frac{\partial}{\partial \mathbf{p}_i} e^{-\beta H}\right) \cdot \frac{\partial}{\partial \mathbf{r}_i} \psi_\alpha \right] \ . \quad (203)$$

Integrating by parts with respect to the derivatives of $e^{-\beta H}$ gives:

$$\begin{aligned} Tr \psi_\beta \{e^{-\beta H}, \psi_\alpha\} &= Tr\, e^{-\beta H} \sum_i \left[-\frac{\partial}{\partial \mathbf{r}_i} \cdot (\psi_\beta \frac{\partial \psi_\alpha}{\partial \mathbf{p}_i}) + \frac{\partial}{\partial \mathbf{p}_i} \cdot (\psi_\beta \frac{\partial \psi_\alpha}{\partial \mathbf{r}_i}) \right] \\ &= Tr\, e^{-\beta H} \{\psi_\alpha, \psi_\beta\} \ , \end{aligned} \quad (204)$$

where we have canceled some cross terms. Putting this result back in Eq. (202) gives the final clean result:

$$\Gamma_{\alpha\beta}^{(s)} = i\beta^{-1} \langle\{\psi_\alpha, \psi_\beta\}\rangle \ . \quad (205)$$

This is a rather nice general result for classical systems. As discussed in Problem 5.20, this theorem holds for more general definitions of the Poisson brackets.

5.6
Separation of Time Scales: The Markoff Approximation

The basic physical picture we want to associate with our Langevin equation is that there are two time scales in our problem. A short time scale is associated with a set of rapidly decaying variables. A second longer time scale is associated with slowly decaying variables. The main idea is to include in the ψ_i the slowly decaying variables while the noise f_i represents the effects of the rapidly decaying variables. We expect, then, that $\langle f_j f_i(t)\rangle$ decays to zero much faster than $\langle \psi_j \psi_i(t)\rangle$. Consequently, if we are interested in long-time phenomena, then $\langle f_j f_i(t)\rangle$ can be taken as very sharply peaked near $t = 0$. We therefore write,

$$\langle f_j f_i(t)\rangle = 2\Gamma_{ij}\delta(t) \ , \quad (206)$$

where:

$$\Gamma_{ij} = \int_0^\infty dt\, \langle f_j f_i(t)\rangle = \int_0^\infty dt\, \sum_\ell K_{i\ell}^{(d)}(t) S_{\ell j} \ , \quad (207)$$

where in the last step we have used Eq. (159). This Markoffian approximation [29] for the noise immediately implies, using the relationship derived earlier between the autocorrelation function for the noise and the *dynamic* part of the memory function, that:

$$\sum_l K_{il}^{(d)}(t - t') S_{lj} = 2\Gamma_{ij}\delta(t - t') \quad (208)$$

or:

$$K_{ij}^{(d)}(t-t') = 2\sum_{l}\Gamma_{il}S_{lj}^{-1}\delta(t-t') \ . \tag{209}$$

Inserting this result in the Langevin equation Eq. (115) we obtain:

$$\frac{\partial \psi_i(t)}{\partial t} + i\sum_{j}K_{ij}^{(s)}\psi_j(t) + \sum_{l,j}\Gamma_{il}S_{lj}^{-1}\psi_j(t) = f_i(t) \ . \tag{210}$$

Within the Markoffian approximation we completely [30] specify the dynamics of a set of variables ψ_j by giving $K_{ij}^{(s)}$, S_{ij} and Γ_{ij}. Different choices for $K^{(s)}$, S and Γ characterize different types of dynamical variables. These variables are typically of three types: relaxational, hydrodynamical and oscillatory. We have already looked at a simple oscillatory model: the simple harmonic oscillator.

5.7
Example: Brownian Motion

The simplest example of a relaxational process is the Brownian motion of a *large* particle in a solvent. In the case where the mass M of the large particle is much greater than m, the mass of the solvent, one expects that the velocity of the Brownian particle will be slow compared to the solvent motions. In this case it is sensible to choose the velocity of the Brownian particle as our slow variable:

$$\psi_i(t) \rightarrow \mathbf{V}(t) \ . \tag{211}$$

The ingredients specifying the Langevin equation are easily chosen in the Markoffian approximation. Consider first the equilibrium averages (i is now a vector component label):

$$S_{ij} = \langle V^i V^j \rangle = \delta_{ij}\frac{k_B T}{M} \tag{212}$$

using the equipartition theorem, and:

$$\begin{aligned}K_{ij}^{(s)}\frac{k_B T}{M} &= -\langle V^j L V^i \rangle = -\delta_{ij}\langle V^i L V^i \rangle \\ &= \delta_{ij}\langle (LV^i)V^i \rangle = 0 \ . \end{aligned} \tag{213}$$

The key assumption is that we can implement the Markoff approximation and write, using the isotropic symmetry in the system:

$$\Gamma_{ij} = \delta_{ij}\Gamma \ . \tag{214}$$

The Langevin equation, Eq. (210), is then given by:

$$\frac{\partial V^i(t)}{\partial t} + \frac{\Gamma M}{k_B T} V^i(t) = f^i(t) \tag{215}$$

and we have from the second fluctuation-dissipation theorem:

$$\langle f_j(t') f_i(t) \rangle = 2\Gamma \delta_{ij} \delta(t - t') \ . \tag{216}$$

Comparing Eq. (215) with Eq. (1.6), we can identify the friction coefficient:

$$\gamma = M \frac{\Gamma}{k_B T} \tag{217}$$

and the noise has the same variance as in Chapter 1.

If we Fourier transform Eq. (215) over time we obtain:

$$V^i(\omega) = \int_{-\infty}^{\infty} dt \, e^{i\omega t} V^i(t)$$

$$= \frac{f^i(\omega)}{(-i\omega + \gamma)} \ . \tag{218}$$

We see from Eq. (215) that the noise *drives* the slow variable and the statistical properties of V^i are controlled by those of the noise. The average of $V^i(\omega)$ is given by:

$$\langle V^i(\omega) V^j(\omega') \rangle = \frac{\langle f_i(\omega) f_j(\omega') \rangle}{(-i\omega + \gamma)(-i\omega' + \gamma)} \ . \tag{219}$$

After Fourier transforming Eq. (216),

$$\langle f^i(\omega) f^j(\omega') \rangle = 2\Gamma \delta_{ij} 2\pi \delta(\omega + \omega') \ , \tag{220}$$

Eq. (219) can be rewritten as:

$$C_{ij}(\omega) 2\pi \delta(\omega + \omega') = \langle V^i(\omega) V^j(\omega') \rangle$$

$$= \frac{2\Gamma \delta_{ij} 2\pi \delta(\omega + \omega')}{\omega^2 + \gamma^2} \tag{221}$$

or:

$$C_{ij}(\omega) = \frac{2\Gamma \delta_{ij}}{\omega^2 + \gamma^2} \ . \tag{222}$$

If we take the inverse Fourier transform of Eq. (222) we obtain:

$$C_{ij}(t) = \delta_{ij} \frac{k_B T}{M} e^{-\gamma |t|} \tag{223}$$

and the velocity autocorrelation function relaxes exponentially to zero for long times. This is in agreement with our results in Chapter 1 given by Eqs. (1.33) and (1.34).

We see that the relaxation rate is given by the dynamic part of the memory function. We should also note that this approximation is only valid in those situations where there is a clear separation of time scales between the variables of interest ψ_i and the noise f_i.

Note that we could have worked directly in terms of the Laplace transform of the correlation function that satisfies:

$$\sum_k \left[z\delta_{ik} - K_{ik}^{(s)} + i \sum_\ell \Gamma_{i\ell} S_{\ell k}^{-1} \right] C_{kj}(z) = S_{ij} \tag{224}$$

in this Markoffian approximation. In the case of Brownian motion, where Eqs. (212), (213) and (214) hold, this reduces to:

$$[z + i\gamma] C_{ij}(z) = \delta_{ij} k_B T / M \tag{225}$$

and:

$$C_{ij}(z) = \delta_{ij} \frac{k_B T / M}{z + i\gamma} . \tag{226}$$

The complex fluctuation function has a simple pole in the lower half complex plane.

5.8
The Plateau-Value Problem

There is a calculation we can carry out that sheds some light on the role of the second (subtraction) term in Eq. (135). Let us first introduce the so-called force–force correlation function:

$$\gamma_{ij}(t - t') = \langle \tilde{F}_i(t) \tilde{F}_j(t') \rangle , \tag{227}$$

where the force is defined by:

$$\tilde{F}_i(t) = \frac{\partial \psi_i(t)}{\partial t} = iL\psi_i(t) . \tag{228}$$

In terms of Laplace transforms we have:

$$\gamma_{ij}(z) = \langle (iL\psi_j) R(z) iL\psi_i \rangle = -\langle (L\psi_j) R(z) L\psi_i \rangle , \tag{229}$$

which we recognize as the first term in the expression Eq. (135) for the dynamic part of the memory function. Let us write Eq. (135) as:

$$\Gamma_{ij}^{(d)}(z) = \gamma_{ij}(z) + P_{ij}(z) , \tag{230}$$

where:

$$P_{ij}(z) = \sum_{k,\ell} \langle \psi_k R(z)(L\psi_i) \rangle C_{k\ell}^{-1}(z) \langle (L\psi_j) R(z) \psi_\ell \rangle \quad . \tag{231}$$

Now remember Eq. (124):

$$\langle \psi_k R(z)(L\psi_i) \rangle = -\sum_s K_{is}(z) C_{sk}(z) \quad , \tag{232}$$

so:

$$\begin{aligned}
P_{ij}(z) &= -\sum_{k,\ell,s} K_{is}(z) C_{sk}(z) C_{k\ell}^{-1}(z) \langle (L\psi_j) R(z) \psi_\ell \rangle \\
&= -\sum_\ell K_{i\ell}(z) \langle (L\psi_j) R(z) \psi_\ell \rangle \quad .
\end{aligned} \tag{233}$$

It is left to Problem 5.12 to show:

$$C_{AB}(z) = -C_{BA}(-z) \quad . \tag{234}$$

Using this result we have:

$$\langle (L\psi_j) R(z) \psi_\ell \rangle = -\langle \psi_\ell R(-z)(L\psi_j) \rangle \tag{235}$$

Again, using Eq. (124) we have:

$$\langle (L\psi_j) R(z) \psi_\ell \rangle = \sum_k K_{jk}(-z) C_{k\ell}(-z) \tag{236}$$

and:

$$\begin{aligned}
P_{ij}(z) &= -\sum_{k,\ell} K_{i\ell}(z) C_{k\ell}(-z) K_{jk}(-z) \\
&= \sum_{k,\ell} K_{i\ell}(z) C_{\ell k}(z) K_{jk}(-z) \quad .
\end{aligned} \tag{237}$$

It is shown in Problem 5.8 that:

$$\Gamma_{ij}^{(s)} = -\langle \psi_j L \psi_i \rangle = -\Gamma_{ji}^{(s)} \quad , \tag{238}$$

while from Problem 5.13 we find:

$$\Gamma_{ij}^{(d)}(z) = -\Gamma_{ji}^{(d)}(-z) \quad . \tag{239}$$

From these last two results it follows that:

$$\sum_\ell K_{i\ell}(z) S_{\ell j} = -\sum_\ell K_{j\ell}(-z) S_{\ell i} \tag{240}$$

or:

$$K_{ji}(-z) = -\sum_{k,\ell} S_{ik}^{-1} K_{k\ell}(z) S_{\ell j} \ . \tag{241}$$

Putting this result back into Eq. (237) we have:

$$P_{ij}(z) = -\sum_{k,\ell,s,m} K_{i\ell}(z) C_{\ell k}(z) S_{ks}^{-1} K_{sm}(z) S_{mj} \tag{242}$$

and we have the equation relating the memory function and the force–force correlation function:

$$\sum_\ell K_{i\ell}^{(d)}(z) S_{\ell j} = \gamma_{ij}(z) - \sum_{k,\ell,s,m} K_{i\ell}(z) C_{\ell k}(z) S_{ks}^{-1} K_{sm}(z) S_{mj} \ . \tag{243}$$

This equation becomes particularly interesting in the case where the correlation functions are diagonal and the static part of the memory function vanishes. This is the case of the Fourier transform of magnetization–magnetization correlation function for the itinerant magnet. In this case we have for the fluctuation function (suppressing the wavenumber label):

$$C(z) = \frac{1}{z - K^{(d)}(z)} S \ , \tag{244}$$

and Eq. (243) reduces to:

$$K^{(d)}(z) S = \gamma(z) - K^{(d)}(z) \frac{1}{z - K^{(d)}(z)} S S^{-1} K^{(d)}(z) S \ . \tag{245}$$

We can solve this equation for $\gamma(z)$ giving:

$$\gamma(z) = \frac{Sz}{z - K^{(d)}(z)} K^{(d)}(z) \ . \tag{246}$$

Alternatively we can solve for $K^{(d)}(z)$ in terms of $\gamma(z)$ to obtain:

$$K^{(d)}(z) = \frac{z \gamma(z)}{zS + \gamma(z)} \ . \tag{247}$$

We see that we have something interesting here. It is our expectation, if we have picked the correct slow variables, that after a microscopic short time τ_s, we can apply the Markoffian approximation and replace:

$$K^{(d)}(z) \to -i\Gamma S^{-1} \ , \tag{248}$$

where Γ is the physical damping coefficient. This means that in the time regime where $K^{(d)}(z)$ is a constant, the force–force correlation is given by:

$$\gamma(z) = -\frac{i\Gamma z}{z + \Gamma S^{-1}} \ . \tag{249}$$

We see that $\gamma(z)$ vanishes as $z \to 0$. It is easier to see what is happening in the time domain. Taking the inverse Laplace transform we obtain:

$$\gamma(t) = 2\Gamma\delta(t) - \Gamma^2 S^{-1} e^{-\Gamma S^{-1} t} \quad . \tag{250}$$

Writing this in terms of the time $\tau = S/\Gamma$:

$$\gamma(t) = \Gamma \left(2\delta(t) - \frac{1}{\tau} e^{-t/\tau} \right) \quad . \tag{251}$$

we see that we can use γ to estimate Γ for times large compared to τ_s and short compared to τ. Thus there is a range of time, a *plateau*, where one integrates $\gamma(t)$ to determine Γ. This plateau-value problem was identified in the 1940s by Kirkwood [31]. It is investigated in Problem 5.14.

Going back to Eq. (243) we see that there are situations where we can replace $K^{(d)}S$ with $\gamma(z)$. If $\gamma(z)$ can be shown to be small in some perturbation theory (density expansion) or because ψ is conserved, then in the limit:

$$K^{(d)} S = \gamma \quad . \tag{252}$$

This is what happens in developing the Green–Kubo relation for a conserved variables where $\Gamma = -iDq^2$ and we have:

$$\lim_{q \to 0} \frac{1}{q^2} \gamma(t) = 2D\delta(t) \quad . \tag{253}$$

All of this means that we must show great case in treating force–force correlations that have delicate long-time contributions. It is advised that one work with the memory functions that do not have these contributions. Instead, the second term in Eq. (135) serves to subtract off these slow degrees of freedom.

5.9
Example: Hydrodynamic Behavior; Spin-Diffusion Revisited

Let us return to the system we discussed earlier in this chapter: the itinerate isotropic ferromagnet with noninteracting spins. In this case the *slow* variable, the magnetization density, does not directly couple to the other slow variables, like the energy density, in the system. Thus we choose:

$$\psi_i \to M(\mathbf{x}) = \sum_{i=1}^{N} S_i \delta(\mathbf{x} - \mathbf{r}_i) \quad . \tag{254}$$

Let us proceed to fill in the various formal steps in our Langevin equation approach. Consider first the static behavior. Since we assume uncorrelated spins (see Eqs. (3) and (4)), we can work out the static properties explicitly. The average magnetization is zero:

$$\langle M(\mathbf{x}) \rangle = 0 \tag{255}$$

and our system is paramagnetic. The static structure function was shown earlier to be given by the simple result:

$$S(\mathbf{q}) = \bar{S} \ . \tag{256}$$

The magnetic susceptibility is given by:

$$\chi_M = \frac{\partial \langle M(\mathbf{x}) \rangle}{\partial H} = \frac{\beta}{V} \int d^3x \, d^3y \langle \delta M(\mathbf{x}) \delta M(\mathbf{y}) \rangle$$
$$= \beta S_M(q=0) = \frac{\bar{S}}{k_B T} \ , \tag{257}$$

which is just the Curie law [32].

Turning to the static part of the memory function we have to evaluate:

$$\int d^3w \, K^{(s)}(\mathbf{x} - \mathbf{w}) S(\mathbf{w} - \mathbf{y}) = -\langle M(\mathbf{y}) L M(\mathbf{x}) \rangle \ . \tag{258}$$

From the continuity equation we know:

$$iLM(\mathbf{x}) = -\nabla \cdot \mathbf{J}(\mathbf{x}) \tag{259}$$

and the static part of the memory function is zero since:

$$\langle M(\mathbf{y}) \mathbf{J}(\mathbf{x}) \rangle = 0 \tag{260}$$

due to the average over the momentum variable appearing in **J**.

The *kinetic equation*, with $K^{(s)} = 0$, takes the form:

$$zC(\mathbf{x} - \mathbf{y}, z) - \int d^3w \, K^{(d)}(\mathbf{x} - \mathbf{w}, z) C(\mathbf{w} - \mathbf{y}, z) = S(\mathbf{x} - \mathbf{y}) \ , \tag{261}$$

where the dynamic part of the memory function can be written as:

$$\int d^3w \, K^{(d)}(\mathbf{x} - \mathbf{w}) S(\mathbf{w} - \mathbf{y}) = K^{(d)}(\mathbf{x} - \mathbf{y}) \bar{S}$$
$$= -\langle [LM(\mathbf{y})] R(z) [LM(\mathbf{x})] \rangle$$
$$+ \int d^3w_1 d^3w_2 \langle [LM(\mathbf{y})] R(z) M(\mathbf{w}_1) \rangle C^{-1}(\mathbf{w}_1 - \mathbf{w}_2, z)$$
$$\times \langle M(\mathbf{w}_2) R(z) [LM(\mathbf{x})] \rangle \ . \tag{262}$$

The equations above simplify if we Fourier transform over space. The kinetic equation reduces to:

$$[z - K^{(d)}(\mathbf{q}, z)] C(\mathbf{q}, z) = \bar{S} \tag{263}$$

and the dynamic part of the memory function is given by:

$$K^{(d)}(\mathbf{q},z)\bar{S} = -\langle [LM(-\mathbf{q})]R(z)[LM(\mathbf{q})]\rangle$$
$$+\langle [LM(-\mathbf{q})]R(z)M(\mathbf{q})\rangle C^{-1}(\mathbf{q},z)\langle M(-\mathbf{q})R(z)[LM(\mathbf{q})]\rangle \quad . \quad (264)$$

It is at this stage in the development that we recall the results from our study of the plateau-value problem. If we introduce the force–force correlation function:

$$\gamma(q,z) = -\langle [LM(-\mathbf{q})]R(z)[LM(\mathbf{q})]\rangle \quad , \qquad (265)$$

then we showed, Eq. (247), that:

$$K^{(d)}(q,z) = \frac{z\gamma(q,z)}{z\bar{S} + \gamma(q,z)} \quad . \qquad (266)$$

An important ingredient, in evaluating the dynamic part of the memory function, is the conservation law:

$$\frac{\partial M}{\partial t} = iLM = -\nabla \cdot \mathbf{J} \quad . \qquad (267)$$

In terms of Fourier transforms:

$$\frac{\partial M(\mathbf{q})}{\partial t} = iLM(\mathbf{q}) = i\mathbf{q}\cdot \mathbf{J}(\mathbf{q}) \quad . \qquad (268)$$

Inserting this result into the equation for the force–force correlation function we obtain:

$$\gamma(q,z) = \sum_{ij} q_j q_i \langle J_j(-\mathbf{q})R(z)J_i(\mathbf{q})\rangle \qquad (269)$$

$$= \frac{q^2}{d} \langle \mathbf{J}(-\mathbf{q})\cdot R(z)\mathbf{J}(\mathbf{q})\rangle \quad , \qquad (270)$$

where the last step follows for an isotropic system in d spatial dimensions.

In the low-frequency small-wavenumber limit we expect:

$$K^{(d)}(q,z) = -iDq^2 \quad , \qquad (271)$$

which we write more carefully as:

$$-iD = \lim_{z\to 0}\lim_{q\to 0} \frac{1}{q^2} K^{(d)}(q,z) \quad . \qquad (272)$$

Substituting for $K^{(d)}(q,z)$, using Eq. (266), we have:

$$-iD = \lim_{z\to 0}\lim_{q\to 0} \frac{1}{q^2}\gamma(q,z)\bar{S}^{-1} \qquad (273)$$

$$= \lim_{z \to 0} \lim_{q \to 0} \frac{1}{d} \langle \mathbf{J}(-\mathbf{q}) \cdot R(z) \mathbf{J}(\mathbf{q}) \rangle \tag{274}$$

or, in the standard Green–Kubo form for the self-diffusion coefficient,

$$D\bar{S} = \frac{1}{dV} \int_0^\infty dt \, \langle \mathbf{J}_T \cdot \mathbf{J}_T(t) \rangle \, . \tag{275}$$

Thus we have reproduced our earlier hydrodynamic results.

Our goal has been to show how the results from linearized hydrodynamics can be *derived* from a well-defined microscopic analysis. It should be understood that there are circumstances where this development breaks down because the various limits ($\mathbf{q} \to 0$, $\omega \to 0$) do not exist. In particular the assumption given by Eq. (272) needs to be checked in detail. One can have a breakdown [33] in conventional hydrodynamics for low-dimensionality systems, systems near critical points and in systems like liquid crystals with strong fluctuations.

We are now in a position to understand the nature of the separation of time scales in this problem. The point is that $M(\mathbf{q}, t)$ will decay back to equilibrium much more slowly than $\mathbf{J}(\mathbf{q}, t)$. We can define characteristic times for the two fields as:

$$\tau_M(\mathbf{q}) = \int_0^\infty dt \langle M(-\mathbf{q}) M(\mathbf{q}, t) \rangle / \langle M(-\mathbf{q}) M(\mathbf{q}) \rangle \tag{276}$$

and:

$$\tau_J(\mathbf{q}) = \int_0^{+\infty} dt \langle \mathbf{J}(-\mathbf{q}) \cdot \mathbf{J}(\mathbf{q}, t) \rangle / \langle \mathbf{J}(-\mathbf{q}) \cdot \mathbf{J}(\mathbf{q}) \rangle \, . \tag{277}$$

We obtain directly from the inverse Laplace transform of Eq. (263) with Eq. (271) that:

$$C(\mathbf{q}, t) = \bar{S} e^{-Dq^2 t} \, . \tag{278}$$

Putting Eq. (278) into Eq. (276) we have in the small q limit,

$$\tau_M(\mathbf{q}) = \int_0^{+\infty} dt \, e^{-Dq^2 t} = \frac{1}{Dq^2} \, . \tag{279}$$

If we use the Green–Kubo formula (see Problem 5.17) we can show:

$$\tau_J(0) = \frac{Dm}{k_B T} \, . \tag{280}$$

Then we have the ratio of the two times:

$$\frac{\tau_M}{\tau_J} = \frac{k_B T}{Dm} \frac{1}{Dq^2} \equiv \frac{V_0^2}{D^2 q^2} \tag{281}$$

and for sufficiently small q,

$$\tau_M(q) \gg \tau_J(0) \tag{282}$$

and we have our separation of time scales and for many purposes it is appropriate to set q and z equal to zero in $\Gamma_{ij}(\mathbf{q}, z)$.

5.10 Estimating the Spin-Diffusion Coefficient

Let us return to the case of the isotropic Heisenberg model where the basic slow variable is the magnetization density $\mathbf{M}(\mathbf{x}, t)$ defined by Eq. (80). The continuity equation satisfied by $\mathbf{M}(\mathbf{x}, t)$ is given by Eq. (85). The spin-density current is now a matrix. We assume that the spin degrees of freedom are isotropic:

$$\begin{aligned} C_{ij}(\mathbf{x} - \mathbf{y}, t - t') &= \langle M_i(\mathbf{x}, t) M_j(\mathbf{y}, t') \rangle \\ &= \delta_{ij} C(\mathbf{x} - \mathbf{y}, t - t') = \frac{\delta_{ij}}{d} \langle \mathbf{M}(\mathbf{x}, t) \cdot \mathbf{M}(\mathbf{y}, t') \rangle \,. \end{aligned} \tag{283}$$

As in the itinerant magnet example in Section 5.8 we consider the force–force correlation function:

$$\gamma_{ij}(q, z) = -\langle (LM_j(-\mathbf{q}) R(z) (LM_i(\mathbf{q})) \rangle \tag{284}$$

$$= -\frac{\delta_{ij}}{d} \langle [L\mathbf{M}(-\mathbf{q})] \cdot R(z) [L\mathbf{M}(\mathbf{q})] \rangle \,. \tag{285}$$

The continuity equation in this case takes the form:

$$L\mathbf{M}(\mathbf{q}) = \sum_\alpha q_\alpha \mathbf{J}_\alpha(\mathbf{q}) \,, \tag{286}$$

where the current $\mathbf{J}_\alpha(\mathbf{q})$ is the spatial Fourier transform of Eq. (86). The force–force correlation function takes the form:

$$\gamma_{ij}(q, z) = \frac{\delta_{ij}}{d} \sum_{\alpha\beta} q_\alpha q_\beta \langle \mathbf{J}_\beta(-\mathbf{q}) \cdot R(z) \mathbf{J}_\alpha(\mathbf{q}) \rangle \,. \tag{287}$$

$\gamma_{ij}(q, z)$ depends on the lattice structure; however we make the simplifying assumption that the system is isotropic in space and assume:

$$\gamma_{ij}(q, z) = \delta_{ij} \frac{q^2}{d} \sum_\alpha \langle \mathbf{J}_\alpha(-\mathbf{q}) \cdot R(z) \mathbf{J}_\alpha(\mathbf{q}) \rangle \,. \tag{288}$$

Comparing with Eq. (270) it is easy to see, following the development in the itinerant magnet case, that the spin-diffusion coefficient for the Heisenberg model is given by the Green–Kubo formula:

$$DS(0) = \frac{1}{d^2 V} \sum_\alpha \int_0^\infty dt \, \langle \mathbf{J}_{T,\alpha}(t) \cdot \mathbf{J}_{T,\alpha} \rangle ,\qquad(289)$$

where the total current is given by Eq. (87):

$$J_{T,\beta}^i(t) = \frac{1}{2} \sum_{\mathbf{R},\mathbf{R}'} \sum_{jk} \varepsilon_{ijk}[\mathbf{R}-\mathbf{R}']_\beta J(\mathbf{R}-\mathbf{R}') M_\mathbf{R}^j(t) M_{\mathbf{R}'}^k(t) \qquad(290)$$

and $S(0) = \langle |\mathbf{M}(0)|^2 \rangle$. Inserting this result for $\mathbf{J}_T$ into Eq. (289) we see that we have a four-field correlation function. In an effort to estimate the spin-diffusion coefficient we adopt a simple classical approach and assume a decoupling approximation:

$$\langle M_{j'}(\mathbf{R}'',0) M_{k'}(\mathbf{R}''',0) M_j(\mathbf{R},t) M_k(\mathbf{R}',t) \rangle$$
$$= C_{jj'}(\mathbf{R},\mathbf{R}'',t) C_{kk'}(\mathbf{R}',\mathbf{R}''',t) + C_{jk'}(\mathbf{R},\mathbf{R}''',t) C_{kj'}(\mathbf{R}',\mathbf{R}'',t), \quad(291)$$

where:

$$C_{jj'}(\mathbf{R},\mathbf{R}',t) = \langle M_j(\mathbf{R},t) M_{j'}(\mathbf{R}',0) \rangle .\qquad(292)$$

The Green–Kubo relation then takes the form:

$$DS(0) = \frac{1}{d^2 V} \int_0^\infty dt \, \frac{1}{4} \sum_{\mathbf{R},\mathbf{R}',\mathbf{R}'',\mathbf{R}'''} \sum_{ijkj'k'} \sum_\alpha \varepsilon_{ijk}(\mathbf{R}-\mathbf{R}')_\alpha J(\mathbf{R}-\mathbf{R}')$$
$$\times \varepsilon_{ij'k'}(\mathbf{R}''-\mathbf{R}''')_\alpha J(\mathbf{R}''-\mathbf{R}''') \qquad(293)$$
$$\times \left[C_{jj'}(\mathbf{R},\mathbf{R}'',t) C_{kk'}(\mathbf{R}',\mathbf{R}''',t) + C_{jk'}(\mathbf{R},\mathbf{R}''',t) C_{kj'}(\mathbf{R}',\mathbf{R}'',t) \right] .$$

Assuming the system is isotropic in spin-space:

$$C_{jj'}(\mathbf{R},\mathbf{R}',t) = \delta_{jj'} C(\mathbf{R},\mathbf{R}',t) ,\qquad(294)$$

we can do the spin sums,

$$\sum_{ijkj'k'} \varepsilon_{ijk} \varepsilon_{ij'k'} \delta_{jj'} \delta_{kk'} = 2d \qquad(295)$$

$$\sum_{ijkj'k'} \varepsilon_{ijk} \varepsilon_{ij'k'} \delta_{jk'} \delta_{kj'} = -2d ,\qquad(296)$$

to obtain:

$$DS(0) = \frac{1}{2dV} \int_0^\infty dt \sum_{\mathbf{R},\mathbf{R}',\mathbf{R}'',\mathbf{R}'''} \sum_\alpha (\mathbf{R}-\mathbf{R}')_\alpha J(\mathbf{R}-\mathbf{R}') (\mathbf{R}''-\mathbf{R}''')_\alpha J(\mathbf{R}''-\mathbf{R}''')$$

$$\times \left[C(\mathbf{R},\mathbf{R}'',t)C(\mathbf{R}',\mathbf{R}''',t) - C(\mathbf{R},\mathbf{R}''',t)C(\mathbf{R}',\mathbf{R}'',t) \right] \ . \tag{297}$$

Next we introduce the Fourier transform:

$$C(\mathbf{R},\mathbf{R}',t) = \frac{1}{N} \sum_{k} e^{-i\mathbf{k}\cdot(\mathbf{R}-\mathbf{R}')} C(k,t) \ , \tag{298}$$

where N is the number of lattice sites, and we find for the spin-diffusion coefficient:

$$DS(0) = \frac{1}{2dV} \int_0^\infty dt \, \frac{1}{N^2} \sum_{k} \sum_{q} C(k,t) C(q,t) I(\mathbf{k},\mathbf{q}) \ , \tag{299}$$

where:

$$I(\mathbf{k},\mathbf{q}) = \sum_{\mathbf{R},\mathbf{R}',\mathbf{R}'',\mathbf{R}'''} \sum_{\alpha} (\mathbf{R}-\mathbf{R}')_\alpha J(\mathbf{R}-\mathbf{R}')(\mathbf{R}''-\mathbf{R}''')_\alpha J(\mathbf{R}'-\mathbf{R}''')$$
$$\times \left[e^{-i\mathbf{k}\cdot(\mathbf{R}-\mathbf{R}'')} e^{-i\mathbf{k}\cdot(\mathbf{R}'-\mathbf{R}''')} - e^{-i\mathbf{k}\cdot(\mathbf{R}-\mathbf{R}''')} e^{-i\mathbf{k}\cdot(\mathbf{R}'-\mathbf{R}'')} \right] \tag{300}$$

These sums factorize into the product:

$$I(\mathbf{k},\mathbf{q}) = \sum_{\alpha} \left[K_\alpha(\mathbf{k},\mathbf{q}) K_\alpha(-\mathbf{k},-\mathbf{q}) - K_\alpha(\mathbf{k},\mathbf{q}) K_\alpha(-\mathbf{q},-\mathbf{k}) \right] \ , \tag{301}$$

where:

$$K_\alpha(\mathbf{k},\mathbf{q}) = \sum_{\mathbf{R},\mathbf{R}'} (\mathbf{R}-\mathbf{R}')_\alpha J(\mathbf{R}-\mathbf{R}') e^{-i\mathbf{k}\cdot\mathbf{R}} e^{-i\mathbf{q}\cdot\mathbf{R}'} \ . \tag{302}$$

In the sum, let $\mathbf{R} = \mathbf{R}' + \mathbf{R}''$, so:

$$\begin{aligned} K_\alpha(\mathbf{k},\mathbf{q}) &= \sum_{\mathbf{R}',\mathbf{R}''} R''_\alpha J(\mathbf{R}'') e^{-i\mathbf{k}\cdot(\mathbf{R}'+\mathbf{R}'')} e^{-i\mathbf{q}\cdot\mathbf{R}'} \\ &= \sum_{\mathbf{R}''} R''_\alpha J(\mathbf{R}'') e^{-i\mathbf{k}\cdot\mathbf{R}''} N\delta_{\mathbf{k},-\mathbf{q}} \\ &= N\delta_{\mathbf{k},-\mathbf{q}} \sigma_\alpha(\mathbf{k}) \ , \end{aligned} \tag{303}$$

where:

$$\sigma_\alpha(\mathbf{k}) = \sum_{\mathbf{R}''} R''_\alpha J(\mathbf{R}'') e^{-i\mathbf{k}\cdot\mathbf{R}''} \ . \tag{304}$$

We have then that:

$$\begin{aligned} I(\mathbf{k},\mathbf{q}) &= \sum_\alpha K_\alpha(\mathbf{k},\mathbf{q}) \left[K_\alpha(-\mathbf{k},-\mathbf{q}) - K_\alpha(-\mathbf{q},-\mathbf{k}) \right] \\ &= N^2 \delta_{\mathbf{k},-\mathbf{q}} \sum_\alpha \sigma_\alpha(\mathbf{k}) \left[\sigma_\alpha(-\mathbf{k}) - \sigma_\alpha(-\mathbf{q}) \right] \end{aligned}$$

$$= 2N^2 \delta_{\mathbf{k},-\mathbf{q}} \vec{\sigma}(\mathbf{k}) \cdot \vec{\sigma}(-\mathbf{k}) \ . \tag{305}$$

Equation (297) for the Green–Kubo correlation function then reduces to:

$$DS(0) = \frac{1}{dV} \int_0^\infty dt \sum_\mathbf{k} C^2(k,t) \vec{\sigma}(\mathbf{k}) \cdot \vec{\sigma}(-\mathbf{k}) \ . \tag{306}$$

The sum over reciprocal lattice vectors can be written:

$$\sum_\mathbf{k} = N \int \frac{d^3(ka)}{(2\pi)^3} \ , \tag{307}$$

where a is a lattice constant and the integral is over the first Brillouin zone. Equation (306) then takes the form:

$$DS(0) = \frac{n}{3} \int_0^\infty dt \int \frac{d^3(ka)}{(2\pi)^3} C^2(k,t) \vec{\sigma}(\mathbf{k}) \cdot \vec{\sigma}(-\mathbf{k}) \ , \tag{308}$$

where $n = N/V$ is the density. Next note that we can write:

$$\sigma_\alpha(\mathbf{k}) = i \nabla_k^\alpha J(\mathbf{k}) \ , \tag{309}$$

where $J(\mathbf{k})$ is the Fourier transform of the exchange constant:

$$J(\mathbf{k}) = \sum_\mathbf{R} J(\mathbf{R}) e^{-i\mathbf{k}\cdot\mathbf{R}} \ . \tag{310}$$

Then:

$$DS(0) = \frac{n}{3} \int_0^\infty dt \int \frac{d^3(ka)}{(2\pi)^3} C^2(k,t) \left(\vec{\nabla}_k J(\mathbf{k}) \right)^2 \ . \tag{311}$$

In the spirit of the Debye theory of solids [34] we assume that the dominant contribution to the integral comes from small wavenumbers and the hydrodynamic mode. We write:

$$C(k,t) = \tilde{S}(k) e^{-Dk^2 t} \tag{312}$$

and do the time integration to obtain:

$$DS(0) = \frac{n}{3} \int \frac{d^3(ka)}{(2\pi)^3} \tilde{S}^2(k) \frac{\left(\vec{\nabla}_k J(\mathbf{k}) \right)^2}{2Dk^2} \ , \tag{313}$$

which we can solve for the diffusion coefficient:

$$D^2 S(0) = \frac{n}{6} \int \frac{d^3(ka)}{(2\pi)^3} \tilde{S}^2(k) \frac{\left(\vec{\nabla}_k J(\mathbf{k}) \right)^2}{k^2} \ . \tag{314}$$

If we assume a nearest-neighbor model we have for the Fourier transform of the exchange coupling:

$$J(k) = 2J\left(\cos k_x a + \cos k_y a + \cos k_z a\right) . \tag{315}$$

Assuming it is the small k region that is important, we can expand:

$$J(k) = 2J[3 - \tfrac{1}{2}(ka)^2 + \ldots] \tag{316}$$

and find the quantity in the integral:

$$\frac{\left(\vec{\nabla}_k J(\mathbf{k})\right)^2}{k^2} = 4J^2 a^4 \tag{317}$$

and:

$$D^2 S(0) = \frac{4n}{6} J^2 a^4 \int \frac{d^3(ka)}{(2\pi)^3} \tilde{S}^2(k) . \tag{318}$$

It is left to Problem 5.18 to show that $S(0) = n\tilde{S}(0)$ so:

$$D^2 = \frac{2}{3} J^2 a^4 \int \frac{d^3(ka)}{(2\pi)^3} \frac{\tilde{S}^2(k)}{\tilde{S}(0)} . \tag{319}$$

Finally we need an approximate form for the static structure factor. A simple mean-field approximation is given by:

$$\tilde{S}(k) = \frac{\hbar^2}{1 - \beta J(k)\hbar^2} , \tag{320}$$

where we have written explicitly that the structure factor, with our definitions here, has dimensions $\hbar^2$. Again evaluating $\tilde{S}(k)$ in the long-wavelength limit we have the Ornstein–Zernike form:

$$\tilde{S}(k) = \frac{\hbar^2}{r + c(ka)^2} , \tag{321}$$

where:

$$r = 1 - \beta 6 J \hbar^2 \tag{322}$$

$$c = 2\beta J \hbar^2 . \tag{323}$$

Using this approximation in Eq. (319) and doing the remaining integral in Problem 5.15, we find:

$$D^2 = \frac{2}{\pi} J^2 a^4 \hbar^2 \frac{1}{r} \left(\frac{r}{c}\right)^{3/2} . \tag{324}$$

From this expression we see that the scale of D is set by $D_0 = a^2 J\hbar$. We can write:

$$D = D_0 \frac{2}{\pi} \frac{1}{2^{3/4}} \sqrt{y}(y-6)^{1/4} \quad , \tag{325}$$

where $y = k_B T / J\hbar^2$. For large temperatures D grows as $y^{3/4}$ and vanishes as $(T - T_c)^{1/4}$ as $T \to T_c = 6J\hbar^2/k_B$. For a more careful quantum calculation and further discussion see Ref. [16].

5.11
References and Notes

1. The classic early numerical simulations on fluids were by B. J. Alder, T. E. Wainwright, Phys. Rev. Lett. **18**, 988 (1968); Phys. Rev. A **1**, 18 (1970); B. J. Alder, D. M. Gass, T. E. Wainwright, J. Chem. Phys. **53**, 3813 (1970). For an introduction, summary of results, and references see J. P. Boon, S. Yip, *Molecular Hydrodynamics*, Mc Graw-Hill, N.Y. (1979). Definitive simulation for the magnetic models we introduced here are more recent. See, for example, D. P. Landau, M. Krech, *Spin Dynamics Simulations of Classical Ferro-and Antiferromagnetic Model Systems: Comparison with Theory and Experiment*, Preprint.
2. There is an extensive discussion of the importance of conservation laws in equilibrium statistical mechanics in Chapter 1 of ESM.
3. Symmetry principles are discussed in some detail in Chapter 1 in ESM and FOD.
4. There is a detailed discussion of Nambu–Goldstone modes in Chapter 5 of FOD.
5. N. Bloembergen, Physica **15**, 386 (1949), introduced the spin-diffusion coefficient.
6. Ohm's Law was introduced by George Ohm in *Die Galvanische Kette Mathematisch Bearbeitet* (1827).
7. See Ref. [1] in Chapter 1.
8. A. Fick, Ann. Phys.(Leipzig) **94**, 59 (1855).
9. M. S. Green, J. Chem. Phys. **30**, 1280 (1952); J. Chem. Phys. **22**, 398 (1954).
10. R. Kubo, J. Phys. Soc. Jpn. **12**, 570 (1957).
11. In terms of the linear response function Eq. (39) takes the form:

    $$D\chi_M = \lim_{\omega \to 0} \lim_{\mathbf{k} \to 0} \frac{\omega}{k^2} \chi_M''(\mathbf{k}, \omega) \quad .$$

    This is the magnetic analog to the expression Eq. (4.122) for the electrical conductivity. The difference is that in the charged case we have the screened response function and one can not carry out the remaining steps leading to the more standard form given by Eq. (39).
12. One actually has an expression for a kinetic coefficient, which is a transport coefficient times a static susceptibility.
13. One can first choose a trial set $\tilde{\psi}_i$ and compute the equilibrium averages $\langle \tilde{\psi}_i \rangle$. Then choose as the final set with zero average $\psi_i = \tilde{\psi}_i - \langle \tilde{\psi}_i \rangle$.
14. P. A. M. Dirac, *The Principles of Quantum Mechanics*, 4th edn, Oxford University Press, Oxford (1958), Section 21.
15. B. O. Koopman, Proc. Natl. Acad. Sci. U. S. A **17**, 315 (1931).
16. For a treatment of spin diffusion in the Heisenberg paramagnet see H. S. Bennett, P. C. Martin, Phys. Rev. **138**, A608 (1965).
17. See Section 5.9.1 in ESM and Section 1.4.1 in FOD.
18. This trick is due to Professor P. C. Martin. See the discussion in Appendix F of ESM.
19. This is the statement, in keeping with causality, that the field does not depend on the noise at later times.
20. The quantum development is discussed in H. Mori, Prog. Theor. Phys. (Kyoto) **33**, 423 (1965).
21. R. W. Zwanzig, in *Lectures in Theoretical Physics, Vol. 3*, Interscience, N.Y. (1961); Phys. Rev. **24**, 983 (1965).
22. H. Mori, Prog. Theor. Phys. (Kyoto) **33**, 423 (1965).

23 G. F. Mazenko, Phys. Rev. A **7**, 209 (1973); Phys. Rev. A **9**, 360 (1974).
24 L. Sjögren, A. Sjölander, J. Phys. C **12**, 4369 (1979).
25 H. C. Anderson, J. Phys. Chem. B **106**, 8326 (2002); J. Phys. Chem. B **107**, 10226 (2003); J. Phys. Chem. B **107**, 10234 (2003).
26 P. Resibois, M. De Leener, Phys. Rev. **152**, 305 (1966); Phys. Rev. **152**, 318 (1969).
27 K. Kawasaki, Ann. Phys. **61**, 1 (1970).
28 We follow the proof of the second fluctuation dissipation theorem given by H. Mori in Ref. [22].
29 The use of the word Markoffian stems from the ideas associated with Markov process or chains [A. A. Markov, Izv. Akad. Nauk St. Petersburg **6**, 61 (1907)]. These processes, roughly, correspond to sequences where the next step depends only on the previous step. In our context of the generalized Langevin equation the Markoffian approximation suggests that the memory function on a long enough time scale has no memory. We can replace $K^{(d)}(z)$ with its $z = 0$ limit.
30 The statistics governing the higher-order correlation functions are generally unknown. This point is discussed further in Chapter 9.
31 J. G. Kirkwood, J. Chem. Phys. **14**, 180 (1946).
32 See p. 370 in ESM.
33 The nonlinear processes that lead to the breakdown of conventional hydrodynamics are discussed in Chapter 9.
34 See Chapter 6 in ESM.

5.12
Problems for Chapter 5

Problem 5.1: Extend the calculation for the Green–Kubo relation given by Eq. (49) to the quantum regime.

Problem 5.2: Show that the time-correlation function:

$$C_M(\mathbf{x} - \mathbf{y}, t - t') = \langle M(\mathbf{x}, t) M(\mathbf{y}, t') \rangle$$

with $M(\mathbf{x}, t)$, defined by Eq. (2), is proportional to the van Hove self-correlation function.

Problem 5.3: Calculate the effect of the Liouville operator L acting on $\mathbf{M}(\mathbf{R})$ in the example of the Heisenberg model. Use the usual spin commutation relations to evaluate:

$$LM^\alpha(\mathbf{R}) = \frac{1}{\hbar}[H, M^\alpha(\mathbf{R})] \ .$$

Problem 5.4: Consider a classical one-dimensional anharmonic oscillator described by the Hamiltonian:

$$H = \frac{p^2}{2M} + \frac{\kappa}{2}x^2 + \frac{u}{4}x^4 \ ,$$

where M, κ and u are all positive. This system is in thermal equilibrium at temperature T.

Using the memory-function formalism and choosing x and p as your slow variables, compute the memory function explicitly in a perturbation theory expansion in the small parameter u. Keep terms up to but not including terms of $\mathcal{O}(u^2)$. Using these results, compute the shift in the resonant frequency from $\omega_0^2 = \kappa/M$ to:

$$\omega_R^2 = \omega_0^2[1 + Au + \mathcal{O}(u^2)] \ .$$

Determine the coefficient A.

Problem 5.5: Consider the same system as in Problem 5.4. Compute the dynamic part of the memory function to second order in the quartic coupling u.

Problem 5.6: Starting with the generalized Langevin equation, Eq. (137), show that:

$$\psi_i(t) = \sum_{k,\ell} C_{ik}(t) S_{k\ell}^{-1} \psi_\ell + \sum_{k\ell} \int_0^t ds \, C_{ik}(t-s) S_{k\ell}^{-1} f_\ell(s) \ .$$

Problem 5.7: Starting with the kinetic equation for the uncoupled harmonic oscillator example:

$$z C_{\alpha\beta}(z) + \sum_\gamma \frac{i\varepsilon_{\alpha\gamma}}{M_\gamma} C_{\gamma\beta}(z) = \delta_{\alpha\beta} M_\alpha k_B T \ ,$$

solve for the complex correlation function $C_{\alpha\beta}(z)$.

Take the inverse Laplace transform to obtain the time-correlation functions given in this chapter.

Problem 5.8: Show for general A and B, using the properties of the Liouville operator and equilibrium averages, that:

$$\langle ALB \rangle = -\langle (LA)B \rangle \ .$$

This follows essentially from the time-translational invariance of the equilibrium distribution function.

Problem 5.9: Given an initial spin disturbance:

$$M(\mathbf{x}, 0) = M_0 e^{-\frac{1}{2}(x/\ell)^2} \ ,$$

find the spin configuration at time t. Assume the system is dynamically controlled by Eq. (20).

Problem 5.10: Evaluate the equal-time current–current correlation function $\langle J_\alpha^i(\mathbf{x}) J_\beta^j(\mathbf{y}) \rangle$ for the Heisenberg model in the high-temperature limit.

Problem 5.11: Show that the interaction part of the two-particle Liouville operator can be written in the form given by Eq. (100).

Problem 5.12: Show that the classical fluctuation function satisfies the symmetry condition:

$$C_{AB}(z) = -C_{BA}(-z) \ .$$

Problem 5.13: Show that:

$$\Gamma_{ij}^{(d)}(z) = \sum_{\ell} K_{i\ell}^{(d)}(z) S_{\ell j}$$

satisfies the symmetry relation:

$$\Gamma_{ij}^{(d)}(z) = -\Gamma_{ji}^{(d)}(-z) \ .$$

Problem 5.14: Show that the integral:

$$\bar{\Gamma}(\tau_p) = \int_0^{\tau_p} dt \gamma(t) \ ,$$

where the force–force correlation function $\gamma(t)$ is given by Eq. (251), shows a plateau as a function of τ_p.

Problem 5.15: Do the wavenumber integral:

$$D^2 = \frac{2}{3} J^2 a^4 \int \frac{d^3(ka)}{(2\pi)^3} \frac{\tilde{S}^2(k)}{\tilde{S}(0)}$$

with the choice for the static structure factor:

$$\tilde{S}(k) = \frac{\hbar^2}{r + c(ka)^2} \ .$$

Problem 5.16: Show that $D_0 = a^2 J\hbar$ has dimensions of a diffusion coefficient.

Problem 5.17: We can define a relaxation time for the spin-density current given by Eq. (277). Use the Green–Kubo formula, Eq. (275), and the necessary results from equilibrium mechanics to show that $\tau_J(0) = \frac{Dm}{kT}$.

Problem 5.18: Show that the structure factor at zero wavenumber, $S(0) = \langle |M(0)|^2 \rangle$, is related to $\tilde{S}(0)$, defined by Eq. (312), by $S(0) = n\tilde{S}(0)$.

Problem 5.19: If our convention for taking Laplace transforms is given by:

$$\mathcal{L}_z[F(t)] = -i \int_0^\infty dt \, e^{izt} F(t) \ ,$$

show that:

$$\mathcal{L}_z\left[\int_0^t ds\, A(t-s) B(s)\right] = i\mathcal{L}_z[A(t)] \mathcal{L}_z[B(t)] \ .$$

Problem 5.20: Prove the identity:

$$\Gamma_{\alpha\beta}^{(s)} = -\langle \psi_\beta L \psi_\alpha \rangle = i\beta^{-1} \langle \{\psi_\alpha, \psi_\beta\} \rangle \ ,$$

where spin variables ψ_α satisfy the Poisson bracket structure:

$$\{A, B\} = \sum_R \sum_{\alpha\beta\gamma} \varepsilon_{\alpha\beta\gamma} \frac{\partial A}{\partial \psi_\alpha(R)} \frac{\partial B}{\partial \psi_\beta(R)} \psi_\gamma(R) \ .$$

6
Hydrodynamic Spectrum of Normal Fluids

6.1
Introduction

In this chapter we discuss the hydrodynamic spectrum of simple normal fluids near equilibrium. Thus we consider the dynamics of fluid systems like argon. This involves application of the general approach developed in the previous chapter to a fairly complicated system. The real complication, though, is the thermodynamic interrelationships among the slow variables. The first step in the program is to select the set of slow variables.

6.2
Selection of Slow Variables

The case of a normal fluid is algebraically complicated because one has five slow variables corresponding to the five conserved quantities in the problem:

$$\tilde{\psi}_\alpha(\mathbf{x},t) = \{n(\mathbf{x},t), g_i(\mathbf{x},t), \varepsilon(\mathbf{x},t)\} \tag{1}$$

where $n(\mathbf{x},t)$ is the number density, $\mathbf{g}(\mathbf{x},t)$ is the momentum density, and $\varepsilon(\mathbf{x},t)$ is the energy density. Classically, we can write:

$$n(\mathbf{x},t) = \sum_{i=1}^{N} \delta(\mathbf{x} - \mathbf{r}_i) \tag{2}$$

$$\mathbf{g}(\mathbf{x},t) = \sum_{i=1}^{N} \mathbf{p}_i \delta(\mathbf{x} - \mathbf{r}_i) \tag{3}$$

$$\varepsilon(\mathbf{x},t) = \sum_{i=1}^{N} \left[\frac{\mathbf{p}_i^2}{2m} + \frac{1}{2} \sum_{j=1}^{N} V(\mathbf{r}_i - \mathbf{r}_j) \right] \delta(\mathbf{x} - \mathbf{r}_i) \tag{4}$$

where $\mathbf{r}_i$ and $\mathbf{p}_i$ are the position and momentum or the ith particle. Each of these densities satisfies a continuity equation:

$$\frac{\partial n}{\partial t} = iLn = -\nabla \cdot \frac{\mathbf{g}}{m} \tag{5}$$

Nonequilibrium Statistical Mechanics. Gene F. Mazenko
Copyright © 2006 WILEY-VCH Verlag GmbH & Co. KGaA, Weinheim
ISBN: 3-527-40648-4

$$\frac{\partial g_i}{\partial t} = iLg_i = -\sum_{j=1}^{3} \nabla_j \sigma_{ij} \tag{6}$$

$$\frac{\partial \varepsilon}{\partial t} = iL\varepsilon = -\nabla \cdot \mathbf{J}_\varepsilon \;, \tag{7}$$

where L is the Liouville operator for this system, σ_{ij} is the stress tensor and $\mathbf{J}_\varepsilon$ is the energy density current. Explicit expressions for σ_{ij} and $\mathbf{J}_\varepsilon$ were given in the previous chapter.

The next step is to make sure that the average of the slow variables is zero. Since $\langle n \rangle = n_0$ and $\langle \varepsilon \rangle = \varepsilon_0$, where n_0 and ε_0 are the ambient particle and energy densities, we must use δn and $\delta \varepsilon$ when constructing our slow variables. When there is no confusion, we will forgo the subscript 0 on ambient quantities.

It will be convenient to introduce, instead of the energy density, $\varepsilon(\mathbf{x}, t)$, the entropy or heat density:

$$q(\mathbf{x}, t) = \varepsilon(\mathbf{x}, t) - \frac{(\varepsilon_0 + p_0)}{n_0} n(\mathbf{x}, t) \tag{8}$$

where p_0 is the equilibrium pressure. We briefly encountered q in Chapter 2 as the field that couples to temperature fluctuations. The heat density is conserved since $\varepsilon(\mathbf{x}, t)$ and $n(\mathbf{x}, t)$ are separately conserved. The physical interpretation of q follows from the thermodynamic identity:

$$TdS = dE + pdV \tag{9}$$

at constant particle number. If we change from E and V to the intensive variables $\varepsilon = E/V$ and $n = N/V$ we have:

$$-\frac{dV}{V} = \frac{dn}{n} \tag{10}$$

and:

$$dE = d(\varepsilon V) = \varepsilon dV + V d\varepsilon \;. \tag{11}$$

Putting this result in Eq. (9) we have:

$$TdS = Vd\varepsilon + (\varepsilon + p)(-V)\frac{dn}{n} \tag{12}$$

and dividing by the volume:

$$T\frac{dS}{V} = d\varepsilon - \frac{(\varepsilon + p)}{n} dn = dq \;. \tag{13}$$

Therefore, up to a factor of the ambient temperature, $q(\mathbf{x},t)$ can be identified with the fluctuating entropy density.

Our set of slow variables with zero average is given by:

$$\psi_\alpha(\mathbf{x},t) = \{\delta n(\mathbf{x},t), g_i(\mathbf{x},t), \delta q(\mathbf{x},t)\} \ . \tag{14}$$

The merits of this choice will become clear when we look below at the static part of the memory function.

6.3 Static Structure Factor

The second task in working out the hydrodynamic spectrum is to evaluate the set of equilibrium averages:

$$S_{\alpha\beta}(\mathbf{x}-\mathbf{x}') = \langle \psi_\alpha(\mathbf{x})\psi_\beta(\mathbf{x}')\rangle \tag{15}$$

in the long-wavelength limit. In this limit we can connect these averages to thermodynamic derivatives. The reason is that in this limit these correlation functions correspond to the fluctuations in the extensive variables in the GCE. We can easily see how this comes about. We have, after taking the spatial Fourier transform and the long wavelength limit,

$$S_{\alpha\beta}(0) = \lim_{k\to 0} S_{\alpha\beta}(\mathbf{k}) = \frac{1}{V}\langle \delta\Psi_\alpha \delta\Psi_\beta\rangle \ , \tag{16}$$

where:

$$\Psi_\alpha = \int d^3x\ \psi_\alpha(\mathbf{x}) \tag{17}$$

is the total number particles N, the total momentum, or $Q = TS$ depending on the index α.

Let us construct the set of correlation functions $S_{\alpha\beta}(0)$ in terms of thermodynamic derivatives. The density–density correlation function follows from Eq. (16) as:

$$S_{nn}(0) = \frac{1}{V}\langle (\delta N)^2\rangle \ . \tag{18}$$

In the GCE we have:

$$S_{nn}(0) = \frac{1}{V}\frac{\partial}{\partial \alpha}\frac{Tr\ e^{-\beta H+\alpha N}N}{Z_G} \ , \tag{19}$$

where the grand partition function is given by:

$$Z_G = Tr\ e^{-\beta H+\alpha N} \tag{20}$$

and $\alpha = \beta\mu$. Equation (19) can then be written as:

$$S_{nn}(0) = \frac{1}{V}\frac{\partial}{\partial\alpha}\langle N\rangle = \left(\frac{\partial n}{\partial\alpha}\right)_{\beta,V}$$

$$= \left(\frac{\partial n}{\partial p}\right)_{\beta,V}\left(\frac{\partial p}{\partial\alpha}\right)_{\beta,V} . \tag{21}$$

However, in the GCE the pressure is related to the partition function by:

$$\beta p V = \ln Z_G , \tag{22}$$

so when we take the derivative with respect to α we obtain the simple result:

$$\left(\frac{\partial p}{\partial\alpha}\right)_{\beta,V}\beta V = \langle N\rangle \tag{23}$$

or,

$$\left(\frac{\partial p}{\partial\alpha}\right)_{\beta,V} = n\beta^{-1} . \tag{24}$$

Returning to the correlation function of interest we have:

$$S_{nn}(0) = n\beta^{-1}\left(\frac{\partial n}{\partial p}\right)_{\beta,V} = \beta^{-1}n^2\kappa_T , \tag{25}$$

where κ_T is the isothermal compressibility. This result tells us that the weight under the frequency spectrum for the density–density correlation function (as in light scattering) at small wavenumbers is proportional to the isothermal compressibility:

$$\int\frac{d\omega}{2\pi}C_{nn}(\mathbf{0},\omega) = nkT\left(\frac{\partial n}{\partial p}\right)_{\beta,V} . \tag{26}$$

This result is due to Einstein [1] who was inspired by the suggestions by Smoluchowski [2] that fluctuations could scatter light. These ideas were extended to nonzero wavenumbers by Ornstein and Zernike [3]. All of these results suggest large scattering near a critical point where $(\partial n/\partial p)_T$ is large. This leads to an explanation of critical opalescence discovered by Andrews [4] in the 1860s.

The correlation function between the density and the heat density is given by:

$$S_{qn}(0) = S_{nq}(0) = \frac{1}{V}\langle\delta N\delta Q\rangle \tag{27}$$

where:

$$Q = \int d^3x\, q(\mathbf{x}) . \tag{28}$$

The evaluation of $S_{qn}(0)$ and:

$$S_{qq}(0) = \frac{1}{V}\langle (\delta Q)^2\rangle \qquad (29)$$

are simplified by noting the following development. Consider the average of some variable A in the GCE:

$$\bar{A} = \langle A \rangle = \frac{\mathrm{Tr}\, e^{-\beta(H-\mu N)} A}{\mathrm{Tr}\, e^{-\beta(H-\mu N)}} \,. \qquad (30)$$

If we look at the changes in $\bar{A}$ as we vary T and μ, holding the volume V fixed, we have:

$$\delta\bar{A} = \left\langle \left[(H-\mu N)\frac{\delta T}{kT^2} + N\frac{\delta\mu}{kT}\right]\delta A \right\rangle, \qquad (31)$$

where, inside the average, $\delta A = A - \bar{A}$. From the Gibbs–Duhem relation [5] we have:

$$\delta\mu = \frac{\delta p}{n} - \frac{S}{N}\delta T \qquad (32)$$

and Eq. (31) takes the form:

$$\delta\bar{A} = \left\langle \left[(H-\mu N)\frac{\delta T}{kT^2} + N\frac{(\frac{\delta p}{n} - \frac{S}{N}\delta T)}{kT}\right]\delta A \right\rangle$$

$$= \langle N\delta A\rangle\frac{\delta p}{nkT} + \langle (H-\mu N - TS)\delta A\rangle\frac{\delta T}{kT^2} \,. \qquad (33)$$

From the Euler relation [6] we have:

$$\mu + \frac{TS}{N} = \frac{E+pV}{N} = \frac{\varepsilon+p}{n} \qquad (34)$$

and we can identify:

$$H - \mu N - TS = H - \frac{(\varepsilon+p)}{n}N = Q \qquad (35)$$

and Eq. (33) takes the form:

$$\delta\bar{A} = \langle \delta N\delta A\rangle\frac{\delta p}{nkT} + \langle \delta Q\delta A\rangle\frac{\delta T}{kT^2} \,. \qquad (36)$$

This equation was quoted in Chapter 2 in the form:

$$\delta\bar{A} = \chi_{nA}\frac{\delta p}{n} + \chi_{qA}\frac{\delta T}{T} \qquad (37)$$

where the χ's are the appropriate static susceptibilities. This result holds in the quantum regime if the variable A is conserved.

Holding the pressure constant in Eq. (36) we have:

$$\langle \delta Q \delta A \rangle = kT^2 \left(\frac{\partial \bar{A}}{\partial T} \right)_{p,V} . \tag{38}$$

In treating $S_{qn}(0)$ we need:

$$\langle \delta Q \delta N \rangle = kT^2 \left(\frac{\partial N}{\partial T} \right)_{p,V} . \tag{39}$$

However:

$$\left(\frac{\partial N}{\partial T} \right)_{p,V} = \frac{\partial(N, p, V)}{\partial(T, p, N)} \frac{\partial(T, p, N)}{\partial(T, p, V)}$$

$$= - \left(\frac{\partial V}{\partial T} \right)_{N,p} n$$

$$= -V \alpha_T n , \tag{40}$$

where α_T is the isothermal expansion coefficient. Putting Eq. (40) back into Eq. (39) gives:

$$\langle \delta Q \delta N \rangle = -kT^2 n \alpha_T V \tag{41}$$

and:

$$S_{qn}(0) = S_{nq}(0) = -kT^2 n \alpha_T . \tag{42}$$

Next, using Eqs. (29), (38) and (35), we can write:

$$S_{qq}(0) = \frac{kT^2}{V} \left[\left(\frac{\partial E}{\partial T} \right)_{p,V} - \left(\frac{(\varepsilon + p)}{n} \right) \left(\frac{\partial N}{\partial T} \right)_{p,V} \right] . \tag{43}$$

Using the first law of thermodynamics,

$$TdS = dE - \mu dN + pdV , \tag{44}$$

to express:

$$\left(\frac{\partial E}{\partial T} \right)_{p,V} = T \left(\frac{\partial S}{\partial T} \right)_{p,V} + \mu \left(\frac{\partial N}{\partial T} \right)_{p,V} . \tag{45}$$

Then:

$$S_{qq}(0) = \frac{kT^2}{V} \left[T \left(\frac{\partial S}{\partial T} \right)_{p,V} + \left(\mu - \frac{(\varepsilon + p)}{n} \right) \left(\frac{\partial N}{\partial T} \right)_{p,V} \right] . \tag{46}$$

Using Eq. (40) we can evaluate:

$$\left(\frac{\partial S}{\partial T}\right)_{p,V} = \frac{\partial(S,V,p)}{\partial(T,N,p)}\frac{\partial(T,N,p)}{\partial(T,V,p)}$$

$$= \left[\left(\frac{\partial S}{\partial T}\right)_{N,p}\frac{1}{n} - \left(\frac{\partial V}{\partial T}\right)_{N,p}\left(\frac{\partial S}{\partial N}\right)_{T,p}\right]n$$

$$= \frac{C_p}{T} - \alpha_T V \frac{S}{N} n \quad . \tag{47}$$

Here we have used the thermodynamic identities (see Problem 6.1):

$$\left(\frac{\partial S}{\partial N}\right)_{T,p} = \frac{S}{N} \tag{48}$$

and:

$$T\left(\frac{\partial S}{\partial T}\right)_{N,p} = C_p \quad , \tag{49}$$

where C_p is the specific heat at constant pressure. Equation (46) is given then by:

$$S_{qq}(0) = \frac{kT^2}{V}\left[T\frac{C_p}{T} - \alpha_T V \frac{S}{N} n - \left(\mu - \frac{(\varepsilon+p)}{n}\right)\alpha_T V n \left(\frac{\partial N}{\partial T}\right)_{p,V}\right] . \tag{50}$$

Using Euler's equation,

$$TS + \mu N - E - pV = 0 \tag{51}$$

and Eq. (40), we see that the coefficient multiplying $\left(\frac{\partial N}{\partial T}\right)_{p,V}$ vanishes and:

$$S_{qq}(0) = kT^2 \frac{C_p}{V} \quad . \tag{52}$$

The equal-time correlation functions between the momentum density and the other slow variables vanishes due to symmetry:

$$S_{g_i,n}(0) = S_{g_i,q}(0) = 0 \quad . \tag{53}$$

The equal-time correlation functions for the momentum density satisfy:

$$S_{g_i g_j}(0) = \delta_{ij} k_B T m n \quad , \tag{54}$$

which follows from the equipartition theorem.

The matrix $S_{\alpha\beta}(0)$ is summarized in Table 6.1.

Table 6.1 Matrix $S_{\alpha\beta}(0)/n_0 k_B T$; note that $n_0 \kappa_T = (\partial n/\partial p)|_T$, and $\alpha = -(\partial n/\partial T)_p$

	n	q	g_j
n	$n_0 \kappa_T$	$-T\alpha$	0
q	$-T\alpha$	TC_p/N	0
g_i	0	0	$m\delta_{ij}$

6.4
Static Part of the Memory Function

Next, we must work out the static part of the memory function for the fluid system. We have the relationship between the static part of the memory function $K^{(s)}$ and the matrix $\Gamma^{(s)}$:

$$\Gamma^{(s)}_{\alpha\beta}(\mathbf{x}-\mathbf{y}) = \sum_\gamma \int d^3w\, K^{(s)}_{\alpha\gamma}(\mathbf{x}-\mathbf{w}) S_{\gamma\beta}(\mathbf{w}-\mathbf{y})$$
$$= -\langle \psi_\beta(\mathbf{y}) L \psi_\alpha(\mathbf{x}) \rangle \ . \tag{55}$$

We must first work out the matrix $\Gamma^{(s)}_{\alpha\beta}$. A key observation in carrying out this analysis is that $L\psi_\alpha$ has a different signature under time reversal than ψ_α. This means (see Problem 6.3) that $\Gamma^{(s)}_{\alpha\beta}$ is zero unless ψ_α and ψ_β have different signatures under time reversal. Thus the only nonzero matrix elements between the hydrodynamic fields are given by $\Gamma^{(s)}_{ng_i}$, $\Gamma^{(s)}_{g_i n}$, $\Gamma^{(s)}_{qg_i}$ and $\Gamma^{(s)}_{g_i q}$. These first two matrix elements can be evaluated using the theorem proved in Section 5.5. We have,

$$\Gamma^{(s)}_{ng_i}(\mathbf{x}-\mathbf{y}) = i\beta^{-1} \langle \{n(\mathbf{x}), g_i(\mathbf{y})\} \rangle$$
$$= i\beta^{-1} \langle \left(-\nabla^i_x [\delta(\mathbf{x}-\mathbf{y}) n] \right) \rangle \ , \tag{56}$$

where the Poisson bracket between the density and momentum density is evaluated in Appendix B. We then have:

$$\Gamma^{(s)}_{ng_i}(\mathbf{x}-\mathbf{y}) = -i\beta^{-1} n_0 \nabla^i_x \delta(\mathbf{x}-\mathbf{y}) \ . \tag{57}$$

Taking the Fourier transform we have the simple result:

$$\Gamma^{(s)}_{ng_i}(\mathbf{k}) = k_i \beta^{-1} n_0 \ . \tag{58}$$

The transpose satisfies:

$$\Gamma^{(s)}_{g_i n}(\mathbf{x}-\mathbf{x}') = -\langle n(\mathbf{x}') L g_i(\mathbf{x}) \rangle = \langle [Ln(\mathbf{x}')] g_i(\mathbf{x}) \rangle = -\Gamma^{(s)}_{ng_i}(\mathbf{x}'-\mathbf{x}) \ . \tag{59}$$

On Fourier transformation we obtain:

$$\Gamma^{(s)}_{g_i n}(\mathbf{k}) = k_i \beta^{-1} n_0 \ . \tag{60}$$

The other matrix elements we need to evaluate are $\Gamma^{(s)}_{g_i,q}$ and $\Gamma^{(s)}_{q,g_i}$. We will first evaluate $\Gamma^{(s)}_{g_i\varepsilon}$ and come back to obtain:

$$\Gamma^{(s)}_{g_iq} = \Gamma^{(s)}_{g_i\varepsilon} - \frac{(\varepsilon+p)}{n}\Gamma^{(s)}_{g_in} \quad . \tag{61}$$

It is more difficult to evaluate $\Gamma^{(s)}_{g_i\varepsilon}$ than it is $\Gamma^{(s)}_{g_in}$. The Poisson brackets between g_i and ε can not be simply expressed in terms of hydrodynamic fields so we take a different approach. We remember from the continuity equation for the momentum density that Lg_i is proportional to the gradient of the stress tensor:

$$Lg_i(\mathbf{x}) = i\sum_j \nabla_j \sigma_{ij}(\mathbf{x}) \tag{62}$$

and we can write:

$$\begin{aligned}\Gamma_{g_i\varepsilon}(\mathbf{x}-\mathbf{x}') &= -\langle \delta\varepsilon(\mathbf{x}')Lg_i(\mathbf{x})\rangle \\ &= -\langle \delta\varepsilon(\mathbf{x}')i\sum_{j=1}^{3}\nabla_j\sigma_{ij}(\mathbf{x})\rangle \\ &= -i\sum_{j=1}^{3}\nabla_j\langle \delta\varepsilon(\mathbf{x}')\sigma_{ij}(\mathbf{x})\rangle \quad . \end{aligned} \tag{63}$$

Taking the Fourier transform we obtain:

$$\Gamma_{g_i\varepsilon}(\mathbf{k}) = \sum_{j=1}^{3} k_j\langle \delta\varepsilon(-\mathbf{k})\sigma_{ij}(\mathbf{k})\rangle \quad . \tag{64}$$

We are interested in the small $\mathbf{k}$ limit where:

$$\begin{aligned}\lim_{\mathbf{k}\to 0}\langle \delta\varepsilon(-\mathbf{k})\sigma_{ij}(\mathbf{k})\rangle &= \frac{1}{V}\langle \delta E\sigma_{ij}(0)\rangle \\ &= \int \frac{d^3x}{V}\langle \delta E\sigma_{ij}(\mathbf{x})\rangle \quad . \end{aligned} \tag{65}$$

Note, however, that we have the derivative relation in the GCE:

$$\frac{\partial}{\partial \beta}\langle \sigma_{ij}(\mathbf{x})\rangle|_{\alpha=\beta\mu} = -\langle \delta E\sigma_{ij}(\mathbf{x})\rangle \quad , \tag{66}$$

so:

$$\lim_{\mathbf{k}\to 0}\langle \delta\varepsilon(-\mathbf{k})\sigma_{ij}(\mathbf{k})\rangle = -\frac{\partial}{\partial \beta}\langle \sigma_{ij}(\mathbf{x})\rangle|_{\alpha=\beta\mu} \quad . \tag{67}$$

The reason it is convenient to introduce the stress tensor in the way that we have is because the equilibrium average of the stress tensor is, by definition, the pressure [7],

$$p\delta_{ij} = \langle \sigma_{ij}(\mathbf{x})\rangle \quad . \tag{68}$$

Then:
$$\lim_{\mathbf{k}\to 0} \langle \delta\varepsilon(-\mathbf{k})\sigma_{ij}(\mathbf{k})\rangle = -\delta_{ij}\left(\frac{\partial p}{\partial \beta}\right)_\alpha \qquad (69)$$

and:
$$\Gamma_{g_i\varepsilon}(\mathbf{k}) = \sum_{j=1}^{3} k_j(-\delta_{ij})\left(\frac{\partial p}{\partial \beta}\right)_\alpha$$
$$= -k_i\left(\frac{\partial p}{\partial \beta}\right)_\alpha . \qquad (70)$$

In order to further evaluate this quantity we need to do a little thermodynamics. Starting with the Gibbs–Duhem relation:
$$S dT = V dp - N d\mu . \qquad (71)$$

Since $\mu = \alpha k_B T$ we have:
$$d\mu = k_B T d\alpha + \alpha k_B dT \qquad (72)$$

and the Gibbs–Duhem relation can be written as:
$$(S - \alpha k_B N)\, dT = V dp - k_B T N d\alpha . \qquad (73)$$

We obtain immediately that:
$$\left(\frac{\partial p}{\partial T}\right)_\alpha = \frac{1}{V}\left(S - \frac{\mu N}{T}\right)$$
$$= \frac{1}{TV}(TS - \mu N) = \frac{1}{TV}(E + pV)$$
$$= \frac{\varepsilon + p}{T} . \qquad (74)$$

Taking the derivative with respect to β rather than T gives:
$$\left(\frac{\partial p}{\partial \beta}\right)_\alpha = -\beta^{-1}(\varepsilon + p) . \qquad (75)$$

Thus we have the final result:
$$\Gamma_{g_i\varepsilon}^{(s)}(\mathbf{k}) = k_i \beta^{-1}(\varepsilon + p) + \mathcal{O}(k^3) . \qquad (76)$$

In a hydrodynamic theory we keep terms in the memory functions up to and including $\mathcal{O}(k^2)$.

Our final goal in this section is to determine the matrix element:
$$\Gamma_{g_i,q}^{(s)}(\mathbf{k}) = \Gamma_{g_i,\varepsilon}^{(s)}(\mathbf{k}) - \left(\frac{\varepsilon + p}{n}\right)\Gamma_{g_i,n}^{(s)}(\mathbf{k}) . \qquad (77)$$

Inserting our results for $\Gamma^{(s)}_{g_i,\varepsilon}$ and $\Gamma^{(s)}_{g_i,n}$ we obtain:

$$\Gamma^{(s)}_{g_i,q}(\mathbf{k}) = k_i\beta^{-1}(\varepsilon+p) - \frac{(\varepsilon+p)}{n}k_i\beta^{-1}n + \mathcal{O}(k^3)$$
$$= \mathcal{O}(k^3) \ . \tag{78}$$

We see here why we choose the heat mode as one of our slow variables. We have the result then, to $\mathcal{O}(k^3)$, that all of the $\Gamma^{(s)}$ vanish except $\Gamma^{(s)}_{n,g_\alpha}(\mathbf{q})$. The various matrix elements are summarized in Table 6.2.

Table 6.2 Matrix $\Gamma^{(s)}_{\alpha\beta}/n_0 k_B T$

	n	q	g_j
n	0	0	k_j
q	0	0	0
g_i	k_i	0	0

Let us now construct $K^{(s)}_{\alpha\beta}$ using these simple results. We must invert the matrix equation:

$$\Gamma^{(s)}_{\alpha\beta}(\mathbf{k}) = \sum_\gamma K^{(s)}_{\alpha\gamma}(\mathbf{k}) S_{\gamma\beta}(\mathbf{k}) \ . \tag{79}$$

This inversion requires constructing the matrix inverse:

$$\sum_\gamma S_{\alpha\gamma}(\mathbf{k})(S^{-1})_{\gamma\beta}(\mathbf{k}) = \delta_{\alpha\beta} \ . \tag{80}$$

The inverse matrix can be constructed using Cramer's rule and the evaluation of determinants is facilitated by the fact that the momentum index does not couple to the other two variables. We then obtain:

$$(S^{-1})_{nn} = \frac{S_{qq}}{d_S} \tag{81}$$

$$(S^{-1})_{nq} = (S^{-1})_{qn} = -\frac{S_{nq}}{d_S} \tag{82}$$

$$(S^{-1})_{qq} = \frac{S_{nn}}{d_S} \tag{83}$$

$$(S^{-1})_{g_i g_j} = \delta_{ij}\frac{1}{mnk_B T} \ , \tag{84}$$

where the subdeterminant:

$$d_S = S_{nn}S_{qq} - S_{nq}S_{qn} \qquad (85)$$

plays an important role in the development. After inserting the explicit thermodynamic expressions for the $S_{\alpha\beta}$ into d_S we find, after using one of the standard thermodynamic relations connecting quantities in the Helmholtz ensemble to those in the Gibbs ensemble, that (see Problem 6.4):

$$d_S = S_{nn}nk_BT^2C_V/N \ , \qquad (86)$$

where C_V is the specific heat at constant volume and shows up as a natural second derivative in the Helmholtz ensemble. Our matrix inverse elements can then be manipulated into their most useful form. Consider first:

$$\begin{aligned}(S^{-1})_{nn} &= \frac{nk_BT^2C_p/N}{S_{nn}nk_BT^2C_V/N} \\ &= \frac{C_p}{C_V}\frac{1}{nk_BT}\left(\frac{\partial p}{\partial n}\right)_T \\ &= \frac{1}{nk_BT}mc^2 \ , \end{aligned} \qquad (87)$$

where:

$$mc^2 = \frac{C_p}{C_V}\left(\frac{\partial p}{\partial n}\right)_T = \left(\frac{\partial p}{\partial n}\right)_S \qquad (88)$$

and c is the adiabatic speed of sound (see Problem 6.5). The matrix element $(S^{-1})_{qq}$ can be written as:

$$(S^{-1})_{qq} = \frac{N}{nk_BT^2C_V} \ . \qquad (89)$$

The matrix element :

$$(S^{-1})_{nq} = \frac{S_{nq}N}{S_{nn}nk_BT^2C_V} \qquad (90)$$

can be put in a more convenient form if we use the result:

$$\frac{S_{nq}S_{qn}}{S_{nn}} = nk_BT^2(C_p - C_V)/N \ , \qquad (91)$$

which follows from Eqs. (52), (85) and (86). We have then:

$$(S^{-1})_{nq} = -\frac{1}{S_{nq}}\left(\frac{C_p - C_V}{C_V}\right) \ . \qquad (92)$$

6.4 Static Part of the Memory Function

The static part of the memory function is given by:

$$K^{(s)}_{\alpha\beta}(\mathbf{k}) = \sum_{\gamma} \Gamma^{(s)}_{\alpha\gamma}(\mathbf{k}) \left(S^{-1}\right)_{\gamma\beta}(\mathbf{k}) . \tag{93}$$

The left index α is nonzero only for $\alpha = n$ or g_i. Take the case $\alpha = n$ first. Then:

$$\begin{aligned} K^{(s)}_{n\beta}(\mathbf{k}) &= \sum_{\gamma} \Gamma^{(s)}_{n\gamma}(\mathbf{k}) \left(S^{-1}\right)_{\gamma\beta}(\mathbf{k}) \\ &= \Gamma^{(s)}_{ng_i}(\mathbf{k}) \left(S^{-1}\right)_{g_i\beta}(\mathbf{k}) , \end{aligned} \tag{94}$$

using the fact that all $\Gamma^{(s)}$ vanish except those coupling n and g_i. Since $\left(S^{-1}\right)_{g_i\beta}(\mathbf{k})$ is diagonal we have:

$$\begin{aligned} K^{(s)}_{n\beta}(\mathbf{k}) &= \Gamma^{(s)}_{ng_i}(\mathbf{k})\delta_{\beta g_i} \left(S^{-1}\right)_{g_i g_i}(\mathbf{k}) \\ &= \delta_{\beta g_i} k_i n k_B T \frac{1}{m n k_B T} \\ &= \delta_{\beta g_i} \frac{k_i}{m} . \end{aligned} \tag{95}$$

The other nonzero matrix element is given by:

$$\begin{aligned} K^{(s)}_{g_i\beta}(\mathbf{k}) &= \sum_{\gamma} \Gamma^{(s)}_{g_i\gamma}(\mathbf{k}) \left(S^{-1}\right)_{\gamma\beta}(\mathbf{k}) \\ &= \Gamma^{(s)}_{g_i n}(\mathbf{k}) \left(S^{-1}\right)_{n\beta}(\mathbf{k}) \\ &= k_i n k_B T \left(S^{-1}\right)_{n\beta}(\mathbf{k}) . \end{aligned} \tag{96}$$

This leads to the two contributions:

$$\begin{aligned} K^{(s)}_{g_i n}(\mathbf{k}) &= k_i n k_B T \frac{mc^2}{k_B T n} \\ &= k_i m c^2 \end{aligned} \tag{97}$$

and:

$$K^{(s)}_{g_i q}(\mathbf{k}) = k_i n k_B T \left(\frac{C_V - C_p}{C_V}\right) \frac{1}{S_{nq}} . \tag{98}$$

This completes our determination of the static part of the memory function in the long wavelength limit. The results are summarized in Table 6.3.

Table 6.3 Matrix $K^{(s)}_{\alpha\beta}(\mathbf{k})$ for small wavenumbers

	n	q	g_j
n	0	0	k_j/m
q	0	0	0
g_i	$mc^2 k_i$	$\dfrac{(C_p-C_V)}{T\alpha C_V}k_i$	0

6.5
Spectrum of Fluctuations with No Damping

We have shown that $K^{(s)} \approx \mathcal{O}(k)$ with corrections, by symmetry, which are of $\mathcal{O}(k^3)$. From our analysis of spin diffusion we would guess, and we shall show, that $K^{(d)} \approx \mathcal{O}(k^2)$. Let us first drop the damping terms and look at the correlation functions at lowest nontrivial order in **k**.

Table 6.4 Small wavenumber limit of matrix $(z - K^{(s)})_{\alpha\beta}$

	n	q	g_L
n	z	0	$-k/m$
q	0	z	0
g_L	$-K^{(s)}_{gn}$	$-K^{(s)}_{gq}$	z

In this limit, the kinetic equation determining the correlation functions is given by:

$$\sum_{\gamma}\left(z\delta_{\alpha\gamma} - K^{(s)}_{\alpha\gamma}(\mathbf{k})\right) C_{\gamma\beta}(\mathbf{k},z) = S_{\alpha\beta}(\mathbf{k}) \ . \tag{99}$$

We are interested in the pole structure of $C_{\alpha\beta}(z)$, which is, of course, associated with zeros of the determinant of the matrix $(z - K^{(s)})_{\alpha\beta}$ given in Table 6.4:

$$D = \det\left(z\delta_{\alpha\beta} - K^{(s)}_{\alpha\beta}\right)$$
$$= z\left(z^2 - \frac{k_i}{m}K^{(s)}_{g_i n}\right)$$
$$= z(z^2 - c^2 k^2) \ . \tag{100}$$

We see immediately that we have a three-pole structure. D vanishes for $z=0$ and $z = \pm ck$. Thus we expect *very* sharp peaks in the correlation functions at these positions of frequency for fixed wavenumber. In fact for small enough wavenumbers where the *damping* is arbitrarily small we have δ-function peaks.

We can compute the residues or weights for the poles by first noting that there is no coupling at this order in **k** between the heat mode and the other modes. Thus if we set $\alpha = q$ in Eq. (99) we find the very simple result:

$$C_{q\beta}(\mathbf{k},z) = \frac{S_{q\beta}}{z} \ , \tag{101}$$

the Fourier transform is given by:

$$C_{q\beta}(\mathbf{k},\omega) = 2\pi\delta(\omega) S_{q\beta} \tag{102}$$

and there is a single heat mode centered at $\omega = 0$ for $\mathbf{k} \to 0$.

From Eq. (99) we have the set of coupled equations determining the density–density correlation function:

$$zC_{nn} - K_{ng_i}^{(s)} C_{g_in} = S_{nn} \tag{103}$$

$$zC_{g_in} - K_{g_in}^{(s)} C_{nn} = K_{g_iq}^{(s)} \frac{S_{qn}}{z}, \tag{104}$$

where we have used Eq. (101) on the right-hand side of Eq. (104). Solving this linear set of equations for C_{nn} gives:

$$C_{nn} = \frac{1}{zD_1}\left[z^2 S_{nn} + K_{ng_i}^{(s)} K_{g_iq}^{(s)} S_{qn}\right], \tag{105}$$

where D_1 is the determinant:

$$D_1 = z^2 - K_{ng_i}^{(s)} K_{g_in}^{(s)} = z^2 - c^2k^2. \tag{106}$$

Putting the various thermodynamic expressions into Eq. (105) gives:

$$C_{nn}(\mathbf{k},z) = \frac{1}{z\left(z^2 - c^2k^2\right)}\left[z^2 S_{nn} + \frac{k^2}{m} nk_B T\left(\frac{C_V - C_p}{C_V}\right)\right]. \tag{107}$$

We can isolated the residues associated with the three poles by using the identities:

$$\frac{z}{z^2 - c^2k^2} = \frac{1}{2}\left[\frac{1}{z+ck} + \frac{1}{z-ck}\right] \tag{108}$$

and:

$$\frac{c^2k^2}{z^2 - c^2k^2} = -1 + \frac{z^2}{z^2 - c^2k^2}. \tag{109}$$

One has then, after substituting for S_{nn} and extracting an overall factor of $nk_B T$,

$$C_{nn}(\mathbf{k},z) = \beta^{-1} n\left\{\left[\frac{1}{2}\left[\frac{1}{z+ck} + \frac{1}{z-ck}\right]\right]\left[\left(\frac{\partial n}{\partial p}\right)_T - \frac{(C_p - C_V)/C_V}{mc^2}\right]\right.$$
$$\left.+ \frac{1}{z}\frac{(C_p - C_V)}{mC_V}\frac{1}{c^2}\right\}. \tag{110}$$

The coefficients of the poles can be put into a much neater forms. First the coefficient of the sound poles can be written as:

$$\left(\frac{\partial n}{\partial p}\right)_T - \frac{(C_p - C_V)/C_V}{mc^2} = \left(\frac{\partial n}{\partial p}\right)_T - \frac{(C_p - C_V)}{C_V} \frac{C_V}{C_p} \left(\frac{\partial n}{\partial p}\right)_T$$
$$= \left(\frac{\partial n}{\partial p}\right)_T [1 - 1 + \frac{C_V}{C_p}] = \left(\frac{\partial n}{\partial p}\right)_T \frac{C_V}{C_p}, \qquad (111)$$

where we have used Eq. (88) in the first line. The coefficient of the $1/z$ term has the weight:

$$\frac{(C_p - C_V)}{mC_V} \frac{1}{c^2} = \frac{C_p - C_V}{C_V} \left(\frac{C_V}{C_p}\right) \left(\frac{\partial n}{\partial p}\right)_T$$
$$= \left(1 - \frac{C_V}{C_p}\right) \left(\frac{\partial n}{\partial p}\right)_T. \qquad (112)$$

Putting the results from Eqs. (111) and (112) back into Eq. (110) gives:

$$C_{nn}(\mathbf{k}, z) = \beta^{-1} n \left(\frac{\partial n}{\partial p}\right)_T \left[\left(\frac{C_V}{C_p}\right) \frac{1}{2} \left[\frac{1}{z + ck} + \frac{1}{z - ck}\right]\right.$$
$$\left. + \left(1 - \frac{C_V}{C_p}\right) \frac{1}{z}\right]. \qquad (113)$$

The dynamic structure factor, which is related to the imaginary part of $C_{nn}(\mathbf{k}, z)$ with $z = \omega + i\eta$, is given by:

$$C_{nn}(\mathbf{k}, \omega) = \beta^{-1} n \left(\frac{\partial n}{\partial p}\right)_T 2\pi \left[\frac{1}{2} \frac{C_V}{C_p} [\delta(\omega + ck) + \delta(\omega - ck)]\right.$$
$$\left. + (1 - C_V/C_p) \delta(\omega)\right]. \qquad (114)$$

We have two sound poles and a heat mode. The weights are governed by the specific heat ratios. The relative weight of the peaks:

$$\frac{C_V/C_p}{1 - C_V/C_p} \qquad (115)$$

is known as the Landau–Placzek ratio [8]. Notice that the zeroth-order sum rule:

$$\int \frac{d\omega}{2\pi} C_{nn}(\mathbf{k}, \omega) = \beta^{-1} n \left(\frac{\partial n}{\partial p}\right)_T \left[\frac{C_V}{C_p} + 1 - \frac{C_V}{C_p}\right]$$
$$= \beta^{-1} n \left(\frac{\partial n}{\partial p}\right)_T$$

$$= S_{nn}(\mathbf{k} \to 0) \tag{116}$$

is properly satisfied.

The decomposition of density fluctuations into a sum of three peaks is a famous result that merits further comment. The central diffusive Rayleigh peak is centered at $\omega = 0$. The traveling wave Brillouin doublet [9] centered at $\omega = \pm ck$ is associated with scattering from sound waves. Notice that it is the physically correct adiabatic speed of sound that enters. Historically, Newton asserted that the speed of sound is given by the isothermal compressibility. This result is experimentally too low by about 20%. Laplace and Poisson eventually discovered [10] the correct result.

6.6
Dynamic Part of the Memory Function

It is then straightforward to construct all of the other correlation functions at this level. Let us turn instead to the inclusion of the damping terms and, equivalently, the evaluation of the dynamical part of the memory function in the hydrodynamical limit.

At this stage we must be more careful about the *vector* nature of the problem entering through the momentum density. We derived previously the result:

$$\Gamma^{(s)}_{g_i n} = k_i \beta^{-1} n \tag{117}$$

and all other $\Gamma^{(s)}_{\alpha\beta}(\mathbf{k})$ are zero for small $\mathbf{k}$. Notice that this matrix element vanishes unless $\mathbf{g}$ is in the same direction as $\mathbf{k}$. That is, we can break $\mathbf{g}(\mathbf{k}, t)$ up into a longitudinal part:

$$g_L(\mathbf{k}, t) = \hat{k} \cdot \mathbf{g}(\mathbf{k}, t) \tag{118}$$

and a transverse part $\mathbf{g}_T(\mathbf{k}, t)$, where:

$$\mathbf{k} \cdot \mathbf{g}_T(\mathbf{k}, t) = 0 \ . \tag{119}$$

Only the longitudinal part of $\mathbf{g}$ couples to the density,

$$\Gamma^{(s)}_{g_T, n}(\mathbf{k}) = 0 \ . \tag{120}$$

This makes sense if we remember the continuity equation can be written:

$$\frac{\partial n}{\partial t} = -i\mathbf{k} \cdot \mathbf{g}(\mathbf{k}, t) \ . \tag{121}$$

Thus the transverse part of the momentum current decouples from the longitudinal part. This is why we only had a 3×3 matrix instead of a 5×5 matrix in our analysis above.

6.7
Transverse Modes

Let us analyze the transverse current fluctuations first. In this case, the kinetic equation becomes:

$$\sum_\ell [z\delta_{i\ell} - K^{(d)}_{i\ell}(\mathbf{k}, z)] C_{\ell j}(\mathbf{k}, z) = S_{ij}(\mathbf{k}) , \qquad (122)$$

where:

$$C_{ij}(\mathbf{k}, z) = \langle g_j(-\mathbf{k}) R(z) g_i(\mathbf{k}) \rangle = \delta_{ij} C_t(\mathbf{k}, z) \qquad (123)$$

and i, j and ℓ label the two directions transverse to $\mathbf{k}$. If we pick $\mathbf{k}$ in the z-direction, then i, j and ℓ can be either x or y. The corresponding equal-time correlation functions are given by

$$S_{ij}(\mathbf{k}) = \delta_{ij} \beta^{-1} mn . \qquad (124)$$

The dynamic part of the memory function is given by:

$$\sum_k K^{(d)}_{ik}(\mathbf{k}, z) S_{kj} = -\langle L g_j(-\mathbf{k}) R(z) L g_i(\mathbf{k}) \rangle$$
$$+ \sum_{k\ell} \langle L g_j(-\mathbf{k}) R(z) g_k(\mathbf{k}) \rangle C^{-1}_{k\ell}(\mathbf{k}, z) \langle g_\ell(-\mathbf{k}) R(z) L g_i(\mathbf{k}) \rangle . \qquad (125)$$

Following steps similar to those used in the case of spin diffusion we find, after remembering:

$$iLg_i(\mathbf{x}) = -\sum_j \nabla^j_x \sigma_{ij}(\mathbf{x}) \qquad (126)$$

and:

$$iLg_i(\mathbf{k}) = -\sum_j ik_j \sigma_{ij}(\mathbf{k}) , \qquad (127)$$

that in the long-wavelength limit the dynamic part of the memory function is given by:

$$K^{(d)}_{ij}(\mathbf{k}, z) \beta^{-1} mn = \sum_{m,\ell} k_m k_\ell \langle \sigma^{j\ell}(-\mathbf{k}) R(z) \sigma^{im}(\mathbf{k}) \rangle + \mathcal{O}(k^4)$$
$$\equiv -i \sum_{m,\ell} k_m k_\ell \Gamma_{im,j\ell} , \qquad (128)$$

where, in the small $\mathbf{k}$ and z limit:

$$\Gamma_{im,j\ell} = \lim_{z \to 0} \lim_{k \to 0} i \langle \sigma^{j\ell}(-\mathbf{k}) R(z) \sigma^{im}(\mathbf{k}) \rangle$$

$$= \frac{1}{V} \int_0^{+\infty} dt \int d^3x\, d^3x'\, \langle \sigma^{j\ell}(\mathbf{x}')\sigma^{im}(\mathbf{x},t)\rangle \ . \tag{129}$$

For an isotropic system:

$$\Gamma_{im,j\ell} = \Gamma_1 \delta_{im}\delta_{j\ell} + \delta_{ij}\delta_{m\ell}\Gamma_2 + \delta_{i\ell}\delta_{jm}\Gamma_3 \ . \tag{130}$$

Since the stress tensor σ can be constructed to be symmetric, we require $\Gamma_2 = \Gamma_3$. In the case of interest here we have the requirements that m and ℓ are in the z-direction while j and i are transverse. Therefore in this case, we pick out only the component:

$$\Gamma_{im,j\ell} = \Gamma_2 \delta_{ij}\delta_{m\ell} \tag{131}$$

and it is given explicitly by:

$$\begin{aligned}\Gamma_2 &= \int_0^{+\infty} dt \int \frac{d^3x\, d^3x'}{V} \langle \sigma^{zx}(\mathbf{x}')\sigma^{zx}(\mathbf{x},t)\rangle \\ &= \frac{1}{V}\int_0^{+\infty} dt\, \langle \sigma_T^{zx}\sigma_T^{zx}(t)\rangle \end{aligned} \tag{132}$$

and:

$$K_{ij}^{(d)}(\mathbf{k},z)\beta^{-1}mn = \delta_{ij}(-i)\Gamma_2 k^2 \ . \tag{133}$$

In this case the kinetic equation, Eq. (122) is diagonal and the transverse current correlation function is given by:

$$(z + i\Gamma_2 k^2 \beta/mn)C_t(\mathbf{k},z) = \beta^{-1}mn \ , \tag{134}$$

and the appropriate transport coefficient is the shear viscosity:

$$\Gamma_2 \beta \equiv \eta \tag{135}$$

where:

$$\eta = \frac{\beta}{V}\int_0^{+\infty} dt\, \langle \sigma_T^{zx}\sigma_T^{zx}(t)\rangle \tag{136}$$

is the corresponding Green–Kubo formula.

The complex transverse current correlation function is given by:

$$C_t(\mathbf{k},z) = \frac{\beta^{-1}\rho}{z + i\eta k^2/\rho} \tag{137}$$

where $\rho = mn$ and the fluctuation spectrum is given by:

$$C_t(\mathbf{k},\omega) = 2\beta^{-1}\frac{\eta k^2}{\omega^2 + (\eta k^2/\rho)^2} \ . \tag{138}$$

This transverse diffusive spectrum is characteristic of fluids since there are no transverse shear traveling modes. This contrasts with solids where one finds the transverse traveling waves (phonons) associated with broken translational symmetry.

6.8
Longitudinal Modes

We now turn to the damping of the longitudinal modes. This requires treating the dynamic part of the memory function:

$$\sum_\gamma K^{(d)}_{\alpha\gamma}(\mathbf{k},z) S_{\gamma\beta} = \Gamma^{(d)}_{\alpha\beta}(\mathbf{k},z) = -\langle \left[L\psi_\beta(-\mathbf{k}) \right] R(z) L\psi_\alpha(\mathbf{k}) \rangle$$

$$+ \sum_{\gamma,\nu} \langle \left[L\psi_\beta(-\mathbf{k}) \right] R(z) \psi_\mu(\mathbf{k}) \rangle C^{-1}_{\mu\nu}(\mathbf{k},z) \langle \psi_\nu(-\mathbf{k}) R(z) L\psi_\alpha(\mathbf{k}) \rangle, \quad (139)$$

where the indices refer to the three longitudinal variables n, $g_L = \hat{k} \cdot \mathbf{g}$ and q. Note first that:

$$\Gamma^{(d)}_{n\gamma}(\mathbf{k},z) = \Gamma^{(d)}_{\gamma n}(\mathbf{k},z) = 0 \tag{140}$$

for any γ since $Ln = -i\nabla \cdot \mathbf{g}$ and linear terms in $L\psi_\alpha$ do not contribute to $\Gamma^{(d)}$. Thus $\Gamma^{(d)}$ reduces to a 2×2 matrix. Remembering the conservation laws,

$$iLg_\alpha(\mathbf{k}) = -\sum_\beta i k_\beta \sigma_{\alpha\beta}(\mathbf{k}) \tag{141}$$

$$iLq(\mathbf{k}) = -i\mathbf{k} \cdot \mathbf{J}^q(\mathbf{k}) , \tag{142}$$

which introduce the stress tensor and heat current, we are led to the discussion surrounding the plateau-value problem and the small k and z limits. We find:

$$\lim_{z \to 0} \lim_{k \to 0} \frac{1}{k^2} \Gamma^{(d)}_{g_z g_z}(\mathbf{k},z) = -i\gamma \tag{143}$$

with:

$$\gamma = \int_0^{+\infty} \frac{dt}{V} \langle \sigma_T^{zz} \sigma_T^{zz}(t) \rangle \tag{144}$$

$$\lim_{z \to 0} \lim_{k \to 0} \frac{1}{k^2} \Gamma^{(d)}_{g_z q}(\mathbf{k},z) = -i\gamma_{zz,z} \tag{145}$$

with:

$$\gamma_{zz,z} = \int_0^{+\infty} \frac{dt}{V} \langle J_z^q \sigma_T^{zz}(t) \rangle , \tag{146}$$

and:

$$\lim_{z \to 0} \lim_{k \to 0} \frac{1}{k^2} \Gamma^{(d)}_{qq}(\mathbf{k}, z) = -i\gamma_q \tag{147}$$

$$\gamma_q = \int_0^{+\infty} \frac{dt}{V} \langle J_z^q J_z^q(t) \rangle \tag{148}$$

$$= \frac{1}{3V} \int_0^{+\infty} dt \langle \mathbf{J}^q \cdot \mathbf{J}^q(t) \rangle \, . \tag{149}$$

Let us consider the cross term proportional to $\gamma_{zz,z}$. It is left to Problem 6.6 to show under parity inversion:

$$q_T \to +q_T \tag{150}$$

$$\mathbf{J}^q \to -\mathbf{J}^q \tag{151}$$

$$\mathbf{g}_T \to -\mathbf{g}_T \tag{152}$$

then:

$$\sigma_T^{\alpha\mu} \to +\sigma_T^{\alpha\mu} \, . \tag{153}$$

Using these results in Eq. (146) we can show:

$$\gamma_{zz,z} \to -\gamma_{zz,z} = 0 \tag{154}$$

and:

$$\Gamma^{(d)}_{g_z q} = \Gamma^{(d)}_{q g_z} = 0 \, , \tag{155}$$

including terms of $\mathcal{O}(k^2)$. Thus the damping of the heat mode decouples from the sound mode.

The matrix $\Gamma^{(d)}$ can be written in the form:

$$\Gamma^{(d)}_{\alpha\beta}(\mathbf{k}, z) = -ik^2 \gamma_\alpha \delta_{\alpha\beta} \, , \tag{156}$$

where $\gamma_n = 0$, $\gamma_q = \gamma_q$, and $\gamma_{g_L} = \gamma$, is summarized in Table 6.5.

Table 6.5 Matrix $\Gamma^{(d)}_{\alpha\beta}(\mathbf{k}, z)$ in the small k and z limits

	n	q	g_L
n	0	0	0
q	0	$-ik^2\gamma_q$	0
g_L	0	0	$-ik^2\gamma$

We must then invert:

$$\Gamma^{(d)}_{\alpha\beta}(\mathbf{k},z) = \sum_\gamma K^{(d)}_{\alpha\gamma} S_{\gamma\beta} = -ik^2 \gamma_\alpha \delta_{\alpha\beta} \qquad (157)$$

to obtain the hydrodynamic limit for the dynamic part of the memory function:

$$\begin{aligned}K^{(d)}_{\alpha\beta}(\mathbf{k},z) &= \sum_\gamma -ik^2 \gamma_\alpha \delta_{\alpha\gamma} (S^{-1})_{\gamma\beta} \\ &= -ik^2 \gamma_\alpha (S^{-1})_{\alpha\beta} \end{aligned} \qquad (158)$$

One can then easily work out the various nonzero matrix elements for $K^{(d)}$. Clearly:

$$K^{(d)}_{n\beta} = 0 \qquad (159)$$

for all β. For $\alpha = g_L$,

$$\begin{aligned}K^{(d)}_{g_L\beta}(\mathbf{k},z) &= -ik^2 \gamma \delta_{\beta g_L}(S^{-1})_{g_L\beta} \\ &= -ik^2 \gamma \delta_{\beta g_L} \frac{1}{k_B T m n} \equiv -ik^2 \tilde{\gamma} \delta_{\beta g_L} \end{aligned} \qquad (160)$$

For $\alpha = q$, :

$$K^{(d)}_{q\beta}(\mathbf{k},z) = -ik^2 \gamma_q (S^{-1})_{q\beta} , \qquad (161)$$

which has only two nonzero elements.:

$$K^{(d)}_{qq} = -ik^2 \frac{\gamma_q}{mn\beta^{-1} T C_V} = -ik^2 \tilde{\gamma}_q \qquad (162)$$

and:

$$\begin{aligned}K^{(d)}_{qn} &= -ik^2 \gamma_q \left(S^{-1}\right)_{nq} \\ &= ik^2 \gamma_q \frac{(C_p - C_V)}{C_V} \frac{1}{S_{nq}} \end{aligned} \qquad (163)$$

Then the dynamic part of the memory function is summarized in Table 6.6.

6.9
Fluctuation Spectrum Including Damping

We are now in a position to work out the hydrodynamic spectrum for the longitudinal case where we include damping. In this case we must invert the matrix:

$$z C_{\alpha\beta} - \sum_\gamma \left(K^{(s)}_{\alpha\gamma} + K^{(d)}_{\alpha\gamma}\right) C_{\gamma\beta} = S_{\alpha\beta} , \qquad (164)$$

Table 6.6 Matrix $K^{(d)}_{\alpha\beta}(\mathbf{q},z)$ in the hydrodynamic regime

	n	q	g_L
n	0	0	0
q	$\frac{ik^2\gamma_q(C_p-C_V)}{C_V S_{nq}}$	$-ik^2\tilde{\gamma}_q$	0
g_L	0	0	$-ik^2\tilde{\gamma}$

where $K^{(s)}$ and $K^{(d)}$ are given in Tables 6.3 and 6.6. The determinant of the matrix $(z-K)$, summarized in Table 6.7, is easily worked out at the appropriate order and is given by:

$$D(\mathbf{k},z) = (z+ik^2\tilde{\gamma}_q)[z(z+ik^2\tilde{\gamma}) - c^2k^2] - \frac{k}{m} K^{(s)}_{gz,q} K^{(d)}_{qn} \,. \tag{165}$$

Since we have:

$$\begin{aligned}K^{(s)}_{gq} K^{(d)}_{qn} &= -k\left(\frac{C_p-C_V}{C_V}\right) S_{qn}^{-1} ik^2 \frac{\gamma_q}{TC_V} \frac{S_{qn}}{S_{nn}} \\ &= -ik^3 \left(\frac{C_p-C_V}{C_V}\right) \frac{\gamma_q}{mTC_V} \frac{1}{S_{nn}} \,, \end{aligned} \tag{166}$$

we can write:

$$\begin{aligned}D(\mathbf{k},z) &= (z+ik^2\tilde{\gamma}_q)[z(z+ik^2\tilde{\gamma}) - c^2k^2] \\ &\quad + ik^4 \frac{(C_p-C_V)}{C_V} \frac{\gamma_q}{m^2 TC_V S_{nn}} \,. \end{aligned} \tag{167}$$

Table 6.7 Matrix $(z-K)_{\alpha\beta}$

	n	q	g_β
n	z	0	$-k/m$
q	$-K^{(s)}_{qn}$	$z-K^{(d)}_{qq}$	0
g_α	$-K^{(s)}_{gn}$	$-K^{(s)}_{gq}$	$z-K^{(d)}_{gg}$

The zeros of $D(k,z)$ control the poles in the complex correlation functions. So, if we look near the heat mode pole and write:

$$z = -iD_T k^2 + \mathcal{O}(k^3) \,, \tag{168}$$

then Eq. (167) becomes:

$$\begin{aligned}D(\mathbf{k},iD_Tk^2) &= [-iD_Tk^2 + ik^2\tilde{\gamma}_q][-iD_Tk^2(-iD_Tk^2 + ik^2\tilde{\gamma}) - c^2k^2] \\ &\quad + ik^4 \frac{(C_p-C_V)}{C_V} \frac{\gamma_q}{m^2 TC_V S_{nn}} \end{aligned}$$

$$= -c^2k^2(-ik^2)(D_T - \tilde{\gamma}_q) + ik^4 \frac{(C_p - C_V)\gamma_q}{m^2 TC_V^2 S_{nn}} \quad , \tag{169}$$

where we have kept terms of $\mathcal{O}(k^4)$. The position of the pole corresponds to a zero of $D(\mathbf{k}, iD_T k^2)$; this determines D_T:

$$c^2(D_T - \tilde{\gamma}_q) = -\frac{(C_p - C_V)\gamma_q}{m^2 TC_V^2 S_{nn}}$$

or:

$$\begin{aligned}
D_T &= \tilde{\gamma}_q - \frac{1}{c^2}\frac{(C_p - C_V)}{m^2 TC_V^2}\frac{\gamma_q}{S_{nn}} \\
&= \frac{\gamma_q}{mn\beta^{-1} TC_V} - \frac{C_V}{C_p}m\left(\frac{\partial n}{\partial p}\right)_T \frac{(C_p - C_V)}{m^2 TC_V^2}\frac{\gamma_q}{S_{nn}} \\
&= \gamma_q\left[\frac{1}{mnT\beta^{-1} TC_V} - \frac{(C_p - C_V)}{C_p}\left(\frac{\partial n}{\partial p}\right)_T \frac{1}{mTC_V\beta^{-1}n}\left(\frac{\partial p}{\partial n}\right)_T\right] \\
&= \frac{\gamma_q}{mnT\beta^{-1}C_V}\left[1 - \frac{(C_p - C_V)}{C_p}\right] \\
&= \frac{\gamma_q}{mnT\beta^{-1}C_p} \quad . \tag{170}
\end{aligned}$$

The damping of the Rayleigh peak, corresponding to the heat mode, is given by:

$$D_T = \frac{\gamma_q}{mnT\beta^{-1}C_p} \quad , \tag{171}$$

where:

$$\gamma_q = \frac{1}{3V}\int_0^{+\infty} \langle \mathbf{J}_T^q \cdot \mathbf{J}_T^q(t)\rangle \quad . \tag{172}$$

D_T is related to the thermal conductivity by:

$$D_T = \frac{\lambda}{mnC_p} \quad , \tag{173}$$

So:

$$\lambda = mnC_p D_T = \frac{\gamma_q}{T\beta^{-1}}$$

$$\lambda = \frac{1}{3k_B T^2 V}\int_0^{+\infty} dt\langle \mathbf{J}_T^q \cdot \mathbf{J}_T^q(t)\rangle \tag{174}$$

is the usual Green–Kubo formula for the thermal conductivity.

The damping of the sound mode can be obtained by putting:

$$z = \pm ck - \frac{i}{2}k^2\Gamma , \qquad (175)$$

in the determinant $D(\mathbf{k}, z)$ where the sound attenuation Γ is to be determined:

$$D(\mathbf{k}, \pm ck - \frac{i}{2}k^2\Gamma) = \left(\pm ck - \frac{i}{2}k^2\Gamma + ik^2\tilde{\gamma}_q\right)$$
$$\times \left[\left(\pm ck - \frac{i}{2}k^2\Gamma\right)\left(\pm ck - \frac{i}{2}k^2\Gamma + ik^2\tilde{\gamma}\right) - c^2k^2\right]$$
$$+ ik^4\frac{(C_p - C_V)\gamma_q}{m^2 T C_V^2 S_{nn}} \qquad (176)$$

and we again work to $\mathcal{O}(k^4)$. Looking first at the term:

$$\left(\pm ck - \frac{i}{2}k^2\Gamma\right)\left[\pm ck + ik^2(\tilde{\gamma} - \Gamma/2)\right] - c^2k^2$$
$$= c^2k^2 \pm ick^3(\tilde{\gamma} - \Gamma/2) \mp \frac{i}{2}ck^3\Gamma + \mathcal{O}(k^4) - c^2k^2$$
$$= \pm ick^3(\tilde{\gamma} - \Gamma/2 - \Gamma/2) \qquad (177)$$

so we can drop terms of $\mathcal{O}(k^5)$ in Eq. (176) and obtain:

$$D(\mathbf{k}, \pm ck - \frac{i}{2}k^2\Gamma) = \pm ck(\pm ick^3)(\tilde{\gamma} - \Gamma)$$
$$+ \frac{ik^4(C_p - C_V)\gamma_q}{m^2 T C_V^2 S_{nn}} \qquad (178)$$
$$= ik^4\left[c^2(\tilde{\gamma} - \Gamma) + \frac{(C_p - C_V)}{m^2 T C_V^2}\frac{\gamma_q}{S_{nn}}\right] . \qquad (179)$$

The zeros of $D(\mathbf{k}, \pm ck - \frac{i}{2}k^2\Gamma)$ determine the sound attenuation coefficient:

$$\Gamma = \tilde{\gamma} + \frac{1}{c^2}\frac{(C_p - C_V)}{m^2 T C_V^2}\frac{\gamma_q}{S_{nn}} . \qquad (180)$$

This expression can be reduced considerably by substituting for c^2 (Eq. (88)), S_{nn} [Eq. (25)], γ_q [Eq. (171)] and $\tilde{\gamma}_q$ [Eq. (160)] with the results:

$$\Gamma = \frac{\beta\gamma}{mn} + \frac{C_V}{C_p}m\left(\frac{\partial n}{\partial p}\right)_T\frac{(C_p - C_V)}{m^2 T C_V^2}$$
$$\times \left(\frac{\partial p}{\partial n}\right)_T\frac{mn}{k_B T n}T\beta^{-1}C_p D_T$$

$$= \frac{\beta\gamma}{mn} + \left(\frac{C_p - C_V}{C_V}\right) D_T \ . \tag{181}$$

It is conventional to define the *bulk viscosity* ζ via:

$$\left(\frac{4}{3}\eta + \zeta\right) / mn = \frac{\beta\gamma}{mn} \ , \tag{182}$$

and so we have the Green–Kubo formula:

$$\left(\frac{4}{3}\eta + \zeta\right) = \frac{1}{k_B T} \int_0^{+\infty} \frac{dt}{V} \langle \sigma_T^{zz} \sigma_T^{zz}(t) \rangle \ . \tag{183}$$

Thus we have characterized the pole structure for the longitudinal modes: there is a diffusive heat mode at $z = -iD_T k^2$ and two damped traveling modes at $z = \pm ck - ik\Gamma/2$ where D_T and Γ are identified with the basic transport coefficients.

We must still determine the residues for the poles in each correlation function. This requires computing the numerator divided by the determinant $D(k,z)$ when inverting Eq. (164) for the correlation functions. We find for the density–density fluctuation function:

$$C_{nn}(\mathbf{k}, z) = \beta^{-1} n \left(\frac{\partial n}{\partial p}\right)_T$$

$$\times \left[\frac{C_V}{C_p} \frac{(z + ik^2(\Gamma + D_T(C_p/C_V - 1)))}{z^2 - c^2 k^2 + izk^2\Gamma} \right.$$

$$\left. + (1 - \frac{C_V}{C_p}) \frac{1}{z + ik^2 D_T} \right] \tag{184}$$

$$C_{qq}(\mathbf{k}, z) = \frac{\beta^{-1} mn C_p T}{z + ik^2 D_T} \tag{185}$$

$$C_{nq}(\mathbf{k}, z) = \beta^{-1} T (\partial n/\partial T)_p \left[\frac{ik^2 D_T}{z^2 - c^2 k^2 + izk^2\Gamma} + \frac{1}{z + ik^2 D_T} \right] \tag{186}$$

$$C_t(\mathbf{k}, z) = \frac{\beta^{-1} mn}{z + ik^2 \nu/mn} \ . \tag{187}$$

The longitudinal current correlations follow from:

$$z C_{n\gamma}(\mathbf{k}, z) - \mathbf{k} \cdot C_{\mathbf{g}\gamma}(\mathbf{k}, z) = S_{n\gamma}(\mathbf{k}) \ , \tag{188}$$

where $\gamma = n, g_L$ or q. Since,

$$z C_{nn}(\mathbf{k}, z) - \mathbf{k} \cdot C_{\mathbf{g}n}(\mathbf{k}, z) = S_{nn}(\mathbf{k}) \tag{189}$$

$$z^2 C_{nn}(\mathbf{k}, z) - \mathbf{k} \cdot C_{\mathbf{gg}}(\mathbf{k}, z) \cdot \mathbf{k} = S_{nn}(\mathbf{k}) \tag{190}$$

$$C_{g_L}(\mathbf{k},z) = \frac{1}{k^2}[z^2 C_{nn}(\mathbf{k},z) - S_{nn}(\mathbf{k})] \ . \tag{191}$$

The transport coefficients are:

$$D_T = \frac{\lambda}{mnC_p} \ , \tag{192}$$

where D_T is the thermal diffusivity,

$$\Gamma = \left(\frac{4}{3}\eta + \zeta\right)/mn + \left(\frac{C_p - C_V}{C_V}\right) D_T \tag{193}$$

is the sound attenuation,

$$\eta = \frac{1}{k_B T \mathcal{V}} \int_0^{+\infty} dt \langle \sigma_T^{zx} \sigma_T^{zx}(t) \rangle \tag{194}$$

is the shear viscosity,

$$\lambda = \frac{1}{3 k_B T^2 \mathcal{V}} \int_0^{+\infty} dt \langle \mathbf{J}_T^q \cdot \mathbf{J}_T^q(t) \rangle \tag{195}$$

is the thermal conductivity, and:

$$\left(\frac{4}{3}\eta + \zeta\right) = \frac{1}{k_B T \mathcal{V}} \int_0^{+\infty} dt \langle \sigma_T^{zz} \sigma_T^{zz}(t) \rangle \tag{196}$$

defines the bulk viscosity. These expressions for the transport coefficients were first derived by Green [11] and Mori [12]. A schematic drawing, Fig. 6.1, of the density–density correlation function shows the three-peak structure, two sound peaks at $\omega = \pm ck$ and a heat mode at the origin. The widths are proportional to Γk^2 and $D_T q^2$ and the area ratio is $(C_V/C_p)/(1 - C_V/C_p) =$

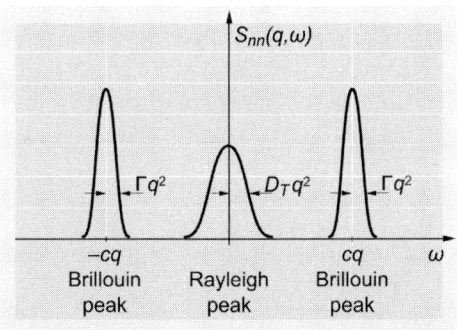

Fig. 6.1 Schematic of the three peak hydrodynamic spectrum for the density–density correlation function.

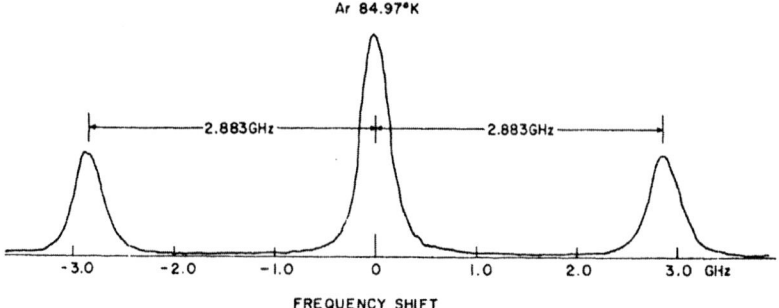

Fig. 6.2 Early light-scattering spectrum for liquid argon. Source: Ref. [16].

$C_p/C_V - 1$. Clearly, in a scattering experiment one can measure the position and widths of all the lines and extract c, Γ, D_T and $C_p/C_V - 1$.

The full theoretical treatment of the density–density spectrum is due to Landau and Placzek [8]. A detailed theoretical analysis was given by Frenkel [13]. Experimentally the existence of the Brillouin doublet was established using light scattering by Gross [14]. However the resolution was not good enough to resolve the width of the Brillouin peaks. The first measurements of the widths of the Brillouin lines in liquids is due to Mash, Starunov and Fabelinskii [15]. For an example of an experimental light scattering spectrum, see Fig. 6.2, which is for liquid argon from Ref. [16].

6.10
References and Notes

1 A. Einstein, Ann. Phys. **38**, 1275 (1910).
2 M. Von Smoluchowski, Ann. Phys. **25**, 205 (1908).
3 L. F. Ornstein, F. Zernike, Proc. Acad. Sci. Amsterdam **17**, 793 (1914).
4 T. Andrews, Philos. Trans. R. Soc. **159**, 575 (1869).
5 The Gibbs–Duhem relation is discussed in Section 2.3.3 in ESM.
6 For a discussion of the Euler relation see Section 2.3.3 in ESM.
7 See Section 1.14.4 in ESM.
8 L. D. Landau, G. Placzek, Phys. Z. Sowjetunion **5**, 172 (1934).
9 L. Brillouin, C. R. Acad. Sci. **158**, 1331 (1914); Ann. Phys. **17**, 88 (1922).
10 S. G. Brush, *Statistical Physics and the Atomic Theory of Matter*, Princeton University Press, Princeton (1983).
11 M. S. Green, J. Chem. Phys. **22**, 398 (1954).
12 H. Mori, Phys. Rev. **112**, 1829 (1958).
13 J. Frenkel, *Kinetic Theory of Liquids*, Oxford University Press, London (1946).
14 E. Gross, Nature **126**, 201, 400, 603 (1930).
15 D. I. Mash, V. S. Starunov, I. L. Fabelinskii, Zh. Eksperim. Teor. Fiz., **47**, 783 (1964). [Enf. Trans. Soviet Phys. JETP **20**, 523 (1965)].
16 P. S. Fleury, J. P. Boon, Phys. Rev. **186**, 244 (1969), see Fig. 3.

6.11
Problems for Chapter 6

Problem 6.1: Show, for a simple fluid system, that we have the thermodynamic identity:

$$\left(\frac{\partial S}{\partial N}\right)_{T,p} = \frac{S}{N} ,$$

where S is the entropy, N the number of particles, T the temperature and p the pressure.

Problem 6.2: In the absence of damping, use Eq. (99) to determine the correlation functions $C_{g_i n}(\mathbf{k}, z)$ and $C_{g_i g_j}(\mathbf{k}, z)$.

Problem 6.3: Show, in the contribution to the static part of the memory Function, that the matrix:

$$\Gamma^{(s)}_{\alpha\beta}(\mathbf{x} - \mathbf{y}) = -\langle \psi_\beta(\mathbf{y}) L \psi_\alpha(\mathbf{x}) \rangle$$

vanishes unless ψ_α and ψ_β have different signatures under time reversal.

Problem 6.4: Show that the determinant:

$$d_s = S_{nn} S_{qq} - S_{nq} S_{qn}$$

can be written in the form given by Eq. (86).

Problem 6.5: Show that:

$$\left(\frac{\partial p}{\partial n}\right)_S = \frac{C_p}{C_V} \left(\frac{\partial p}{\partial n}\right)_T .$$

Problem 6.6: Show using symmetry that:

$$\gamma_{zz,z} = \int_0^{+\infty} \frac{dt}{V} \langle J_z^q \sigma_T^{zz}(t) \rangle = 0 .$$

Problem 6.7: Show using Eq. (136) and Eq. (196) that the bulk viscosity is positive.

Problem 6.8: If one works with the undamped continuity equations:

$$\frac{\partial \rho}{\partial t} = -\nabla \cdot \mathbf{g}$$

and

$$\frac{\partial \mathbf{g}}{\partial t} = -\nabla p ,$$

where ρ is the mass density, $\mathbf{g}$ the momentum density, and $p = p(\rho)$ the pressure, show that one is led to sound waves with the isothermal speed of sound.

7
Kinetic Theory

7.1
Introduction

Kinetic theory is the theoretical approach, which attempts to carry out a complete dynamical treatment of large fluid systems starting at the atomistic level. This is to be contrasted with a hydrodynamical point of view where we organize things in terms of the slow variables. The kinetic theory of gases has an rich history. A brief description of the early history is given in ESM [1] while a much more extensive discussion is given by Brush [2]. The history and controversy associated with kinetic theory are connected with the establishment of the existence of atoms and the use of probability theory in physics. Indeed much of the seminal work [3] in kinetic theory was carried out in the mid-nineteenth century well before the size of atoms was established [4] definitively, as discussed in Chapter 1, in the early part of the twentieth century. In this chapter we first build up an understanding of traditional kinetic theory using elementary ideas of geometry and probability theory. This discussion introduces the concepts of mean-free path and mean-free time. This leads us naturally to a discussion of the Boltzmann equation and the approach of dilute fluid systems to equilibrium. Analysis of the nonlinear Boltzmann equation allows for a dynamical understanding of the Maxwell velocity probability distribution function. Next, we restrict our discussion to situations near equilibrium and the linearized Boltzmann equation. We study, as a first approximation, the single relaxation time approximation and traditional transport in driven systems. In the second half of this chapter we show how kinetic theory in the linear- response regime can be made systematic via the use of phase-space time-correlation functions in the memory-function approach.

7.2 Boltzmann Equation

7.2.1 Ideal Gas Law

In this section we give a derivation of the Boltzmann equation. We begin with an elementary example that sets up some of the ideas we will use later. The appropriate probability distribution for treating dilute-gas kinetics is related to the phase-space density [5],

$$\hat{f}(\mathbf{x},\mathbf{p}) = \sum_{i=1}^{N} \delta(\mathbf{x} - \mathbf{r}_i)\delta(\mathbf{p} - \mathbf{p}_i) \;, \qquad (1)$$

for a system of N particles with phase-space coordinates $\{\mathbf{r}_i, \mathbf{p}_i\}$. Momentum integrals of $\hat{f}$ give back observables of interest. The particle number density is given by:

$$\hat{n}(\mathbf{x},t) = \int d^3p\, \hat{f}(\mathbf{x},\mathbf{p}) = \sum_{i=1}^{N} \delta(\mathbf{x} - \mathbf{r}_i) \;, \qquad (2)$$

the momentum density is given by:

$$\hat{\mathbf{g}}(\mathbf{x},t) = \int d^3p\, \mathbf{p}\, \hat{f}(\mathbf{x},\mathbf{p},t) \qquad (3)$$

and the kinetic energy current density is given by:

$$\hat{\mathbf{J}}_K(\mathbf{x},t) = \int d^3p\, \varepsilon_0(p) \frac{\mathbf{p}}{m}\, \hat{f}(\mathbf{x},\mathbf{p},t) \;, \qquad (4)$$

where the particles have mass m and:

$$\varepsilon_0(p) = \frac{\mathbf{p}^2}{2m} \;. \qquad (5)$$

We are interested in determining the nonequilibrium average [6] of $\hat{f}(\mathbf{x},\mathbf{p},t)$,

$$f(\mathbf{x},\mathbf{p},t) = \langle \hat{f}(\mathbf{x},\mathbf{p},t)\rangle_{NE} \;. \qquad (6)$$

This quantity is a probability distribution function and is traditionally called the *singlet* probability distribution function. The reason is that it monitors the statistical properties of a typical single particle [7]. If we can determine the singlet distribution function we can obtain the nonequilibrium averages:

$$n(\mathbf{x},t) = \langle \hat{n}(\mathbf{x},t)\rangle_{NE} = \int d^3p\, f(\mathbf{x},\mathbf{p},t) \;, \qquad (7)$$

$$\mathbf{g}(\mathbf{x},t) = \int d^3p\, \mathbf{p}\, f(\mathbf{x},\mathbf{p},t) \qquad (8)$$

and:

$$J_K(\mathbf{x}, t) = \int d^3p\, \varepsilon_0(p) \frac{\mathbf{p}}{m} f(\mathbf{x}, \mathbf{p}, t) \ . \tag{9}$$

What is the equilibrium form for $f(\mathbf{x}, \mathbf{p}, t)$? This is worked out in Appendix C with the result that:

$$f_{EQ}(\mathbf{x}, \mathbf{p}) = f_0(\mathbf{p}) = n\Phi(\mathbf{p}) \ , \tag{10}$$

where $n = N/V = \langle \hat{n}(\mathbf{x}) \rangle$ is the average density and:

$$\Phi(\mathbf{p}) = \frac{e^{-\beta p^2/2m}}{(2\pi m k_B T)^{3/2}} \tag{11}$$

is the normalized Maxwell velocity probability distribution for a system at equilibrium with temperature T.

Let us investigate the usefulness of the singlet distribution function in the simplest kinetic context. We want to use [8] geometry and elementary probability theory to compute the pressure acting on a wall due to a dilute gas. Before proceeding further it is useful to point out that, in certain circumstances, it is more convenient to use velocity rather than momentum as labels on the singlet distribution function. It is easy to see (Problem 7.1) that these two distributions are simply related,

$$f(\mathbf{x}, \mathbf{v}) = \left\langle \sum_{i=1}^{N} \delta(\mathbf{x} - \mathbf{r}_i)\delta(\mathbf{v} - \mathbf{v}_i) \right\rangle_{NE} = m^3 f(\mathbf{x}, \mathbf{p}) \ , \tag{12}$$

and we will freely move back and forth between the two as is convenient.

The singlet distribution function is normalized such that:

$$\int_V d^3x\, d^3v\, f(\mathbf{x}, \mathbf{v}, t) = N \tag{13}$$

where N is the total number of molecules in the volume V. More locally $f(\mathbf{x}, \mathbf{v}, t)d^3x\, d^3v$ gives the number of molecules in the region of phase-space between $\mathbf{x} + d\mathbf{x}$ and $\mathbf{x}$ and $\mathbf{v} + d\mathbf{v}$ and $\mathbf{v}$.

Consider a very dilute gas colliding with a wall where the collisions are assumed to be elastic. What is the pressure on the wall? Pressure is the momentum transferred by the molecules to a unit area of wall per unit of time. Consider, as shown in Fig. 7.1, a small flat area of wall of area δA and look at the momentum transferred in time δt. All the molecules with speed v that make a collision with the area δA during time δt lie inside the cylinder shown in Fig. 7.1. The cylinder with base δA has a side parallel to $\mathbf{v}$ with a length $v\delta t$. The number of molecules n_V with velocity $\mathbf{v}$ in the cylinder is given by:

$$n_V = \int_{\delta v} d^3x\, f(\mathbf{x}, \mathbf{v}, t) \ , \tag{14}$$

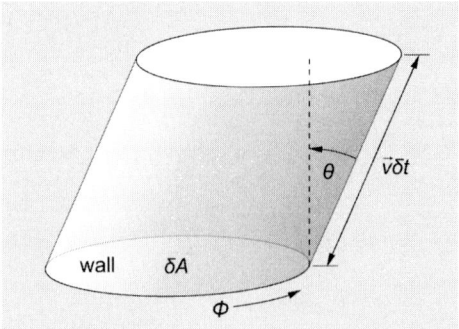

Fig. 7.1 Collision cylinder for particles with velocity **v** that collide with a wall during a time δt.

where the volume of the cylinder is given by:

$$\delta \mathcal{V} = \delta A \delta t \, v \cos \theta \ . \tag{15}$$

The momentum exchange during a collision with the wall is:

$$\Delta p_z = 2mv \cos \theta \tag{16}$$

and if the molecules are to hit the wall, $v_\perp = v \cos \theta > 0$.

The total momentum P_V transferred to the area δA of the wall by molecules that move with velocity **v** in time δt is:

$$\begin{aligned} P_V &= 2mv \cos \theta \, n_V \\ &= 2mv \cos \theta \int_{\delta \mathcal{V}} d^3 x f(\mathbf{x}, \mathbf{v}, t) \ . \end{aligned} \tag{17}$$

The total momentum P transferred to the wall by molecules of all possible velocities is obtained by integration over v, θ and ϕ where ϕ is the azimuthal angle:

$$P = \int_0^\infty dv \, v^2 \int_0^{\pi/2} d\theta \, \sin \theta \int_0^{2\pi} d\phi \, 2mv \cos \theta \int_{\delta \mathcal{V}} d^3 x f(\mathbf{x}, \mathbf{v}, t) \ . \tag{18}$$

The integration extends over $0 \leq \theta \leq \pi/2$, since for $\theta > \pi/2$ the molecules travel away from the wall. Let us assume that the gas is spatially homogeneous, then:

$$f(\mathbf{x}, \mathbf{v}, t) = n f(\mathbf{v}, t) \ , \tag{19}$$

and Eq. (18) reduces to:

$$P = \int_0^\infty dv \, v^2 \int_0^{\pi/2} d\theta \, \sin \theta \int_0^{2\pi} d\phi \, 2mv \cos \theta \int_{\delta \mathcal{V}} d^3 x \, n f(\mathbf{v}, t)$$

$$= \int_0^\infty dv\, v^2 \int_0^{\pi/2} d\theta \sin\theta \int_0^{2\pi} d\phi\, 2mv \cos\theta\, \delta \mathcal{V} n f(\mathbf{v},t)$$

$$= \int_0^\infty dv\, v^2 \int_0^{\pi/2} d\theta \sin\theta \int_0^{2\pi} d\phi\, 2mv \cos\theta\, \delta A \delta t\, v \cos\theta\, n f(\mathbf{v},t)$$

$$= n\delta A \delta t \int_0^\infty dv\, v^4 \int_0^{\pi/2} d\theta \sin\theta \cos^2\theta\, 4\pi m\, f(\mathbf{v},t) \quad . \tag{20}$$

Letting $u = \cos\theta$, $du = -\sin\theta\, d\theta$, and assuming the system is isotropic in velocity space we have:

$$P = n\delta A \delta t 4\pi m \int_0^\infty dv\, v^4 f(v,t) \int_1^0 -du\, u^2$$

$$= n\delta A \delta t 4\pi m \int_0^\infty dv\, v^4 f(v,t) \frac{1}{3} \quad . \tag{21}$$

The pressure is given by:

$$p = \frac{P}{\delta A \delta t} = n \frac{4\pi m}{3} \int_0^\infty dv\, v^4 f(v,t) \quad . \tag{22}$$

In equilibrium we have the Maxwell velocity distribution:

$$f_{EQ}(v,t) = \Phi(\mathbf{v}) = \left(\frac{m}{2\pi kT}\right)^{3/2} e^{-\frac{1}{2}\beta mv^2} \quad . \tag{23}$$

Letting $v = \sqrt{\frac{2kT}{m}} x$ we obtain:

$$p = n\frac{4\pi m}{3} \left(\frac{m}{2\pi kT}\right)^{3/2} \left(\frac{2\pi kT}{m}\right)^{5/2} \int_0^\infty dx\, x^4 e^{-x^2}$$

$$= \frac{8}{3} \frac{n}{\sqrt{\pi}} kT \int_0^\infty dx\, x^4 e^{-x^2} \quad . \tag{24}$$

From the standard tables we have:

$$\int_0^\infty dx\, x^4 e^{-x^2} = \frac{3\sqrt{\pi}}{8} \quad , \tag{25}$$

which leads to the final result:

$$p = nkT \quad , \tag{26}$$

which is the ideal gas law.

We can use the same reasoning to compute the number of molecules that strike the area δA in time δt. The first step is to compute the number of molecules with a given velocity $\mathbf{v}$ that collide with δA in time δt. This is given by:

$$N_V = \int_{\delta \mathcal{V}} d^3 x f(\mathbf{x},\mathbf{v},t) \quad . \tag{27}$$

The total number of molecules striking δA in time δt is given by:

$$\tilde{N} = \int_0^\infty v^2 dv \int_0^{\pi/2} d\theta \sin\theta \int_0^{2\pi} d\phi\, N_V \;. \tag{28}$$

Again, assuming the gas is homogeneous in space:

$$N_V = \delta V n f(\mathbf{v}, t) = \delta A \delta t\, v \cos\theta\, n f(\mathbf{v}, t) \tag{29}$$

and the total number is given by:

$$\begin{aligned}\tilde{N} &= \int_0^\infty v^2 dv \int_0^{\pi/2} d\theta \sin\theta \int_0^{2\pi} d\phi\, \delta A \delta t\, v \cos\theta\, n f(\mathbf{v}, t) \\ &= \delta A \delta t\, n \int_0^\infty v^3 dv \int_0^{\pi/2} d\theta \sin\theta \cos\theta \int_0^{2\pi} d\phi\, f(\mathbf{v}, t) \;. \end{aligned} \tag{30}$$

If we further assume the system is in thermal equilibrium, then we can use Eq. (23) and evaluate the remaining integrals (see Problem 7.2) to obtain:

$$\tilde{N} = a\delta A \delta t \;, \tag{31}$$

where a, the rate of effusion, is given by:

$$a = n \left(\frac{kT}{2\pi m}\right)^{1/2}, \tag{32}$$

and is the number of particles that escape from a hole in a wall enclosing a sample per unit time per unit area.

7.2.2
Mean-Free Path

Having considered collisions between molecules and a wall we now move on to consider binary collisions between molecules. The problem we address was formulated by Clausius [9]. In looking at the kinetics of dilute gases, one finds that disturbances are not communicated across a sample at thermal speeds. If one had thermal ballistic motion, disturbances would travel across a sample with speed $v_0^2 = k_B T/m$. Instead, thermal transport is much slower due to collisions. Collisions lead to a type of random walk of particles across the sample, with steps on the order of the mean-free path – the average distance between collisions. Thus we have diffusion as opposed to ballistic motion. We can estimate this mean-free path using elementary geometrical and probabilistic arguments.

Let us assume that the range of interactions between particles is r_0 and is short compared to the distance between molecules in the gas. To a first approximation we can think of the molecules as spheres of diameter r_0.

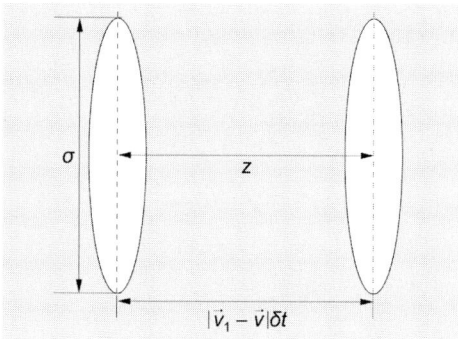

Fig. 7.2 Collision cylinder for particles with velocities **v** and $\mathbf{v}_1$ that collide during a time interval δt.

We carry out this determination of the mean-free path using the idea of a collision cylinder. Suppose a molecule with velocity **v** collides with another particle with velocity $\mathbf{v}_1$ at time t. We describe the collision by choosing a coordinate system (see Fig. 7.2) with its origin at the center of the first molecule, and the z-axis is drawn in the direction of the relative velocity vector $\mathbf{v}_1 - \mathbf{v}$. The molecules collide with each other only if the distance between their centers is smaller than r_0. This means that the center of the molecule with velocity $\mathbf{v}_1$ must at time t lie inside the cylinder if a collision is to take place in the succeeding time interval δt. The height of this collision cylinder (see Fig. 7.2) is:

$$h = |\mathbf{v}_1 - \mathbf{v}|\delta t \quad . \tag{33}$$

The cross sectional area is simply πr_0^2, so the volume of the collision cylinder is:

$$V_c = \pi r_0^2 h = \pi r_0^2 |\mathbf{v}_1 - \mathbf{v}|\delta t \quad . \tag{34}$$

We can use the collision cylinder to compute the number of binary collisions that take place in a small column of gas in time δt between molecules with velocity **v** and $\mathbf{v}_1$. In a small volume d^3x in the gas, there are $f(\mathbf{x}, \mathbf{v}, t)d^3x\, d^3v$ molecules with velocity **v** located at position **x**. To each of these molecules there is attached a collision cylinder appropriate for collisions with molecules of velocity $\mathbf{v}_1$ within a time interval δt. The number of such collision cylinders is therefore $f(\mathbf{x}, \mathbf{v}, t)d^3x\, d^3v$. The total volume occupied by the collision cylinder is the number of cylinders times the volume per cylinder:

$$V_c^T = f(\mathbf{x}, \mathbf{v}, t)d^3x\, d^3v\, V_c = f(\mathbf{x}, \mathbf{v}, t)d^3x\, d^3v\, \pi r_0^2 |\mathbf{v}_1 - \mathbf{v}|\delta t \quad . \tag{35}$$

To compute the number of $(\mathbf{v}, \mathbf{v}_1)$ collisions, we must compute the number of molecules with velocity $\mathbf{v}_1$ that are present in collision cylinders at the beginning of the time interval. Going forward we assume that the gas is sufficiently dilute that collision cylinders contain at most one molecule with velocity $\mathbf{v}_1$ and these molecules lead to $(\mathbf{v}, \mathbf{v}_1)$ collisions.

The number of molecules with velocity $\mathbf{v}_1$ present in the $(\mathbf{v}, \mathbf{v}_1)$ collision cylinders at the instant t, N_T, is equal to the numbers of $\mathbf{v}_1$-molecules per unit volume,

$$N(\mathbf{v}_1) = f(\mathbf{x}, \mathbf{v}_1, t)\, d^3v_1 \; , \tag{36}$$

multiplied by the total volume, V_c^T, of $(\mathbf{v}, \mathbf{v}_1)$ cylinders:

$$N_T = N(\mathbf{v}_1) V_c^T \; . \tag{37}$$

In summary the number of collisions occurring within time interval δt between molecules in the velocity range $\mathbf{v}$ to $\mathbf{v} + d\mathbf{v}$ and the molecules in the range $\mathbf{v}_1$ to $\mathbf{v}_1 + d\mathbf{v}_1$ in volume d^3x of gas centered about position $\mathbf{x}$ is given by:

$$\begin{aligned} N_T &= f(\mathbf{x}, \mathbf{v}_1, t)\, d^3v_1 f(\mathbf{x}, \mathbf{v}, t) d^3x\, d^3v\, \pi r_0^2 |\mathbf{v}_1 - \mathbf{v}| \delta t \\ &= f(\mathbf{x}, \mathbf{v}, t) f(\mathbf{x}, \mathbf{v}_1, t) \pi r_0^2 |\mathbf{v}_1 - \mathbf{v}|\, d^3v\, d^3v_1\, d^3x\, \delta t \; . \end{aligned} \tag{38}$$

Notice that this expression is symmetric in $\mathbf{v}$ and $\mathbf{v}_1$ [10].

The total number of collisions suffered by molecules of velocity $\mathbf{v}$ in spatial volume d^3x in time δt comes from integrating over the velocities of all their collision partners:

$$\begin{aligned} & N(\mathbf{x}, \mathbf{v}, t)\, d^3v\, d^3x \delta t \\ &= f(\mathbf{x}, \mathbf{v}, t) \pi r_0^2 d^3x \delta t\, d^3v \int d^3v_1\, f(\mathbf{x}, \mathbf{v}_1, t) |\mathbf{v}_1 - \mathbf{v}| \; . \end{aligned} \tag{39}$$

The total number of collisions suffered by molecules of all velocities in spatial volume d^3x and in time δt is given by:

$$N(\mathbf{x}, t) d^3x \delta t = d^3x \delta t \int d^3v \pi r_0^2 \int d^3v_1\, f(\mathbf{x}, \mathbf{v}_1, t) |\mathbf{v}_1 - \mathbf{v}| \; . \tag{40}$$

Assuming the gas is homogeneous in space, where Eq. (19) holds, we have:

$$N(\mathbf{v}, t) = n^2 f(\mathbf{v}, t) \pi r_0^2 \int d^3v_1\, f(\mathbf{v}_1, t) |\mathbf{v}_1 - \mathbf{v}| \tag{41}$$

and the number of collisions per unit volume per unit time is:

$$\begin{aligned} N(t) &= \int d^3v N(\mathbf{v}, t) \\ &= n^2 \pi r_0^2 \int d^3v f(\mathbf{v}, t) \int d^3v_1\, f(\mathbf{v}_1, t) |\mathbf{v}_1 - \mathbf{v}| \; . \end{aligned} \tag{42}$$

The average collision rate $\nu(t)$ is the number of collisions per unit volume per unit time divided by the number of molecules per unit volume,

$$\nu(t) = \frac{N(t)}{n} = n\pi r_0^2 \int d^3v f(\mathbf{v},t) \int d^3v_1 \, f(\mathbf{v}_1,t)|\mathbf{v}_1 - \mathbf{v}| \;. \tag{43}$$

The average time between collisions, the mean-free time, is:

$$\tau = 1/\nu \;. \tag{44}$$

The mean-free path is then given by:

$$\ell = u\tau \;, \tag{45}$$

where u is the average speed:

$$u = \langle v \rangle = \int d^3v \, vf(\mathbf{v},t) \;. \tag{46}$$

If we further restrict ourselves to thermal equilibrium then $\nu(t)$ is independent of time and $f(v)$ is the Maxwell velocity distribution. In Problem 7.3 we show that the average speed is given by:

$$u = \left(\frac{8kT}{\pi m}\right)^{1/2} \;. \tag{47}$$

The collision rate in equilibrium can then be evaluated using:

$$\nu = n\pi r_0^2 \left(\frac{m}{2\pi kT}\right)^3 \int d^3v \int d^3v_1 \, e^{-\frac{1}{2}\beta m(v^2+v_1^2)}|\mathbf{v}_1 - \mathbf{v}| \;. \tag{48}$$

The remaining integrals are evaluated in Problem 7.4, with the result:

$$\tau^{-1} = \nu = 4nr_0^2 \sqrt{\frac{\pi k_B T}{m}} \;. \tag{49}$$

The mean-free path is given by:

$$\ell = \frac{u}{\nu} = \frac{1}{\sqrt{2}n\pi r_0^2} \;. \tag{50}$$

As expected the mean-free path is inversely proportional to density and molecular diameter.

7.2.3
Boltzmann Equation: Kinematics

The elementary ideas developed in the previous sections makes the assumption that we know the singlet distribution function in equilibrium. Here, we develop some of the ideas we need to determine the singlet distribution under a variety of nonequilibrium conditions. The first step is to look at the equation of motion for the phase-space density. In Chapter 5 we established that:

$$\frac{\partial}{\partial t}\hat{f}(\mathbf{x},\mathbf{p},t) = iL\hat{f}(\mathbf{x},\mathbf{p},t) \ , \tag{51}$$

where L is the Liouville operator. Going further, we found that:

$$\frac{\partial}{\partial t}\hat{f}(\mathbf{x},\mathbf{p},t) = -\frac{\mathbf{p}\cdot\nabla_x}{m}\hat{f}(\mathbf{x},\mathbf{p}) + \hat{f}_I(\mathbf{x},\mathbf{p},t) \tag{52}$$

where, for pair interactions, the nonlinear interaction contribution is given by:

$$\hat{f}_I(\mathbf{x},\mathbf{p},t) = -\int d^3x_1 d^3p_1 iL_I(\mathbf{x}-\mathbf{x}_1,\mathbf{p},\mathbf{p}_1)\hat{f}(\mathbf{x}_1,\mathbf{p}_1)\hat{f}(\mathbf{x},\mathbf{p}) \ , \tag{53}$$

where we define the interaction part of the two-body Liouville operator,

$$L_I(\mathbf{x}-\mathbf{x}_1,\mathbf{p},\mathbf{p}_1) = i\nabla_x V(\mathbf{x}-\mathbf{x}_1)\cdot(\nabla_\mathbf{p}-\nabla_{\mathbf{p}_1}) \ . \tag{54}$$

It is left to Problem 7.5 to show that if we have an external force $\mathbf{F}_E$ that occurs in Newton's law in the form:

$$\frac{d\mathbf{p}_i(t)}{dt} = \mathbf{F}_E(\mathbf{r}_i,\mathbf{p}_i,t) - \sum_{j(\neq i)=1}^N \nabla_{r_i} V(\mathbf{r}_i-\mathbf{r}_j) \ , \tag{55}$$

then the equation of motion for the phase-space density takes the form:

$$\frac{\partial}{\partial t}\hat{f}(\mathbf{x},\mathbf{p},t) = -\frac{\mathbf{p}\cdot\nabla_x}{m}\hat{f}(\mathbf{x},\mathbf{p},t) - \nabla_p\cdot\left(\mathbf{F}_E(\mathbf{x},\mathbf{p},t)\hat{f}(\mathbf{x},\mathbf{p},t)\right) + \hat{f}_I(\mathbf{x},\mathbf{p},t) \ . \tag{56}$$

If we have uniform applied electric and magnetic fields then:

$$\mathbf{F}_E = q\mathbf{E} + \frac{q}{c}(\mathbf{v}\times\mathbf{B}) \tag{57}$$

and because:

$$\nabla_p\cdot(\mathbf{p}\times\mathbf{B}) = 0 \ , \tag{58}$$

then we have the equation of motion:

$$\left[\frac{\partial}{\partial t} + \frac{\mathbf{p}\cdot\nabla_x}{m} + \mathbf{F}_E(\mathbf{x},\mathbf{p},t)\cdot\nabla_p\right]\hat{f}(\mathbf{x},\mathbf{p},t) = \hat{f}_I(\mathbf{x},\mathbf{p},t) \ . \tag{59}$$

After taking the average over a nonequilibrium ensemble we have:

$$\left[\frac{\partial}{\partial t} + \frac{\mathbf{p}\cdot\nabla_x}{m} + \mathbf{F}_E(\mathbf{x},\mathbf{p},t)\cdot\nabla_p\right]f(\mathbf{x},\mathbf{p},t) = f_I(\mathbf{x},\mathbf{p},t) \ . \tag{60}$$

7.2.4
Boltzmann Collision Integral

Let us turn next to the treatment of collisions in the Boltzmann kinetic equation. The effect of intermolecular collisions is to modify the count of molecules entering and leaving the region in phase space $\mathbf{x}$ to $\mathbf{x} + d\mathbf{x}$ and $\mathbf{v}$ to $\mathbf{v} + d\mathbf{v}$ in time interval δt. The number of particles in this volume is effected by collisions in two ways. Some streaming molecules are kicked out while others are knocked into the volume of interest. We can write the contribution to the kinetic equation in the form:

$$f_I(\mathbf{x}, \mathbf{p}, t) d^3x d^3v \delta t = (J_+ - J_-) d^3x d^3v \delta t \quad . \tag{61}$$

This is the net change in the number of particles in $d^3x d^3v \delta t$ due to collisions. In this equation $J_+ d^3x d^3v \delta t$ is the number of particles entering the region of phase space in time δt via collisions, and $J_- d^3x d^3v \delta t$ the number of particles leaving the region of phase space in time δt via collisions.

Before determining J_+ and J_- we need some background on two-body dynamics. We assume we have elastic collisions between two particles of equal mass m as shown schematically in Fig. 7.3. We assume that well before the collision at time t at position $\mathbf{x}$, the particles have velocities $\mathbf{v}_1$ and $\mathbf{v}_2$. Well after the collision, the particles have the final velocities $\mathbf{v}'_1$ and $\mathbf{v}'_2$. These asymptotic velocities are connected by conservation of momentum and kinetic energy (for short-ranged interactions):

$$\mathbf{v}_1 + \mathbf{v}_2 = \mathbf{v}'_1 + \mathbf{v}'_2 \tag{62}$$

$$\mathbf{v}_1^2 + \mathbf{v}_2^2 = (\mathbf{v}'_1)^2 + (\mathbf{v}'_2)^2 \quad . \tag{63}$$

If we introduce center of mass (COM) variables:

$$\mathbf{V} = \frac{1}{2}(\mathbf{v}_1 + \mathbf{v}_2) \tag{64}$$

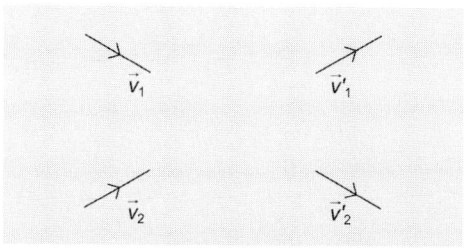

Fig. 7.3 Before and after collision kinematics for two particles with velocities $\mathbf{v}_1$ and $\mathbf{v}_2$ before the collision and $\mathbf{v}'_1$ and $\mathbf{v}'_2$ after the collision.

$$\mathbf{v} = \mathbf{v}_1 - \mathbf{v}_2 \ , \tag{65}$$

then clearly:

$$\mathbf{V} = \mathbf{V}' \ . \tag{66}$$

Since we have the inverse relations:

$$\mathbf{v}_1 = \mathbf{V} + \frac{\mathbf{v}}{2} \tag{67}$$

$$\mathbf{v}_2 = \mathbf{V} - \frac{\mathbf{v}}{2} \ , \tag{68}$$

we have for conservation of kinetic energy:

$$\mathbf{v}_1^2 + \mathbf{v}_2^2 = 2\mathbf{V}^2 + \frac{1}{2}\mathbf{v}^2 = 2(\mathbf{V}')^2 + \frac{1}{2}(\mathbf{v}')^2 \ . \tag{69}$$

Since $\mathbf{V}^2 = (\mathbf{V}')^2$ we have that the relative speed is unchanged by the collision: $\mathbf{v}^2 = (\mathbf{v}')^2$. The angle between $\mathbf{v}$ and $\mathbf{v}'$ is defined by:

$$\mathbf{v} \cdot \mathbf{v}' = v^2 \cos\theta \ . \tag{70}$$

If we limit ourselves at first to the case of hard spheres, then we have the collision diagram shown in Fig. 7.4. We can then connect the before and after velocities in the COM using:

$$\mathbf{v}' = \mathbf{v} - 2\hat{\rho}(\hat{\rho} \cdot \mathbf{v}) \ , \tag{71}$$

where $\hat{\rho}$ is the unit vector connecting the centers of the two colliding particles at contact. The scattering angle θ is related to the angle between $\hat{\rho}$ and $\mathbf{v}$,

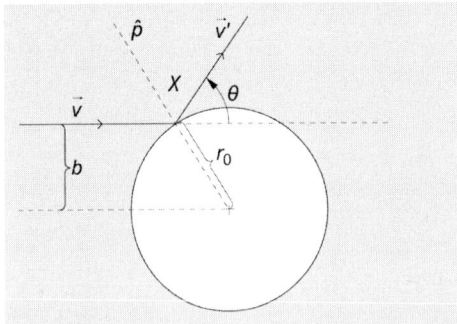

Fig. 7.4 Scattering kinematics for particles with initial relative velocity $\mathbf{v}$, final velocity $\mathbf{v}'$, impact parameter b, r_0 is the hard-sphere diameter and $\hat{\rho}$ is the unit vector connecting the centers of the particles at contact.

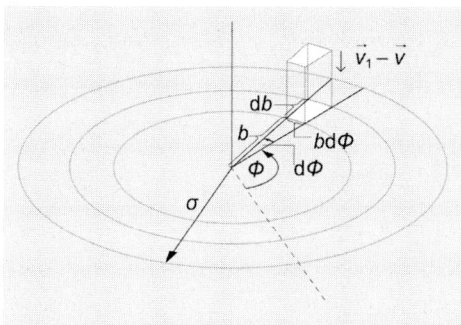

Fig. 7.5 Kinematics for two-particle collisions in terms of a collision cylinder.

$\cos \chi = \hat{\rho} \cdot \mathbf{v}$ by $2\chi + \theta = \pi$. We also have $\sin \chi = b/r_0$ where b is the impact parameter.

Let us determine J_-, the number of molecules that leave the volume $d^3x d^3v$ in time δt via collisions. Consider collisions between molecules with velocity $\mathbf{v}$ in volume d^3x centered at position $\mathbf{x}$ and molecules that move with a different velocity $\mathbf{v}_1$ such that the impact parameter of the $(\mathbf{v}, \mathbf{v}_1)$ collision falls in the range of impact parameter b to $b + db$, as shown in Fig. 7.5. The azimuthal angle of the collision is assumed to be confined to the range ϕ to $\phi + d\phi$ about a plane fixed in space and containing the relative velocity $\mathbf{v}_1 - \mathbf{v}$. A collision takes place if the centers of the two molecules are located inside the collision cylinder shown in Fig. 7.5. Notice that the problem here is very similar to the determination of the collision cylinder in determining the mean collision rate. The difference is the base we use for the collision cylinder. In Fig. 7.5 the base is the full collisional cross sectional area πr_0^2. In the current case, for reasons that become clear when we treat J_+, the base is $db\, b\, d\phi$ as shown in Fig. 7.5.

The number of molecules moving with velocity $\mathbf{v}$ in the region d^3x is $f(\mathbf{x}, \mathbf{v}, t) d^3x d^3v$ and the total volume of collision cylinders for the $(\mathbf{v}, \mathbf{v}_1)$ collision is:

$$f(\mathbf{x}, \mathbf{v}, t) d^3x d^3v\, b\, db\, d\phi\, |\mathbf{v}_1 - \mathbf{v}| \delta t \ . \tag{72}$$

The number of molecules with velocity $\mathbf{v}_1$ in the collision cylinder is:

$$f(\mathbf{x}, \mathbf{v}_1, t) d^3v_1 \tag{73}$$

so the number of collisions under consideration is:

$$\mathcal{N}(\mathbf{v}, \mathbf{v}_1) = f(\mathbf{x}, \mathbf{v}_1, t) d^3v_1 f(\mathbf{x}, \mathbf{v}, t) d^3x d^3v\, b\, db\, d\phi\, |\mathbf{v}_1 - \mathbf{v}| \delta t \ . \tag{74}$$

The molecular flux J_- comes from integrating over all velocities $\mathbf{v}_1$, impact parameters b and azimuthal angles ϕ:

$$J_- d^3x d^3v \delta t = \left[\int d^3v_1 \int_0^{r_0} db \int_0^{2\pi} b d\phi \, |\mathbf{v}_1 - \mathbf{v}| f(\mathbf{x}, \mathbf{v}, t) f(\mathbf{x}, \mathbf{v}_1, t) \right] \times d^3x d^3v \delta t \, , \tag{75}$$

where r_0 denotes the range of forces. In this case, in contrast with our treatment of J_+ below, we can do the integral over b and ϕ to find that $J_- = \mathcal{N}(\mathbf{x}, \mathbf{v}, t)$ where $\mathcal{N}(\mathbf{x}, \mathbf{v}, t)$ is given in the treatment of the mean collision rate by Eq. (39).

Next we need to evaluate J_+, the number of particles scattered into the phase-space volume $d^3v d^3x$ in time δt. This is the process where two particles with velocities $\mathbf{v}'$ and $\mathbf{v}'_1$ are scattered into $\mathbf{v}$ and $\mathbf{v}_1$. We easily have that the number of particles going from $(\mathbf{v}', \mathbf{v}'_1)$ into $(\mathbf{v}, \mathbf{v}_1)$ in the volume $d^3v' d^3v'_1 d^3x \delta t$, is, in complete analogy with Eq. (74),

$$\mathcal{N}(\mathbf{v}', \mathbf{v}'_1) = f(\mathbf{x}, \mathbf{v}'_1, t) d^3v'_1 f(\mathbf{x}, \mathbf{v}', t) d^3x d^3v' b db d\phi \, |\mathbf{v}'_1 - \mathbf{v}'| \delta t \, . \tag{76}$$

We show in Problem 7.6 that:

$$d^3v' d^3v'_1 = d^3v d^3v_1 \, , \tag{77}$$

and we know that:

$$|\mathbf{v}'_1 - \mathbf{v}'| = |\mathbf{v}_1 - \mathbf{v}| \, , \tag{78}$$

since the magnitude of the relative velocity is preserved in an elastic collision. We have then:

$$\mathcal{N}(\mathbf{v}', \mathbf{v}'_1) = f(\mathbf{x}, \mathbf{v}'_1, t) f(\mathbf{x}, \mathbf{v}', t) b db d\phi \, |\mathbf{v}_1 - \mathbf{v}| d^3v d^3v_1 d^3x \delta t \, . \tag{79}$$

Then after integrating over $\mathbf{v}_1$, b and ϕ we arrive at the expression for J_+:

$$J_+ d^3x d^3v \delta t = \left[\int d^3v_1 \int_0^{r_0} b db \int_0^{2\pi} d\phi \, |\mathbf{v}_1 - \mathbf{v}| f(\mathbf{x}, \mathbf{v}', t) f(\mathbf{x}, \mathbf{v}'_1, t) \right] \times d^3x d^3v \delta t \, . \tag{80}$$

Since $\mathbf{v}'$ and $\mathbf{v}'_1$ depend on $\mathbf{v}$, $\mathbf{v}_1$ and b we can not carry out the integral over b. Putting together the results for J_- and J_+ in Eq. (61) we have the famous expression for the Boltzmann collision integral:

$$f_I(\mathbf{x}, \mathbf{v}, t) = \int d^3v_1 \int_0^{r_0} b db \int_0^{2\pi} d\phi \, |\mathbf{v}_1 - \mathbf{v}| \left(f' f'_1 - f f_1 \right) \, , \tag{81}$$

where we have used the convenient notation $f = f(\mathbf{x}, \mathbf{v}, t)$, $f_1 = f(\mathbf{x}, \mathbf{v}_1, t)$, $f' = f(\mathbf{x}, \mathbf{v}', t)$ and $f'_1 = f(\mathbf{x}, \mathbf{v}'_1, t)$. Putting Eq. (81) back into Eq. (60) the Boltzmann equation [11] is then given by:

$$\frac{\partial f}{\partial t} + \mathbf{v} \cdot \vec{\nabla} f = f_I ,\qquad(82)$$

where, here, we assume there is zero external force acting.

It is conventional to write the collisional contribution to the Boltzmann equation in term of the differential cross section σ. We show in Problem 7.7 that we can replace the integral over the impact parameter with a properly weighted average over the scattering angles:

$$\int b\,db\,d\phi \ldots = \int d\Omega\, \sigma(|\mathbf{v}_1 - \mathbf{v}|, \Omega) \ldots ,\qquad(83)$$

where:

$$d\Omega = \sin\theta\, d\theta d\phi \qquad(84)$$

is the usual differential for the solid angle in spherical coordinates. Then we have for the Boltzmann collision integral:

$$f_I(\mathbf{x}, \mathbf{v}, t) = \int d^3 v_1 \int d\Omega\, \sigma(|\mathbf{v}_1 - \mathbf{v}|, \Omega)\, |\mathbf{v}_1 - \mathbf{v}|\, (f' f'_1 - f f_1) . \qquad(85)$$

7.2.5
Collisional Invariants

It is important to consider momentum integrals of the collision integral defined by:

$$M_\chi(\mathbf{x}, t) \equiv \int d^3 v\, \chi(\mathbf{v}) f_I(\mathbf{x}, \mathbf{v}, t)$$

$$= \int d^3 v\, \chi(\mathbf{v}) \int d^3 v_1 \int d\Omega\, \sigma(|\mathbf{v}_1 - \mathbf{v}|, \Omega)\, |\mathbf{v}_1 - \mathbf{v}|\, (f' f'_1 - f f_1) . \quad(86)$$

If we exchange $\mathbf{v}$ and $\mathbf{v}_1$ in the integral we have:

$$M_\chi(\mathbf{x}, t) = \int d^3 v_1\, \chi(\mathbf{v}_1) \int d^3 v$$
$$\times \int d\Omega\, \sigma(|\mathbf{v} - \mathbf{v}_1|, \Omega) |\mathbf{v} - \mathbf{v}_1|\, (f'_1 f' - f_1 f) . \qquad(87)$$

We see that the integrand is the same as before the exchange except $\chi(\mathbf{v}) \to \chi(\mathbf{v}_1)$. Next we make the change of variables $\mathbf{v} \to \mathbf{v}'$ and $\mathbf{v}_1 \to \mathbf{v}'_1$ in Eq. (86):

$$M_\chi(\mathbf{x}, t) = \int d^3 v'\, \chi(\mathbf{v}') \int d^3 v'_1$$

$$\times \int d\Omega\, \sigma(|\mathbf{v}'_1 - \mathbf{v}'|, \Omega)\, |\mathbf{v}'_1 - \mathbf{v}'|\, (ff_1 - f'f'_1) \quad . \tag{88}$$

Remembering that $d^3v'd^3v'_1 = d^3v d^3v_1$ and $|\mathbf{v}'_1 - \mathbf{v}'| = |\mathbf{v}_1 - \mathbf{v}|$ we have:

$$M_\chi(\mathbf{x}, t) = -\int d^3v\, \chi(\mathbf{v}') \int d^3v_1$$
$$\times \int d\Omega\, \sigma(|\mathbf{v}_1 - \mathbf{v}|, \Omega)\, |\mathbf{v}_1 - \mathbf{v}|\, (f'f'_1 - ff_1) \quad . \tag{89}$$

The final change of variables of interest is $\mathbf{v} \to \mathbf{v}'_1$ and $\mathbf{v}_1 \to \mathbf{v}'$:

$$M_\chi(\mathbf{x}, t) = \int d^3v'_1\, \chi(\mathbf{v}'_1) \int d^3v'$$
$$\times \int d\Omega\, \sigma(|\mathbf{v}' - \mathbf{v}'_1|, \Omega)\, |\mathbf{v}' - \mathbf{v}'_1|\, (f_1 f - f'_1 f') \quad . \tag{90}$$

Using the same relations as used in the last transformation this reduces to:

$$M_\chi(\mathbf{x}, t) = -\int d^3v\, \chi(\mathbf{v}'_1) \int d^3v_1$$
$$\times \int d\Omega\, \sigma(|\mathbf{v}_1 - \mathbf{v}|, \Omega)\, |\mathbf{v}_1 - \mathbf{v}|\, (f'f'_1 - ff_1) \quad . \tag{91}$$

We find a key result when we add the four equivalent results and divide by 4:

$$M_\chi(\mathbf{x}, t) = \frac{1}{4} \int d^3v d^3v_1 \int d\Omega \sigma(|\mathbf{v}_1 - \mathbf{v}|, \Omega)\, |\mathbf{v}_1 - \mathbf{v}|$$
$$\times [\chi + \chi_1 - \chi' - \chi'_1]\, (f'f'_1 - ff_1) \quad . \tag{92}$$

This tells us that if we have a function of velocity that satisfies

$$\chi(\mathbf{v}) + \chi(\mathbf{v}_1) = \chi(\mathbf{v}') + \chi(\mathbf{v}'_1) \quad , \tag{93}$$

then the matrix element vanishes:

$$M_\chi(\mathbf{x}, t) = 0 \quad . \tag{94}$$

Clearly such *collisional invariants* correspond to conservation laws:

$$\chi(\mathbf{v}) = 1 \quad \text{conservation of particle number} \tag{95}$$
$$\chi(\mathbf{v}) = \mathbf{v} \quad \text{conservation of momentum} \tag{96}$$
$$\chi(\mathbf{v}) = v^2 \quad \text{conservation of kinetic energy} \quad . \tag{97}$$

Returning to the Boltzmann equation given by Eq. (82) we see that we obtain the hydrodynamic equations if we multiply by $\chi_i(\mathbf{p})$ and integrate over all $\mathbf{p}$. For example with $\chi_1(\mathbf{p}) = 1$, we have:

$$\frac{\partial n}{\partial t} = -\vec{\nabla} \cdot \frac{\mathbf{g}}{m} \quad , \tag{98}$$

which is just the continuity equation for particle number. Similarly, for conservation of momentum:

$$\frac{\partial g_i}{\partial t} = -\int d^3p\, p_i \sum_{j=1}^{3} v_j \nabla_j f(\mathbf{x}, \mathbf{p}, t) \tag{99}$$

$$= -\sum_{j=1}^{3} \nabla_j \sigma_{ij}, \tag{100}$$

where:

$$\sigma_{ij} = \int d^3p\, \frac{p_i p_j}{m} f(\mathbf{x}, \mathbf{p}, t) \tag{101}$$

is the kinetic part of the stress tensor. Finally, choosing $\chi(v) = \frac{m}{2}v^2$ and integrating, we obtain:

$$\frac{\partial \varepsilon_K}{\partial t} = -\vec{\nabla} \cdot \mathbf{J}_K, \tag{102}$$

where the kinetic energy current is given by:

$$\mathbf{J}_K = \int d^3p\, \varepsilon_0(p) \frac{\mathbf{p}}{m} f(\mathbf{x}, \mathbf{p}, t). \tag{103}$$

As we shall see, in traditional kinetic theory we extract transport coefficients using the currents σ_{ij} and $\mathbf{J}_K$.

7.2.6
Approach to Equilibrium

Boltzmann developed an ingenious argument [11] to show that the Boltzmann equation drives an arbitrary velocity distribution to the Maxwell velocity distribution. To begin, we introduce a function similar to the entropy function defined in equilibrium. This is the Boltzmann H-function [12]:

$$H(t) = \int d^3v\, f(\mathbf{v}, t) \ln f(\mathbf{v}, t), \tag{104}$$

where for simplicity we assume the system is uniform in space and f is independent of $\mathbf{x}$. Now take the time derivative of H:

$$\frac{dH(t)}{dt} = \int d^3v \left(\frac{\partial f}{\partial t}(1 + \ln f(\mathbf{v}, t)) \right). \tag{105}$$

Next use the Boltzmann equation, Eq. (82) absent the gradient term, to obtain:

$$\frac{dH(t)}{dt} = \int d^3v (1 + \ln f(\mathbf{v}, t)) f_I(\mathbf{v}, t) = M_{\chi_H}(t) \tag{106}$$

where we have a matrix element of the collision integral, using the notation introduced in Eq. (86), with:

$$\chi_H = 1 + \ln f(\mathbf{v}, t) \ . \tag{107}$$

If we then use the result Eq. (92) with $\chi = \chi_H$ we have:

$$\begin{aligned}\frac{dH(t)}{dt} &= \frac{1}{4} \int d^3v d^3v_1 \int d\Omega \, \sigma(|\mathbf{v}_1 - \mathbf{v}|, \Omega) \, |\mathbf{v}_1 - \mathbf{v}| \, (f'f_1' - ff_1) \\ &\quad \times \left[1 + \ln f + 1 + \ln f_1 - 1 - \ln f' - 1 - \ln f_1'\right] \\ &= \frac{1}{4} \int d^3v d^3v_1 \int d\Omega \, \sigma(|\mathbf{v}_1 - \mathbf{v}|, \Omega) \, |\mathbf{v}_1 - \mathbf{v}| \\ &\quad \times (f'f_1' - ff_1) \ln\left(\frac{ff_1}{f'f_1'}\right) \ . \end{aligned} \tag{108}$$

Consider the inequality (see Problem 7.8):

$$(x - y) \ln\left(\frac{y}{x}\right) \leq 0 \ , \tag{109}$$

where the equality holds only for $x = y$. Let us choose $x = f'f_1'$ and $y = ff_1$ in Eq. (108). Then, since the rest of the integral in Eq. (108) is manifestly positive, we have the inequality:

$$\frac{dH(t)}{dt} \leq 0 \ . \tag{110}$$

Thus $H(t)$ decreases monotonically with time, and we can conclude, independent of the initial distribution function, the system will be driven to a final state where the distribution function satisfies:

$$f_0(\mathbf{v}) f_0(\mathbf{v}_1) = f_0(\mathbf{v}') f_0(\mathbf{v}_1') \ ; \tag{111}$$

we have added the subscript on f to indicate the distribution corresponds to the stationary state. Taking the logarithm, we have:

$$\ln f_0(\mathbf{v}) + \ln f_0(\mathbf{v}_1) = \ln f_0(\mathbf{v}') + \ln f_0(\mathbf{v}_1') \ , \tag{112}$$

which is of the form of a conservation law like Eq. (93).

The general solution to Eq. (112) is a sum of all of the invariants:

$$\begin{aligned}\ln f_0(\mathbf{v}) &= a_1 \chi_1(\mathbf{v}) + a_2 \chi_2(\mathbf{v}) + a_3 \chi_3(\mathbf{v}) + a_4 \chi_4(\mathbf{v}) + a_5 \chi_5(\mathbf{v}) \\ &= a + \mathbf{b} \cdot \mathbf{v} + c v^2 \ , \end{aligned} \tag{113}$$

where we have the five constants a, $\mathbf{b}$, and c to determine. These are determined by the conditions:

$$\int d^3v f_0(\mathbf{v}) = n \tag{114}$$

$$\int d^3v\, \mathbf{v} f_0(\mathbf{v}) = n\mathbf{u} \tag{115}$$

$$\int d^3v\, \frac{m}{2}(\mathbf{v}-\mathbf{u})^2 f_0(\mathbf{v}) = n\mathbf{u}kT\ , \tag{116}$$

where the density n, average velocity $\mathbf{u}$ and temperature T may be functions of position. It is easy to show that the general solution in equilibrium is of the form:

$$f_0(\mathbf{v}) = n\left(\frac{m}{2\pi kT}\right)^{3/2} e^{-\beta\frac{m}{2}(\mathbf{v}-\mathbf{u})^2}\ . \tag{117}$$

7.2.7
Linearized Boltzmann Collision Integral

In most situations where we can make progress, we must assume that we are near equilibrium. Thus we make the assumption that the singlet distribution function is close to its equilibrium value. Thus we write:

$$f(\mathbf{x},\mathbf{p},t) = f_0(\mathbf{x},\mathbf{p}) + g(\mathbf{x},\mathbf{p},t)\ , \tag{118}$$

where g is treated as small. We can then expand the Boltzmann collision integral in powers of g:

$$\begin{aligned} f(\mathbf{x},\mathbf{v},t)_I &= \int d^3v_1 \int d\Omega\, \sigma(|\mathbf{v}_1-\mathbf{v}|,\Omega)\, |\mathbf{v}_1-\mathbf{v}|\, (f'f_1' - ff_1)\\ &= \int d^3v_1 \int d\Omega\, \sigma(|\mathbf{v}_1-\mathbf{v}|,\Omega)\, |\mathbf{v}_1-\mathbf{v}|\\ &\quad \times [(f_0'+g')(f_{0,1}'+g_1') - (f_0+g)(f_{0,1}+g_1)]\end{aligned} \tag{119}$$

$$\begin{aligned} &= \int d^3v_1 \int d\Omega\, \sigma(|\mathbf{v}_1-\mathbf{v}|,\Omega)\, |\mathbf{v}_1-\mathbf{v}|\\ &\quad \times [(f_0'f_{0,1}' - f_0 f_{0,1}) + g'f_{0,1}' + g_1'f_0' - gf_{0,1} - g_1 f_0 + \ldots]\end{aligned} \tag{120}$$

The zeroth-order term in g vanishes due to Eq. (111) and at linear order we have:

$$\begin{aligned} f(\mathbf{x},\mathbf{v},t)_I &= \int d^3v_1 \int d\Omega\, \sigma(|\mathbf{v}_1-\mathbf{v}|,\Omega)\, |\mathbf{v}_1-\mathbf{v}| \int d^3v_2\\ &\quad \times [f_{0,1}'\delta(\mathbf{v}_2-\mathbf{v}') + f_0'\delta(\mathbf{v}_2-\mathbf{v}_1') - f_{0,1}\delta(\mathbf{v}_2-\mathbf{v})\\ &\quad\ - f_0\delta(\mathbf{v}_2-\mathbf{v}_1)]\, g(\mathbf{x},\mathbf{v}_2,t)\end{aligned} \tag{121}$$

$$\equiv \int d^3v_2\, K_B(\mathbf{v},\mathbf{v}_2) g(\mathbf{x},\mathbf{v}_2,t)\ , \tag{122}$$

where the linearized Boltzmann collision integral is given by:

$$K_B(\mathbf{v},\mathbf{v}_2) = \int d^3v_1 \int d\Omega\, \sigma(|\mathbf{v}_1-\mathbf{v}|,\Omega)\, |\mathbf{v}_1-\mathbf{v}|$$

$$\times \left[f'_{0,1} \delta(\mathbf{v}_2 - \mathbf{v}') + f'_0 \delta(\mathbf{v}_2 - \mathbf{v}'_1) - f_{0,1} \delta(\mathbf{v}_2 - \mathbf{v}) \right.$$
$$\left. - f_0 \delta(\mathbf{v}_2 - \mathbf{v}_1) \right] \, . \tag{123}$$

It is useful to consider the quantity:

$$K_B(\mathbf{v}, \mathbf{v}_2) f_0(\mathbf{v}_2) = \int d^3 v_1 \int d\Omega \, \sigma(|\mathbf{v}_1 - \mathbf{v}|, \Omega) \, |\mathbf{v}_1 - \mathbf{v}|$$
$$\times \left[f'_{0,1} \delta(\mathbf{v}_2 - \mathbf{v}') + f'_0 \delta(\mathbf{v}_2 - \mathbf{v}'_1) - f_{0,1} \delta(\mathbf{v}_2 - \mathbf{v}) \right.$$
$$\left. - f_0 \delta(\mathbf{v}_2 - \mathbf{v}_1) \right] f_0(\mathbf{v}_2) \, . \tag{124}$$

Using $f'_0 f'_{0,1} = f_0 f_{0,1}$ and the δ-functions we find:

$$K_B(\mathbf{v}, \mathbf{v}_2) f_0(\mathbf{v}_2) = \int d^3 v_1 \int d\Omega \, \sigma(|\mathbf{v}_1 - \mathbf{v}|, \Omega) \, |\mathbf{v}_1 - \mathbf{v}|$$
$$\times f_0 f_{0,1} \left[\delta(\mathbf{v}_2 - \mathbf{v}') + \delta(\mathbf{v}_2 - \mathbf{v}'_1) - \delta(\mathbf{v}_2 - \mathbf{v}) \right.$$
$$\left. - \delta(\mathbf{v}_2 - \mathbf{v}_1) \right] \, . \tag{125}$$

It is useful to show that $K_B(\mathbf{v}, \mathbf{v}_2) f_0(\mathbf{v}_2)$ is symmetric under exchange $\mathbf{v} \leftrightarrow \mathbf{v}_2$. We focus on the first of four contributions to K_B in Eq. (125):

$$K_B^{(1)}(\mathbf{v}, \mathbf{v}_2) f_0(\mathbf{v}_2) = \int d^3 v_1 \int d\Omega \, \sigma(|\mathbf{v}_1 - \mathbf{v}|, \Omega) |\mathbf{v}_1 - \mathbf{v}|$$
$$\times f_0(\mathbf{v}) f_0(\mathbf{v}_1) \delta(\mathbf{v}_2 - \mathbf{v}') \, . \tag{126}$$

Inserting the identity:

$$\int d^3 v_3 \delta(\mathbf{v}_3 - \mathbf{v}'_1) = 1 \, , \tag{127}$$

we have:

$$K_B^{(1)}(\mathbf{v}, \mathbf{v}_2) f_0(\mathbf{v}_2) = \int d^3 v_1 \int d^3 v_3 \int d\Omega \, \sigma(|\mathbf{v}_1 - \mathbf{v}|, \Omega) \, |\mathbf{v}_1 - \mathbf{v}|$$
$$\times f_0(\mathbf{v}) f_0(\mathbf{v}_1) \delta(\mathbf{v}_2 - \mathbf{v}') \delta(\mathbf{v}_3 - \mathbf{v}'_1)$$
$$= \int d^3 v_1 \int d^3 v_3 \int d\Omega \, \sigma(|\mathbf{v}'_1 - \mathbf{v}'|, \Omega) \, |\mathbf{v}'_1 - \mathbf{v}'| f_0(\mathbf{v}')$$
$$\times f_0(\mathbf{v}'_1) \delta(\mathbf{v}_2 - \mathbf{v}') \delta(\mathbf{v}_3 - \mathbf{v}'_1)$$
$$= \int d^3 v_1 \int d^3 v_3 \int d\Omega \, \sigma(|\mathbf{v}_3 - \mathbf{v}_2|, \Omega)$$
$$\times |\mathbf{v}_3 - \mathbf{v}_2| f_0(\mathbf{v}_3) f_0(\mathbf{v}_2) \delta(\mathbf{v}_2 - \mathbf{v}') \delta(\mathbf{v}_3 - \mathbf{v}'_1) \tag{128}$$

The key observation is that:

$$\delta(\mathbf{v}_2 - \mathbf{v}') \delta(\mathbf{v}_3 - \mathbf{v}'_1) = \delta(\mathbf{v}'_2 - \mathbf{v}) \delta(\mathbf{v}'_3 - \mathbf{v}_1) \, , \tag{129}$$

where we have introduced the collision pair $(\mathbf{v}_2, \mathbf{v}_3) \to (\mathbf{v}'_2, \mathbf{v}'_3)$. We then have:

$$K_B^{(1)}(\mathbf{v}, \mathbf{v}_2) f_0(\mathbf{v}_2) = \int d^3 v_1 \int d^3 v_3 \int d\Omega\, \sigma(|\mathbf{v}_3 - \mathbf{v}_2|, \Omega)\, |\mathbf{v}_3 - \mathbf{v}_2| \\ \times f_0(\mathbf{v}_3) f_0(\mathbf{v}_2) \delta(\mathbf{v}'_2 - \mathbf{v}) \delta(\mathbf{v}'_3 - \mathbf{v}_1) \ . \tag{130}$$

Next, do the integral over $\mathbf{v}_1$ with the result:

$$K_B^{(1)}(\mathbf{v}, \mathbf{v}_2) f_0(\mathbf{v}_2) = \int d^3 v_3 \int d\Omega\, \sigma(|\mathbf{v}_3 - \mathbf{v}_2|, \Omega) \\ \times |\mathbf{v}_3 - \mathbf{v}_2| f_0(\mathbf{v}_3) f_0(\mathbf{v}_2) \delta(\mathbf{v}'_2 - \mathbf{v}) \ . \tag{131}$$

Comparing this result with Eq. (126) we see that we have the symmetry:

$$K_B^{(1)}(\mathbf{v}, \mathbf{v}_2) f_0(\mathbf{v}_2) = K_B^{(1)}(\mathbf{v}_2, \mathbf{v}) f_0(\mathbf{v}) \ . \tag{132}$$

It is left to Problem 7.9, using similar manipulations, to show that the three other contributions, and the full K_B, also obey the symmetry.

7.2.8
Kinetic Models

It is very difficult to make progress solving the linearized Boltzmann equation. Here we describe a practical method for both extracting the physics contained in the linearized Boltzmann equation and for establishing a systematic accurate solution. We focus here on the collisional contribution to the kinetic equation, which can be written in the general form:

$$f_I(\mathbf{x}, \mathbf{p}, t) = -\int d^3 p'\, K_B(\mathbf{p}, \mathbf{p}') f(\mathbf{x}, \mathbf{p}', t) \ , \tag{133}$$

where $K_B(\mathbf{p}, \mathbf{p}')$ is the linearized Boltzmann collision operator. It is an integral operator. For our purposes here, this operator has two sets of properties. It satisfies the symmetry:

$$\Gamma_B(\mathbf{p}, \mathbf{p}') = K_B(\mathbf{p}, \mathbf{p}') f_0(\mathbf{p}') = K_B(\mathbf{p}', \mathbf{p}) f_0(\mathbf{p}) \tag{134}$$

and certain momentum integrals, related to the conservation laws, have zero matrix elements:

$$\int d^3 p\, \psi_i(\mathbf{p}) K_B(\mathbf{p}, \mathbf{p}') = 0 \ . \tag{135}$$

If the particles of interest satisfy conservation of particles, momentum and energy then Eq. (135) is true for the unnormalized set:

$$\psi_i = \{1, \mathbf{p}, \varepsilon_0(p)\} \ . \tag{136}$$

In the case of electrical transport there is positive charge in addition to the carriers and typically in solids there is scattering from the lattice. Thus, the only conservation law for the charge carriers is conservation of particle number.

K_B is a rather complicated [13] function of momenta and depends on the details of the interactions in the system. Here, we discuss a method for extracting information from the associated kinetic equation that depends on the general properties listed above. This is the method of kinetic modeling [14].

The first step in this approach is to construct a set of momentum polynomials, $\psi_i(\mathbf{p})$, that are complete and orthonormal with respect to the Maxwell distribution,

$$\Phi(\mathbf{p}) = \frac{e^{-\beta p^2/2m}}{(2\pi mkT)^{3/2}} , \qquad (137)$$

as a weight. Orthonormality means:

$$\int d^3p\, \Phi(\mathbf{p})\psi_i(\mathbf{p})\psi_j(\mathbf{p}) = \delta_{ij} , \qquad (138)$$

while completeness is assumed to take the form:

$$\sum_{i=1}^{\infty} \psi_i(\mathbf{p})\psi_i(\mathbf{p}')\Phi(\mathbf{p})\Phi(\mathbf{p}') = \delta(\mathbf{p}-\mathbf{p}')\Phi(\mathbf{p}) . \qquad (139)$$

The polynomials are labeled by a set of integers and clearly can be constructed [15] in terms of a product of Hermit polynomials. Ideally the $\psi_i(\mathbf{p})$ would be eigenfunctions of K_B. Generally we do not know [16] these eigenfunctions.

It is shown in Problem 7.10 that we can choose as the first five such polynomials the set:

$$\psi_1 = 1 \qquad (140)$$

$$\psi_2 = p_x/p_0 \qquad (141)$$

$$\psi_3 = p_y/p_0 \qquad (142)$$

$$\psi_4 = p_z/p_0 \qquad (143)$$

$$\psi_5 = \sqrt{\frac{2}{3}}(\varepsilon_0(p) - \bar{\varepsilon}_0)/(k_B T) , \qquad (144)$$

where:

$$p_0 = \sqrt{mk_B T} \qquad (145)$$

and:
$$\bar{\varepsilon}_0 = \frac{3}{2}k_B T \tag{146}$$

Clearly one can continue on constructing higher-order polynomials.

In kinetic modeling one assumes that one can expand the collision integral in terms of these polynomials:

$$K_B(\mathbf{p},\mathbf{p}')\Phi(\mathbf{p}') = \sum_{i,j=1}^{\infty} k_{ij}\psi_i(\mathbf{p})\psi_j(\mathbf{p}')\Phi(\mathbf{p})\Phi(\mathbf{p}') \ , \tag{147}$$

where we have used the symmetry given by Eq. (134). We assume for a given problem (interaction potential) that we can evaluate the set of matrix elements:

$$k_{ij} = \int d^3p\, d^3p'\, \psi_i(\mathbf{p}) K_B(\mathbf{p},\mathbf{p}')\Phi(\mathbf{p})\psi_j(\mathbf{p}') \ . \tag{148}$$

This may be very difficult in practice.

The key assumption in the kinetic modeling approach is that Eq. (147) can be divided up into a block with $i, j \leq N$ and the rest:

$$\begin{aligned}K_B(\mathbf{p},\mathbf{p}')\Phi(\mathbf{p}) &= \sum_{i=1}^{N}\sum_{j=1}^{N} k_{ij}\psi_i(\mathbf{p})\psi_j(\mathbf{p}')\Phi(\mathbf{p})\Phi(\mathbf{p}') \\ &+ \left(\sum_{i=1}^{N}\sum_{j=N+1}^{\infty} + \sum_{i=N+1}^{\infty}\sum_{j=1}^{N} + \sum_{i=N+1}^{\infty}\sum_{j=N+1}^{\infty}\right) \\ &\times k_{ij}\psi_i(\mathbf{p})\psi_j(\mathbf{p}')\Phi(\mathbf{p})\Phi(\mathbf{p}') \ . \end{aligned} \tag{149}$$

We next make two rather bold assumptions or approximations:

- If either i or j or both are greater than N then we assume:

$$k_{ij} = k_i \delta_{ij} \ . \tag{150}$$

- For $i > N$ then k_i is independent of i:

$$k_i = k_{N+1,N+1} \ . \tag{151}$$

Applying these assumptions to the expansion, Eq. (149), for the collision integral we obtain:

$$\begin{aligned}K_B(\mathbf{p},\mathbf{p}')\Phi(\mathbf{p}) &= \sum_{i=1}^{N}\sum_{j=1}^{N} k_{ij}\psi_i(\mathbf{p})\psi_j(\mathbf{p}')\Phi(\mathbf{p})\Phi(\mathbf{p}') \\ &+ \sum_{i=N+1}^{\infty}\sum_{j=N+1}^{\infty} \delta_{ij} k_{N+1,N+1}\psi_i(\mathbf{p})\psi_j(\mathbf{p}')\Phi(\mathbf{p})\Phi(\mathbf{p}')\end{aligned}$$

$$\begin{aligned}
&= \sum_{i=1}^{N}\sum_{j=1}^{N} k_{ij}\psi_i(\mathbf{p})\psi_j(\mathbf{p}')\Phi(\mathbf{p})\Phi(\mathbf{p}') \\
&\quad + k_{N+1,N+1}\sum_{i=N+1}^{\infty}\psi_i(\mathbf{p})\psi_i(\mathbf{p}')\Phi(\mathbf{p})\Phi(\mathbf{p}') \\
&= \sum_{i=1}^{N}\sum_{j=1}^{N} k_{ij}\psi_i(\mathbf{p})\psi_j(\mathbf{p}')\Phi(\mathbf{p})\Phi(\mathbf{p}') \\
&\quad + k_{N+1,N+1}\left(\sum_{i=1}^{\infty}-\sum_{i=1}^{N}\right)\psi_i(\mathbf{p})\psi_i(\mathbf{p}')\Phi(\mathbf{p})\Phi(\mathbf{p}') \\
&= \sum_{i=1}^{N}\sum_{j=1}^{N}\left(k_{ij}-k_{N+1,N+1}\delta_{ij}\right)\psi_i(\mathbf{p})\psi_j(\mathbf{p}')\Phi(\mathbf{p})\Phi(\mathbf{p}') \\
&\quad + k_{N+1,N+1}\delta(\mathbf{p}-\mathbf{p}')\Phi(\mathbf{p}) \ ,
\end{aligned} \qquad (152)$$

where we have used the completeness relation Eq. (139). This defines the collision integral for the Nth-order collision model. If we define:

$$\gamma_{ij} = k_{ij} - k_{N+1,N+1}\delta_{ij} \qquad (153)$$

then the Nth-order model is given by:

$$\begin{aligned}
K_B(\mathbf{p},\mathbf{p}')\Phi(\mathbf{p}') &= \sum_{i=1}^{N}\sum_{j=1}^{N}\gamma_{ij}\psi_i(\mathbf{p})\psi_j(\mathbf{p}')\Phi(\mathbf{p})\Phi(\mathbf{p}') \\
&\quad + k_{N+1,N+1}\delta(\mathbf{p}-\mathbf{p}')\Phi(\mathbf{p}) \ .
\end{aligned} \qquad (154)$$

7.2.9
Single-Relaxation-Time Approximation

The simplest model is given by $N = 1$ where the collision operator is given by:

$$\begin{aligned}
K_B(\mathbf{p},\mathbf{p}')\Phi(\mathbf{p}') &= (k_{11}-k_{22})\psi_1(\mathbf{p})\psi_1(\mathbf{p}')\Phi(\mathbf{p})\Phi(\mathbf{p}') \\
&\quad + k_{22}\delta(\mathbf{p}-\mathbf{p}')\Phi(\mathbf{p})
\end{aligned} \qquad (155)$$

or:

$$K_B(\mathbf{p},\mathbf{p}') = (k_{11}-k_{22})\psi_1(\mathbf{p})\Phi(\mathbf{p})\psi_1(\mathbf{p}') + k_{22}\delta(\mathbf{p}-\mathbf{p}') \ . \qquad (156)$$

The Boltzmann collision integral, Eq. (133), then takes the form:

$$\begin{aligned}
f_I(\mathbf{x},\mathbf{p},t) &= -\int d^3p' K_B(\mathbf{p},\mathbf{p}')f(\mathbf{x},\mathbf{p}',t) \\
&= -(k_{11}-k_{22})\psi_1(\mathbf{p})\Phi(\mathbf{p})\int d^3p'\,\psi_1(\mathbf{p}')f(\mathbf{x},\mathbf{p}',t)
\end{aligned}$$

$$-k_{22} f(\mathbf{x}, \mathbf{p}, t) \ . \tag{157}$$

Remember that $\psi_1(\mathbf{p}) = 1$. If we enforce conservation of particles, then $k_{11} = 0$ and, remembering the normalization:

$$\int d^3 p' \psi_1(\mathbf{p}') f(\mathbf{x}, \mathbf{p}', t) = n(\mathbf{x}, t) \ , \tag{158}$$

we have for the collisional contribution:

$$f_I(\mathbf{x}, \mathbf{p}, t) = -\frac{1}{\tau} [f(\mathbf{x}, \mathbf{p}, t) - n(\mathbf{x}, t) \Phi(\mathbf{p})] \ , \tag{159}$$

where we have identified the relaxation time $\tau^{-1} = k_{22}$. This approximation, including driving forces, take the form of a kinetic equation:

$$\left[\frac{\partial}{\partial t} + \frac{\mathbf{p} \cdot \nabla_x}{m} + F_E(\mathbf{x}, \mathbf{p}, t) \cdot \nabla_p \right] f(\mathbf{x}, \mathbf{p}, t)$$
$$= -\frac{1}{\tau} [f(\mathbf{x}, \mathbf{p}, t) - n(\mathbf{x}) \Phi(\mathbf{p})] \ . \tag{160}$$

We will look at this equation from various different points of view. First we look at initial-value problems.

Let us look at the simple example where we have a uniform, $n(\mathbf{x}) = n$, initial configuration that depends only on momentum:

$$f(\mathbf{x}, \mathbf{p}, t = 0) = f_{in}(\mathbf{p}) \tag{161}$$

Assume the system is in zero external field, $F_E = 0$. In this case, because the initial distribution is uniform in space, $f_0(p) = n\Phi(p)$, the kinetic equation takes the form:

$$\frac{\partial}{\partial t} f(\mathbf{p}, t) = -\frac{1}{\tau} [f(\mathbf{p}, t) - f_0(\mathbf{p})] \ . \tag{162}$$

We can look for a solution of the form:

$$f(\mathbf{p}, t) = e^{-t/\tau} \phi(\mathbf{p}, t) \ , \tag{163}$$

where:

$$\phi(\mathbf{p}, t = 0) = f_{in}(\mathbf{p}) \ . \tag{164}$$

Inserting Eq. (163) into Eq. (162) we obtain:

$$-\frac{1}{\tau} f(\mathbf{x}, \mathbf{p}, t) + e^{-t/\tau} \frac{\partial}{\partial t} \phi(\mathbf{p}, t) = -\frac{1}{\tau} [f(\mathbf{x}, \mathbf{p}, t) - f_0(\mathbf{p})] \ . \tag{165}$$

Canceling terms, this reduces to:

$$\frac{\partial}{\partial t} \phi(\mathbf{p}, t) = e^{t/\tau} \frac{1}{\tau} f_0(\mathbf{p}) \ . \tag{166}$$

Integrating over time we obtain:

$$\phi(\mathbf{p},t) - \phi(\mathbf{p},t=0) = \int_0^t \frac{dt'}{\tau} e^{t'/\tau} f_0(\mathbf{p}) \,, \tag{167}$$

which leads to:

$$\phi(\mathbf{p},t) = f_{\text{in}}(\mathbf{p}) + (e^{t/\tau} - 1) f_0(\mathbf{p}) \,. \tag{168}$$

Putting this back into Eq. (163), we have the final solution:

$$f(\mathbf{p},t) = e^{-t/\tau} f_{\text{in}}(\mathbf{p}) + (1 - e^{-t/\tau}) f_0(\mathbf{p}) \,. \tag{169}$$

As time evolves, the average phase-space density evolves from the initial value $f_{\text{in}}(\mathbf{p})$ to the final equilibrium form $f_0(\mathbf{p})$ on a time scale determined by the relaxation time τ.

Let us next consider the initial-value problem where we must solve the kinetic equation:

$$\left[\frac{\partial}{\partial t} + \frac{\mathbf{p} \cdot \nabla_x}{m} \right] f(\mathbf{x},\mathbf{p},t) = -\frac{1}{\tau} [f(\mathbf{x},\mathbf{p},t) - n(\mathbf{x},t)\Phi(\mathbf{p})] \tag{170}$$

subject to the initial condition:

$$f(\mathbf{x},\mathbf{p},t=t_0) = n_0(\mathbf{x})\Phi(\mathbf{p}) \,, \tag{171}$$

where $n_0(\mathbf{x})$ is given. We easily solve this problem using the Fourier–Laplace transform:

$$f(\mathbf{k},\mathbf{p},z) = -i \int_0^\infty dt\, e^{izt} \int d^3x\, e^{-i\mathbf{k}\cdot\mathbf{x}} f(\mathbf{x},\mathbf{p},t) \,. \tag{172}$$

Taking the Fourier–Laplace transform of the kinetic equation, Eq. (170), we find after some rearrangements:

$$\left[z - \frac{\mathbf{k} \cdot \mathbf{p}}{m} + i\tau^{-1} \right] f(\mathbf{k},\mathbf{p},z) = n_0(\mathbf{k})\Phi(\mathbf{p}) + i\tau^{-1} n(\mathbf{k},z)\Phi(\mathbf{p})$$

$$= \left(n_0(\mathbf{k}) + i\tau^{-1} n(\mathbf{k},z) \right) \Phi(\mathbf{p}) \,. \tag{173}$$

Dividing by $z - \frac{\mathbf{k}\cdot\mathbf{p}}{m} + i\tau^{-1}$, we have:

$$f(\mathbf{k},\mathbf{p},z) = \left(n_0(\mathbf{k}) + i\tau^{-1} n(\mathbf{k},z) \right) \frac{\Phi(\mathbf{p})}{z - \frac{\mathbf{k}\cdot\mathbf{p}}{m} + i\tau^{-1}} \,. \tag{174}$$

We obtain a *trap* for the unknown density by integrating over all momenta:

$$n(\mathbf{k},z) = \left(n_0(\mathbf{k}) + i\tau^{-1} n(\mathbf{k},z) \right) I(\mathbf{k},z) \,, \tag{175}$$

where:

$$I(\mathbf{k},z) = \int d^3p \frac{\Phi(\mathbf{p})}{z - \frac{\mathbf{k}\cdot\mathbf{p}}{m} + i\tau^{-1}} \quad . \tag{176}$$

Solving for the transformed density:

$$n(\mathbf{k},z) = n_0(\mathbf{k}) \frac{I(\mathbf{k},z)}{1 - i\tau^{-1}I(\mathbf{k},z)} \quad . \tag{177}$$

We can not invert this transform analytically but we can look at the relaxation of the density in the hydrodynamic limit. It is left as a problem, 7.11, to show that an expansion in powers of the wavenumber gives the result at second order:

$$I(\mathbf{k},z) = \frac{1}{z + i\tau^{-1}} + \frac{k_B T}{m} \frac{k^2}{[z + i\tau^{-1}]^3} + \cdots \quad . \tag{178}$$

Expanding the coefficient of $n_0(\mathbf{k})$ in Eq. (177) in powers of k we obtain:

$$n(\mathbf{k},z) = \frac{n_0(\mathbf{k})}{z}\left(1 + \frac{k_B T}{m} \frac{k^2}{z[z + i\tau^{-1}]}\right) + \cdots \quad . \tag{179}$$

To this order in k, we can rewrite this in the form:

$$n(\mathbf{k},z) = \frac{n_0(\mathbf{k})}{z - \frac{k_B T}{m} \frac{k^2}{[z+i\tau^{-1}]}} \quad . \tag{180}$$

We can write this in the form:

$$n(\mathbf{k},z) = \frac{n_0(\mathbf{k})}{z + iD(z)k^2} \quad , \tag{181}$$

where we have a frequency-dependent diffusion coefficient:

$$D(z) = \frac{k_B T}{m}\tau \frac{1}{1 - iz\tau} \quad . \tag{182}$$

In the low-frequency limit $z\tau \ll 1$, and we are in the diffusion limit where we can replace $D(z)$ with $D = D(0)$ and invert the Laplace transform to obtain:

$$n(\mathbf{k},t) = n_0(\mathbf{k})e^{-Dk^2 t} \quad , \tag{183}$$

where the diffusion constant is given by $D = \frac{k_B T}{m}\tau$.

7.2.10
Steady-State Solutions

Let us look next at the case of a *uniform* applied force. If the system and the associated average phase-space density is initially spatially uniform, then it will remain uniform and the kinetic equation takes the form:

$$\frac{\partial}{\partial t}f(\mathbf{p},t) = -\mathbf{F}(t)\cdot\nabla_p f(\mathbf{p},t) - \frac{1}{\tau}[f(\mathbf{p},t) - n\Phi(\mathbf{p})] \ . \tag{184}$$

Let us assume that the force is initially zero and the system is in equilibrium:

$$f(\mathbf{p}, t=0) = f_0(\mathbf{p}) \tag{185}$$

and we turn on the force in the fashion:

$$\mathbf{F}(t) = \left(1 - e^{-t/t_0}\right)\mathbf{F} \ . \tag{186}$$

Let us write:

$$g(\mathbf{p},t) = f(\mathbf{x},\mathbf{p},t) - f_0(\mathbf{p}) \tag{187}$$

with:

$$g(\mathbf{p}, t=0) = 0 \ . \tag{188}$$

The kinetic equation, Eq. (184), takes the form:

$$\frac{\partial}{\partial t}g(\mathbf{p},t) = -\mathbf{F}(t)\cdot\nabla_p[f_0(\mathbf{p}) + g(\mathbf{p},t)] - \frac{1}{\tau}g(\mathbf{p},t) \ . \tag{189}$$

Introduce:

$$g(\mathbf{p},t) = e^{-t/\tau}\phi(\mathbf{p},t) \ , \tag{190}$$

where:

$$\phi(\mathbf{p}, t=0) = g(\mathbf{p}, t=0) = 0 \ , \tag{191}$$

into the kinetic equation as before. Again we can cancel off the term proportional to τ^{-1} and obtain:

$$\frac{\partial}{\partial t}\phi(\mathbf{p},t) = -\mathbf{F}(t)\cdot\nabla_p\left[e^{t/\tau}f_0(\mathbf{p}) + \phi(\mathbf{p},t)\right] \ . \tag{192}$$

Let us first consider the case where the external force is weak and we can determine ϕ as a power-series expansion in $\mathbf{F}$. At first order in $\mathbf{F}$ we can drop the second-order contribution, $\phi \approx F$, in the kinetic equation to obtain:

$$\frac{\partial}{\partial t}\phi(\mathbf{p},t) = -\mathbf{F}(t)\cdot\nabla_p e^{t/\tau}f_0(\mathbf{p}) \ . \tag{193}$$

Remembering the initial condition, $\phi(p, t=0) = 0$, we integrate Eq. (193) to obtain:

$$\phi(\mathbf{p},t) = -\int_0^t dt' e^{t'/\tau}\mathbf{F}(t')\cdot\nabla_p f_0(\mathbf{p}) \ . \tag{194}$$

The average phase-space density is given by:

$$f(\mathbf{p},t) = f_0(\mathbf{p}) - \mathcal{F}(t) \cdot \nabla_p f_0(\mathbf{p}) ,\qquad(195)$$

where:

$$\mathcal{F}(t) = e^{-t/\tau} \int_0^t dt' e^{t'/\tau} \mathbf{F}(t') .\qquad(196)$$

It is left as a problem (Problem 7.12) to show that:

$$\mathcal{F}(t) = \left[\tau\left(1 - e^{-t/\tau}\right) - \left(\tau^{-1} - t_0^{-1}\right)\left(e^{-t/t_0} - e^{-t/\tau}\right)\right]\mathbf{F} .\qquad(197)$$

For short times we easily find:

$$\mathcal{F}(t) = \frac{1}{2}\frac{t^2}{t_0}\mathbf{F} + \dots ,\qquad(198)$$

while for long times $t \gg \tau, t_0$, there is exponential convergence to the constant result:

$$\lim_{t\to\infty} \mathcal{F}(t) = \tau\mathbf{F} \qquad(199)$$

and the long-time solution at linear order is given by:

$$g(\mathbf{p}) = f(\mathbf{p}) - f_0(\mathbf{p}) = -\tau\mathbf{F} \cdot \nabla_p f_0(\mathbf{p}) .\qquad(200)$$

This agrees with the steady-state solution of Eq. (184) where we drop the time-derivative and expand $f = f_0 + g + \dots$.

7.3
Traditional Transport Theory

7.3.1
Steady-State Currents

A key application of kinetic theory is to transport theory: the nonequilibrium steady-state currents driven by external forces. Currents of interest can be derived as momentum integrals of the singlet distribution function. The momentum current, for example, is given by:

$$\mathbf{g}(\mathbf{x},t) = \int d^3p\, \mathbf{p} f(\mathbf{p},t) .\qquad(201)$$

In this section we assume the system is spatially homogeneous and in steady state. In a charged system, where the momentum and energy of the electrons alone are not conserved, the charge current is given by:

$$\mathbf{J} = q \int d^3p \frac{\mathbf{p}}{m} f(\mathbf{p}) ,\qquad(202)$$

where the mobile particles carry charge q. For our model to make sense we must maintain overall charge neutrality. Thus there must be an inert background of *positive* charge. This is known as a *jellium* model [17].

Another current of interest is the kinetic energy current:

$$J_K = \int d^3p \, \varepsilon_0(\mathbf{p}) \frac{\mathbf{p}}{m} f(\mathbf{p}) \;, \tag{203}$$

where:

$$\varepsilon_0(p) = \frac{p^2}{2m} \;. \tag{204}$$

We will also be interested in the heat current defined by:

$$J_q = \int d^3p \, \tilde{\varepsilon}_0(p) \frac{\mathbf{p}}{m} f(\mathbf{p}) \;, \tag{205}$$

where:

$$\tilde{\varepsilon}_0(p) = \varepsilon_0(p) - \frac{3}{2} k_B T \;. \tag{206}$$

Note that we can write:

$$J_q = J_K - \frac{3}{2} \frac{k_B T}{q} \mathbf{J} \;. \tag{207}$$

The relevance of the heat current will become clearer when we discuss systems driven by temperature gradients.

Clearly, the average currents for a system in equilibrium, due to the symmetry $\Phi(\mathbf{p}) = \Phi(-\mathbf{p})$, are zero:

$$\langle \mathbf{J} \rangle_{EQ} = \langle \mathbf{J}_K \rangle_{EQ} = \langle \mathbf{J}_q \rangle_{EQ} = 0 \;. \tag{208}$$

These results are useful if we are interested in transport, where the average currents are driven by the applied force. Suppose some average current is given by:

$$J_X = \int d^3p \, X(p) \frac{\mathbf{p}}{m} f(\mathbf{p}) \;, \tag{209}$$

where $X(p) = X(-p)$. Then for long times and at linear order we find, using Eq. (200),

$$J_X(t) = \int d^3p \, X(p) \frac{\mathbf{p}}{m} \left[f_0(\mathbf{p}) - \tau \mathbf{F} \cdot \nabla_p f_0(\mathbf{p}) \right] \;.$$

The leading order contribution to the current vanishes by symmetry and the first-order contribution is given by by:

$$J_X^i = -\tau \sum_j F_j \int d^3p \, X(p) \frac{p_i}{m} \nabla_p^j f_0(\mathbf{p}) \;. \tag{210}$$

Integrating by parts we have:

$$J_X^i = \tau \sum_j F_j \int d^3p\, f_0(\mathbf{p}) \nabla_p^j \left(X(p) \frac{p_i}{m} \right) \ . \tag{211}$$

For a case of particular interest we consider the application of an electric field to a set of electrons, $\mathbf{F}_E = -e\mathbf{E}$, and look at the average electric current, $X(p) = -e$, then:

$$\begin{aligned} J^i &= \tau \sum_j (-eE_j) \int d^3p\, f_0(\mathbf{p}) \nabla_p^j \left((-e) \frac{p_i}{m} \right) \\ &= \frac{\tau e^2}{m} \sum_j E_j \delta_{ij} n \\ &= \sigma E_i \ , \end{aligned} \tag{212}$$

where we write things in the form of Ohm's law with the electrical conductivity given by:

$$\sigma = \frac{\tau e^2 n}{m} \ . \tag{213}$$

This is the Drude formula found in Chapters 2 and 4. This equation can be used to empirically determine the relaxation time τ. Using measured conductivities, one can estimate τ as in Section 2.1.5. τ is on the order of 10^{-14} s at room temperatures for metallic systems in the solid state.

It is appropriate to consider here the contrast between this method of determining a transport coefficient with the Green–Kubo approach we developed in detail in earlier chapters. Here we must solve the kinetic equation before we can identify the expression for the transport coefficient and there are various approximations we make along the way. In the last portion of this chapter we show how we can organize kinetic theory to be compatible with the Green–Kubo formulation.

We can also determine the average kinetic-energy current. With $X(p) = \mathbf{p}^2/2m$ we have in Eq. (211)

$$\begin{aligned} J_K^i &= \tau \sum_j F_j \int d^3p\, f_0(\mathbf{p}) \nabla_p^j \left(\frac{\mathbf{p}^2}{2m} \frac{p_i}{m} \right) \\ &= \tau \sum_j F_j \int d^3p\, f_0(\mathbf{p}) \left(\frac{\mathbf{p}^2}{2m^2} \delta_{ij} + \frac{p_i}{m^2} p_j \right) \\ &= \tau \sum_j F_j \int d^3p\, f_0(\mathbf{p}) \delta_{ij} \left(\frac{\mathbf{p}^2}{2m^2} + \frac{\mathbf{p}^2}{3m^2} \right) \\ &= \frac{\tau F_i}{m} \frac{5}{3} \int d^3p\, f_0(\mathbf{p}) \frac{\mathbf{p}^2}{2m} \ . \end{aligned} \tag{214}$$

We have in the momentum average:

$$\int d^3p \, f_0(\mathbf{p}) \frac{\mathbf{p}^2}{2m} = n \int d^3p \, \frac{\mathbf{p}^2}{2m} \Phi(\mathbf{p}) = n\frac{3}{2}kT \,, \tag{215}$$

using the equipartition theorem in the last step. We have then:

$$J_K^i = \frac{\tau F_i}{m} \frac{5}{2} nkT \,. \tag{216}$$

Assuming an electric field coupled to a set of conduction electrons we have the energy current:

$$J_K = -\frac{\tau e E}{m} \frac{5}{2} kT \,. \tag{217}$$

This says that an applied electric field induces a flow of heat in a system of conduction electrons. We will return to this later.

We can treat the nonlinear corrections to the linear response results just discussed. Let us focus on the stationary solution where $\mathbf{F}$ is a constant and one can drop the time derivative term in the kinetic equation to obtain the steady-state result:

$$0 = -\mathbf{F} \cdot \nabla_p f(\mathbf{p}) - \frac{1}{\tau}[f(\mathbf{p}) - f_0(\mathbf{p})] \,. \tag{218}$$

Let $\mathbf{p}_0 = \tau \mathbf{F}$ so we can write:

$$(1 + \mathbf{p}_0 \cdot \nabla_p) f(\mathbf{p}) = f_0(\mathbf{p}) \,. \tag{219}$$

With $\mathbf{p}_0 = p_0 \hat{z}$ then:

$$\left(1 + p_0 \frac{\partial}{\partial p_z}\right) f(\mathbf{p}) = f_0(\mathbf{p}) \,. \tag{220}$$

Next, we can assume:

$$f(\mathbf{p}) = e^{-p_z/p_0} \phi(\mathbf{p}) \,, \tag{221}$$

so that the kinetic equation, Eq. (220), is reduced to:

$$p_0 \frac{\partial}{\partial p_z} \phi(\mathbf{p}) = e^{p_z/p_0} f_0(\mathbf{p}) \,. \tag{222}$$

Integrating this equation:

$$\phi(\mathbf{p}) = \frac{1}{p_0} \int_{-\infty}^{p_z} dz \, e^{z/p_0} f_0(p_x, p_y, z) \tag{223}$$

and putting this back into Eq. (221) we obtain:

$$f(\mathbf{p}) = \frac{1}{p_0} \int_{-\infty}^{p_z} dz\, e^{(z-p_z)/p_0} f_0(p_x, p_y, z) \ . \qquad (224)$$

Letting $z = p_z + p_0 y$ we obtain:

$$f(\mathbf{p}) = \int_{-\infty}^{0} dy\, e^{y} f_0(p_x, p_y, p_z + p_0 y) \ . \qquad (225)$$

Let $y \to -y$ in the integral to obtain the final result:

$$f(\mathbf{p}) = \int_{0}^{\infty} dy\, e^{-y} f_0(p_x, p_y, p_z - \tau F y) \ . \qquad (226)$$

It is left to Problem 7.13 to check the normalization of $f(\mathbf{p})$, recover the linear response result for small $\mathbf{F}$, and find the nonlinear contributions to the electrical and heat currents.

7.3.2
Thermal Gradients

Besides the application of a direct mechanical force one can have thermal perturbations corresponding to a weak thermal gradient in the system. This is treated in the theory by using the idea of *local* equilibrium. This takes advantage of the fact that the steady-state solution of the nonlinear Boltzmann equation are of the form given by Eq. (117) where n, and T and $\mathbf{u}$ may be slowly varying functions of position. With a temperature that is slowly varying in space, we replace f_0 in the kinetic equation with:

$$f_0[\mathbf{p}, T(\mathbf{x})] = \frac{n}{[2\pi m k_B T(\mathbf{x})]^{3/2}} e^{-\beta(\mathbf{x}) \mathbf{p}^2/2m} \ . \qquad (227)$$

In the absence of applied external forces one has the kinetic equation:

$$\left(\frac{\partial}{\partial t} + \frac{\mathbf{p} \cdot \nabla_x}{m}\right) f(\mathbf{x}, \mathbf{p}, t) = -\frac{1}{\tau}[f(\mathbf{x}, \mathbf{p}, t) - f_0(\mathbf{p}, T(\mathbf{x}))] \ . \qquad (228)$$

Looking for a stationary solution, the kinetic equation reduces to:

$$\frac{\mathbf{p} \cdot \nabla_x}{m} f(\mathbf{x}, \mathbf{p}) = -\frac{1}{\tau}[f(\mathbf{x}, \mathbf{p}) - f_0(\mathbf{p}, T(\mathbf{x}))] \ . \qquad (229)$$

If we write:

$$g(\mathbf{x}, \mathbf{p}) = f(\mathbf{x}, \mathbf{p}) - f_0(\mathbf{p}, T(\mathbf{x})) \ , \qquad (230)$$

then the kinetic equation takes the form:

$$g(\mathbf{x}, \mathbf{p}) = -\tau \frac{\mathbf{p} \cdot \nabla_x}{m} f(\mathbf{x}, \mathbf{p}) \qquad (231)$$

and for weak gradients we can treat $g(\mathbf{x}, \mathbf{p})$ as a perturbation. We have then, to lowest order:

$$g(\mathbf{x}, \mathbf{p}) = -\tau \frac{\mathbf{p} \cdot \nabla_x}{m} f_0(\mathbf{x}, \mathbf{p})$$

$$= -\tau \left(\frac{\mathbf{p} \cdot \nabla_x}{m} T(\mathbf{x}) \right) \frac{\partial}{\partial T(\mathbf{x})} f_0(\mathbf{x}, \mathbf{p}) \ . \tag{232}$$

We assume that the variation in temperature is small. In particular, the temperature gradient is small. To linear order in the temperature gradient we can replace the temperature by its average value, $T(\mathbf{x}) \approx T$, with corrections that depend on the gradient. Then, to linear order, Eq. (232) reads:

$$\begin{aligned}
g(\mathbf{x}, \mathbf{p}) &= -\tau \left(\frac{\mathbf{p} \cdot \nabla_x}{m} T(\mathbf{x}) \right) \frac{\partial}{\partial T} f_0(\mathbf{p}) \\
&= -\tau \left(\frac{\mathbf{p} \cdot \nabla_x}{m} T(\mathbf{x}) \right) \frac{\partial}{\partial T} n \frac{e^{-\beta p^2/2m}}{(2\pi m k_B T)^{3/2}} \\
&= -\tau \left(\frac{\mathbf{p} \cdot \nabla_x}{m} T(\mathbf{x}) \right) f_0(\mathbf{p}) \left(-\frac{3}{2T} + \frac{p^2}{2m} \frac{1}{k_B T^2} \right) \\
&= -\tau \mathbf{v} \cdot \nabla_x T(\mathbf{x}) \frac{f_0(\mathbf{p})}{k_B T^2} \left(\varepsilon_0(\mathbf{p}) - \frac{3}{2} k T \right) \\
&= \beta \tau \tilde{\varepsilon}_0(\mathbf{p}) \mathbf{v} \cdot \left(-\frac{\vec{\nabla}_x T(\mathbf{x})}{T} \right) f_0(\mathbf{p}) \ .
\end{aligned} \tag{233}$$

Remember Eq. (200), that gives the linear change in the average phase-space density for an applied mechanical force. We see Eqs. (200) and (233) are of the same form if we identify the effective driving *forces*:

$$\mathbf{F}_e = \mathbf{E} \ , \tag{234}$$

$$\mathbf{F}_q = -\frac{\vec{\nabla}_x T(\mathbf{x})}{T} \tag{235}$$

and the associated effective currents:

$$\mathbf{v}_e = -e\mathbf{v} \tag{236}$$

$$\mathbf{v}_q = \tilde{\varepsilon}_0(p)\mathbf{v} \ . \tag{237}$$

Then both types of perturbations can be written in the same form:

$$g_\alpha = \beta \tau \mathbf{v}_\alpha \cdot \mathbf{F}_\alpha f_0(\mathbf{p}) \ , \tag{238}$$

where $\alpha = e$ and q.

Thus we can write the electric current:

$$\mathbf{J} = \int d^3p\, \mathbf{v}_e g_\alpha(\mathbf{p}) \;, \qquad (239)$$

and the heat density current:

$$\mathbf{J}_q = \int d^3p\, \mathbf{v}_q g_\alpha(\mathbf{p}) \;. \qquad (240)$$

This more symmetric notation will be useful to us later.

The average heat-current density due to a temperature gradient is given then by:

$$\mathbf{J}_q = \int d^3p\, \mathbf{v}_q g_q(\mathbf{p}) \;, \qquad (241)$$

where, setting $\alpha = q$ in Eq. (238),

$$g_q = \beta \tau \mathbf{v}_q \cdot \mathbf{F}_q f_0(\mathbf{p}) \;. \qquad (242)$$

Since $\mathbf{F}_q$ is proportional to the temperature gradient we can write average heat-current density in the standard form:

$$J_q^i = -\sum_j \lambda_{ij} \nabla_x^j T(\mathbf{x}) \;,$$

where the thermal conductivity is given by:

$$\lambda_{ij} = \frac{\beta \tau}{T} \int d^3p\, \tilde{\varepsilon}_0^2(\mathbf{p}) v_i v_j f_0(\mathbf{p}) \;. \qquad (243)$$

The remaining integral is evaluated in Problem 7.32 with the result:

$$\lambda_{ij} = \lambda \delta_{ij} \;, \qquad (244)$$

where:

$$\lambda = \frac{7}{2} \frac{\tau n k_B^2 T}{m} \;. \qquad (245)$$

Thus we obtain Fourier's law:

$$\mathbf{J}_q = -\lambda \vec{\nabla} T(\mathbf{x}) \;, \qquad (246)$$

with the thermal conductivity given by Eq. (245).

If we use the mean-free time, Eq. (49), to estimate τ, we find that the thermal conductivity is given by:

$$\lambda = \frac{7}{8} \frac{k_B}{\sqrt{\pi}} \frac{v_0}{r_0^2} \;, \qquad (247)$$

where $mv_0^2 = k_BT$ defines the thermal speed. This can be compared with the lowest order result from the Boltzmann equation given by:

$$\lambda = \frac{75}{64} \frac{k_B\, v_0}{\sqrt{\pi}\, r_0^2} \ . \tag{248}$$

See Chapter 12 in Ref. [18].

If we compare our expression for the electrical conductivity with that for the thermal conductivity we see that both depend on the relaxation time τ. Thus it is interesting to consider the ratio of the two where this dependence cancels. This ratio is given in the single relaxation time approximation by:

$$\frac{\lambda}{\sigma} = \frac{7\tau k_B^2 T n m}{2m\tau e^2 n} = \frac{7}{2}\frac{k_B^2 T}{e^2} \ , \tag{249}$$

which is linear in temperature. If we plot the Lorentz number, $\frac{\lambda}{T\sigma}$ it should be a constant. This is the old and famous Wiedemann-Franz (1853) result. If one generalizes the results to include corrections due to quantum statistics one can show [19]:

$$\frac{\lambda}{T\sigma} = \frac{\pi^2}{3}\left(\frac{k_B}{e}\right)^2 \ , \tag{250}$$

which agrees well with experiments.

It is interesting to consider the case where we have both an applied electric field and a temperature gradient. In this case the steady-state part of the singlet distribution that contributes to the currents is:

$$g(\mathbf{p}) = \sum_\gamma g_\gamma(\mathbf{p}) = \sum_\gamma \beta\tau \mathbf{v}_\gamma \cdot \mathbf{F}_\gamma f_0(\mathbf{p}) \ , \tag{251}$$

where we have used Eq. (238). The steady-state currents are given by:

$$\begin{aligned}
J_\alpha^i &= \int d^3p\, v_\alpha^i g(\mathbf{p}) \\
&= \int d^3p\, v_\alpha^i \sum_\gamma \beta\tau \mathbf{v}_\gamma \cdot \mathbf{F}_\gamma f_0(\mathbf{p}) \\
&= \sum_{\gamma,j} \mathcal{L}_{\alpha\gamma}^{ij} F_\gamma^j \ ,
\end{aligned} \tag{252}$$

where the response function is given by:

$$\mathcal{L}_{\alpha\gamma}^{ij} = \beta\tau \int d^3p\, v_\alpha^i v_\gamma^j f_0(\mathbf{p}) \ . \tag{253}$$

Assuming our system is isotropic we have:

$$\mathcal{L}_{\alpha\gamma}^{ij} = \delta_{ij} \mathcal{L}_{\alpha\gamma} \ , \tag{254}$$

$$J_\alpha^i = \sum_\gamma \mathcal{L}_{\alpha\gamma} F_\gamma^i \tag{255}$$

and:

$$\mathcal{L}_{\alpha\gamma} = \frac{\beta\tau}{3} \int d^3p\, v_\alpha \cdot v_\gamma f_0(\mathbf{p}) \ . \tag{256}$$

We have already shown that:

$$\mathcal{L}_{ee} = \sigma \tag{257}$$

$$\mathcal{L}_{qq} = \lambda T \ . \tag{258}$$

For the off-diagonal components we have:

$$\mathcal{L}_{eq} = \mathcal{L}_{qe} = \frac{\beta\tau}{3} \int d^3p\, v_e \cdot v_q f_0(\mathbf{p}) \tag{259}$$

$$= -\frac{e\tau}{m} k_B T \tag{260}$$

and the momentum integral leading to the final result is worked out in Problem 7.33. The symmetry between off-diagonal elements of $\mathcal{L}$ was discovered [20] by Lars Onsager in 1931. Clearly our analysis here is much less general than that we gave in our discussion of linear response theory in Chapter 2.

For a given temperature gradient we can adjust the applied electric field such that there is no net electric current. This is known as the Seebeck effect [21], and gives the relation between the applied electric field and the temperature gradient:

$$\mathbf{E} = Q \vec{\nabla} T \ . \tag{261}$$

Setting Eq. (255), with $\alpha = e$, to zero, allows one to compute the Seebeck coefficient as:

$$Q = -\frac{e\tau n k_B}{m\sigma} = -\frac{e\tau n k_B m}{m n e^2 \tau} = -\frac{k_B}{e} \ . \tag{262}$$

There are many variations one can play on this theme. One can for example look for the form of the thermal conductivity for the case where the is no net charge flow.

7.3.3
Shear Viscosity

We can also compute the shear viscosity within the single-relaxation time approximation. In this case an off-diagonal component of the momentum current (stress tensor) is driven by a gradient of the (small) local inhomogeneous flow velocity $\mathbf{u}(\mathbf{x})$. Thus we assume, in analogy with the assumption of an inhomogeneous temperature in treating the thermal conductivity, that we have inhomogeneous flow velocity, $\mathbf{u}(\mathbf{x})$. In this case, the local equilibrium solution for the singlet distribution has the form $f_0[\mathbf{p} - m\mathbf{u}(\mathbf{x})]$. The induced current in this case are the off-diagonal components of the stress tensor with the constitutive relation:

$$\sigma_{xy} = -\eta \left(\frac{\partial u_x}{\partial y} + \frac{\partial u_y}{\partial x} \right) \tag{263}$$

defining the shear viscosity. We calculate the off-diagonal component of the stress tensor using:

$$\sigma_{xy} = \int d^3 p \, \frac{p_x p_y}{m} f(\mathbf{x}, \mathbf{p}) \ . \tag{264}$$

In complete analogy to Eq. (229), we have in the steady-state limit for the single-relaxation time approximation:

$$\frac{\mathbf{p} \cdot \nabla}{m} f(\mathbf{x}, \mathbf{p}) = -\frac{1}{\tau} [f(\mathbf{x}, \mathbf{p}) - f_0(\mathbf{p} - m\mathbf{u}(\mathbf{x}))] \ . \tag{265}$$

The solution for f can be expressed in the form:

$$f(\mathbf{x}, \mathbf{p}) = f_0(\mathbf{p} - m\mathbf{u}(\mathbf{x})) + g_s(\mathbf{x}, \mathbf{p}) \ , \tag{266}$$

where g_s is treated as a perturbation. To lowest order we have:

$$g_s(\mathbf{x}, \mathbf{p}) = -\tau \frac{\mathbf{p} \cdot \nabla}{m} f_0(\mathbf{p} - m\mathbf{u}(\mathbf{x})) \ . \tag{267}$$

Using the chain rule for differentiation and expanding to lowest order in $\mathbf{u}$:

$$g_s(\mathbf{x}, \mathbf{p}) = -\tau \sum_{ij} \frac{p_i}{m} \nabla_p^j [f_0(\mathbf{p} - m\mathbf{u}(\mathbf{x}))] \nabla_x^i [p_j - m u_j(\mathbf{x})]$$

$$= \tau \sum_{ij} p_i \nabla_p^j f_0(\mathbf{p}) \nabla_x^i u_j(\mathbf{x}) \ . \tag{268}$$

g_s can be written in the same basic form as for the electric and thermal gradient forces:

$$g_s(\mathbf{p}) = -\beta \frac{\tau}{m} \sum_{ij} p_i p_j f_0(\mathbf{p}) \nabla_x^i u_j$$

$$= -\beta \frac{\tau}{2m} \sum_{ij} p_i p_j f_0(\mathbf{p}) \left(\nabla_x^i u_j + \nabla_x^j u_i \right)$$

$$= -\beta m \tau \sum_{ij} \frac{v_i v_j}{2} F_{ij} \ , \tag{269}$$

where:

$$F_{ij} = \left(\nabla_x^i u_j + \nabla_x^j u_i\right) \ . \tag{270}$$

An off-diagonal component of the average stress tensor is given by:

$$\sigma_{xy} = \int d^3 p \frac{p_x p_y}{m} \left(f_0(\mathbf{p} - m\mathbf{u}(\mathbf{x})) + g_s(\mathbf{x}, \mathbf{p})\right) \ . \tag{271}$$

It is left to Problem 7.21 to show that the local-equilibrium contribution, $f_0(\mathbf{p} - m\mathbf{u}(\mathbf{x}))$, gives no contribution to the average stress tensor up to second order. Therefore, to first order in the driving force, we have for the stress tensor:

$$\begin{aligned}
\sigma_{xy} &= \int d^3 p \frac{p_x p_y}{m} g_s(\mathbf{p}) \\
&= \int d^3 p \frac{p_x p_y}{m} (-\beta m \tau) \sum_{ij} \frac{v_i v_j}{2} F_{ij} f_0(\mathbf{p}) \\
&= (-\beta \tau) \int d^3 p\, p_x p_y \frac{1}{2}(v_x v_y + v_y v_x) F_{xy} f_0(\mathbf{p}) \\
&= -\eta F_{xy} \ ,
\end{aligned} \tag{272}$$

where the shear viscosity is given by:

$$\begin{aligned}
\eta &= \frac{\beta \tau}{m^2} \int d^3 p\, p_x^2 p_y^2 f_0(\mathbf{p}) \\
&= \frac{\beta \tau}{m^2} n \langle p_x^2 \rangle \langle p_y^2 \rangle \ .
\end{aligned} \tag{273}$$

Using the equipartition function $\langle p_x^2 \rangle = m k_B T$ the shear viscosity is given by:

$$\eta = \tau n k_B T \ . \tag{274}$$

The mean-free time determined earlier can be used to estimate τ:

$$\tau^{-1} = 4 n r_0^2 \sqrt{\frac{\pi k_B T}{m}} \ . \tag{275}$$

Using Eq. (275) back in Eq. (274), the shear viscosity is given by:

$$\eta = \frac{1}{4 r_0^2} \sqrt{\frac{m k_B T}{\pi}} = \frac{m v_0}{4 \sqrt{\pi} r_0^2} \ , \tag{276}$$

where $m v_0^2 = k_B T$. Notice that the viscosity, like the thermal conductivity, is independent of the density in the low-density limit. We will return to this expression at the end of this chapter and compare with a more direct kinetic theory calculation.

It is shown in Problem 7.34 that the electric and thermal gradient diving forces do not drive the off-diagonal components of the stress tensor at linear order.

7.3.4
Hall Effect

Suppose we have applied steady electric and magnetic fields and have a steady-state solution to the Boltzmann equation of the form:

$$f = f_0 + g , \qquad (277)$$

where:

$$g = -\tau \mathbf{F} \cdot \nabla_p f \qquad (278)$$

and **F** is the electric plus Lorentz force:

$$\mathbf{F} = -e \left(\mathbf{E} + \frac{\mathbf{v}}{c} \times \mathbf{B} \right) . \qquad (279)$$

Writing out Eq. (278) in terms of g we have:

$$\begin{aligned} g &= e\tau \left(\mathbf{E} + \frac{\mathbf{v}}{c} \times \mathbf{B} \right) \cdot \nabla_p (f_0 + g) \\ &= e\tau (\mathbf{E} + \frac{\mathbf{v}}{c} \times \mathbf{B}) \cdot \left[-\beta \frac{\mathbf{p}}{m} f_0 + \nabla_p g \right] . \end{aligned} \qquad (280)$$

Since:

$$(\mathbf{v} \times \mathbf{B}) \cdot \mathbf{v} = 0 , \qquad (281)$$

we have the equation for g:

$$g = e\tau \mathbf{E} \cdot \nabla_p (f_0 + g) + e\tau \left(\frac{\mathbf{v}}{c} \times \mathbf{B} \right) \cdot \nabla_p g . \qquad (282)$$

The electric current is given then by multiplying by $-ev_i$ and integrating over all momenta:

$$\begin{aligned} J_i &= \int d^3p (-ev_i) g(\mathbf{p}) \\ &= -e^2\tau \sum_j E_j \int d^3p \frac{p_i}{m} \nabla_p^j f_0(\mathbf{p}) - e^2\tau \sum_j E_j \int d^3p \frac{p_i}{m} \nabla_p^j g(\mathbf{p}) \\ &\quad - \frac{e^2\tau}{c} \int d^3p\, v_i (\mathbf{v} \times \mathbf{B}) \cdot \vec{\nabla}_p g(\mathbf{p}) . \end{aligned} \qquad (283)$$

In the middle term we have:

$$\int d^3p\, p_i \nabla_p^j g(\mathbf{p}) = -\int d^3p\, \delta_{ij} g(\mathbf{p}) = -\delta_{ij} \int d^3p [f(\mathbf{p}) - f_0(\mathbf{p})]$$

7.3 Traditional Transport Theory

$$= -\delta_{ij}(n-n) = 0 \; , \tag{284}$$

and we have the electric current in the form:

$$J_i = \sigma E_i + J_i^{(1)} \; , \tag{285}$$

where $\sigma = \frac{ne^2\tau}{m}$ is the Drude result, and:

$$\begin{aligned} J_i^{(1)} &= -\frac{e^2\tau}{c} \int d^3p \, v_i \sum_{jk\ell} \varepsilon_{jk\ell} v_k B_\ell \nabla_p^j g(\mathbf{p}) \\ &= \frac{e^2\tau}{cm^2} \int d^3p \, g(\mathbf{p}) \sum_{jk\ell} \varepsilon_{jk\ell} \nabla_p^j (p_i p_k) B_\ell \\ &= \frac{e^2\tau}{cm^2} \int d^3p \, g(\mathbf{p}) \sum_{jk\ell} \varepsilon_{jk\ell} B_\ell \left(\delta_{ij} p_k + p_i \delta_{jk} \right) \\ &= \frac{e^2\tau}{cm^2} \sum_{k\ell} \varepsilon_{ik\ell} B_\ell \int d^3p \, p_k \, g(\mathbf{p}) \\ &= \frac{e^2\tau}{cm} \sum_{k\ell} \varepsilon_{ik\ell} B_\ell \left(-\frac{J_k}{e} \right) \\ &= -\frac{e\tau}{cm} \sum_{k\ell} \varepsilon_{ik\ell} J_k B_\ell \; . \end{aligned} \tag{286}$$

Putting this back into Eq. (285) we have the equation for the electric current:

$$\mathbf{J} = \sigma \mathbf{E} - \frac{e\tau}{cm} \mathbf{J} \times \mathbf{B} \; . \tag{287}$$

Let us interpret this result in the context of the following system configuration. Apply an electric field E_x along the x-direction. This drives a current along the x-direction:

$$J_x = \sigma E_x \; . \tag{288}$$

Now apply a magnetic field **B** in the z-direction. The Lorentz force will cause the electrons to move in the −y-direction. Thus **B** induces an electric field in the y-direction that balances the Lorentz force in the the y-direction. This is known as the Hall effect [22]. Thus we expect:

$$E_y = R_H J_x B \; , \tag{289}$$

where the Hall coefficient R_H is the constant of proportionality. Applying this geometry to Eq. (287) we find:

$$J_x \hat{x} = \sigma \mathbf{E} - \frac{e\tau}{cm} \mathbf{J} \times \mathbf{B}$$

$$= \sigma \mathbf{E} - \frac{e\tau}{cm} J_x B(\hat{x} \times \hat{z})$$
$$= \sigma \mathbf{E} + \frac{e\tau}{cm} J_x B \hat{y} \; . \tag{290}$$

The x-component reduces to Eq. (288) while the y component is given by:

$$\sigma E_y = -\frac{e\tau}{cm} J_x B \; . \tag{291}$$

This gives the Hall coefficient:

$$R_H = \frac{E_y}{J_x B} = -\frac{\tau e}{mc\sigma} = -\frac{\tau e m}{mcne^2\tau} = -\frac{1}{nec} \; . \tag{292}$$

Measurement of the Hall coefficient gives a method for determining [23] the density of charge carried in the current.

If we introduce the cyclotron frequency:

$$\omega_c = \frac{eB}{mc} \tag{293}$$

and the dimensionless magnetic field:

$$\mathbf{H} = \frac{\tau e}{mc} \mathbf{B} = \tau \omega_c \hat{z} \; , \tag{294}$$

then Eq. (287) can be written in the simple form:

$$\mathbf{J} = \sigma \mathbf{E} - \mathbf{J} \times \mathbf{H} \; . \tag{295}$$

It is left to Problem 7.15 to show that one can solve for $\mathbf{J}$ with the result:

$$\mathbf{J} = \frac{\sigma}{1 + \mathbf{H}^2} \left[\mathbf{E} + \mathbf{H} \times \mathbf{E} + (\mathbf{E} \cdot \mathbf{H}) \mathbf{H} \right] \; . \tag{296}$$

This can be written in the Ohm's law form:

$$J_i = \sum_j \sigma_{ij} E_j \; , \tag{297}$$

where we have the conductivity matrix:

$$\sigma_{ij} = \frac{\sigma}{1 + \mathbf{H}^2} \left[\delta_{ij} + \sum_j \varepsilon_{i\ell j} H_\ell + H_i H_j \right] \; . \tag{298}$$

7.4
Modern Kinetic Theory

Having established the traditional approach to kinetic theory, we now want to discuss a more modern and systematic approach. We will use the memory function approach developed in Chapter 5 to develop this theory. The appropriate dynamical variable in this case, ψ in Chapter 5, is the phase-space density given by:

$$\hat{f}(1) = \hat{f}(\mathbf{x}_1, \mathbf{p}_1) = \sum_{i=1}^{N} \delta(\mathbf{x}_1 - \mathbf{r}_i) \delta(\mathbf{p}_1 - \mathbf{p}_i) \quad . \tag{299}$$

The correlation function of interest in this case is the Laplace transform of the phase-space correlation function:

$$\begin{aligned} C(12;z) &= -i \int_0^\infty dt e^{izt} \langle \delta\hat{f}(2)\delta\hat{f}(1;t) \rangle \\ &= \langle \delta\hat{f}(2) R(z) \delta\hat{f}(1) \rangle \quad , \end{aligned} \tag{300}$$

where, as before, the resolvant operator is given by:

$$R(z) = (z+L)^{-1} \quad . \tag{301}$$

One of the motivations for computing $C(12;z)$ is that we then determine the dynamic structure factor that is measured in scattering experiments. The connection to conventional transport kinetic theory developed in the first half of this chapter is via linear response theory. Earlier we computed the nonequilibrium average: $f(1) = \langle \hat{f}(1) \rangle_{NEQ}$. Suppose we have a disturbance generated by a potential $U(1)$ that couples linearly to the phase-space density in the governing Hamiltonian,

$$H_T = H_0 - \int d1 \hat{f}(1) U(1, t_1) \quad , \tag{302}$$

then classically we have in linear response theory:

$$f(1) = \int_{-\infty}^{t_1} dt_2 \beta \int d^3 x_2 \int d^3 p_2\, C(12; t_1 - t_2) U(2; t_2) \quad . \tag{303}$$

In our memory-function development in Chapter 5 we showed that $C(12;z)$ satisfies the kinetic equation:

$$zC(12;z) - \int d3\, K(13;z) C(32;z) = S(12) \quad , \tag{304}$$

where $\int d3 \equiv \int d^3 x_3 d^3 p_3$ and the equal-time correlation function is given by:

$$S(12) = \langle \delta\hat{f}(2) \delta\hat{f}(1) \rangle \quad . \tag{305}$$

The memory function in Eq. (304) is given by:

$$K(13;z) = K(\mathbf{x}_1 - \mathbf{x}_3, \mathbf{p}_1, \mathbf{p}_3; z) = K^{(s)}(13) + K^{(c)}(13;z) \quad , \tag{306}$$

where the static part of the memory function is given by:

$$\int d3\, K^{(s)}(13) S(32) = -\langle \delta \hat{f}(2) L \delta \hat{f}(1) \rangle \qquad (307)$$

while the *collisional* part of the memory function is given by:

$$\int d3\, K^{(c)}(13;z) S(32) = -\langle [L\delta\hat{f}(2)] R(z) L\delta\hat{f}(1) \rangle$$
$$+ \int d3 d4\, \langle [\delta\hat{f}(3)] R(z) L\delta\hat{f}(1) \rangle$$
$$\times C^{-1}(34;z) \langle [L\delta\hat{f}(2)] R(z) \delta\hat{f}(4) \rangle \ . \qquad (308)$$

The elements of Eq. (304) can be worked out more explicitly. First let us concentrate on the static properties. The first thing is to compute the equilibrium average of the field itself. We have from our earlier work:

$$\langle \hat{f}(1) \rangle = f_0(1) = n\Phi(\mathbf{p}) \ , \qquad (309)$$

which is just the Maxwell result. Next we need the static correlation function. We find:

$$S(12) = \langle \delta\hat{f}(2) \delta\hat{f}(1) \rangle = \langle \hat{f}(2) \hat{f}(1) \rangle - \langle \hat{f}(2) \rangle \langle \hat{f}(1) \rangle$$
$$= \left\langle \sum_{i=1}^{N} \delta(\mathbf{x}_1 - \mathbf{r}_i) \delta(\mathbf{p}_1 - \mathbf{p}_i) \sum_{j=1}^{N} \delta(\mathbf{x}_2 - \mathbf{r}_j) \delta(\mathbf{p}_2 - \mathbf{p}_j) \right\rangle - f_0(1) f_0(2)$$
$$= \left\langle \sum_{i=1}^{N} \delta(\mathbf{x}_1 - \mathbf{r}_i) \delta(\mathbf{p}_1 - \mathbf{p}_i) \delta(\mathbf{x}_2 - \mathbf{r}_i) \delta(\mathbf{p}_2 - \mathbf{p}_i) \right\rangle$$
$$+ \left\langle \sum_{i \neq j=1}^{N} \delta(\mathbf{x}_1 - \mathbf{r}_i) \delta(\mathbf{p}_1 - \mathbf{p}_i) \delta(\mathbf{x}_2 - \mathbf{r}_j) \delta(\mathbf{p}_2 - \mathbf{p}_j) \right\rangle - f_0(1) f_0(2)$$
$$= \delta(\mathbf{x}_1 - \mathbf{x}_2) \delta(\mathbf{p}_1 - \mathbf{p}_2) f_0(1)$$
$$+ f_0(1) f_0(2) g(\mathbf{x}_1 - \mathbf{x}_2) - f_0(1) f_0(2) \ , \qquad (310)$$

where we show in Problem 7.22 that:

$$f_0(1) f_0(2) g(\mathbf{x}_1 - \mathbf{x}_2)$$
$$= \left\langle \sum_{i \neq j=1}^{N} \delta(\mathbf{x}_1 - \mathbf{r}_i) \delta(\mathbf{p}_1 - \mathbf{p}_i) \delta(\mathbf{x}_2 - \mathbf{r}_j) \delta(\mathbf{p}_2 - \mathbf{p}_j) \right\rangle \qquad (311)$$

and the radial distribution function is defined by:

$$n^2 g(\mathbf{x}_1 - \mathbf{x}_2) = \left\langle \sum_{i \neq j=1}^{N} \delta(\mathbf{x}_1 - \mathbf{r}_i) \delta(\mathbf{x}_2 - \mathbf{r}_j) \right\rangle \ . \qquad (312)$$

We write finally:
$$S(12) = \delta(\mathbf{x}_1 - \mathbf{x}_2)\delta(\mathbf{p}_1 - \mathbf{p}_2)f_0(1) + f_0(1)f_0(2)h(\mathbf{x}_1 - \mathbf{x}_2) , \quad (313)$$
where:
$$h(\mathbf{x}_1 - \mathbf{x}_2) = g(\mathbf{x}_1 - \mathbf{x}_2) - 1 \quad (314)$$
is the *hole function*, which vanishes for $|\mathbf{x}_1 - \mathbf{x}_2|$ large. For a brief discussion of the evaluation of h in a low-density expansion see p. 299 in ESM.

We turn next to the static part of the memory function given by:
$$\int d3 \, K^{(s)}(13)S(32) = -\langle \delta \hat{f}(2) L \delta \hat{f}(1)\rangle . \quad (315)$$

Using the identity derived in Chapter 5 we have:
$$\int d3 \, K^{(s)}(13)S(32) = -i\beta^{-1}\langle \{\hat{f}(1), \hat{f}(2)\}\rangle , \quad (316)$$

which requires us to work out the Poisson bracket between $\hat{f}(1)$ and $\hat{f}(2)$. This is left to Problem 7.23, where we show:
$$\{\hat{f}(1), \hat{f}(2)\} = (\nabla_{\mathbf{p}_1}\cdot\nabla_{\mathbf{x}_2} - \nabla_{\mathbf{p}_2}\cdot\nabla_{\mathbf{x}_1})\left[\delta(\mathbf{x}_1-\mathbf{x}_2)\delta(\mathbf{p}_1-\mathbf{p}_2)\hat{f}(1)\right]. \quad (317)$$

We next need to take the equilibrium average of this result to obtain:
$$\begin{aligned}\langle\{\hat{f}(1),\hat{f}(2)\}\rangle &= (\nabla_{\mathbf{p}_1}\cdot\nabla_{\mathbf{x}_2} - \nabla_{\mathbf{p}_2}\cdot\nabla_{\mathbf{x}_1})[\delta(\mathbf{x}_1-\mathbf{x}_2)\delta(\mathbf{p}_1-\mathbf{p}_2)f_0(1)]\\ &= -\nabla_{\mathbf{x}_1}\cdot(\nabla_{\mathbf{p}_1}+\nabla_{\mathbf{p}_2})[\delta(\mathbf{x}_1-\mathbf{x}_2)\delta(\mathbf{p}_1-\mathbf{p}_2)f_0(1)]\\ &= -\nabla_{\mathbf{x}_1}\cdot[\delta(\mathbf{x}_1-\mathbf{x}_2)\delta(\mathbf{p}_1-\mathbf{p}_2)\nabla_{\mathbf{p}_1}f_0(1)]\\ &= \beta\frac{\mathbf{p}_1\cdot\nabla_{\mathbf{x}_1}}{m}[\delta(\mathbf{x}_1-\mathbf{x}_2)\delta(\mathbf{p}_1-\mathbf{p}_2)f_0(1)] . \end{aligned} \quad (318)$$

Our equation for the static part of the memory function is given by:
$$\int d^3x_3 d^3p_3 \, K^{(s)}(\mathbf{x}_1-\mathbf{x}_3;\mathbf{p}_1\mathbf{p}_3)S(\mathbf{x}_3-\mathbf{x}_2;\mathbf{p}_3\mathbf{p}_2)$$
$$= -i\frac{\mathbf{p}_1\cdot\nabla_{\mathbf{x}_1}}{m}[\delta(\mathbf{x}_1-\mathbf{x}_2)\delta(\mathbf{p}_1-\mathbf{p}_2)f_0(1)] . \quad (319)$$

Because of the translational invariance it is useful to Fourier transform over space and obtain:
$$\int d^3p_3 \, K^{(s)}(\mathbf{k};\mathbf{p}_1\mathbf{p}_3)S(\mathbf{k};\mathbf{p}_3\mathbf{p}_2) = \frac{\mathbf{k}\cdot\mathbf{p}_1}{m}\delta(\mathbf{p}_1-\mathbf{p}_2)f_0(1) . \quad (320)$$

This is still an integral equation for $K^{(s)}$. The Fourier transform of the static correlation function is given by:
$$S(\mathbf{k};\mathbf{p}_1,\mathbf{p}_2) = \delta(\mathbf{p}_1-\mathbf{p}_2)f_0(1) + f_0(1)f_0(2)h(\mathbf{k}) , \quad (321)$$

where:
$$h(\mathbf{k}) = \int d^3x_1\, e^{-i\mathbf{k}\cdot(\mathbf{x}_1-\mathbf{x}_2)} h(\mathbf{x}_1 - \mathbf{x}_2) \ . \tag{322}$$

If we integrate $S(\mathbf{k};\mathbf{p}_1,\mathbf{p}_2)$ over $\mathbf{p}_1$ and $\mathbf{p}_2$, we find that the static structure factor is given by:
$$nS(k) = \int d^3p_1 d^3p_2 S(\mathbf{k};\mathbf{p}_1,\mathbf{p}_2) = n + n^2 h(k) \ . \tag{323}$$

Using Eq. (321) in Eq. (320) gives:
$$\int d^3p_3\, K^{(s)}(\mathbf{k};\mathbf{p}_1\mathbf{p}_3) S(\mathbf{k};\mathbf{p}_3\mathbf{p}_2) = K^{(s)}(\mathbf{k};\mathbf{p}_1,\mathbf{p}_2) f_0(2)$$
$$+ \int d^3p_3\, K^{(s)}(\mathbf{k};\mathbf{p}_1\mathbf{p}_3) f_0(3) f_0(2) h(\mathbf{k})$$
$$= \frac{\mathbf{k}\cdot\mathbf{p}_1}{m} \delta(\mathbf{p}_1 - \mathbf{p}_2) f_0(1) \ . \tag{324}$$

Next we integrate this equation over $\mathbf{p}_2$ and use the fact that:
$$\int d^3p_2\, f_0(2) = n \ , \tag{325}$$

to obtain:
$$\int d^3p_3\, K^{(s)}(\mathbf{k};\mathbf{p}_1\mathbf{p}_3) f_0(3)[1 + nh(\mathbf{k})] = \frac{\mathbf{k}\cdot\mathbf{p}_1}{m} f_0(1) \tag{326}$$

or:
$$\int d^3p_3\, K^{(s)}(\mathbf{k};\mathbf{p}_1\mathbf{p}_3) f_0(3) = \frac{\mathbf{k}\cdot\mathbf{p}_1}{m} \frac{f_0(1)}{[1 + nh(\mathbf{k})]} \ . \tag{327}$$

Putting this result back into Eq. (324) we obtain:
$$K^{(s)}(\mathbf{k};\mathbf{p}_1\mathbf{p}_2) = \frac{\mathbf{k}\cdot\mathbf{p}_1}{m} \delta(\mathbf{p}_1 - \mathbf{p}_2) - \frac{\mathbf{k}\cdot\mathbf{p}_1}{m} f_0(1) \frac{h(\mathbf{k})}{[1 + nh(\mathbf{k})]} \ . \tag{328}$$

It is conventional to introduce the direct correlation function [24]:
$$c_D(k) = \frac{h(\mathbf{k})}{[1 + nh(\mathbf{k})]} \ , \tag{329}$$

and obtain finally:
$$K^{(s)}(\mathbf{k};\mathbf{p}_1\mathbf{p}_2) = \frac{\mathbf{k}\cdot\mathbf{p}_1}{m} \delta(\mathbf{p}_1 - \mathbf{p}_2) - \frac{\mathbf{k}\cdot\mathbf{p}_1}{m} f_0(1) c_D(k) \ . \tag{330}$$

7.4.1
Collisionless Theory

Let us suppose that the collision part of the memory function is zero, $K^{(c)} = 0$, and we can investigate the physics in the kinetic equation associated with the static part of the memory function alone. This development is relevant during the short-time high-frequency regime or where the system is weakly interacting. The kinetic equation in this case takes the form:

$$\left(z - \frac{\mathbf{k}\cdot\mathbf{p}_1}{m}\right) C(\mathbf{k},\mathbf{p}_1\mathbf{p}_2,z) + \int d^3p_3 \, \frac{\mathbf{k}\cdot\mathbf{p}_1}{m} f_0(1) c_D(k) C(\mathbf{k},\mathbf{p}_3\mathbf{p}_2,z)$$
$$= S(\mathbf{k},\mathbf{p}_1\mathbf{p}_2) \ . \qquad (331)$$

This is an integral equation in the momentum variables. Fortunately we can solve this equation exactly because it is of separable form. To see this divide Eq. (331) by $z - \frac{\mathbf{k}\cdot\mathbf{p}_1}{m}$ and integrate over $\mathbf{p}_1$. This leads to the equation:

$$\int d^3p_3 \, C(\mathbf{k},\mathbf{p}_3\mathbf{p}_2,z) \left[1 + c_D(k) I_1(\mathbf{k},z)\right] = I_s(\mathbf{k},\mathbf{p}_2,z) \ , \qquad (332)$$

where:

$$I_1(\mathbf{k},z) = \int d^3p_3 \, \frac{\mathbf{k}\cdot\mathbf{p}_3}{m} \frac{f_0(3)}{z - \frac{\mathbf{k}\cdot\mathbf{p}_3}{m}} \qquad (333)$$

and:

$$I_s(\mathbf{k};\mathbf{p}_2,z) = \int d^3p_3 \, \frac{S(\mathbf{k},\mathbf{p}_3\mathbf{p}_2)}{z - \frac{\mathbf{k}\cdot\mathbf{p}_3}{m}} \ . \qquad (334)$$

Putting the expression for the static correlation function, given by Eq. (321), into Eq. (334) we obtain:

$$I_s(\mathbf{k};\mathbf{p}_2,z) = \frac{f_0(2)}{z - \frac{\mathbf{k}\cdot\mathbf{p}_2}{m}} + f_0(2) h(\mathbf{k}) I_0(\mathbf{k},z) \ , \qquad (335)$$

where:

$$I_0(\mathbf{k},z) = \int d^3p_3 \, \frac{f_0(3)}{z - \frac{\mathbf{k}\cdot\mathbf{p}_3}{m}} \ . \qquad (336)$$

Notice that:

$$I_1(\mathbf{k},z) = -n + z I_0(\mathbf{k},z) \ . \qquad (337)$$

Putting Eq. (335) back into Eq. (332) we find:

$$\int d^3p_3 \, C(\mathbf{k},\mathbf{p}_3\mathbf{p}_2,z)$$
$$= \frac{1}{1 + c_D(k) I_1(\mathbf{k},z)} \left(\frac{f_0(2)}{z - \frac{\mathbf{k}\cdot\mathbf{p}_2}{m}} + f_0(2) h(\mathbf{k}) I_0(\mathbf{k},z)\right) \ . \qquad (338)$$

Putting this back into Eq. (331), we have:

$$C(\mathbf{k}, \mathbf{p}_1\mathbf{p}_2, z) = \frac{S(\mathbf{k}, \mathbf{p}_1\mathbf{p}_2)}{z - \frac{\mathbf{k}\cdot\mathbf{p}_1}{m}}$$
$$- \frac{\mathbf{k}\cdot\mathbf{p}_1}{m} f_0(1) c_D(\mathbf{k}) \frac{1}{1 + c_D(\mathbf{k}) I_1(\mathbf{k}, z)}$$
$$\times \left(\frac{f_0(2)}{z - \frac{\mathbf{k}\cdot\mathbf{p}_2}{m}} + f_0(2) h(\mathbf{k}) I_0(\mathbf{k}, z) \right) \quad . \tag{339}$$

It is shown in Problem 7.24 that this can be written in a form symmetric in $\mathbf{p}_1$ and $\mathbf{p}_2$:

$$C(\mathbf{k}, \mathbf{p}_1\mathbf{p}_2, z) = \frac{f_0(1)}{z - \frac{\mathbf{k}\cdot\mathbf{p}_1}{m}} \delta(\mathbf{p}_1 - \mathbf{p}_2)$$
$$+ \frac{f_0(1) f_0(2) c_D(k)}{(z - \frac{\mathbf{k}\cdot\mathbf{p}_1}{m})(z - \frac{\mathbf{k}\cdot\mathbf{p}_2}{m})} \frac{1}{1 + c_D(\mathbf{k}) I_1(\mathbf{k}, z)}$$
$$\times \left[(z - \frac{\mathbf{k}\cdot(\mathbf{p}_1 + \mathbf{p}_2)}{m})[1 + zh(\mathbf{k}) I_0(\mathbf{k}, z)] \right.$$
$$\left. \times + \frac{\mathbf{k}\cdot\mathbf{p}_1}{m} \frac{\mathbf{k}\cdot\mathbf{p}_2}{m} h(\mathbf{k}) I_0(\mathbf{k}, z) \right] \quad . \tag{340}$$

The solution for the density–density correlation function is given by integrating Eq. (338) over $\mathbf{p}_2$ to obtain:

$$C_{nn}(\mathbf{k}, z) = \frac{1}{1 + c_D(\mathbf{k}) I_1(\mathbf{k}, z)} [I_0(\mathbf{k}, z) + nh(\mathbf{k}) I_0(\mathbf{k}, z)]$$
$$= S(k) \frac{I_0(\mathbf{k}, z)}{1 + c_D(\mathbf{k}) I_1(\mathbf{k}, z)} \quad , \tag{341}$$

where the static structure factor, from Eq. (323), is $S(k) = 1 + nh(k)$. We must still evaluate the integrals I_0 and I_1.

7.4.2
Noninteracting Gas

Let us look first at the case of noninteracting particles where $c_D(\mathbf{k}) = 0$ and the phase-space fluctuation function is given by:

$$C^{(0)}(\mathbf{k}, \mathbf{p}_1\mathbf{p}_2, z) = \frac{f_0(1)}{z - \frac{\mathbf{k}\cdot\mathbf{p}_1}{m}} \delta(\mathbf{p}_1 - \mathbf{p}_2) \quad . \tag{342}$$

Then in particular, the density–density fluctuation function is given by:

$$C_{nn}^{(0)}(\mathbf{k},z) = \int d^3p_1 d^3p_2 C^{(0)}(\mathbf{k},\mathbf{p}_1\mathbf{p}_2,z)$$
$$= \int d^3p_1 \frac{f_0(1)}{z - \frac{\mathbf{k}\cdot\mathbf{p}_1}{m}}$$
$$= I_0(\mathbf{k},z) \ . \tag{343}$$

The correlation function (Fourier transform) is given by:

$$C_{nn}^{(0)}(\mathbf{k},\omega) = -2\, Im\, C_{nn}^{(0)}(\mathbf{k},\omega+i\eta)$$
$$= -2\, Im \int d^3p_1 \frac{f_0(1)}{\omega + i\eta - \frac{\mathbf{k}\cdot\mathbf{p}_1}{m}}$$
$$= -2 \int d^3p_1 f_0(1)(-\pi)\delta\left(\omega - \frac{\mathbf{k}\cdot\mathbf{p}_1}{m}\right)$$
$$= \int_{-\infty}^{\infty} dt \int d^3p_1 f_0(1) e^{it\left(\omega - \frac{\mathbf{k}\cdot\mathbf{p}_1}{m}\right)} \ . \tag{344}$$

The integrals over momentum are separable. We need, for example,

$$\int dp_x e^{-\beta p_x^2/2m} e^{-itk_x p_x/m} = \sqrt{\frac{2\pi m}{\beta}} e^{-t^2 k_x^2/2m\beta} \ , \tag{345}$$

which leads to:

$$C_{nn}^{(0)}(\mathbf{k},\omega) = n\left(\frac{\beta}{2\pi m}\right)^{3/2} \int_{-\infty}^{\infty} dt\, e^{it\omega} \left(\sqrt{\frac{2\pi m}{\beta}}\right)^3 e^{-t^2(k_x^2+k_y^2+k_z^2)/2m\beta}$$
$$= n \int_{-\infty}^{\infty} dt\, e^{it\omega} e^{-t^2 k^2 v_0^2/2}$$
$$= n \frac{\sqrt{2\pi}}{kv_0} e^{-\frac{\omega^2}{2k^2 v_0^2}} \ , \tag{346}$$

where $k_B T = mv_0^2$. We see that the density–density correlation function is a Gaussian in frequency for a system of free particles. This is to be contrasted with the hydrodynamical spectrum, which consists of three Lorentzians. The hydrodynamical spectrum is valid in the regime where $k\ell \ll 1$ where ℓ is the mean free path. The free particle spectrum is valid in the opposite limit $k\ell \gg 1$. This corresponds to probing regions $\approx 1/k$, sufficiently small that, on average, the particles appear to be free.

7.4.3
Vlasov Approximation

Let us recall the collisionless solution for $C_{nn}(\mathbf{k},z)$ given by Eq. (341):

$$C_{nn}(\mathbf{k},z) = \frac{S(\mathbf{k}) I_0(\mathbf{k},z)}{1 + c_D(\mathbf{k}) I_1(\mathbf{k},z)} \ . \tag{347}$$

This approximation is related to the linearized Vlasov approximation [25], which is useful in treating charged systems. In the case of neutral systems it leads to a discussion of collisionless sound seen [26] at high frequencies. For a discussion of charged systems from this point of view see Ref. [27].

In order to go further we need some understanding of the direct correlation function. This quantity was discussed in Chapter 4 of ESM [28]. In the low-density limit it is easy to show that the pair correlation function is given by:

$$g(r) = e^{-\beta V(r)} , \qquad (348)$$

where $V(r)$ is the pair potential acting between two particles. The hole function is given then by:

$$h(r) = e^{-\beta V(r)} - 1 \qquad (349)$$

and the Fourier transform:

$$h(\mathbf{k}) = \int d^3r\, e^{-i\mathbf{k}\cdot\mathbf{r}} h(\mathbf{r}) \qquad (350)$$

$$= \int d^3r\, e^{-i\mathbf{k}\cdot\mathbf{r}} \left(e^{-\beta V(r)} - 1 \right) . \qquad (351)$$

We can evaluate this explicitly for a system of hard spheres where:

$$V(r) = \begin{matrix} 0 & r > r_0 \\ \infty & r < r_0 \end{matrix} , \qquad (352)$$

where r_0 is a hard-sphere diameter. We find in Problem 7.25 that for this case:

$$h(\mathbf{k}) = -\frac{4\pi}{k^3}\left[\sin(kr_0) - kr_0\cos(kr_0)\right] . \qquad (353)$$

We have from Eq. (329) that to lowest order in the density:

$$c_D(\mathbf{k}) = h(\mathbf{k}) . \qquad (354)$$

We can also expand (see Problem 7.35) the direct correlation function as a power series in the potential. To first order in V, independent of the density,

$$c_D(\mathbf{k}) = h(\mathbf{k}) = \int d^3r\, e^{-i\mathbf{k}\cdot\mathbf{r}}(-\beta V(r)) = -\beta V(k) . \qquad (355)$$

As a final example, we have that in the case of a Coulomb gas, due to the long range nature [29] of the interactions, that:

$$c_D(\mathbf{k}) = \lim_{k \to 0} -\beta V(k) = -\beta 4\pi e^2/k^2 . \qquad (356)$$

Armed with this information, let us look at the denominator in Eq. (347) for small wavenumbers. We have, expanding in powers of k,

$$I_1(\mathbf{k},z) = \int d^3p \frac{\mathbf{k}\cdot\mathbf{p}}{m} \frac{f_0(p)}{z - \frac{\mathbf{k}\cdot\mathbf{p}}{m}}$$

$$= \frac{1}{z^2} \int d^3p \left(\frac{\mathbf{k}\cdot\mathbf{p}}{m}\right)^2 f_0(p) + \mathcal{O}(k^4) \quad . \tag{357}$$

It is shown in Problem 7.26 that:

$$\int d^3p \left(\frac{\mathbf{k}\cdot\mathbf{p}}{m}\right)^2 f_0(p) = k^2 \frac{k_B T}{m} = (kv_0)^2 \quad . \tag{358}$$

To lowest order in k,

$$I_0(\mathbf{k},z) = \int d^3p \frac{f_0(p)}{z - \frac{\mathbf{k}\cdot\mathbf{p}}{m}} = \frac{n}{z} \quad . \tag{359}$$

The density–density fluctuation function is then given by:

$$C_{nn}(\mathbf{k},z) = \frac{S(k)\frac{n}{z}}{1 + c_D(k)n(kv_0)^2/z^2}$$

$$= nS(k)\frac{z}{z^2 + c_D(k)n(kv_0)^2} \quad . \tag{360}$$

Since we have assumed large z and small k, the poles at $z = \pm\sqrt{-c_D(k)n(kv_0)^2}$ are unphysical. Remember, however, that for a plasma, for small wavenumbers the direct correlation function is given by Eq. (356). Putting this into Eq. (360) we have:

$$C_{nn}(\mathbf{k},z) = nS(k)\frac{z}{z^2 - \omega_P^2} \quad , \tag{361}$$

where the plasma frequency is given by:

$$\omega_P^2 = \frac{4\pi e^2 n}{m} \quad . \tag{362}$$

We can then write:

$$C_{nn}(\mathbf{k},z) = nS(k)\frac{1}{2}\left[\frac{1}{z-\omega_P} + \frac{1}{z+\omega_P}\right] \quad . \tag{363}$$

The correlation function is given by:

$$C_{nn}(\mathbf{k},\omega) = -2\,\text{Im}\,C_{nn}(\mathbf{k},\omega+i\eta)$$

$$= nS(k)\pi\left[\delta(\omega-\omega_P) + \delta(\omega+\omega_P)\right] \quad . \tag{364}$$

In a plasma the dynamics are dominated by plasma modes or plasma oscillations:

$$C_{nn}(\mathbf{k},t) = nS(k)\,\cos(\omega_p t) \ . \tag{365}$$

7.4.4 Dynamic Part of Memory Function

Let us now move on to discuss the dynamic or collisional part of the memory function. We have from Eq. (308):

$$\int d3\, K^{(c)}(13;z)S(32) = -\langle [L\delta\hat{f}(2)]R(z)L\delta\hat{f}(1)\rangle$$
$$+ \int d3d4\, \langle [\delta\hat{f}(3)]R(z)L\delta\hat{f}(1)\rangle C^{-1}(34;z)\langle [L\delta\hat{f}(2)]R(z)\delta\hat{f}(4)\rangle. \tag{366}$$

We have, from Eq. (5.101), the identity:

$$L\hat{f}(1) = -L_0(1)\hat{f}(1) - \int d1'\, L_I(11')\hat{f}(1)\hat{f}(1') \ , \tag{367}$$

where the kinetic part of the single-particle Liouville operator is:

$$L_0(1) = -i\frac{\mathbf{p}_1 \cdot \nabla_{x_1}}{m} \tag{368}$$

and:

$$L_I(11') = i\nabla_{x_1} V(\mathbf{x}_1 - \mathbf{x}_{1'}) \cdot (\nabla_{p_1} - \nabla_{p_{1'}}) \ . \tag{369}$$

We recall from our general discussion of memory functions that the linear term $-L_0(1)\hat{f}(1)$ will not contribute to the dynamic part of the memory function, so:

$$\int d3\, K^{(c)}(13;z)S(32) = -\int d1d1'\, L_I(11')L_I(22')$$
$$\times \Bigg[\langle \delta[\hat{f}(2)\hat{f}(2')]R(z)\delta[\hat{f}(1)\hat{f}(1')]\rangle$$
$$- \int d3d4\, \langle [\delta\hat{f}(3)]R(z)\delta[\hat{f}(1)\hat{f}(1')]\rangle$$
$$\times C^{-1}(34;z)\langle \delta[\hat{f}(2)\hat{f}(2')]R(z)\delta\hat{f}(4)\rangle \Bigg]. \tag{370}$$

To simplify the notation we define the new correlation function:

$$\begin{aligned} G(11',22';z) &= \langle \delta[\hat{f}(2)\hat{f}(2')]R(z)\delta[\hat{f}(1)\hat{f}(1')]\rangle \\ &\quad - \int d3\,d4 \,\langle [\delta\hat{f}(3)]R(z)\delta[\hat{f}(1)\hat{f}(1')]\rangle \\ &\quad \times C^{-1}(34;z)\langle \delta[\hat{f}(2)\hat{f}(2')]R(z)\delta\hat{f}(4)\rangle \quad . \end{aligned} \quad (371)$$

The dynamic part of the memory function is written then in the form:

$$\int d3\, K^{(c)}(13;z)S(32) = -\int d1'd2'\, L_I(11')L_I(22')G(11',22';z) \quad . \quad (372)$$

There is still an integral equation to solve for $K^{(c)}$. Fortunately, we can easily solve for $K^{(c)}(13;z)$. Inserting:

$$S(32) = \delta(32)f_0(2) + f_0(3)f_0(2)h(\mathbf{x}_3 - \mathbf{x}_2) \quad (373)$$

on the left in Eq. (372) we have:

$$\begin{aligned} \int d3\, K^{(c)}(13;z)S(32) &= K^{(c)}(12;z)f_0(2) \\ &\quad + \int d3\, K^{(c)}(13;z)f_0(3)f_0(2)h(\mathbf{x}_3 - \mathbf{x}_2) \quad . \end{aligned} \quad (374)$$

We see that we must treat the integral $\int d^3 p_3 K^{(c)}(13;z)f_0(3)$. This quantity vanishes due to conservation of particle number reflected in the quantity:

$$\int d^3 p_2 \int d2'\, L_I(22')G(11',22';z) = 0 \quad . \quad (375)$$

We have then:

$$K^{(c)}(12;z)f_0(2) = -\int d1'd2'\, L_I(11')L_I(22')G(11',22';z) \quad . \quad (376)$$

We need to evaluate $K^{(c)}$ in some physically sensible approximation.

7.4.5
Approximations

One can evaluate the dynamic part of the memory function in perturbation theory in the coupling and in the density. It is not difficult to show that in these limits the correlation function $G(11',22';z)$ has nice [30,31] properties. In particular, to lowest order in the potential (see Problem 7.30),

$$\begin{aligned} G_{00}(11',22';z) &= [z - L_0(1) - L_0(1')]^{-1} \left[\delta(12)\delta(1'2') + \delta(12')\delta(1'2)\right] \\ &\quad \times f_0(1)f_0(1') \quad . \end{aligned} \quad (377)$$

If we put this result back into Eq. (376), we obtain the dynamic part of the memory function for a weakly interacting system studied in detail in Refs. [32,33]. Here we move on to the low-density case. It was shown by Mazenko [30] that one can develop a general and systematic analysis of $G(11', 22'; z)$. One finds to lowest order in the density (see Problem 7.31):

$$G_0(11', 22'; z) = [z - L(11')]^{-1} [\delta(12)\delta(1'2') + \delta(12')\delta(1'2)] \\ \times f_0(11') , \qquad (378)$$

where:

$$L(11') = L_0(1) + L_0(1') + L_I(11') \qquad (379)$$

is the full two-body Liouville operator,

$$\delta(12) = \delta(\mathbf{x}_1 - \mathbf{x}_2) \delta(\mathbf{p}_1 - \mathbf{p}_2) , \qquad (380)$$

and:

$$f_0(11') = n^2 \left(\frac{\beta}{2\pi m}\right)^3 e^{-\beta H_2(11')} , \qquad (381)$$

where:

$$H_2(11') = \frac{\mathbf{p}_1^2}{2m} + \frac{\mathbf{p}_{1'}^2}{2m} + V(\mathbf{x}_1 - \mathbf{x}_{1'}) \qquad (382)$$

is just the Hamiltonian for a two-particle system. We can rewrite this result for G_0, remembering:

$$L(11') f_0(11') = 0 \qquad (383)$$

$$G_0(11', 22'; z) = f_0(11')[z + L(22')]^{-1} [\delta(12)\delta(1'2') + \delta(12')\delta(1'2)] . \qquad (384)$$

If we take the inverse Laplace transform, we have in the time regime:

$$G_0(11', 22'; t) = f_0(11') e^{itL(22')} \\ \times [\delta(12)\delta(1'2') + \delta(12')\delta(1'2)] . \qquad (385)$$

If we then use the property that $e^{itL(22')}$ propagates particles 2 and 2' forward in time, then for any function of the phase-space coordinates of particles 2 and 2' we have:

$$e^{itL(22')} g(2, 2') = g[2(t), 2'(t)] \qquad (386)$$

and:

$$G_0(11', 22'; t) = f_0(11')$$

$$\times \left[\delta(12(t))\delta(1'2'(t)) + \delta(12'(t))\delta(1'2(t))\right] \quad . \tag{387}$$

$G_0(11', 22'; t)$ has the physical interpretation as the probability of finding two particles at the points 1 and $1'$ at time t given that the two particles are at 2 and $2'$ at time $t = 0$. This amounts to solving the two-body problem given the initial phase-space coordinates $\mathbf{x}_2, \mathbf{p}_2, \mathbf{x}_{2'}, \mathbf{p}_{2'}$. The low-density memory function is given then by:

$$\begin{aligned} K_0^{(c)}(12; z) f_0(2) &= -\int d1' d2' L_I(11') L_I(22') [z - L(11')]^{-1} \\ &\quad \times \left[\delta(12)\delta(1'2') + \delta(12')\delta(1'2)\right] f_0(11') \quad . \end{aligned} \tag{388}$$

This expression has been studied rather extensively. Here we point out two particular special cases. It can be shown [34] that:

$$\lim_{z \to i0^+} \lim_{k \to 0} K_0^{(c)}(\mathbf{k}, \mathbf{pp}'; z) f_0(\mathbf{p}') = i K_B(\mathbf{pp}') f_0(\mathbf{p}') \quad , \tag{389}$$

where K_B is the linearized Boltzmann collision operator found earlier [Eq. (125)].

A second interesting case, where we can investigate the low-density memory function, is the case of hard spheres. It has been shown [15] that for hard spheres:

$$\begin{aligned} K_0^{(c)}(\mathbf{k}, \mathbf{pp}'; z) f_0(\mathbf{p}') &= K_1^{(c)}(\mathbf{k}, \mathbf{pp}'; z) f_0(\mathbf{p}') \\ &\quad + K_2^{(c)}(\mathbf{k}, \mathbf{pp}'; z) f_0(\mathbf{p}') \quad , \end{aligned} \tag{390}$$

where:

$$K_2^{(c)}(\mathbf{k}, \mathbf{pp}'; z) f_0(\mathbf{p}') = \frac{\mathbf{k} \cdot \mathbf{p}'}{m} c_D^{(0)}(k) f_0(\mathbf{p}) f_0(\mathbf{p}') \quad , \tag{391}$$

where $c_D^{(0)}(k)$ is the low-density expression for the direct correlation function for a system of hard spheres, and we also have:

$$\begin{aligned} K_1^{(c)}(\mathbf{k}, \mathbf{pp}'; z) f_0(\mathbf{p}') &= i \int d^3 p_1 d\Omega_r \sigma \frac{\hat{r} \cdot (\mathbf{p} - \mathbf{p}_1)}{m} \theta[\hat{r} \cdot (\mathbf{p}_1 - \mathbf{p})] \\ &\quad \times f_0(\mathbf{p}) f_0(\mathbf{p}_1) \left[\delta(\mathbf{p}' - \mathbf{p}^*) - \delta(\mathbf{p}' - \mathbf{p}) \right. \\ &\quad \left. + e^{i\mathbf{k} \cdot \hat{r} r_0} \delta(\mathbf{p}' - \mathbf{p}_1^*) - e^{-i\mathbf{k} \cdot \hat{r} r_0} \delta(\mathbf{p}' - \mathbf{p}_1)\right] \quad , \end{aligned} \tag{392}$$

where the post-collision momentum is given by:

$$\mathbf{p}^* = \mathbf{p} - 2\hat{r}(\hat{r} \cdot \mathbf{p}) \tag{393}$$

and satisfies $(\mathbf{p}^*)^2 = \mathbf{p}^2$. We see in this case that the dynamic part of the memory function is explicitly frequency independent for hard spheres and low densities. This gives some justification for the use of the *local* time approximation in treating $K^{(d)}$ in fluids. If we include higher order terms in the density expansion, the memory function will become frequency dependent even for hard spheres.

Next we note that the low-density memory function differs from the Boltzmann result (where $\sigma = \pi r_0^2$ for hard spheres) only through the term $K_2^{(c)}$ and the phase factors $e^{i\mathbf{k}\cdot\hat{r}_0}$ in $K_1^{(c)}$. In the small k limit they coincide, as they must. The small k corrections are due to the finite size of the particles: the collisions occur when the centers of the particles are separated by a distance r_0.

Once we have an approximation for $K^{(c)}$, we must still solve the associated kinetic equation given by Eq. (304). For the approximation given by Eq. (390), one can solve for the correlation functions using the method of kinetic models introduced in Section 7.2.8. This calculation has been carried out by Mazenko, Wei and Yip [15] and leads to results for $C_{nn}(\mathbf{k},\omega)$ valid over the whole $\mathbf{k}$ and ω plane, but restricted to low-density systems. These calculations compare favorably with neutron-scattering experiments on dilute systems and show the transition from the hydrodynamical ($k\ell \ll 1$) to the kinetics of free particles regime ($k\ell \gg 1$).

7.4.6
Transport Coefficients

Given a kinetic equation like Eq. (304), how do we calculate the transport coefficients? This is discussed in detail by Forster and Martin [32]. We will give a brief overview of the method. This analysis is facilitated by the introduction of a linear vector space spanning momentum space. We assume there exist orthonormal and complete momentum states:

$$\langle \mathbf{p}_1 | \mathbf{p}_2 \rangle = \delta(\mathbf{p}_1 - \mathbf{p}_2) \ , \tag{394}$$

where:

$$\langle \mathbf{p}_1 | L_0(\mathbf{k}) | \mathbf{p}_2 \rangle = \frac{\mathbf{k}\cdot\mathbf{p}_1}{m} \langle \mathbf{p}_1 | \mathbf{p}_2 \rangle \tag{395}$$

$$K^{(s)}(\mathbf{k},\mathbf{p}_1\mathbf{p}_2) = \langle \mathbf{p}_1 | K^{(s)}(\mathbf{k}) | \mathbf{p}_2 \rangle \tag{396}$$

$$K^{(c)}(\mathbf{k},\mathbf{p}_1\mathbf{p}_2;z) = \langle \mathbf{p}_1 | K^{(c)}(\mathbf{k},z) | \mathbf{p}_2 \rangle \tag{397}$$

$$S(\mathbf{k},\mathbf{p}_1\mathbf{p}_2) = \langle \mathbf{p}_1 | S(\mathbf{k}) | \mathbf{p}_2 \rangle \tag{398}$$

$$C(\mathbf{k},\mathbf{p}_1\mathbf{p}_2;z) = \langle \mathbf{p}_1 | C(\mathbf{k},z) | \mathbf{p}_2 \rangle \ , \tag{399}$$

where $L_0(\mathbf{k})$, $K^{(s)}(\mathbf{k})$, $K^{(c)}(\mathbf{k},z)$, $S(\mathbf{k})$ and $C(\mathbf{k},z)$ are now interpreted as operators. It is clear then that the kinetic equation can be written:

$$\int d^3 p_3 \langle \mathbf{p}_1 | [z - K(\mathbf{k},z)] | \mathbf{p}_3 \rangle \langle \mathbf{p}_3 | C(\mathbf{k},z) | \mathbf{p}_2 \rangle = \langle \mathbf{p}_1 | S(\mathbf{k}) | \mathbf{p}_2 \rangle \ . \tag{400}$$

Since the momentum states are assumed to be complete,

$$\int d^3 p_3 | \mathbf{p}_3 \rangle \langle \mathbf{p}_3 | = 1 \ , \tag{401}$$

Eq. (400) can be written as an operator equation:

$$[z - K(\mathbf{k},z)] C(\mathbf{k},z) = S(\mathbf{k}) \ , \tag{402}$$

which has the formal solution:

$$C(\mathbf{k},z) = [z - K(\mathbf{k},z)]^{-1} S(\mathbf{k}) \ . \tag{403}$$

If we want to compute the shear viscosity, then we need to calculate the transverse current fluctuation function:

$$C_t(\mathbf{k},z) = \int d^3 p_1 \int d^3 p_2 p_1^x p_2^x C(\mathbf{k}, \mathbf{p}_1 \mathbf{p}_2; z) \ , \tag{404}$$

where we assume that $\mathbf{k}$ is in the z-direction. This can then be written as:

$$C_t(\mathbf{k},z) = \int d^3 p_1 \int d^3 p_2 p_1^x p_2^x \langle \mathbf{p}_1 | [z - K(\mathbf{k},z)]^{-1} S(\mathbf{k}) | \mathbf{p}_2 \rangle \ . \tag{405}$$

Looking at the quantity on the right-hand side,

$$\int d^3 p_2 p_2^x \langle \mathbf{p}_3 | S(\mathbf{k}) | \mathbf{p}_2 \rangle = \int d^3 p_2 p_2^x [\delta(\mathbf{p}_2 - \mathbf{p}_3) f_0(\mathbf{p}_2) + h(k) f_0(\mathbf{p}_2) f_0(\mathbf{p}_3)]$$
$$= p_3^x f_0(\mathbf{p}_3) \tag{406}$$

and Eq. (405) reduces to:

$$C_t(\mathbf{k},z) = \int d^3 p_1 \int d^3 p_2 p_1^x p_2^x f_0(\mathbf{p}_2) \langle \mathbf{p}_1 | [z - K(\mathbf{k},z)]^{-1} | \mathbf{p}_2 \rangle \ . \tag{407}$$

We now need to compare this expression with the hydrodynamical result:

$$C_t(\mathbf{k},z) = \frac{k_B T m n}{z + i \eta k^2 / m n} \ , \tag{408}$$

valid for small k and z. The viscosity η can be extracted from the correlation function using the limiting process we discussed in treating Green–Kubo functions. As shown in detail in Problem 7.27 we have the identity:

$$\eta = \lim_{z \to 0} \lim_{k \to 0} \frac{iz}{k_B T k^2} [z C_t(\mathbf{k},z) - m n k_B T] \ . \tag{409}$$

Inserting Eq. (407) into this expression we obtain:

$$\eta = \lim_{z \to 0} \lim_{k \to 0} \frac{iz}{k_B T k^2} \int d^3 p_1 \int d^3 p_2$$
$$\times p_1^x p_2^x f_0(\mathbf{p}_2) <\mathbf{p}_1 | z[z - K(\mathbf{k},z)]^{-1} - 1 | \mathbf{p}_2 > , \qquad (410)$$

where we have used the result:

$$\int d^3 p_1 \int d^3 p_2 p_1^x p_2^x f_0(\mathbf{p}_2) <\mathbf{p}_1|\mathbf{p}_2> = mnk_B T . \qquad (411)$$

Next, if we use the operator identity:

$$z[z - K(\mathbf{k},z)]^{-1} = 1 + K(\mathbf{k},z)[z - K(\mathbf{k},z)]^{-1} , \qquad (412)$$

we find:

$$\eta = \lim_{z \to 0} \lim_{k \to 0} \frac{iz}{k_B T k^2} \int d^3 p_1 \int d^3 p_2$$
$$\times p_1^x p_2^x f_0(\mathbf{p}_2) \langle \mathbf{p}_1 | K(\mathbf{k},z)[z - K(\mathbf{k},z)]^{-1} | \mathbf{p}_2 \rangle . \qquad (413)$$

Using the operator identity, Eq. (412), one more time we obtain:

$$\eta = \lim_{z \to 0} \lim_{k \to 0} \frac{i}{k_B T k^2} \int d^3 p_1 \int d^3 p_2 p_1^x p_2^x f_0(\mathbf{p}_2)$$
$$\times \langle \mathbf{p}_1 | K(\mathbf{k},z) \left[1 + [z - K(\mathbf{k},z)]^{-1} K(\mathbf{k},z) \right] | \mathbf{p}_2 \rangle . \qquad (414)$$

We have then from the first term:

$$\eta_1 = \lim_{z \to 0} \lim_{k \to 0} \frac{i}{k_B T k^2} \int d^3 p_1 \int d^3 p_2 p_1^x p_2^x f_0(\mathbf{p}_2) \langle \mathbf{p}_1 | K(\mathbf{k},z) | \mathbf{p}_2 \rangle$$
$$= \lim_{z \to 0} \lim_{k \to 0} \frac{i}{k_B T k^2} \int d^3 p_1 \int d^3 p_2 p_1^x p_2^x f_0(\mathbf{p}_2)$$
$$\times \left[\frac{\mathbf{k} \cdot \mathbf{p}_1}{m} \delta(\mathbf{p}_1 - \mathbf{p}_2) - \frac{\mathbf{k} \cdot \mathbf{p}_1}{m} c_D(k) f_0(\mathbf{p}_1) - K^{(d)}(\mathbf{k}, \mathbf{p}_1 \mathbf{p}_2; z) \right]$$
$$= \lim_{z \to 0} \lim_{k \to 0} \frac{-i}{k_B T k^2} \int d^3 p_1 \int d^3 p_2 p_1^x p_2^x K^{(d)}(\mathbf{k}, \mathbf{p}_1 \mathbf{p}_2; z) f_0(\mathbf{p}_2) . \qquad (415)$$

This contribution is clearly of second order in the density and can be dropped consistently in an expansion in the density at lowest order.

The second contribution to the viscosity from Eq. (414) is given by:

$$\eta_2 = \lim_{z \to 0} \lim_{k \to 0} \frac{i}{k_B T k^2} \int d^3 p_1 \int d^3 p_2 p_1^x p_2^x f_0(\mathbf{p}_2)$$
$$\times \langle \mathbf{p}_1 | K(\mathbf{k},z)[z - K(\mathbf{k},z)]^{-1} K(\mathbf{k},z) | \mathbf{p}_2 \rangle . \qquad (416)$$

To lowest order in the density we can replace the K's in the numerator by $L_0(\mathbf{k})$. Since $\mathbf{k}$ is taken to be in the z-direction we have:

$$\begin{aligned}\eta_2 &= \lim_{z\to 0}\lim_{k\to 0}\frac{i}{k_B T m^2}\int d^3p_1\int d^3p_2 p_1^x p_1^z p_2^x p_2^z f_0(\mathbf{p}_2)\\ &\quad \times \langle \mathbf{p}_1|[z-K(\mathbf{k},z)]^{-1}|\mathbf{p}_2\rangle\\ &= \frac{i}{k_B T m^2}\int d^3p_1\int d^3p_2 p_1^x p_1^z p_2^x p_2^z f_0(\mathbf{p}_2)\\ &\quad \times \langle \mathbf{p}_1|[K(0,i0^+)]^{-1}|\mathbf{p}_2\rangle \ . \end{aligned} \quad (417)$$

From Eq. (389) we have that $K(0,i0^+) = iK_B$ to lowest order in the density. We then find:

$$\begin{aligned}\eta &= \eta_2\\ &= \frac{1}{k_B T m^2}\int d^3p_1\int d^3p_2\, p_1^x p_1^z p_2^x p_2^z f_0(\mathbf{p}_2)\langle \mathbf{p}_1|(-K_B)^{-1}|\mathbf{p}_2\rangle \ . \end{aligned} \quad (418)$$

To go further one must develop methods for handling the inverse of the Boltzmann collision operator. Let us rewrite our expression for the shear viscosity in the form:

$$\eta = \frac{1}{k_B T m^2}\langle T|(K_B)^{-1}f_0|T\rangle \ , \quad (419)$$

where we have used the completeness of the momentum states $|\mathbf{p}>$ and defined:

$$\langle T|\mathbf{p}\rangle = p^x p^y \quad (420)$$

and:

$$f_0|\mathbf{p}\rangle = f_0(\mathbf{p})|\mathbf{p}\rangle \ . \quad (421)$$

It will turn out to be useful to deal with the symmetric operator:

$$\Gamma_B = K_B f_0 \ . \quad (422)$$

As a first step in this process, let us define the operator:

$$A = K_B^{-1} f_0 \ , \quad (423)$$

which enters the equation for η. Multiply on the left by K_B to obtain:

$$K_B A = f_0 \ . \quad (424)$$

We then write this in the form:

$$K_B f_0 f_0^{-1} A = \Gamma_B f_0^{-1} A = f_0 \ . \quad (425)$$

Multiplying from the left by Γ_B^{-1}, then f_0 gives the symmetric form:

$$A = f_0 \Gamma_B^{-1} f_0 , \tag{426}$$

and:

$$\eta = \frac{1}{k_B T m^2} \langle T | f_0 \Gamma_B^{-1} f_0 | T \rangle \tag{427}$$

$$= \frac{1}{k_B T m^2} \langle T' | \Gamma_B^{-1} | T' \rangle , \tag{428}$$

where $\langle T' | \mathbf{p} \rangle = p^x p^y f_0(\mathbf{p})$.

The next step is to construct the eigenvalues and and eigenfunctions of Γ_B:

$$\Gamma_B | \lambda \rangle = \Gamma_\lambda | \lambda \rangle . \tag{429}$$

This has been accomplished for Maxwell molecules [16]. Assuming that this set is complete and orthonormal we can write:

$$\eta = \frac{1}{k_B T m^2} \sum_\lambda \langle T' | \lambda \rangle \frac{1}{\Gamma_\lambda} \langle \lambda | T' \rangle . \tag{430}$$

The simplest approximation, accurate to about 5%, is to assume that $|T\rangle$ is an eigenvector of Γ_B, and:

$$\eta = \frac{1}{k_B T m^2} \langle T' | T \rangle \frac{1}{\Gamma_T} \langle T | T' \rangle , \tag{431}$$

where $\Gamma_T = \langle T | \Gamma_B | T \rangle$. The evaluation of the shear viscosity reduces to the evaluation of momentum integrals:

$$\langle T | f_0 | T \rangle = \int d^3 p_1 \int d^3 p_2 \, p_1^x p_1^z p_2^x p_2^z f_0(\mathbf{p}_2) \delta(\mathbf{p}_1 - \mathbf{p}_2) \tag{432}$$

and:

$$\langle T | K_B f_0 | T \rangle = \int d^3 p_1 \int d^3 p_2 \, p_1^x p_1^z p_2^x p_2^z K_B(\mathbf{p}_1 \mathbf{p}_2) f_0(\mathbf{p}_2) . \tag{433}$$

This evaluation of $\langle T | f_0 | T \rangle$ is not difficult and is evaluated in Problem 7.29. The evaluation of the matrix element of the linearized Boltzmann collision integral is involved and depends on the potential of interaction. If we restrict ourselves to the case of hard spheres the analysis is discussed in detail in Ref. [15]. The final result for the shear viscosity is given by:

$$\eta = \frac{5 m v_0}{16 \sqrt{\pi} r_0^2} . \tag{434}$$

Clearly this result is within a numerical factor of the result given by Eq. (276). One can then proceed to compute systematic corrections to Eq. (434). Clearly we can carry out a similar analysis for the thermal conductivity and bulk viscosity. It should be clear that the key building blocks in carrying out a detained analysis of kinetic theory is the set of matrices $\langle i|K^{(c)}f_0|j\rangle$. These can be used to numerically solve the associated eigenvalue problem or one can adopt the kinetic model approach discussed earlier in this chapter. We also see that a given approximation for $K^{(d)}$ implies not just a contribution from Eq. (417) but the full set of contributions given by Eq. (414).

These methods have been extended [30, 35] to higher order in the density where one generates long-time tails in the correlation functions that enter the Green–Kubo formulae. We will treat these long-time tails using less involved methods in Chapter 9.

7.5 References and Notes

1. See footnote 68 on p. 101 in ESM.
2. S. Brush, *Kinetic Theory, Vol. 1*, Pergamon, N.Y. (1965).
3. Seminal work in kinetic theory was carried out between 1859 and 1877 by J. C. Maxwell and L. Boltzmann. For a detailed discussion see S. Brush, *Statistical Physics and the Atomic Theory of Matter*, Princeton University Press, Princeton (1983).
4. J. Perin, *Atoms*, Ox Bow Press, Woodbridge, Conn. (1990).
5. This quantity was introduced in Chapter 5.
6. One of the advantages of this approach is that one need not be precise about the nature of the nonequilibrium average at this stage in the calculation. Eventually, when one wants precision, the definiteness of linear response theory becomes an asset.
7. It is clear that the two-particle distribution function is of the form:

$$f(\mathbf{x}_1, \mathbf{p}_1; \mathbf{x}_2, \mathbf{p}_2) = \left\langle \sum_{i\neq j=1}^{N} \delta(\mathbf{x}_1 - \mathbf{r}_i) \right. \\ \left. \times \delta(\mathbf{p}_1 - \mathbf{p}_i)\delta(\mathbf{x}_2 - \mathbf{r}_j)\delta(\mathbf{p}_2 - \mathbf{p}_j) \right\rangle_{NE}$$

8. Our development parallels that in J. Kestin, J. R. Dorfman, *A Course in Statistical Thermodynamics*, Academic, N.Y. (1971), Chapters 12, 13.
9. R. Clausius, Ann. Phys. ser. 2, **105**, 239 (1865); see the discussion on p. 54 in S. Brush, *Statistical Physics and the Atomic Theory of Matter*, Princeton University Press, Princeton (1983).
10. In a more careful treatment N_T is proportional to a two-particle distribution $f(\mathbf{x}_1, \mathbf{p}_1; \mathbf{x}_2, \mathbf{p}_2)$. We have followed Boltzmann and adopted the famous Stosszahlansatz or molecular chaos assumption where the two-particle distribution factorizes into a product of singlet-distribution function $f(\mathbf{x}_1, \mathbf{p}_1; \mathbf{x}_2, \mathbf{p}_2) = f(\mathbf{x}_1, \mathbf{p}_1)f(\mathbf{x}_2, \mathbf{p}_2)$.
11. L. Boltzmann, Math.-Naturwiss. Kl. **66**, 275 (1872). English translation in S. Brush, *Kinetic Theory, Vol. 2*, Pergamon, N.Y. (1966).
12. Boltzmann actually used the letter E rather than H. H was introduced later in S. H. Burbury, Philos. Mag. Ser. 5 **30**, 298 (1890).
13. A detailed discussion is given in S. Chapman, T. G. Cowling, *The Mathematical Theory of Nonuniform Gases*, Cambridge University Press, N.Y. (1953).
14. P. L. Bhatnagar, E. P. Gross, M. Krook, Phys. Rev. **94**, 511 (1954); E. P. Gross, E. A. Jackson, Phys. Fluids **2**, 432 (1959).
15. G. F. Mazenko, T. Wei, S. Yip, Phys. Rev. A **6**, 1981 (1972).

16 The exception is for so-called *Maxwell molecules*. In this case, where the interaction potential goes as r^{-4}, one can solve for the eigenvalues and eigenfunctions explicitly. J. C. Maxwell, *Collected Papers, Vol. 2*, p. 42; Philos. Trans. R. Soc. **157**, 49 (1867). See p. 175 in Ref. [13].

17 In this case the negative carriers move in a fixed background of positive charge, which provides overall change neutrality.

18 T. M. Reed, K. E. Gubbins, *Applied Statistical Mechanics*, Mc Graw-Hill, N.Y. (1973).

19 See the discussions in N. Ashcroft, N. D. Mermin, *Solid State Physics*, Holt, Rinehardt and Winston, N.Y. (1976), Chapters 1, 13.

20 L. Onsager, Phys. Rev. **37**, 405 (1931); Phys. Rev. **38**, 2265 (1931).

21 For a discussion of a variety of thermoelectric, thermomagnetic and galvanomagnetic effects, like the Seebeck effect, see H. B. Callen, Phys. Rev. **73**, 1349 (1948).

22 E. H. Hall, Am. J. Math. **2**, 287 (1879).

23 See the discussion in Ashcroft and Mermin, Ref. [19], p. 15.

24 L. S. Ornstein, F. Zernike, Proc. Acad. Sci. Amsterdam **17**, 793 (1914); see p. 305 in ESM.

25 E. E. Salpeter, Phys. Rev. **120**, 1528 (1960); D. C. Montgomery, D. A. Tidman, *Plasma Kinetic Theory*, McGraw-Hill, N.Y. (1964).

26 M. Nelkin, S. Ranganthan, Phys. Rev. **164**, 222 (1967).

27 H. Gould, G. F. Mazenko, Phys. Rev. A **15**, 1274 (1977); Phys. Rev. Lett. **35**, 1455 (1975).

28 See the discussion of the direct correlation function in Section 4.7.1 in ESM.

29 The direct correlation function for a plasma is given on p. 316 in ESM.

30 G. F. Mazenko, Phys. Rev. A **9**, 360 (1974).

31 H. C. Anderson, J. Phys. Chem. B **107**, 10234 (2003).

32 D. Forster, P. C. Martin, Phys. Rev. A **12**, 1575 (1970).

33 A. Z. Akcasu, J. J. Duderstadt, Phys. Rev. **188**, 479 (1969).

34 G. F. Mazenko, Phys. Rev. A **5**, 2545 (1972).

35 J. R. Dorfmann, E. G. D. Cohen, Phys. Rev. Lett. **25**, 1257 (1970); Phys. Rev. A **6**, 776 (1972).

7.6
Problems for Chapter 7

Problem 7.1: Show that the singlet distribution function $f(\mathbf{x}, \mathbf{p})$, defined by Eqs.(1) and (6), is related to $f(\mathbf{x}, \mathbf{v})$, defined by Eq. (12), by $f(\mathbf{x}, \mathbf{v}) = m^3 f(\mathbf{x}, \mathbf{p})$ where m is the mass and we work in three dimensions.

Problem 7.2: Evaluate the integral:

$$a = n \int_0^\infty v^3 dv \int_0^{\pi/2} d\theta \, \sin\theta \, \cos\theta \int_0^{2\pi} d\phi \, \Phi(\mathbf{v}) \ ,$$

giving the effusion rate for a dilute gas in thermal equilibrium. $\Phi(\mathbf{v})$ is the Maxwell velocity distribution given by Eq. (23).

Problem 7.3: Evaluate the average speed u in a classical gas using:

$$u = \int d^3 v \, v \Phi(\mathbf{v}) \ ,$$

where $\Phi(\mathbf{v})$ is the Maxwell velocity distribution given by Eq. (23).

Problem 7.4: Finish the calculation of the collision rate for a dilute gas in thermal equilibrium by evaluating the velocity integrals:

$$I = \int d^3v \int d^3v_1 \, e^{-\frac{1}{2}\beta m(v^2+v_1^2)} |\mathbf{v}_1 - \mathbf{v}| \ .$$

Problem 7.5: Starting with the phase-space density $\hat{f}(\mathbf{x}, \mathbf{p}, t)$, defined by Eq. (1), and assuming that the individual momenta satisfy the equation of motion given by Eq. (55), show that the equation of motion satisfied by $\hat{f}(\mathbf{x}, \mathbf{p}, t)$ is given by Eq. (56).

Problem 7.6: Consider a collision process between two particles with initial velocities $\mathbf{v}$ and $\mathbf{v}_1$ that scatter into final velocities $\mathbf{v}'$ and $\mathbf{v}_1'$. Show that:

$$d^3v \, d^3v_1 = d^3v' \, d^3v_1' \ .$$

Problem 7.7: Show how the change of integration variables given by Eq. (83) follow from the two-body kinematics.

Problem 7.8: Show that the inequality:

$$(x-y)\ln\left(\frac{y}{x}\right) \leq 0$$

holds.

Problem 7.9: Show that the linearized Boltzmann collision operator, defined by Eq. (125), obeys the symmetry:

$$K_B(p, p') f_0(p') = K_B(p', p) f_0(p) \ .$$

Problem 7.10: Consider the set of polynomials given by Eqs.(140)–(144). Show that the constants N_i and $\bar{\varepsilon}_0$ can be chosen such that the polynomials are orthonormal, as in Eq. (138).

Problem 7.11: Show that the expansion of $I(k, z)$, defined by Eq. (176), in powers of wavenumber leads to the result given by Eq. (178).

Problem 7.12: If we turn on a time-dependent force of the form given by Eq. (186), then in linear response theory we need to evaluate the integral:

$$\mathcal{F}(t) = e^{-t/\tau} \int_0^t dt' e^{t'/\tau} \mathbf{F}(t') \ .$$

Show that:

$$\mathcal{F}(t) = \left[\tau\left(1 - e^{-t/\tau}\right) - \left(\tau^{-1} - t_0^{-1}\right)\left(e^{-t/t_0} - e^{-t/\tau}\right)\right] \mathbf{F} \ .$$

Problem 7.13: Starting with $f(p)$, given by Eq. (226), show that $f(p)$ is properly normalized, leads to the correct linear response result for small $\mathbf{F}$, and find the nonlinear contributions to the electrical and heat currents.

Problem 7.14: Evaluate the momentum integrals of the form:

$$\bar{\varepsilon}_0^n = \int d^3p\, \varepsilon_0^n(p) f_0(p)$$

for $n = 1, 2, 3$.

Problem 7.15: Solve the transport equation:

$$\mathbf{J} = \sigma \mathbf{E} - \mathbf{J} \times \mathbf{H}$$

to obtain the current $\mathbf{J}$ as a function of $\mathbf{E}$, $\mathbf{H}$ and σ.

Problem 7.16: Show that the expression for the rate of effusion given by Eq. (32) has the correct units.

Problem 7.17: Estimate the mean-free path and time assuming a temperature $T \approx 300$ K, pressure $p \approx 1$ atm and hard-core diameter $r_0 \approx 4 \times 10^{-8}$ cm.

Problem 7.18: One can work out the general solution to the case where a set of electrons is driven by a static inhomogeneous temperature. The appropriate kinetic equation is given by:

$$\frac{\mathbf{p} \cdot \nabla_x}{m} f(\mathbf{x}, \mathbf{p}, t) = -\frac{1}{\tau}[f(\mathbf{x}, \mathbf{p}, t)] - f_0[\mathbf{p}, T(\mathbf{x})] \ .$$

Show that this equation has the general solution:

$$f(\mathbf{x}, \mathbf{p}) = \int_0^\infty ds\, e^{-s} f_0[\mathbf{p}, T(\mathbf{x} - \frac{s\tau\mathbf{p}}{m})] \ .$$

Show that the number of particles is conserved in this quantity. Investigate the perturbation theory expansion in a weak temperature gradient. If $T(x) = T_0 + \frac{x}{L}(T_L - T_0)$, then show that to linear order $\nabla_x T = \frac{(T_L - T_0)}{L}$, we recover Eq. (246).

Problem 7.19: Given the solution :

$$f(\mathbf{x}, \mathbf{p}) = \int_0^\infty ds\, e^{-s} f_0[\mathbf{p}, T(\mathbf{x} - \frac{s\tau\mathbf{p}}{m})]$$

in the problem for a system driven in local equilibrium by a temperature gradient. Assume:

$$T(x) = T_0 + \frac{x}{L}(T_L - T_0)$$

in a system of length L in the x-direction. Write:

$$T(x) = T + \delta T(x) \ ,$$

where:

$$T = \frac{1}{L}\int_0^L dx\, T(x)$$

is the average temperature, and expand $f(\mathbf{x}, \mathbf{p})$ in a power series in $\delta T(x)$. Compute the averages:

$$n(x) = \int d^3p\, f(\mathbf{x}, \mathbf{p})$$

$$\mathbf{g}(x) = \int d^3p\, \mathbf{p}\, f(\mathbf{x}, \mathbf{p})$$

$$K(x) = \int d^3p\, \frac{\mathbf{p}^2}{2m} f(\mathbf{x}, \mathbf{p})$$

$$\mathbf{J}_K(x) = \int d^3p\, \frac{\mathbf{p}}{m} \frac{\mathbf{p}^2}{2m} f(\mathbf{x}, \mathbf{p})\ .$$

Problem 7.20: Starting with the single-relaxation approximation, Eq. (184), for a system in an AC electric field find the frequency-dependent conductivity $\sigma(\omega)$ relating the electrical current and the applied field.

Problem 7.21: In the case of a system with inhomogeneous flow, show that the leading order contribution to the off-diagonal stress tensor,

$$\sigma_{ij}^{(0)} = \int d^3p\, \frac{p_i p_j}{m} f_0[\mathbf{p} - m\mathbf{u}(x)]\ ,$$

$i \ne j$, is second order in the small flow velocity $\mathbf{u}(x)$ and does not contribute to the shear viscosity.

Problem 7.22: Show that:

$$f_0(1) f_0(2) g(\mathbf{x}_1 - \mathbf{x}_2) = \left\langle \sum_{i \ne j=1}^{N} \delta(\mathbf{x}_1 - \mathbf{r}_i)\, \delta(\mathbf{p}_1 - \mathbf{p}_i)\, \delta(\mathbf{x}_2 - \mathbf{r}_j)\, \delta(\mathbf{p}_2 - \mathbf{p}_j) \right\rangle ,$$

where the radial distribution function is defined by:

$$n^2 g(\mathbf{x}_1 - \mathbf{x}_2) = \left\langle \sum_{i \ne j=1}^{N} \delta(\mathbf{x}_1 - \mathbf{r}_i)\, \delta(\mathbf{x}_2 - \mathbf{r}_j) \right\rangle\ .$$

Problem 7.23: Show that the Poisson bracket between phase-space densities is given by:

$$\{\hat{f}(1), \hat{f}(2)\} = (\nabla_{\mathbf{p}_1} \cdot \nabla_{\mathbf{x}_2} - \nabla_{\mathbf{p}_2} \cdot \nabla_{\mathbf{x}_1}) \left[\delta(\mathbf{x}_1 - \mathbf{x}_2)\, \delta(\mathbf{p}_1 - \mathbf{p}_2)\, \hat{f}(1)\right]\ .$$

Problem 7.24: Show that Eq. (339) can be put into the form symmetric under exchange of $\mathbf{p}_1$ and $\mathbf{p}_2$.

Problem 7.25: Show that for a low-density fluid of hard spheres that the Fourier transform of the hole function is given by:

$$h(\mathbf{k}) = -\frac{4\pi}{k^3}\left[\sin(kr_0) - kr_0\cos(kr_0)\right] \quad .$$

Problem 7.26: Show that:

$$\int d^3p \left(\frac{\mathbf{k}\cdot\mathbf{p}}{m}\right)^2 f_0(p) = k^2\frac{k_BT}{m}n \quad .$$

Problem 7.27: Given the hydrodynamical form for the transverse correlation function, Eq. (408), show:

$$\eta = \lim_{z\to 0}\lim_{k\to 0}\frac{iz}{k_BTk^2}\left[zC_T(\mathbf{k},z) - mnk_BT\right] \quad .$$

Problem 7.28: Write down the $N = 6$ kinetic model for the collision integral $K_B(\mathbf{p},\mathbf{p}')$ for the case where we preserve the five conservation laws. Put this result into the kinetic equation and assume that the system is spatially uniform. Solve for $f(\mathbf{p},t)$, assuming that initially $f(\mathbf{p},t=0) = f_{\text{in}}(\mathbf{p})$.

Problem 7.29: Evaluate the momentum integral:

$$\langle T'|T\rangle = \langle T|f_0|T\rangle = \int d^3p_1 \int d^3p_2\, p_1^x p_1^z p_2^x p_2^z f_0(\mathbf{p}_2)\delta(\mathbf{p}_1-\mathbf{p}_2) \quad ,$$

which enters the determination of the shear viscosity.

Problem 7.30: Evaluate $G(11';22';z)$, defined by Eq. (371), for an ideal gas. Note that terms cancel between the two pieces in Eq. (371).

Problem 7.31: Evaluate $G(11';22';z)$, defined by Eq. (371), in a density expansion to first order in the density.

Problem 7.32: Evaluate the integral for the thermal conductivity given by Eq. (243).

Problem 7.33: Evaluate the remaining integral for the *off-diagonal* response functions given by Eq. (259).

Problem 7.34: Do the electric and thermal gradient forces induce a contribution to the off-diagonal components of the average stress tensor? Explain.

Problem 7.35: Expand the direct correlation function to first order in a weak interaction potential.

8
Critical Phenomena and Broken Symmetry

8.1
Dynamic Critical Phenomena

8.1.1
Order Parameter as a Slow Variable

In our development in Chapters 5 and 6 we have established the typical or conventional hydrodynamic picture. How do these ideas break down? One of the most obvious places to look for a breakdown of conventional hydrodynamics is near a second-order phase transition [1]. The key point in this case is that there is a length in the problem, the correlation length ξ, which becomes infinite as one goes to the transition. This length measures the correlations of the order parameter associated with the phase transition. Since these correlated regions can involve very many degrees of freedom we expect the time evolution of the order parameter to be slow. Thus we potentially have a new mechanism for introducing a slow variable into the dynamics of a macroscopic system: the kinetics of the order parameter near a critical point. In this chapter we analyze these kinetics using the linearized Langevin equation approach developed earlier. This leads to a mean-field analysis of the problem. A more sophisticated treatment of dynamic critical phenomena is given in Chapter 10 after we have discussed nonlinear Langevin methods in Chapter 9.

The first step in analyzing critical phenomena is to identify the associated order parameter [2]. The order parameter field $\psi(\mathbf{x})$ has the following properties [3]:

- The average of the order parameter vanishes at temperatures above some transition temperature T_c. The zero average of the order parameter is usually due to some symmetry principle [4]. Below the transition temperature the average value of the order parameter is nonzero:

$$\langle \psi(\mathbf{x}) \rangle = \begin{matrix} 0 & T > T_c \\ M & T < T_c \end{matrix} , \qquad (1)$$

where, near the critical point, M vanishes as $(T_c - T)^\beta$ where β is one of a handful of critical exponents [5] that characterize the analytic properties of a critical point.

- The order-parameter equal-time or static correlation function is defined by:

$$S(\mathbf{x} - \mathbf{y}) = \langle \delta\psi(\mathbf{x})\delta\psi(\mathbf{y})\rangle \ . \tag{2}$$

We are interested in the Fourier transform, the order parameter structure factor $S(q, T)$. This quantity becomes large as one approaches the critical point and additional critical indices, η and γ, are defined by:

$$S^{-1}(\mathbf{q}, T_c) \approx q^{2-\eta} \tag{3}$$

and:

$$S(0, T) \approx |T - T_c|^{-\gamma} \ . \tag{4}$$

These expressions indicate the singular nature of the critical point as probed by order parameter fluctuations. An important quantity in measuring the growing correlations as one approaches the critical point is the correlation length, which can be defined by:

$$\xi^2 = \frac{\int d^d x\, x^2 S(x, T)}{\int d^d x\, S(x, T)} \tag{5}$$

and ξ grows as $T \to T_c$ as:

$$\xi \approx |T - T_c|^{-\nu} \ . \tag{6}$$

It turns out that in three dimensions the index η is typically small [6]. In the approximation where we set $\eta = 0$, one is led to the *Ornstein–Zernike* result [7]:

$$S(\mathbf{q}, T) = \frac{c}{q^2 + \xi^{-2}} \tag{7}$$

for T near T_c. Inverting the Fourier transform in three dimensions gives the order parameter correlation function:

$$S(\mathbf{x} - \mathbf{y}, T) = c\frac{e^{-|\mathbf{x}-\mathbf{y}|/\xi}}{4\pi|\mathbf{x} - \mathbf{y}|} \tag{8}$$

for $T \geq T_c$ where c is a positive temperature-independent constant as $T \to T_c$. Notice that Eq. (7) implies that:

$$S(\mathbf{0}, T) = c\xi^2 \tag{9}$$

and $\gamma = 2\nu$.

Since $S(\mathbf{q}, T_c)$ diverges as $\mathbf{q} \to 0$, we see that one of our hydrodynamic assumptions [8] breaks down. We note that ξ introduces a large length scale in the problem and even if $\mathbf{q}$ is very small it is inappropriate to replace $S(\mathbf{q})$ by $S(\mathbf{0})$ except if $q\xi \ll 1$.

8.1.2
Examples of Order Parameters

One can identify, for essentially all known second-order phase transitions, the associated order parameter [9]. Here we will concentrate on the variety of order parameters associated with the Heisenberg model for magnetic systems. It will turn out that many physical systems have order parameters that are similar in nature to one of the order parameters in this model.

Suppose we have a set of spins $\mathbf{S}(\mathbf{R})$ setting on a lattice at site $\mathbf{R}$. We assume that the dynamics of these spins is governed by the Hamiltonian of the form:

$$H = -\frac{1}{2} \sum_{\mathbf{R},\mathbf{R}'} \sum_k J^{(k)}(\mathbf{R} - \mathbf{R}') S_k(\mathbf{R}) S_k(\mathbf{R}') , \qquad (10)$$

where the *exchange interaction*, J, is assumed to depend on the distance between the two spins. In the simplest models with short-range interactions it is sufficient to consider only nearest-neighbor interactions and write:

$$H = - \sum_{\langle \mathbf{R},\mathbf{R}' \rangle_{NN}} \sum_k J^{(k)} S_k(\mathbf{R}) S_k(\mathbf{R}') . \qquad (11)$$

There are then three important cases. First we have the case where the system is isotropic and the exchange interaction is the same in all directions: $J^{(k)} = J$. In this case one has the isotropic Heisenberg model with Hamiltonian:

$$H = - \sum_{\langle \mathbf{R},\mathbf{R}' \rangle_{NN}} J \mathbf{S}(\mathbf{R}) \cdot \mathbf{S}(\mathbf{R}') . \qquad (12)$$

When the system orders it will be equally likely along any of the directions in three-dimensional space. For an isotropic ferromagnet the order parameter is a three-dimensional vector. The second case is where there is an easy plane for rotation:

$$J^{(x)} = J^{(y)} > J^{(z)} . \qquad (13)$$

In this case it is energetically favorable for the spins to line up in the xy plane and the ordering will be in this plane. The associated order parameter is a two-component vector corresponding to ordering in the plane. This model is called an XY or planar model.

Finally one can have the situation where there is an easy axis:

$$J^{(z)} > J^{(x)}, J^{(y)} \ . \tag{14}$$

In this case, the order parameter is a scalar, and there is ordering parallel or antiparallel to the z-direction. This corresponds to an Ising model. Thus we can have order parameters with differing numbers of ordering components denoted by n. For $n = 3$ we have the isotropic Heisenberg model, for $n = 2$ we have an XY model and for $n = 1$ we have an Ising model. We know that the critical phenomena of a given system depend [10] on the value of n.

There are other considerations that enter into the selection of an order parameter that are more important in the case of dynamic critical phenomena compared to static critical phenomena. Consider the simple situation of just two Ising spins in one dimension governed by the Hamiltonian:

$$H_{12} = -JS_1 S_2 \ . \tag{15}$$

Let us assume that $J > 0$. There are then two possible energetically distinct configurations:

- If the spins are parallel ↑↑ or ↓↓, then:

$$H_{12}^p = -JS_1 S_2 < 0 \ . \tag{16}$$

- If the spins are antiparallel ↓↑ or ↑↓ then:

$$H_{12}^A = -JS_1 S_2 > 0 \ . \tag{17}$$

Since a system, at low temperature, acts to lower its energy for $J > 0$ the system *wants* to line up parallel. Clearly, if $J < 0$ then:

$$H_{12}^p > 0 \ , \tag{18}$$

$$H_{12}^A < 0 \tag{19}$$

and the spins want to line up antiparallel.

The case $J < 0$ corresponds to a system that aligns antiferromagnetically at low temperatures. A possible alignment at zero temperature is shown in Fig. 8.1. We see that there is now no net magnetization:

$$\langle \mathbf{M}_T \rangle = \left\langle \sum_{\mathbf{R}} \mathbf{S}(\mathbf{R}) \right\rangle = 0 \ . \tag{20}$$

We note, however, that there is order at low temperatures. Let us define a quantity called the staggered magnetization:

$$\mathbf{N}(\mathbf{x}) = \sum_{\mathbf{R}} \eta(\mathbf{R}) \mathbf{S}(\mathbf{R}) \delta(\mathbf{x} - \mathbf{R}) \ . \tag{21}$$

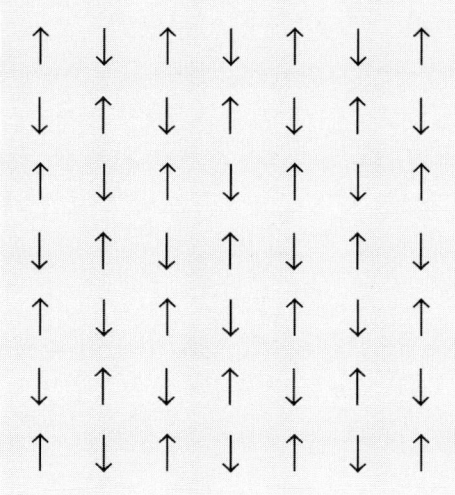

Fig. 8.1 Zero-temperature ordering of an Ising antiferromagnet on a square lattice.

For a square or cubic lattice, the $\eta(\mathbf{R})$ are defined as follows. Divide the lattice up into two interpenetrating sublattices. If a given spin is on one sublattice then all of its nearest neighbors are on the other sublattice. Then arbitrarily assign $\eta(\mathbf{R}) = +1$ on one sublattice and $\eta(\mathbf{R}) = -1$ on the other sublattice, as shown [11] in Fig. 8.2. Finally we plot in Fig. 8.3 the value of $\eta(\mathbf{R})\mathbf{S}(\mathbf{R})$ at the various lattice sites at low temperatures. The -1 values of $\eta(\mathbf{R})$ flip the orientation of the spin at that site and the staggered magnetization appears ordered. It is then clear that there is a net total staggered magnetization:

$$\langle \mathbf{N}_T \rangle = \left\langle \sum_{\mathbf{R}} \eta(\mathbf{R})\mathbf{S}(\mathbf{R}) \right\rangle \neq 0 \ . \tag{22}$$

The staggered magnetization is the ordered or symmetry breaking field (order parameter) for an antiferromagnet. Thus we have:

$$\langle \mathbf{N}(\mathbf{x}) \rangle = \begin{matrix} 0 & T \geq T_N \\ N\hat{z} & T < T_N \end{matrix} \tag{23}$$

where T_N is the Néel transition temperature and near and below the transition $N \approx (T_c - T)^\beta$. Note that ferro- and antiferromagnets may be in the same static universality class and share the same values, for example, for β.

With respect to their equilibrium critical properties ferro and antiferromagnets with the same symmetry of the order parameter (same value of n) fall into the same universality classes. This need not be the case with respect to dynamic critical phenomena. The reason has to do with conservation laws.

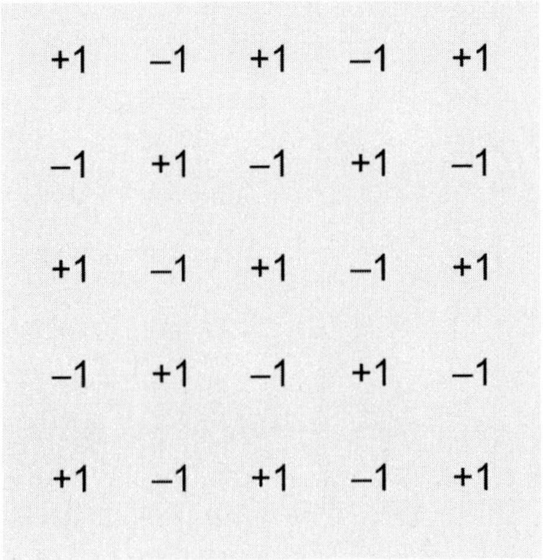

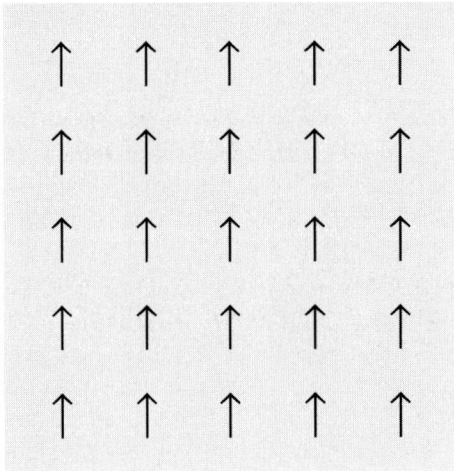

Fig. 8.2 One choice for the phase factors $\eta(\mathbf{R})$ for a two-dimensional square lattice.

Fig. 8.3 Ordered configuration for $\eta(\mathbf{R})S(\mathbf{R})$ for an antiferromagnet on a square lattice.

We have seen previously that for the isotropic case the magnetization is conserved. If there is an easy plane:

$$J^{(x)} = J^{(y)} \neq J^{(z)} , \qquad (24)$$

then only the magnetization in the z-direction is conserved. If there are three unequal exchange couplings:

$$J^{(x)} \neq J^{(y)} \neq J^{(z)} , \qquad (25)$$

then the magnetization is not conserved. It is easy to see, even in the isotropic case, that the staggered magnetization is not conserved:

$$\{\mathbf{N}_T, H\} = \sum_{\langle \mathbf{R},\mathbf{R}'\rangle_{NN}} J(\mathbf{R}-\mathbf{R}')[\eta(\mathbf{R})-\eta(\mathbf{R}')][\mathbf{S}(\mathbf{R})\times\mathbf{S}(\mathbf{R}')] \quad . \qquad (26)$$

Whether the order parameter is conserved or not will clearly matter in treating dynamic critical phenomena.

8.1.3
Critical Indices and Universality

We have introduced above a number of critical indices that characterize the properties of the order parameter near the critical point. Equations(1), (6), (3) and (4) define β, ν, η and γ. In the 1950s, 60s and 70s there was extensive experimental, numerical and theoretical study [12] of the critical properties (indices) of a wide variety of systems. One of the organizing ideas to come out of this work is that there are similarities between systems with seemingly very different microscopic interaction mechanisms. This idea can be summarized in the hypothesis of *universality* [13]:

All phase transition problems can be divided into a small number of different classes depending upon the dimensionality d of the system and the symmetries n of the ordered state. Within each class, all phase transitions have identical behavior in the critical region; only the names of the variables are changed.

As a first step one must find the appropriate values of d and n. The appropriate spatial dimensionality is usually three, but some layered or surface compounds are more appropriately [14] taken as two-dimensional. The n-dependence tells one whether one has an order parameter that corresponds to an Ising, XY or Heisenberg model. As an example, in superfluid helium the order parameter is a complex creation operator that has two components ($n = 2$). Therefore the critical properties of the λ-transition are the same as for an XY magnet. The liquid–gas transition and Ising magnets are in the same universality class. There are additional classes beyond the $\mathcal{O}(n)$ symmetric models. One can have systems with cubic symmetry, others with uniaxial symmetry, etc., or more exotic symmetries as for certain liquid crystals [15].

8.1.4
The Scaling Hypothesis

Closely related to the proposition of universality is the idea of scaling. Universality implies that the microscopic details of a system are unimportant near the critical point. The critical indices cannot depend on the lattice separation a in a ferromagnet or the hard-core separation distance r_0 in a fluid. If these lengths are to be removed from the problem, then there must be one length that is much longer and which dominates these microscopic lengths. This length is clearly the correlation length ξ, and it dominates all other microscopic lengths in the problem, $a/\xi \ll 1$, near T_c. According to the *scaling hypothesis*, due to Widom [16] and Kadanoff [17], the divergence of ξ is responsible for the singular dependence on $T - T_c$ of physical quantities, and, as far as the singular dependence is concerned, ξ is the only relevant length. This scaling assumption can help correlate a large amount of information.

Consider first the order-parameter correlation function. We note that this quantity is a function, in general, of $\mathbf{q}$, T and the temperature-independent variables such as, for a magnet, the lattice spacing a and the nearest neighbor interaction J. We can, in principle, eliminate the temperature in terms of the correlation length and write:

$$S = S(\mathbf{q}, \xi, a, J) \ . \tag{27}$$

For simplicity, we assume that the field conjugate to the order parameter, like an external magnetic field for a ferromagnet, is set to its critical value. According to the scaling hypothesis $S(\mathbf{q}, \xi, a, J)$ becomes a homogeneous function in the critical or *scaling* region:

$$S(q, \xi) = \xi^x f(q\xi) \ , \tag{28}$$

where the index x is unknown. Scaling laws of this type are particularly useful in analyzing experimental data. First one collects scattering data as a function of $\mathbf{q}$ for a given temperature T_1. Then collect another set of data as a function of $\mathbf{q}$ for another temperature T_2, and so on. According to our scaling assumption, if we plot $S(\mathbf{q})/\xi^x$ versus $q\xi$ then we can find a temperature-dependent length ξ and an exponent x such that all of the data collapses on a single curve.

The scaling assumption has immediate consequences concerning the critical indices. If we hold ξ fixed and set $\mathbf{q} = 0$ in Eq. (28), we can use Eq. (4) to obtain:

$$S(0, \xi) = \xi^x f(0) \approx |T - T_c|^{-\gamma} \ . \tag{29}$$

If we remember Eq. (6), we can write:

$$|T - T_c| \approx \xi^{-1/\nu} \ , \tag{30}$$

and, after inserting this in Eq. (29), identify:

$$x = \gamma/\nu \ . \tag{31}$$

Similarly if we set $T = T_c$, then, for small q, Eq. (28) reads:

$$S(q, T_c) = \lim_{\xi \to \infty} \xi^x f(q\xi) \ . \tag{32}$$

Since $S(q, T_c)$ is finite for $q \neq 0$, the limit in Eq. (32) must exist and we require:

$$f(q\xi, 0) \to (q\xi)^{-x} \qquad (33)$$

as $q\xi \to \infty$. This means that Eq. (32) reduces to:

$$S(q, T_c) \approx q^{-x} \ . \qquad (34)$$

If we compare this with Eq. (3) we can identify:

$$x = 2 - \eta \ . \qquad (35)$$

Combining this result for x with Eq. (31) gives the relation:

$$\gamma = (2 - \eta)\nu \ . \qquad (36)$$

Thus, the scaling hypothesis allows us to find relations among the various critical indices. This suggests that the critical indices are not all independent.

There is a simple set of approximate theories one can use to check the scaling hypothesis. Mean-field theory [18] is a collection of theories like the Landau theory [19], Curie-Weiss theory [20], and van der Waals' theory [21]. These theories all assume a certain level of analyticity near the critical point and give for all (n, d), $\gamma = 1$, $\nu = 1/2$ and $\eta = 0$. It turns out that these results hold in $d = 4$ dimensions for reasons that become apparent in a renormalization group analysis, as discussed in Chapter 10. This set of indices do obey scaling. Another important theory was Onsager's solution [22] of the two-dimensional Ising model where we know $\gamma = 7/4$, $\nu = 1$, and $\eta = 1/4$. These values are quite different from the mean-field theory values, but do also obey the scaling relations.

While the scaling hypothesis played an important role in the development of our modern theory of critical phenomena, it was still far from a complete theory. It reduced the number of independent indices to two, but it did not tell us anything about how to calculate them. More fundamentally, we did not know how to establish the scaling hypothesis from first principles. The renormalization group theory led to the solution of this problem.

8.1.5
Conventional Approximation

With this background on static critical phenomena we turn to the study of dynamic critical phenomena. Let us proceed with the assumption that the order parameter ψ is the slow variable, which is not coupled to any other variable and we can use our Langevin equation approach. We begin by assuming that the order parameter is not conserved.

As usual in this approach, the first step after identifying the slow variables is to determine their static correlations. We assume that in this case, the static order parameter structure factor $S(\mathbf{q})$ is given approximately by the Ornstein–Zernike form expressed by Eq. (7).

We turn next to the evaluation of the static part of the memory function. If we have a single-component order parameter there is no nonzero Poisson bracket we can form and:

$$K^{(s)}(\mathbf{q}) = 0 \quad . \tag{37}$$

Turning to the dynamic part of the memory function, assuming that the order parameter is not conserved, the simplest assumption, known as the *conventional* [23] or *van Hove approximation*, is that for small frequencies and wavenumbers:

$$\Gamma^{(d)}(\mathbf{q}, z) = K^{(d)}(q, z) S(q) = -i\Gamma \quad , \tag{38}$$

where Γ is assumed to be a constant. The dynamic part of the memory function is given in this approximation by:

$$K^{(d)}(\mathbf{q}, z) = -i\Gamma S^{-1}(\mathbf{q}) \quad , \tag{39}$$

and the simplest assumption is that the kinetic coefficient Γ shows a weak temperature dependence near the critical point.

Inserting Eqs. (37) and (39) into Eq. (5.121), the equation of motion for the order parameter fluctuation function is given by:

$$[z + i\Gamma S^{-1}(\mathbf{q})] C(\mathbf{q}, z) = S(\mathbf{q}) \quad . \tag{40}$$

Inverting the Laplace transform, we obtain the order parameter time-correlation function:

$$C(\mathbf{q}, t) = S(\mathbf{q}) e^{-\Gamma S^{-1}(\mathbf{q}) t} \quad . \tag{41}$$

If we are at the transition, where $T = T_c$ and $S(\mathbf{q}) = c/q^2$, we have:

$$C(\mathbf{q}, t) = \frac{c}{q^2} e^{-\frac{\Gamma}{c} q^2 t} \quad . \tag{42}$$

We are now in a position to see why we must treat the order parameter as a slow variable. The order parameter time-correlation function, at the critical point, looks as though the order parameter is conserved since its decay time is given by:

$$\tau(\mathbf{q}) = \frac{c}{\Gamma q^2} \quad , \tag{43}$$

which goes to infinity as $q \to 0$.

If we are near T_c, where the correlation length ξ is large, the static susceptibility in Eq. (41) can be replaced by:

$$S(\mathbf{q}) = \frac{c}{q^2 + \xi^{-2}} \tag{44}$$

and the relaxation time is given by:

$$\tau(\mathbf{q}, \xi) = \frac{c}{\Gamma(q^2 + \xi^{-2})} \,, \tag{45}$$

and at zero wavenumber,

$$\tau(\mathbf{0}, \xi) = \frac{c\xi^2}{\Gamma} \tag{46}$$

is large. The existence of this large time is known as *critical slowing down* [24].

More generally, in the critical regime, one has, using Eq. (35) in Eq. (28), the scaling form:

$$S(\mathbf{q}, \xi) = q^{-2+\eta} f(q\xi) \,. \tag{47}$$

In the Ornstein–Zernike case, $\eta = 0$, the scaling function is given by:

$$f(x) = (1 + x^{-2})^{-1} \,. \tag{48}$$

In the case of dynamic critical phenomena, one can generalize the static scaling given by Eq. (47) by taking into account the existence of the large dominant time τ given by Eq. (46). It is then conventional to introduce the dynamic scaling form [25] for the order parameter fluctuation function:

$$C(\mathbf{q}, z) = \frac{S(\mathbf{q}, \xi)}{\omega_c(\mathbf{q}, \xi)} F[z/\omega_c(\mathbf{q}, \xi)] \,, \tag{49}$$

where ω_c is the characteristic frequency with the assumed scaling form:

$$\omega_c(\mathbf{q}, \xi) = q^z f_D(q\xi) \,, \tag{50}$$

where z is the dynamic critical index and f_D and F are scaling functions. In our simple example, using Eq. (40), we have:

$$C(\mathbf{q}, z) = \frac{S(\mathbf{q})}{z + i\Gamma S^{-1}(\mathbf{q})}$$

$$= \frac{S(\mathbf{q})}{\omega_c(\mathbf{q}, \xi)} \frac{1}{[\Omega + i]} \,, \tag{51}$$

where the dimensionless frequency is given by $\Omega = z/\omega_c$, and the characteristic frequency can be identified as:

$$\omega_c = \Gamma S^{-1}(\mathbf{q}, \xi) \tag{52}$$

$$= q^2 \frac{\Gamma}{c}[1 + (q\xi)^{-2}] \tag{53}$$

and the dynamic critical index is given [26] by:

$$z = 2 \ . \tag{54}$$

The scaling function is given by:

$$f(x) = \frac{\Gamma}{c}(1 + x^{-2}) \ . \tag{55}$$

This is the simple behavior that one expects from a nonconserved Ising system.

If the order parameter is conserved, as for certain isotropic ferromagnets, then for small q and z, Eq. (38) is replaced by:

$$\Gamma^{(d)}(\mathbf{q}, z) = -iq^2 D \ , \tag{56}$$

where D is a diffusion coefficient, and the dynamic part of the memory function is given by:

$$K^{(d)}(\mathbf{q}, z) = -iDq^2 S^{-1}(\mathbf{q}) \ . \tag{57}$$

The order parameter fluctuation function, assuming again that the static part of the memory function vanishes, then satisfies:

$$[z + iq^2 DS^{-1}(\mathbf{q})]C(\mathbf{q}, z) = S(\mathbf{q}) \ . \tag{58}$$

Inverting the Laplace transform,

$$C(\mathbf{q}, t) = S(\mathbf{q}) e^{-DS^{-1}(\mathbf{q})q^2 t} \ . \tag{59}$$

The most interesting consequence of this result comes if we assume we are precisely at the transition temperature $T = T_c$ where $S(\mathbf{q}) = c/q^2$. In this limit,

$$C(\mathbf{q}, t) = \frac{c}{q^2} e^{-Dq^4 t/c} \ , \tag{60}$$

and the Fourier transform becomes:

$$C(\mathbf{q}, \omega) = \frac{2Dq^2/c}{\omega^2 + (Dq^4/c)^2} \ . \tag{61}$$

Equation (58) can be written in the dynamic scaling form with characteristic frequency:

$$\omega_c(\mathbf{q}, \xi) = DS^{-1}(\mathbf{q})q^2$$

$$= q^4 \frac{D}{c}(1+(q\xi)^{-2}) \ , \tag{62}$$

and we can identify the dynamic critical index:

$$z = 4 \tag{63}$$

and the dynamic scaling function:

$$F(\Omega) = \frac{1}{i+\Omega} \ . \tag{64}$$

is the same as before in the nonconserved case.

What about our assumption, Eq. (38), that $\Gamma(\mathbf{q},z)$ is regular as $q,z \to 0$? This does not hold in general. There are strong nonlinear processes driven by the large correlation length ξ near the critical point that lead to anomalous behavior in $\Gamma(\mathbf{q},z)$ as $\xi \to \infty$, $q,z \to 0$. Thus, for example, the shear viscosity in a fluid near its liquid–gas critical point shows *critical* anomalies. We can not understand these effects using our linearized Langevin approach. In general, as we see in Chapter 10, we must treat nonlinear fluctuations in the order parameter, as well as couplings to other slow degrees of freedom.

8.2
More on Slow Variables

In the previous section we assumed that the order parameter is a scalar variable that does not couple to any other slow variables in the system. This is most appropriate to the case of a nonconserved Ising system. More generally, we must allow for coupling between the order parameter and other slow modes in the system. In this way, the universality classes in dynamic critical phenomena are more diverse than for static critical phenomena.

At this stage we understand that slow modes exist for two basic reasons: conservation laws and critical slowing down. In the next section we find another source of slow variables: Nambu–Goldstone modes in systems with a broken continuous symmetry. These variables must also be included in our list of slow variables.

A key point is not just that we identify the slow variables but that we must also be able to decide how they couple to other slow modes. This is tied up with the construction of the static part of the memory function for the system where we include all of the slow modes in our set ψ_α. Slow modes ψ and ϕ are coupled if the static part of the memory function $\Gamma^{(s)}_{\psi\phi}$ is nonzero. From our theorem from Chapter 5 for the static part of the memory function $\Gamma^{(s)}_{\psi\phi}$ is nonzero if the average of the Poisson brackets between ψ and ϕ is nonzero. Thus generally we must construct the Poisson brackets among all of the slow modes to see if they are coupled.

To see how these Poisson brackets enter the discussion, let us look at a magnetic system. Suppose our fields of interest are the magnetization density **M** and the staggered magnetization density **N**. Notice that the definitions of these quantities and their Poisson brackets are independent of the choice of Hamiltonian and the explicit choice of an order parameter.

Starting with the fundamental Poisson brackets [27] for a set of classical spins:

$$\{S_i(\mathbf{R}), S_j(\mathbf{R}')\} = \delta_{\mathbf{R},\mathbf{R}'} \sum_k \varepsilon_{ijk} S_k(\mathbf{R}) \quad , \tag{65}$$

and defining the magnetization density:

$$\mathbf{M}(\mathbf{x}) = \sum_{\mathbf{R}} \delta(\mathbf{x} - \mathbf{R}) \mathbf{S}(\mathbf{R}) \tag{66}$$

and the staggered magnetization density:

$$\mathbf{N}(\mathbf{x}) = \sum_{\mathbf{R}} \delta(\mathbf{x} - \mathbf{R}) \eta(\mathbf{R}) \mathbf{S}(\mathbf{R}) \quad , \tag{67}$$

then it is shown in Appendix D that:

$$\{M_i(\mathbf{x}), M_j(\mathbf{y})\} = \delta(\mathbf{x} - \mathbf{y}) \sum_k \varepsilon_{ijk} M_k(\mathbf{x}) \tag{68}$$

$$\{N_i(\mathbf{x}), N_j(\mathbf{y})\} = \delta(\mathbf{x} - \mathbf{y}) \sum_k \varepsilon_{ijk} M_k(\mathbf{x}) \tag{69}$$

$$\{M_i(\mathbf{x}), N_j(\mathbf{y})\} = \delta(\mathbf{x} - \mathbf{y}) \sum_k \varepsilon_{ijk} N_k(\mathbf{x}) \quad . \tag{70}$$

We will need these Poisson bracket relations as we go along.

When we turn to the static part of the memory function we see that it depends on the average of the Poisson bracket [28] between the slow variables. If we focus on the magnetization densities then we have from Eq. (5.205):

$$\begin{aligned}\Gamma_{ij}^{(s)}(\mathbf{x} - \mathbf{y}) &= i\beta^{-1} \langle \{M_i(\mathbf{x}), M_j(\mathbf{y})\} \rangle \\ &= i\beta^{-1} \delta(\mathbf{x} - \mathbf{y}) \sum_k \varepsilon_{ijk} \langle M_k(\mathbf{x}) \rangle \quad . \end{aligned} \tag{71}$$

This vanishes unless the system has a nonzero average magnetization. Thus we have new contributions to the static part of the memory function in the ordered phase of a ferromagnetic or if we apply an ordering magnetic field. In either case $\langle M_k(\mathbf{x}) \rangle \neq 0$ and certain components of $\Gamma^{(s)}$ are nonzero.

8.3
Spontaneous Symmetry Breaking and Nambu–Goldstone Modes

Let us now turn our attention to the ordered ferromagnetic state where $T < T_c$. In this case, the dynamics are more complicated, but also extremely interesting. We must first discuss the static properties of the ordered state that breaks a continuous symmetry. Thus, we consider cases where the order parameter has at least two components.

The first thing we note is that in the ordered state the average magnetization is no longer zero in zero external field. There is a net magnetization that points, we assume, in the z-direction:

$$\langle \mathbf{M}(\mathbf{x}) \rangle = M\hat{z} \ . \tag{72}$$

Recall that, in an isotropic system, one has a system described by a Hamiltonian that is rotationally invariant - no direction is preferred. Nevertheless, at some temperature (the critical temperature T_c) the system chooses to be in a thermodynamic state that breaks the rotational symmetry. This phenomenon is known as *spontaneous symmetry breaking* [29]

We next have to say something about the static correlation functions in the ordered phase. We first realize, once there is a preferred direction, that fluctuations along and perpendicular to the direction of ordering may be different. We define the longitudinal correlation function:

$$S_L(\mathbf{x} - \mathbf{x}') = \langle \delta M^z(\mathbf{x}) \delta M^z(\mathbf{x}') \rangle \ , \tag{73}$$

where we now subtract off the average value of M^z explicitly, and the transverse correlation functions are defined by:

$$S_T(\mathbf{x} - \mathbf{x}') = \langle M^x(\mathbf{x}) M^x(\mathbf{x}') \rangle = \langle M^y(\mathbf{x}) M^y(\mathbf{x}') \rangle \ . \tag{74}$$

The cross correlation functions are zero by symmetry. We are interested in the behavior of S_L and S_T in the ordered phase. It turns our that the behavior of S_T is largely determined by the existence of Nambu–Goldstone (NG) modes in the system. We give first a simple argument leading to the identification of NG modes.

Consider the thermodynamic free-energy density of an isotropic magnet $f(B)$ as a function of an applied conjugate external magnetic field $\mathbf{B}$. Also suppose the system is symmetric and f depends only on the magnitude of $\mathbf{B}$ (See Problem 8.3). For a paramagnet this leads to the usual result:

$$f(B) = f(0) - \frac{1}{2}\chi B^2 + \ldots \ , \tag{75}$$

where χ is the magnetic susceptibility. More generally, the average magnetization density is given by:

$$m_\alpha = -\frac{\partial f}{\partial B_\alpha} = -f' \hat{B}_\alpha \equiv M \hat{B}_\alpha \ . \tag{76}$$

In this case, the magnetic susceptibility is a matrix:

$$\chi_{\alpha\beta} = \frac{\partial m_\alpha}{\partial B_\beta} = -\frac{\partial}{\partial B_\beta}(f'\hat{B}_\alpha)$$

$$= -f''\hat{B}_\alpha\hat{B}_\beta - \frac{f'}{B}(\delta_{\alpha\beta} - \hat{B}_\alpha\hat{B}_\beta) \quad . \tag{77}$$

The longitudinal part of the free energy can be read off as:

$$\chi_L = -f'' \; , \tag{78}$$

while the transverse part has the form:

$$\chi_T = -\frac{f'}{B} = \frac{M}{B} \; . \tag{79}$$

In the ordered ferromagnetic regime, as $B \to 0$ the average magnetization density reduces to the spontaneous magnetization $M \to M_0$ and χ_T blows up. Since:

$$\beta^{-1}\chi_T = \lim_{q \to 0} S_T(q) \tag{80}$$

in the classical limit, we have that the transverse component of the static correlation function blows up as B and q go to zero in the ferromagnetic phase. One can show [30] rigorously that if the interactions are short ranged then:

$$S_T(q) \geq \frac{A}{q^2} \; , \tag{81}$$

where A is a constant as $q \to 0$. For our purposes here, we assume that in the long-wavelength limit we can write:

$$S_T(\mathbf{q}) = \frac{1}{\rho_0 q^2} \tag{82}$$

and to lowest order we can treat ρ_0 as weakly temperature dependent near the phase transition. This is the famous Nambu–Goldstone boson or mode. Let us see how this physics influences the dynamics. We will use our linearized Langevin description.

8.4
The Isotropic Ferromagnet

Let us determine the magnetization–magnetization time-correlation functions in the ordered phase, $T < T_c$, in the conventional approximation for an isotropic ferromagnet. We first summarize our results for the static correlation functions in the ordered phase $T < T_c$. Assuming the magnetization is ordered in the z-direction we have:

$$S_{ij}(\mathbf{q}) = \delta_{ij} S_i(\mathbf{q}) \tag{83}$$

and:

$$S_T(\mathbf{q}) = S_x(\mathbf{q}) = S_y(\mathbf{q}) = \frac{1}{\rho_0 q^2} \tag{84}$$

for small q. The simplest assumption is that the longitudinal component of the order parameter correlation function, $S_L(\mathbf{q} = 0) = S_z(\mathbf{q} = 0)$ is a weakly temperature-dependent constant [31]. We next look at the static part of the memory function. Using the theorem given by Eq. (71) and the Poisson brackets given by Eq. (68) we have:

$$\Gamma_{ij}^{(s)}(\mathbf{q}) = \sum_l K_{il}^{(s)}(\mathbf{q}) S_{lj}(\mathbf{q}) = i\beta^{-1} \sum_l \varepsilon_{ijl} \langle M_l(\mathbf{x}) \rangle \ . \tag{85}$$

In the ordered phase we have assumed:

$$\langle M_\ell(\mathbf{x}) \rangle = \delta_{\ell,z} M. \tag{86}$$

We can then solve Eq. (85) for $K^{(s)}$,

$$K_{ij}^{(s)}(\mathbf{q}) = i\beta^{-1} \varepsilon_{ijz} M S_j^{-1}(\mathbf{q}) \ . \tag{87}$$

Turning to the dynamic part of the memory function, we assume, in the absence of any additional information, that we can use a form of the conventional approximation, remembering that the order parameter in this case is conserved,

$$\Gamma_{ij}^{(d)}(\mathbf{q}, z) = \sum_l K_{il}^{(d)}(\mathbf{q}, z) S_{lk}(\mathbf{q}) = -iq^2 \delta_{ij} D_i \tag{88}$$

or, using Eq. (83),

$$K_{ij}^{(d)}(\mathbf{q}, z) = -iq^2 \delta_{ij} D_i S_j^{-1}(\mathbf{q}) \ , \tag{89}$$

where the diffusion coefficient D_i may be different in the longitudinal and transverse directions. Armed with these results the fluctuation function:

$$C_{ij}(\mathbf{q}, z) = -i \int_0^{+\infty} dt\, e^{+izt} \int d^d x\, e^{+i\mathbf{q}\cdot(\mathbf{x}-\mathbf{y})} \langle \delta M_i(\mathbf{x}, t) \delta M_j(\mathbf{y}, 0) \rangle \tag{90}$$

satisfies, as usual, the kinetic equation:

$$\sum_l [z\delta_{il} - K_{il}^{(s)}(\mathbf{q}) - K_{il}^{(d)}(\mathbf{q}, z)] C_{lj}(\mathbf{q}, z) = S_{ij}(\mathbf{q}) \ . \tag{91}$$

Putting in the results for the memory functions, Eqs. (87) and (89), we have:

$$zC_{ij}(\mathbf{q},z) - i\beta^{-1}M\sum_{l}\varepsilon_{ilz}S_l^{-1}(\mathbf{q})C_{lk}(\mathbf{q},z) + iq^2 D_i S_i^{-1}(\mathbf{q})C_{ij}(\mathbf{q},z) = S_{ij}(\mathbf{q}) \quad (92)$$

Let us look at this first for the case of longitudinal fluctuations, setting $i = z = L$ in Eq. (92) gives:

$$zC_{zj}(\mathbf{q},z) + iq^2 D_L S_L^{-1}(\mathbf{q})C_{zj}(\mathbf{q},z) = \delta_{jz}S_L(\mathbf{q}) \quad (93)$$

or:

$$C_{zj}(\mathbf{q},z) = \frac{S_L(\mathbf{q})\delta_{jz}}{z + iq^2 D_L S_L^{-1}(\mathbf{q})} \quad , \quad (94)$$

and one has, for the longitudinal component, diffusive behavior just as in the paramagnetic phase. In some sense this is the continuation of the paramagnetic behavior into the ordered region. Dynamical scaling near $T = T_c$ gives $z = 4$ as we found for $T \geq T_c$.

Consider next the transverse case. Taking the xx matrix element of Eq. (92) we find:

$$[z + iq^2 D_T S_T^{-1}(\mathbf{q})]C_{xx}(q,z) - i\beta^{-1}MS_T^{-1}(\mathbf{q})C_{yx}(q,z) = S_T(\mathbf{q}) \quad , \quad (95)$$

which couples C_{xx} to C_{yx}. Taking the $y-x$ matrix element of Eq. (92) gives:

$$[z + iq^2 D_T S_T^{-1}(\mathbf{q})]C_{yx}(z) + i\beta^{-1}MS_T^{-1}(\mathbf{q})C_{xx}(z) = 0 \quad . \quad (96)$$

Thus Eqs. (95) and (96) form a closed set of equations. The determinant of coefficients for this coupled set of equations is given by:

$$D(\mathbf{q},z) = [z + iq^2 D_T S_T^{-1}(\mathbf{q})]^2 - \beta^{-2}M^2 S_T^{-2}(\mathbf{q}) \quad . \quad (97)$$

One therefore has poles at the complex frequencies:

$$z = \pm\beta^{-1}MS_T^{-1}(q) - iq^2 D_T S_T^{-1}(\mathbf{q}) \quad . \quad (98)$$

Since the transverse correlation functions support NG modes,

$$S_T(\mathbf{q}) = \frac{1}{\rho_0 q^2} \quad , \quad (99)$$

we have the dispersion relation:

$$z = \pm cq^2 - iq^4 D_T \rho_0 \quad , \quad (100)$$

where:

$$c = \beta^{-1}M\rho_0 \quad . \quad (101)$$

8.4 The Isotropic Ferromagnet

The fluctuation function is given then by solving Eqs. (95) and (96) to obtain:

$$
\begin{aligned}
C_{xx}(\mathbf{q},z) &= C_T(\mathbf{q},z) \\
&= \frac{S_T(q)(z+iq^4\rho_0 D_T)}{(z-cq^2+iq^4\rho_0 D_T)(z+cq^2+iq^4\rho_0 D_T)} \\
&= \frac{S_T(q)}{2}\left[\frac{1}{z-cq^2+iq^4\rho_0 D_T}+\frac{1}{z+cq^2+iq^4\rho_0 D_T}\right]. \quad (102)
\end{aligned}
$$

The associated correlation function is just the imaginary part and given by:

$$
C_T(\mathbf{q},\omega) = \frac{S_T(\mathbf{q})}{2}\left[\frac{q^4\rho_0 D_T}{(\omega-cq^2)^2+(\rho_0 D_T q^4)^2}+\frac{q^4\rho_0 D_T}{(\omega+cq^2)^2+(\rho_0 D_T q^4)^2}\right]. \quad (103)
$$

We see that below the transition temperature the transverse correlation function is characterized by two *spin-wave* peaks. The spin-wave frequency is given by:

$$
\omega_s(q) = cq^2 = M\beta^{-1}\rho_0 q^2 , \quad (104)
$$

while the spin-wave damping is given by:

$$
\begin{aligned}
W(q) &= q^2 D_T S_T^{-1}(q) \\
&= D_T \rho_0 q^4 . \quad (105)
\end{aligned}
$$

If we use ω_s as the characteristic frequency used in dynamical scaling, we have:

$$
\omega_s(q) \approx Mq^2 . \quad (106)
$$

If we assume $M \approx |T-T_c|^\beta \approx \xi^{-\beta/\nu}$ and use the reasonably accurate values for ferromagnets of $\beta = 1/3$, $\nu = 2/3$, then we have:

$$
\omega_s(q) \approx \xi^{-1/2}q^2 = q^z/\sqrt{(q\xi)} , \quad (107)
$$

and $z = 5/2$. This is a pretty good estimate of the measured [32] index for a ferromagnet. The damping contribution to the characteristic frequency has, as for the longitudinal component, a dynamic critical index $z = 4$. The discrepancy in values of z calls into question the conventional approximation, which assumes that D_T and D_L are regular in temperature as $T \to T_c$. In Problem 8.9 we discuss the temperature dependence required for D_T if cq^2 and $q^4 D_T \rho_0$ have the same scaling index. Clearly these results require D_T (and D_L) to show a strong temperature dependence. The source of this strong temperature dependence is discussed in Chapter 10.

The structure of the correlation function is changed qualitatively by the broken symmetry. An important comment is that the spin-wave spectrum is a property of the ordered phase, not just a property of the critical point.

8.5
Isotropic Antiferromagnet

We now want to turn to the dynamics of isotropic antiferromagnets. The order parameter, the total staggered magnetization $\mathbf{N}_T$, is not conserved. Thus above T_N there is nothing particularly special about the staggered magnetization autocorrelation function. In the conventional approximation we obtain (see Problem 8.5):

$$C_N(\mathbf{q},\omega) = \frac{2\Gamma_N}{\omega^2 + (\Gamma_N S_N^{-1}(\mathbf{q}))^2} \tag{108}$$

where Γ_N is a kinetic coefficient. It is left as an exercise to show that the dynamic scaling analysis of Eq. (108) as $T \to T_N^+$ leads to a dynamic scaling index $z = 2$.

Let us move on to the interesting case of the ordered phase, $T < T_N$. We will concentrate on the possibility of NG modes.

Following the treatment for the ferromagnet (now the field $\mathbf{H}_A$ conjugate to $\mathbf{N}$ is unphysical) we again argue that for small q we have the NG mode:

$$N^T(\mathbf{q}) = \frac{1}{\rho_0 q^2}, \tag{109}$$

where ρ_0 is assumed to be weakly temperature dependent near the transition. Thus the transverse component of the staggered field shows the NG mode behavior. The transverse correlation function for the magnetization does *not* show a q^{-2} behavior below T_N! It is regular for small wavenumbers:

$$\lim_{q \to 0} S_M^T(\mathbf{q}) = \frac{1}{\rho_M}. \tag{110}$$

For simplicity we assume that (see Problem 8.6):

$$\lim_{q \to 0} S_{MN}^{ij}(\mathbf{q}) = 0. \tag{111}$$

Let us now turn to a discussion of the dynamics for $T < T_N$. The first order of business is the choice of slow variables ψ_i to be used in the Langevin equation. We said previously that we should include in ψ_i all of the *slow* variables. In our problem here, the magnetization is conserved and is certainly

8.5 Isotropic Antiferromagnet

a slow variable. We also want to include the staggered magnetization in our set. We therefore analyze our generalized Langevin equation in the case with:

$$\psi_i \to \psi_\alpha^i(\mathbf{x}) \,, \tag{112}$$

where i is a vector label and $\alpha = N$ or M. Remembering Eq. (111) we have for the slow variable structure factor:

$$S_{\alpha\beta}^{ij}(\mathbf{q}) = \delta_{ij}\delta_{\alpha\beta} S_\alpha^i(\mathbf{q}) \,. \tag{113}$$

Next we need to evaluate the static part of the memory function. From our general theorem we have:

$$\sum_{\gamma,k} \int d^3 z \, K_{\alpha\gamma}^{(s)ik}(\mathbf{x}-\mathbf{z}) S_{\gamma\beta}^{kj}(\mathbf{z}-\mathbf{y}) = -i\beta^{-1} \langle \{\psi_\alpha^i(\mathbf{x}), \psi_\beta^j(\mathbf{y})\} \rangle \,. \tag{114}$$

The Poisson brackets among the various fields are worked out in Appendix D:

$$\{M_i(\mathbf{x}), N_j(\mathbf{y})\} = \sum_k \varepsilon_{ijk} N_k(\mathbf{x}) \delta(\mathbf{x}-\mathbf{y}) \tag{115}$$

$$\{N_i(\mathbf{x}), N_j(\mathbf{y})\} = \sum_k \varepsilon_{ijk} M_k(\mathbf{x}) \delta(\mathbf{x}-\mathbf{y}) \tag{116}$$

$$\{M_i(\mathbf{x}), M_j(\mathbf{y})\} = \sum_k \varepsilon_{ijk} M_k(\mathbf{x}) \delta(\mathbf{x}-\mathbf{y}) \,. \tag{117}$$

Since $\langle \mathbf{M} \rangle = 0$, we see that $\Gamma^{(s)}$ is zero unless $\alpha \neq \beta$ and i and j are not equal and in the set x,y. Therefore the only nonzero elements are:

$$K_{NM}^{(s)xy}(\mathbf{q}) S_M^y(\mathbf{q}) = -i\beta^{-1} N \tag{118}$$

$$K_{NM}^{(s)yx}(\mathbf{q}) S_M^x(\mathbf{q}) = +i\beta^{-1} N \tag{119}$$

$$K_{MN}^{(s)xy}(\mathbf{q}) S_N^y(\mathbf{q}) = -i\beta^{-1} N \tag{120}$$

$$K_{MN}^{(s)yx}(\mathbf{q}) S_N^x(\mathbf{q}) = +i\beta^{-1} N \,, \tag{121}$$

where $N = \langle N_z \rangle$. We can rewrite these equations in the compact form:

$$K_{\alpha\beta}^{ij(s)}(\mathbf{q}) = -i\beta^{-1} \varepsilon_{ijz} N (1-\delta_{\alpha\beta}) [S_\beta^i(\mathbf{q})]^{-1} \,. \tag{122}$$

We assume the simplest forms for the dynamic parts of the memory function compatible with the conservation law satisfied by the magnetization density:

$$K^{ij(d)}_{\alpha\beta}(\mathbf{q}) = -i\delta_{\alpha\beta}\delta_{i,j}\Gamma^i_\alpha(\mathbf{q})(S^i_\beta(\mathbf{q}))^{-1} \tag{123}$$

$$\Gamma^i_M(\mathbf{q}) = q^2 D^i_M \tag{124}$$

$$\Gamma^i_N(\mathbf{q}) = \Gamma^i_N . \tag{125}$$

Our kinetic equation takes the form:

$$zC^{ij}_{\alpha\beta}(\mathbf{q},z) - \sum_{\gamma k} K^{ik}_{\alpha\gamma}(\mathbf{q})C^{kj}_{\gamma\beta}(\mathbf{q},z) = S^{ij}_{\alpha\beta}(\mathbf{q}) . \tag{126}$$

Let us focus here on the transverse components that we expect are influenced by NG modes. Taking the xx matrix element of Eq. (126), we obtain:

$$zC^{xx}_{\alpha\beta}(\mathbf{q},z) + i\beta^{-1}NS^T_{-\alpha}(\mathbf{q})^{-1}C^{yx}_{-\alpha\beta}(\mathbf{q},z) + i\Gamma^T_\alpha(\mathbf{q})S^T_\alpha(\mathbf{q})^{-1}C^{xx}_{\alpha\beta}(\mathbf{q},z)$$
$$= \delta_{\alpha\beta}S^T_\alpha(\mathbf{q}) , \tag{127}$$

where $-\alpha$ means M if $\alpha = N$ or N if $\alpha = M$. Taking the NN matrix element of Eq. (127) gives:

$$zC^{xx}_{NN}(\mathbf{q},z) + i\beta^{-1}NS^T_M(\mathbf{q})^{-1}C^{yx}_{MN}(\mathbf{q},z) + i\Gamma^T_N(\mathbf{q})S^T_N(\mathbf{q})^{-1}C^{xx}_{NN}(\mathbf{q},z)$$
$$= S^T_N(q)$$

or:

$$(z + i\Gamma^T_N S^T_N(\mathbf{q})^{-1})C^{xx}_{NN}(\mathbf{q},z) + i\beta^{-1}NS^T_M(\mathbf{q})^{-1}C^{yx}_{MN}(\mathbf{q},z) = S^T_N(\mathbf{q}) . \tag{128}$$

So C^{xx}_{NN} is coupled to C^{yx}_{MN}. An equation for this quantity is given by taking the MN, yx matrix elements of Eq. (127) to obtain:

$$zC^{yx}_{MN}(\mathbf{q},z) - i\beta^{-1}NS^T_N(\mathbf{q})^{-1}C^{xx}_{NN}(\mathbf{q},z) + i\Gamma^T_M(\mathbf{q})S^T_M(\mathbf{q})^{-1}C^{yx}_{MN}(\mathbf{q},z) = 0$$

or:

$$-i\beta^{-1}NS^T_N(\mathbf{q})^{-1}C^{xx}_{NN}(\mathbf{q},z) + (z + iD^T_M q^2 S^T_M(\mathbf{q})^{-1})C^{yx}_{MN}(\mathbf{q},z) = 0 . \tag{129}$$

Equations (128) and (129) form a simple 2×2 set of equations. The pole structure is determined by the zeros of the determinant:

$$\mathcal{D} = \left[z + i\Gamma^T_N S^T_N(\mathbf{q})^{-1}\right]\left[z + iD^T_M q^2 S^T_M(\mathbf{q})^{-1}\right]$$
$$- \beta^{-2} N^2 S^T_N(\mathbf{q})^{-1} S^T_M(\mathbf{q})^{-1} \tag{130}$$

Setting this to zero gives:

$$z^2 + iz\left(\Gamma_N^T S_N^T(\mathbf{q})^{-1} + D_M^T q^2 S_M^T(\mathbf{q})^{-1}\right)$$
$$-\Gamma_N^T D_M^T q^2 S_M^T(\mathbf{q})^{-1} S_N^T(\mathbf{q})^{-1} - \beta^{-2} N^2 S_N^T(\mathbf{q})^{-1} S_M^T(\mathbf{q})^{-1} = 0. \quad (131)$$

Substituting the small wavenumber forms for the structure factors gives the result:

$$z^2 + izq^2\left(\Gamma_N^T \rho_0 + D_M^T \rho_M\right) - q^4 \Gamma_N^T D_M^T \rho_0 \rho_M - \beta^{-2} N^2 \rho_0 \rho_M q^2 = 0. \quad (132)$$

Let us define:

$$\Gamma = \Gamma_N^T \rho_0 + D_M^T \rho_M \quad (133)$$

and:

$$c^2(\mathbf{q}) = \beta^{-2} N^2 \rho_0 \rho_M + q^2 \Gamma_N^T D_M^T \rho_0 \rho_M, \quad (134)$$

then Eq. (132) takes the form:

$$z^2 + iz\Gamma q^2 - c^2(\mathbf{q}) q^2 = 0. \quad (135)$$

Solving this quadratic equation, we find that the poles are located at complex frequencies:

$$z = -\frac{1}{2} i \Gamma q^2 \pm \frac{1}{2}\left[-\Gamma q^2 + 4c^2(q) q^2\right]^{1/2}. \quad (136)$$

We can rewrite the argument of the square root as:

$$4c^2(q) q^2 - \Gamma^2 q^4 = 4c^2(0) q^2 + 4q^4 \Gamma_N^T D_M^T \rho_0 \rho_M - \left(\Gamma_N^T \rho_0 + D_M^T \rho_M\right)^2 q^4$$
$$= 4c^2 q^2 - \left(\Gamma_N^T \rho_0 - D_M^T \rho_M\right)^2 q^4, \quad (137)$$

where $c = c(0)$. We can then evaluate the pole positions as $q \to 0$:

$$z = -\frac{i\Gamma q^2}{2} \pm \frac{1}{2}\left(4c^2 q^2 - \left(\Gamma_N^T \rho_0 - D_M^T \rho_M\right)^2 q^4\right)^{1/2}$$
$$= \frac{-i\Gamma q^2}{2} \pm cq\left(1 - \frac{q^2}{4c^2}\left(\Gamma_N^T \rho_0 - D_M^T \rho_M\right)^2\right)^{1/2}$$
$$= \frac{-i\Gamma q^2}{2} \pm cq\left(1 - \frac{q^2}{8c^2}\left(\Gamma_N^T \rho_0 - D_M^T \rho_M\right)^2 + \ldots\right). \quad (138)$$

The spin-wave spectrum corresponds to poles at:

$$z = \pm cq - i\Gamma \frac{q^2}{2} + \mathcal{O}(q^3), \quad (139)$$

where the spin-wave speed is given by:

$$c = \beta^{-1} N (\rho_0 \rho_M)^{1/2} ,\qquad (140)$$

and the spin-wave damping by:

$$\Gamma = \Gamma_N^T \rho_0 + D_M^T \rho_M .\qquad (141)$$

Inverting the matrix equation given by Eqs. (128) and (129), we find after some algebra the fluctuation function for the transverse staggered fields:

$$C_{NN}^{xx}(\mathbf{q}, z) = S_N^T(\mathbf{q}) \frac{(z + i D_M^T q^2 \rho_M)}{(z - cq + \frac{i\Gamma q^2}{2})(z + cq + \frac{i\Gamma q^2}{2})} .\qquad (142)$$

Then, near each pole we can write:

$$C_{NN}^T(\mathbf{q}, z) = S_N^T(\mathbf{q}) \frac{1}{2} \left[\frac{1}{z - cq + i\Gamma q^2/2} + \frac{1}{z + cq + i\Gamma q^2/2} \right] \qquad (143)$$

plus terms higher order in q. It is clear that the transverse magnetization correlation functions also shows a spin-wave spectrum (the two correlation functions C_N^T and C_M^T share the same denominator and therefore the same poles). It is interesting to note that the spectrum of C_N^T now looks very much like that for the Brillouin peaks in the density–density correlation function in a fluid. In the fluid one has a dispersion relation:

$$z = \pm cq + i\Gamma q^2/2 ,\qquad (144)$$

where c is the speed of sound and Γ the sound attenuation. In the spin-wave case, things look identical. Remember, however, that the sound wave spectrum was generated by the conservation laws (the density and momentum is conserved). The staggered magnetization is not conserved. We see then that the Nambu–Goldstone mode associated with the spontaneous symmetry breaking has the dynamical effect of simulating a conservation law.

If we look at dynamic scaling in the case where we are near the critical point, we have, assuming Γ is regular in temperature near T_N,

$$C_{NN}^T(\mathbf{q}, z) = \frac{S_N^T(\mathbf{q})}{\omega_c(\mathbf{q})} \frac{1}{2} \left[\frac{1}{\Omega - 1} + \frac{1}{\Omega + 1} \right] ,\qquad (145)$$

where the characteristic frequency is given by:

$$\omega_c(\mathbf{q}) = cq = \beta^{-1} N \sqrt{\rho_0 \rho_M} q .$$

Using the result $N \approx \xi^{-1/2}$, which is the same as for the ferromagnetic case, we have:

$$\omega_c(\mathbf{q}) \approx \xi^{-1/2} q \approx q^{3/2} \frac{1}{\sqrt{q\xi}} \qquad (146)$$

and we can identify the dynamic critical index as $z = 3/2$. This result is in good agreement with experiment [33] for isotropic antiferromagnets. However, it disagrees with the conventional theory, $z = 2$, found for $T > T_N$ and for the longitudinal component C_{NN}^L.

We must conclude, as for the case of the isotropic ferromagnet, that the conventional theory is not a quantitative theory near the critical point. We need a self-consistent theory for all the exponents, z, $β$, and $ν$. We *know* that $β$, and $ν$ require a nontrivial treatment of critical fluctuations, so also z requires a renormalization group treatment of fluctuations. This is developed in Chapter 10.

8.6
Summary

One can associate with every broken continuous symmetry, when there are short-range interactions in the disordered state, a set of Nambu–Goldstone modes (variables) that must be included in the set of thermodynamic state variables and the set of hydrodynamic variables. Typically these variables generate traveling modes like spin waves in magnets, second sound in helium and transverse sound in solids. These modes also form the lowest lying excitations [34] in the low-temperature quantum analysis of such systems.

8.7
References and Notes

1 In Chapter 1 of FOD the nature of the critical points in a variety of systems is discussed.

2 The identification of order parameter for a variety of systems is discussed in Chapter 1 of FOD.

3 For simplicity at the beginning we assume a scalar order parameter as one would find for an Ising model. We also assume for simplicity that the thermodynamically conjugate field is set to its critical value. For magnetic systems this means the externally applied magnetic field is set to zero.

4 For example, the average of the order parameter, the magnetization density, vanishes for a ferromagnet in the paramagnetic phase due to rotational invariance. This assumes zero external applied field.

5 See the discussion in Sections 4.1 and 4.2 in FOD.

6 See Table 4.4 in FOD.

7 L. S. Ornstein, F. Zernike, Proc. Acad. Sci. Amsterdam **17**, 793 (1914).

8 As part of the Markovian approximation in the hydrodynamic limit we set $S^{-1}(q) = S^{-1}(0)$ and assume that $S^{-1}(0)$ is a constant.

9 See Table 1.2 in FOD.

10 The dependence of the critical properties as a function of n is discussed in Chapter 5 in ESM.

11 If the square lattice points are specified by $\mathbf{R} = an_x\hat{x} + an_y\hat{y}$ where a is the lattice spacing and n_x and n_y are integers then one can choose $η = (-1)^{n_x+n_y}$.

12 See Sections 4.1 and 4.2 in FOD.

13 L. P. Kadanoff, W. Götze, D. Hamblen, R. Hecht, E. A. S. Lewis, V. V. Palciauskas,

M. Rayl, J. Swift, D. Aspnes, J. Kane, Rev. Mod. Phys. **39**, 395 (1967).

14 In certain anisotropic materials the correlation length in a plane may be much longer than in the third dimension. As long as the correlation length in the third direction is small compared with the lattice spacing in that direction the system can be taken as quasi-two-dimensional.

15 Liquid crystals with biaxial symmetry are discussed in Chapter 10 of FOD.

16 B. Widom, J. Chem. Phys. **43**, 3989 (1965).

17 L. P. Kadanoff, Physics (N.Y.) **2**, 263 (1966).

18 See Section 4.3.2 in FOD.

19 See Section 2.7 in ESM.

20 P. Weiss, J. Phys. **6**, 661 (1907); see Section 5.9.2 in ESM.

21 See Section 2.9 in ESM. Johannes Diderik Van der Waals, *Over der Continuiteit van den Gas-en vloeisoftoestand*, Leiden, Sijthoff (1873). [Transl.: The Continuity of the Liquid and Gaseous States of Matter], in Physical Memoirs, Vol. 1, Part 3. Taylor and Francis (The Physical Society), London (1890).

22 L. Onsager, Phys. Rev. **65**, 117 (1944); see also B. Kaufmann, Phys. Rev. **76**, 1232 (1949) and B. M. Mc Coy, T. T. Wu, *The Two-Dimensional Ising Model*, Harvard University Press, Cambridge, (1973).

23 L. van Hove, Phys. Rev. **95**, 1374 (1954).

24 P. C. Hohenberg, B. I. Halperin, Rev. Mod. Phys. **49**, 435 (1977).

25 R. A. Ferrell, N. Menyhárd, H. Schmidt, F. Schwable, P. Szépfalusy, Phys. Rev. Lett. **18**, 891 (1967); R. A. Ferrell, N. Menyhárd, H. Schmidt, F. Schwable, P. Szépfalusy, Ann. Phys. (N.Y.) **47**, 565 (1968); B. I. Halperin, P. C. Hohenberg, Phys. Rev. Lett. **19**, 700 (1967); B. I. Halperin, P. C. Hohenberg, Phys. Rev. **177**, 952 (1969).

26 If we use the exact static structure factor in Eq. (52) then $z = 2 - \eta$.

27 The classical Poisson brackets are replaced by commutation relations in quantum-mechanical systems.

28 Remember that the Poisson brackets between the density and the momentum density lead to the sound modes in a simple fluid.

29 Spontaneous symmetry breaking is discussed in some detail in Section 1.3.2 in FOD.

30 See the discussion in Section 5.3 in FOD on Goldstone's theorem.

31 See the discussion in Section 5.2.1 in FOD.

32 The dynamic critical index for a ferromagnet is measured in O. W. Dietrich, J. Als-Nielsen, L. Passell, Phys. Rev. B **14**, 4923 (1976).

33 See the discussion on p. 466 in Ref. [24].

34 In the low-temperature quantum analysis of such systems the NG modes are the lowest lying excitations. See the discussion in Section 5.6 in FOD.

8.8
Problems for Chapter 8

Problem 8.1: Consider the anisotropic Heisenberg Hamiltonian given by Eq. (10). If there is an easy plane:

$$J^{(x)} = J^{(y)} \neq J^{(z)} ,$$

show that only the magnetization in the z-direction is conserved. Show, in the isotropic case, that the staggered magnetization is not conserved.

Problem 8.2: Take the inverse Fourier transform of the Ornstein–Zernike result for the order parameter static correlation function:

$$S(q) = \frac{c}{q^2 + \xi^{-2}}$$

to obtain $S(x)$ in d dimensions.

Problem 8.3: After minimizing the Landau free energy density:

$$f = \frac{r}{2}\mathbf{m}^2 + \frac{u}{4}\mathbf{m}^4 - \mathbf{m}\cdot\mathbf{B} \ ,$$

where r, u are positive, $\mathbf{m}$ is the fluctuating magnetization density and $\mathbf{B}$ the applied external magnetic field, show that $f = f(B)$. Expand f in powers of B in the paramagnetic phase and identify the magnetic susceptibility.

Problem 8.4: Find the magnetization–magnetization correlation functions for the case where the classical Hamiltonian is of the form:

$$H = \int d^d x \left[\frac{r}{2}\mathbf{M}^2(\mathbf{x}) - \mathbf{M}(\mathbf{x})\cdot\mathbf{B}\right] \ ,$$

where r is a positive constant and $\mathbf{B}$ an applied constant magnetic field. Use the Poisson bracket relations given by Eq. (68) to generate the dynamics.

Problem 8.5: Using the conventional approximation, find an approximate solution for the staggered magnetization fluctuation function in the disordered $T > T_N$ regime. From this, find the correlation function $C_N(\mathbf{q},\omega)$. Write your final result in dynamic scaling form and extract the dynamic critical index z.

Problem 8.6: Suppose an antiferromagnet can be described by the Landau free energy density:

$$f = \frac{r_M}{2}\mathbf{M}^2 + \frac{r}{2}\mathbf{N}^2 + \frac{u}{4}\mathbf{N}^4 + \frac{v}{2}\mathbf{N}^2\mathbf{M}^2 - \mathbf{M}\cdot\mathbf{H} \ ,$$

where r_M, u and v are positive. Compute within Landau theory the cross-susceptibility:

$$\chi_{MN} = \frac{\partial N_i}{\partial H_j}$$

in zero external field.

Problem 8.7: Starting with Eq. (126) for an antiferromagnet, work out the longitudinal and transverse correlation functions for the magnetization density. We expect the transverse components to share the spin-wave spectrum with that for the staggered field.

Problem 8.8: Suppose one has an effective Hamiltonian governing the behavior of an n-component vector order parameter, M_α, $\alpha = 1, 2 \cdots, n$, of the $\mathcal{O}(n)$ symmetric form:

$$\mathcal{H} = \int d^d x \left[\frac{r}{2}\mathbf{M}^2 + \frac{u}{4}(\mathbf{M}^2)^2 + \frac{c}{2}\sum_{i\alpha}(\nabla_i M_\alpha)^2\right] \ ,$$

where $r = r_0(T - T_c)$ and r_0, c and u are temperature independent. Determine the average order parameter and the fluctuation spectra in the approximation:

$$\frac{\delta\mathcal{H}}{\delta M_\alpha(\mathbf{x})}\bigg|_{M=\bar{M}} = 0$$

$$C_{\alpha\beta}^{-1}(\mathbf{x}-\mathbf{y}) = \frac{\delta^2 \mathcal{H}}{\delta M_\alpha(\mathbf{x})\delta M_\alpha(\mathbf{y})}\bigg|_{M=\bar{M}} .$$

Work in the ordered regime where $T < T_c$ and determine $C_T(q) = 1/\rho_0 q^2$ and obtain an estimate for ρ_0 in terms of the parameters of the model.

Problem 8.9: If the spin-wave speed and damping contributions to the isotropic ferromagnet's characteristic frequency scale in the same manner, what is the dependence of the transverse diffusion coefficient D_T on ξ. Assume ρ_0 is a constant.

Problem 8.10: One can also include the conjugate field B in the scaling relations and write the free energy in the form:

$$F \approx \xi^{x-2x_1} f(B\xi^{x_1}) ,$$

where x and x_1 are constant exponents. Noting the definitions of the critical indices for the magnetization:

$$M(B, T_c) \approx B^{1/\delta}$$

and:

$$M(0, T) \approx \varepsilon^\beta ,$$

find expressions for x and x_1 in terms of indices δ, β, γ and ν. Find a relation connecting γ, β, and δ.

9
Nonlinear Systems

9.1
Historical Background

From our work in the previous chapter we discovered that linearized hydrodynamics breaks down near a second-order phase transition or in the ordered phase of a system with a broken continuous symmetry. In these systems there are long-range correlations and the hydrodynamic assumption that there are no large lengths in the problem breaks down. It turns out however that the validity of linearized hydrodynamics can be questioned even in the case of systems, like normal fluids, where there are no broken symmetries or large correlation lengths. In the hydrodynamic description of fluid flow, one can have large flows that take the system out of the linear regime. In this case we know that there are convective nonlinearities that enter into a nonlinear hydrodynamic description. Even when one is not externally driving such systems one can look at the nonlinear response to internal fluctuations. Let us review some of the history in this area.

It was appreciated through the pioneering work [1] of the 1950s that a coherent theory for the dynamics of fluids can be conveniently organized through the study of equilibrium-averaged time-correlation functions. In particular, the Green–Kubo formula related transport coefficients, λ, to time integrals over current–current correlation functions:

$$\lambda = \int_0^\infty dt \langle J_\lambda(t) J_\lambda(0) \rangle \ . \tag{1}$$

For almost 100 years there was a single theory [2], developed around the Boltzmann equation, which described the dynamics of strongly interacting fluids. The basic ingredient of this theory was, of course, short-ranged uncorrelated two-body collisions. Such processes lead to correlation functions that decay exponentially with time:

$$\langle J_\lambda(t) J_\lambda(0) \rangle = \langle J_\lambda^2(0) \rangle e^{-t/\tau_\lambda} \ , \tag{2}$$

and transport coefficients are simply related to the decay rate τ_λ through the Green–Kubo formula. Putting Eq. (2) into Eq. (1) and doing the time integral

Nonequilibrium Statistical Mechanics. Gene F. Mazenko
Copyright © 2006 WILEY-VCH Verlag GmbH & Co. KGaA, Weinheim
ISBN: 3-527-40648-4

we find:

$$\lambda = \langle J_\lambda^2(0)\rangle \tau_\lambda \ . \tag{3}$$

During the 1960s, more careful studies of these time-correlation functions went beyond the Boltzmann approximation and led to several surprises.

- One of the theoretical developments inspired by the newly discovered Green–Kubo equations was to develop a low-density, n, expansion [3]:

$$\lambda = \lambda_B(1 + An + Bn^2 + \ldots) \ , \tag{4}$$

where λ_B is the Boltzmann expression for the transport coefficient. It was found via direct calculation [4] that the coefficient of the second-order term, B, blows up. It was only after resuming higher-order terms [5] in the expansion that it was understood that we should write:

$$B(n) = B_0 + B_1 \ln(n\sigma^3) \ , \tag{5}$$

where σ is a hard-sphere diameter.

- The pioneering molecular dynamics simulations of Alder and Wainwright [6] found that the time correlation functions were not exponentially decaying with time but showed power-law decays known as *long-time tails*. A typical current–current time-correlation function was found to go for long times as:

$$\langle J_\lambda(t)J_\lambda(0)\rangle \approx A_\lambda t^{-d/2} \ , \tag{6}$$

where d is the spatial dimensionality of the system. Putting this result back into the Green–Kubo formula for the transport coefficient λ in the long-time regime where it is appropriate, one obtains:

$$\begin{aligned}\lambda(\tau) &= \lambda_0 + \int_{\tau_0}^\tau dt\, A_\lambda t^{-d/2} \\ &= \lambda_0 + \frac{2A_\lambda}{d-2}\left[\frac{1}{\tau_0^{(d-2)/2}} - \frac{1}{\tau^{(d-2)/2}}\right] \ ,\end{aligned} \tag{7}$$

where λ_0 is the contribution to the transport coefficient from times less than τ_0. Then, for $d > 2$, as τ increases $\lambda(\tau)$ approaches the physical transport coefficient:

$$\lambda = \lim_{\tau\to\infty} \lambda(\tau) \ . \tag{8}$$

We then obtain the remarkable result that for $d \leq 2$ that the transport coefficient and conventional hydrodynamics do not exist. For $d = 2$, $\lambda(\tau) \approx \ln(\tau/\tau_0)$. Thus the effect producing the long-time tails grows stronger as one lowers the spatial dimensionality.

Remarkably, after careful study [7], it was found that the origins of both effects, divergence of the coefficient B and long-time tails, is due to related collective or hydrodynamic effects on a semimicroscopic level. Such effects have since come to be called *mode-coupling effects*. We discuss the theoretical origins of mode-coupling theory at the end of this chapter and see how long-time tails come out of the theory.

9.2
Motivation

The basic assumption of linearized hydrodynamics, as indicated above and in previous chapters, is that the dynamic part of the memory function is proportional to q^2 (in almost all cases for a conserved slow mode) with a coefficient that is regular in the small $\mathbf{q}$ and ω limit. We have developed the important idea that in this picture we have identified all of the slow variables [8] in our set ψ_i. This picture held intact for nearly 100 years. Eventually, however, as discussed in the previous section, in the mid 1960s it became recognized that it is not always sufficient to identify the slow variables (in terms of conserved variables, order parameters or Nambu–Goldstone modes) and write down the linearized Langevin equation as we have discussed in detail previously. What can go wrong with our picture? It has now been recognized in a number of different contexts that *nonlinear interactions* can lead to a breakdown of conventional linearized hydrodynamics. If $\psi(\mathbf{x})$ is a slow variable, should not $\psi^2(\mathbf{x})$ also be treated as a slow variable? If, roughly speaking, the slow mode goes as:

$$\psi(\mathbf{q},t) = \psi(\mathbf{q}) e^{-Dq^2 t} \quad , \tag{9}$$

then the Fourier transform of $\psi^2(\mathbf{x},t)$ is:

$$\int d^d x \, e^{+i\mathbf{q}\cdot\mathbf{x}} \psi^2(\mathbf{x},t) = \int \frac{d^d k}{(2\pi)^d} \psi(\mathbf{k},t) \psi(\mathbf{q}-\mathbf{k},t)$$

$$= \int \frac{d^d k}{(2\pi)^d} \psi(\mathbf{k}) \psi(\mathbf{q}-\mathbf{k}) e^{-D[k^2+(\mathbf{q}-\mathbf{k})^2]t} \quad . \tag{10}$$

Let us look at the long distance behavior and let $\mathbf{q} \to 0$ to obtain:

$$\int d^d x \, \psi^2(\mathbf{x},t) = \int \frac{d^d k}{(2\pi)^d} \psi^2(\mathbf{k}) e^{-2Dk^2 t} \quad . \tag{11}$$

The point is that the contribution to the integral is dominated by the contributions at small wavenumbers $\mathbf{k}$ and we can write:

$$\int d^d x \, \psi^2(\mathbf{x},t) = \tilde{K}_d \, \psi^2(0) \int_0^\Lambda k^{d-1} dk \, e^{-2Dk^2 t} \quad , \tag{12}$$

where Λ bounds the low-wavenumber behavior from above, and:

$$\tilde{K}_d = \int \frac{d\Omega_d}{(2\pi)^d} \qquad (13)$$

is proportional to the d-dimensional angular average. Making the change of variables from k to $x = \sqrt{2Dt}k$ we easily obtain:

$$\begin{aligned}
\int d^d x \, \psi^2(\mathbf{x}, t) &= \frac{\psi^2(\mathbf{0})}{(2Dt)^{d/2}} \tilde{K}_d \int_0^{\sqrt{2Dt\Lambda}} x^{d-1} dx \, e^{-x^2} \\
&= \frac{\psi^2(\mathbf{0})}{(2\pi Dt)^{d/2}} \tilde{K}_d \int_0^{\infty} x^{d-1} dx \, e^{-x^2} \\
&= \frac{\psi^2(\mathbf{0})}{(2\pi Dt)^{d/2}} \tilde{K}_d \tfrac{1}{2} \Gamma(d/2) \, , \qquad (14)
\end{aligned}$$

where the last result holds in the long-time limit and we notice that it is independent of the cutoff Λ. We see that the Fourier transform of $\psi^2(\mathbf{x})$ decays algebraically with time, *not* exponentially as expected for variables that are not part of the slow set. Thus it seems wise to treat all products of $\psi(\mathbf{x})$ as *slow variables*. These nonlinear quantities are slower in lower dimension. We shall see that there are situations where these nonlinearities qualitatively affect the long-time and distance behavior of our system.

9.3
Coarse-Grained Variables and Effective Hamiltonians

In order to develop a method for treating the nonlinear dynamics of a set of fluctuating fields, we need to develop a bit of formal structure. We again consider a set of variables ψ_i ($\psi_i = \{n, \mathbf{g}, q\}$ for a fluid, $\mathbf{M}$ for a ferromagnet and $\mathbf{M}$ and $\mathbf{N}$ for antiferromagnets.) We want to derive Langevin-like equations as before, but we want to allow for all possible nonlinear couplings among the ψ_i's. The end result of this development is a set of nonlinear field theories with many of the same technical difficulties (renormalization, etc.) found in other field theories. We shall avoid some of these problems by defining our *fields* on a lattice. This means that all of our fields can be labeled by a set of discrete labels. Thus in the microscopic realization of our model, we assume there are a set of slow variables $\psi_i(t)$. A variable that keeps track of these variables *and their products* is:

$$g_\phi(t) = \prod_i \delta[\phi_i - \psi_i(t)] \equiv \delta[\phi - \psi(t)] \, . \qquad (15)$$

Here, $g_\phi(t)$ maps the dynamical variables $\psi_i(t)$ onto the set of functions ϕ_i. If we take the thermal average of $g_\phi(t)$ over the space that includes the variables

$\psi_i(t)$, then we have a quantity:

$$W_\phi = \langle g_\phi(t) \rangle , \qquad (16)$$

which gives the equilibrium probability distribution governing the variables ϕ_i. It is useful, when doing multiple integrals over ϕ, to introduce the notation:

$$\mathcal{D}(\phi) = \left(\prod_i d\phi_i \right) \qquad (17)$$

and:

$$\int \mathcal{D}(\phi) g_\phi(t) = \prod_i \int d\phi_i \delta[\phi_i - \psi_i(t)] = 1 . \qquad (18)$$

This result guarantees that W_ϕ is normalized:

$$\int \mathcal{D}(\phi) W_\phi = 1 . \qquad (19)$$

Note that the variable $g_\phi(t)$ can generate all nonlinear equal-time couplings among the variables ψ_i. We see, for example, that:

$$\int \mathcal{D}(\phi) \phi_i \phi_j g_\phi(t) = \psi_i(t) \psi_j(t) . \qquad (20)$$

We shall also deal with the equal-time correlation functions:

$$\begin{aligned} S_{\phi\phi'} &= \langle g_\phi g_{\phi'} \rangle = \delta(\phi - \phi') \langle g_\phi \rangle \\ &= \delta(\phi - \phi') W_\phi . \end{aligned} \qquad (21)$$

Clearly there is no new information in $S_{\phi\phi'}$.

In principle, we must carry out a detailed microscopic analysis to obtain W_ϕ. Since we expect W_ϕ to be positive, we can write:

$$W_\phi = \frac{e^{-\beta \mathcal{H}_\phi}}{Z} , \qquad (22)$$

where $\mathcal{H}_\phi$ is a coarse-grained or effective Hamiltonian or free energy. In practice, in a variety of situations one can write down the effective Hamiltonian from general considerations [9] without having to carry out a detailed microscopic calculation. Let us discuss a few examples.

In the case of a normal fluid or thermodynamic systems with no long-range correlations, Landau and Lifshitz [10] have shown that the fluctuations are governed by a Gaussian probability distribution or a quadratic effective Hamiltonian:

$$\mathcal{H}_\phi = \frac{1}{2} \int d^d x \sum_{ij} \phi_i(\mathbf{x}) M_{ij} \phi_j(\mathbf{x}) , \qquad (23)$$

where the matrix M_{ij} consists of thermodynamic derivatives. It is easy to see that this matrix can be determined from the standard relation for Gaussian integrals [11]:

$$S_{ij}(\mathbf{k}) = \langle \phi_j(-\mathbf{k})\phi_i(\mathbf{k}) \rangle = k_B T M_{ij}^{-1}. \tag{24}$$

We explored this statement in detail in Chapter 6, where we showed that $S_{nn}(0)$, $S_{nq}(0)$, and $S_{qq}(0)$ could be written in terms of thermodynamic derivatives. Thus we constructed $(M^{-1})_{ij}$ for the set of variables n and q. If one chooses as fundamental variables the fluctuating temperature $T(\mathbf{x})$ or the particle density $n(\mathbf{x})$ then the matrix M is diagonal and one has:

$$\mathcal{H}_\phi = \frac{1}{2} \int d^d x \left[\frac{C_V}{VT} [\delta T(\mathbf{x})]^2 + \frac{1}{n\kappa_T} [\delta n(\mathbf{x})]^2 \right], \tag{25}$$

where C_V is the specific heat at constant volume and κ_T is the isothermal compressibility. One also has a diagonal form if one chooses pressure and entropy density as the independent fluctuating fields:

$$\mathcal{H}_\phi = \frac{1}{2} \int d^d x \left[\left(\frac{\partial n}{\partial p} \right)_s (\delta p(\mathbf{x}))^2 + \frac{TV}{C_p} (\delta s(\mathbf{x}))^2 \right]. \tag{26}$$

Nonlinear models in the case of critical phenomena are well developed. In Wilson's original [12] renormalization group treatment of second-order phase transitions he focused on the evolution of effective Hamiltonians as one coarse-grained or *integrated out* short-distance information to obtain $\mathcal{H}_\phi$ where the order parameter ϕ was restricted to small wavenumbers. A key idea that came out of this development was that the details of the effective Hamiltonian were not crucial, but its general form could be determined from principles of symmetry and analyticity, as in the original development of Landau [13]. Many critical systems could be understood in terms of the Landau–Ginzburg–Wilson (LGW) effective Hamiltonian:

$$\mathcal{H}_\phi = \frac{1}{2} \int d^d x \left[\frac{c}{2} (\nabla \vec{\phi})^2 + V(\vec{\phi}) \right], \tag{27}$$

where c is a positive constant and V is a potential of the general form:

$$V(\vec{\phi}) = \frac{r}{2} \vec{\phi}^2 + \frac{u}{4} (\vec{\phi}^2)^2, \tag{28}$$

where u is positive but r is proportional to $T - T_c^{(0)}$ and changes sign at the mean-field theory [14] transition temperature $T_c^{(0)}$. It is crucial that $u > 0$. If $u = 0$ then we have a Gaussian probability distribution and a quadratic effective Hamiltonian that leads to mean-field or *linear* critical properties. Terms cubic or higher order in the effective Hamiltonian are called nonlinear, since they make nonlinear contributions to the conjugate force $\delta \mathcal{H}/\delta \phi$.

In the case of simple fluids, where one wants to include flow in the description, the slow variables are the mass density ρ, the momentum density $\mathbf{g}$ and the energy density ε. In the simplest circumstances we can ignore the energy density and work [15] with the effective Hamiltonian:

$$\mathcal{H}[\rho, \mathbf{g}] = \int d^d x \left[\frac{1}{2} \frac{\mathbf{g}^2(\mathbf{x})}{\rho(\mathbf{x})} + f[\rho(\mathbf{x})] \right] , \tag{29}$$

where the first term represents the kinetic energy.

In the case of more complex fluids, one introduces effective Hamiltonians almost out of necessity, since a more microscopic treatment is very unwieldy and probably unnecessary if one is interested in long-time, long-distance behavior. In the case of a nematic liquid crystal it is the director field $\hat{n}(\mathbf{x})$ that corresponds to the Nambu–Goldstone (NG) mode in the ordered phase. It is widely accepted [16] that the appropriate effective Hamiltonian is the Frank effective Hamiltonian:

$$\mathcal{H}[\hat{n}] = \frac{1}{2} \int d^d x \left[K_1 (\nabla \cdot \hat{n})^2 + K_2 (\hat{n} \cdot (\nabla \times \hat{n}))^2 + K_3 (\hat{n} \times (\nabla \times \hat{n}))^2 \right] , \tag{30}$$

where the K_i are elastic constants. In the case of smectic A liquid crystals [17] the NG mode is the displacement field $u(\mathbf{x})$ and the Landau effective Hamiltonian in this case is given by:

$$\mathcal{H}_u = \frac{1}{2} \int d^d x \left[BE[u] + K_1 (\nabla^2 u)^2 \right] , \tag{31}$$

where B is a constant and:

$$E[u] = \frac{\partial u}{\partial z} + \frac{1}{2} (\nabla u)^2 . \tag{32}$$

In the case of solids one has the elastic contribution [18] to the effective Hamiltonian:

$$\mathcal{H}_U = \frac{1}{2} \int d^d x \, C_{\alpha\beta\gamma\mu} U_{\gamma\alpha} U_{\mu\beta} , \tag{33}$$

where the matrix C is the set of elastic constants appropriate for the symmetry of the lattice and the U are the *strain* fields:

$$U_{\gamma\alpha} = \frac{1}{2} \left(\frac{\partial u_\alpha}{\partial x_\gamma} + \frac{\partial u_\gamma}{\partial x_\alpha} \right) . \tag{34}$$

Finally we have the case of the Ginsburg–Landau free energy [19] for Superconductors, where $\psi(\mathbf{x})$ is the complex order parameter governed by the effective Hamiltonian [20]:

$$\mathcal{H}[\psi, \mathbf{A}] = \frac{1}{2} \int d^d x \left[V[\psi(\mathbf{x})] + \frac{\hbar^2}{2m^*} |(\nabla - iq\mathbf{A}(\mathbf{x})\psi(\mathbf{x})|^2 \right] + \mathcal{H}_{EM}[\mathbf{A}, \phi] ,\quad (35)$$

where q is the charge, $\mathbf{A}$ and ϕ are the electromagnetic potential and $\mathcal{H}_{EM}$ is the associated effective Hamiltonian that generates the macroscopic Maxwell's equations in the static case. Neutral superfluids are described by $\mathcal{H}[\psi, \mathbf{A}]$ with $q = 0$.

9.4
Nonlinear Coarse-Grained Equations of Motion

9.4.1
Generalization of Langevin Equation

We want to work out the equation of motion satisfied by $g_\phi(t)$. Our analysis will follow the same procedure [21] used in Chapter 5 for the set of variables $\psi_i(t)$, except now we replace $\psi_i(t)$ with $g_\phi(t)$. Thus one again has a linear theory, but now linear in $g_\phi(t)$, which treats all equal-time products of $\psi_i(t)$. This is not the most general set of nonlinearities, but appears to be sufficient for our purposes.

The first step in writing a generalized Langevin equation for $g_\phi(t)$ is to introduce the Laplace transform:

$$\begin{aligned} g_\phi(z) &= -i \int_0^{+\infty} dt\, e^{+izt} g_\phi(t) \\ &= R(z) g_\phi ,\end{aligned} \quad (36)$$

where $R(z)$ is again the resolvant operator,

$$R(z) = [z + L]^{-1} ,\quad (37)$$

with L the Liouville operator governing the system of interest. In complete analogy with our work in Chapter 5, we can immediately write down the generalized Langevin equation of the form:

$$z g_\phi(z) - K_{\phi\bar\phi}(z) g_{\bar\phi}(z) = g_\phi + i N_\phi(z) ,\quad (38)$$

where we introduce the notation for repeated barred labels:

$$K_{\phi\bar\phi}(z) g_{\bar\phi} \equiv \int \mathcal{D}(\bar\phi) K_{\phi\bar\phi}(z) g_{\bar\phi}(z) ,\quad (39)$$

and $N_\phi(z)$ is the Laplace transform of the associated noise. Again we demand that the noise satisfy:

$$\langle g_{\phi'} N_\phi(z) \rangle = 0 \tag{40}$$

or:

$$\langle g_{\phi'} N_\phi(t) \rangle = 0 \tag{41}$$

for $t \geq 0$. The correlation function:

$$G_{\phi\phi'}(z) = \langle g_{\phi'} R(z) g_\phi \rangle \tag{42}$$

then satisfies the kinetic equation:

$$z G_{\phi\phi'}(z) - K_{\phi\bar{\phi}} G_{\bar{\phi}\phi'}(z) = S_{\phi\phi'} \;, \tag{43}$$

where the memory function is, as before, a sum of two parts:

$$K_{\phi\phi'}(z) = K^{(s)}_{\phi\phi'} + K^{(d)}_{\phi\phi'}(z) \;, \tag{44}$$

where the static part of the memory function as usual is given by:

$$K^{(s)}_{\phi\bar{\phi}} S_{\bar{\phi}\phi'} = -\langle g_{\phi'} L g_\phi \rangle \tag{45}$$

and the dynamic part of the memory function is given by:

$$\begin{aligned} K^{(d)}_{\phi\bar{\phi}}(z) S_{\bar{\phi}\phi'} &= -\langle (L g_{\phi'}) R(z) (L g_\phi) \rangle \\ &\quad + \langle (L g_{\phi'}) R(z) g_{\bar{\phi}} \rangle G^{-1}_{\bar{\phi}\bar{\phi}'}(z) \langle g_{\bar{\phi}'} R(z) (L g_\phi) \rangle \;. \end{aligned} \tag{46}$$

All of this seems impressively complicated. Let us see how it can be simplified.

9.4.2
Streaming Velocity

We have the now familiar sequence of steps developed in Chapter 5. We have chosen our set of slow variables, $g_\phi(t)$, and its equilibrium average is given by:

$$W_\phi = \langle g_\phi(t) \rangle = e^{-\beta \mathcal{H}_\phi} / Z_\phi \;, \tag{47}$$

where this equation defines the effective Hamiltonian $\mathcal{H}_\phi$ up to a normalization constant given by Z_ϕ. Next, the static structure factor for these variables is given by:

$$S_{\phi\phi'} = \langle g_\phi g_{\phi'} \rangle = \delta(\phi - \phi') W_{\phi'} \;. \tag{48}$$

Consider now the static part of the memory function given by Eq. (45). Since L is a linear operator we have the very useful chain-rule identity:

$$Lg_\phi = -\sum_i \frac{\partial}{\partial \phi_i} g_\phi L\psi_i \;, \qquad (49)$$

which leads to the result:

$$\begin{aligned} K^{(s)}_{\phi\phi'} W_{\phi'} &= \sum_i \frac{\partial}{\partial \phi_i} \langle g_{\phi'} g_\phi L\psi_i \rangle \\ &= \sum_i \frac{\partial}{\partial \phi_i} [\delta(\phi' - \phi)\langle g_\phi L\psi_i \rangle] \;. \end{aligned} \qquad (50)$$

We can use the result from Chapter 5, Eq. (5.205), that:

$$\langle \psi_j L\psi_i \rangle = -i\beta^{-1}\langle \{\psi_i, \psi_j\} \rangle \;, \qquad (51)$$

to obtain:

$$\langle g_\phi L\psi_i \rangle = -i\beta^{-1}\langle \{\psi_i, g_\phi\} \rangle \;. \qquad (52)$$

Since in the Poisson bracket we have a linear operator acting on g_ϕ, we again use the chain-rule to obtain:

$$\{\psi_i, g_\phi\} = -\sum_j \frac{\partial}{\partial \phi_j} g_\phi \{\psi_i, \psi_j\} \qquad (53)$$

and:

$$K^{(s)}_{\phi\phi'} W_{\phi'} = \sum_i \frac{\partial}{\partial \phi_i} \left[\delta(\phi' - \phi)(i\beta^{-1}) \sum_j \frac{\partial}{\partial \phi_j} \langle g_\phi \{\psi_i, \psi_j\} \rangle \right] \;. \qquad (54)$$

Let us define:

$$Q_{ij}[\phi] = \langle g_\phi \{\psi_i, \psi_j\} \rangle / W_\phi \;. \qquad (55)$$

If we have the explicit result for the Poisson brackets expressed in terms of the slow modes:

$$\{\psi_i, \psi_j\} = f_{ij}(\psi) \;, \qquad (56)$$

then:

$$\begin{aligned} Q_{ij}[\phi] &= \langle g_\phi f_{ij}(\psi) \rangle / W_\phi \\ &= f_{ij}(\phi) \langle g_\phi \rangle / W_\phi \\ &= f_{ij}(\phi) \;. \end{aligned} \qquad (57)$$

Putting Eq. (55) back into Eq. (54) we have:

$$K^{(s)}_{\phi\phi'} W_{\phi'} = i \sum_{ij} \frac{\partial}{\partial \phi_i} \left[\delta(\phi' - \phi) \frac{\partial}{\partial \phi_j} \beta^{-1} W_\phi Q_{ij}[\phi] \right] \quad (58)$$

$$= i \sum_{ij} \frac{\partial}{\partial \phi_i} \left[\delta(\phi' - \phi) \left(\frac{\partial W_\phi}{\partial \phi_j} \right) \beta^{-1} Q_{ij}[\phi] + \delta(\phi' - \phi) W_\phi \beta^{-1} \frac{\partial}{\partial \phi_j} Q_{ij}[\phi] \right] .$$

Using the definition of the effective Hamiltonian:

$$\beta \frac{\partial \mathcal{H}_\phi}{\partial \phi_j} = -\frac{1}{W_\phi} \frac{\partial}{\partial \phi_j} W_\phi , \quad (59)$$

we can then cancel a common factor of $W_{\phi'}$ to obtain:

$$K^{(s)}_{\phi\phi'} = i \sum_{ij} \frac{\partial}{\partial \phi_i} \left[-\delta(\phi' - \phi) Q_{ij}[\phi] \frac{\partial \mathcal{H}_\phi}{\partial \phi_j} + \delta(\phi' - \phi) \frac{\partial}{\partial \phi_j} \beta^{-1} Q_{ij}[\phi] \right]$$

$$= -i \sum_i \frac{\partial}{\partial \phi_i} \left[\mathcal{V}_i[\phi] \delta(\phi - \phi') \right] , \quad (60)$$

where:

$$\mathcal{V}_i[\phi] = \sum_j \left(Q_{ij}[\phi] \frac{\partial \mathcal{H}_\phi}{\partial \phi_j} - \beta^{-1} \frac{\partial}{\partial \phi_j} Q_{ij}[\phi] \right) . \quad (61)$$

$\mathcal{V}_i[\phi]$ is a *streaming velocity* in function ϕ space. It satisfies the important *divergence* property:

$$\sum_i \frac{\partial}{\partial \phi_i} [\mathcal{V}_i(\phi) W_\phi] = 0 . \quad (62)$$

Proof: if we insert Eq. (61) into Eq. (62) we find:

$$\sum_i \frac{\partial}{\partial \phi_i} \left[\sum_j \left(Q_{ij}[\phi] \frac{\partial \mathcal{H}_\phi}{\partial \phi_j} - \beta^{-1} \frac{\partial}{\partial \phi_j} Q_{ij}[\phi] \right) W_\phi \right]$$

$$= -\sum_{ij} \frac{\partial}{\partial \phi_i} \left[Q_{ij}[\phi] \beta^{-1} \frac{\partial}{\partial \phi_j} W_\phi + W_\phi \beta^{-1} \frac{\partial}{\partial \phi_j} Q_{ij}[\phi] \right]$$

$$= -\sum_{ij} \frac{\partial}{\partial \phi_i} \frac{\partial}{\partial \phi_j} \left[\beta^{-1} Q_{ij}[\phi] W_\phi \right] . \quad (63)$$

Since Q_{ij} is antisymmetric,

$$Q_{ij}[\phi] = -Q_{ji}[\phi] , \quad (64)$$

the double sum vanishes and the theorem is proven. The usefulness of this theorem will become apparent later. It is left to Problem 9.1 to show that the average of the streaming velocity is zero:

$$\langle \mathcal{V}_i[\phi] \rangle = 0 \ . \tag{65}$$

9.4.3
Damping Matrix

Let us turn next to the dynamic part of the *memory function* given by:

$$K_{\phi\phi'}^{(d)}(z)W_{\phi'} = -\langle (Lg_{\phi'})R(z)(Lg_\phi)\rangle + \langle (Lg_{\phi'})R(z)g_{\bar\phi}\rangle G_{\bar\phi\bar\phi'}^{-1}(z)\langle g_{\bar\phi'}R(z)(Lg_\phi)\rangle \ .$$

Using the identity given by Eq. (49), we see that this can be written in the form,

$$K_{\phi\phi'}^{(d)}(z)W_{\phi'} = \sum_{ij} \frac{\partial}{\partial\phi_i}\frac{\partial}{\partial\phi'_j} T_{\phi\phi'}^{ij}(z) \ , \tag{66}$$

where $T_{\phi\phi'}^{ij}(z)$ is a symmetric but complicated functional of ϕ and ϕ'. We know that any part of $L\psi_i$ that can be expressed in terms of ψ_i, or products of ψ_i at the same time t, will not contribute to $K^{(d)}$. The fundamental assumption at this point is that only fast variables contribute to $K_{\phi\phi'}^{(d)}(z)$ and it can be replaced by its *Markoffian*-type approximation for $T_{\phi\phi'}^{ij}(z)$:

$$T_{\phi\phi'}^{ij}(z) = -i\Gamma_0^{ij}(\phi)\beta^{-1}\delta(\phi-\phi')W_\phi \ , \tag{67}$$

where $\Gamma_0^{ij}(\phi)$ is a set, in general, of field-dependent bare kinetic coefficients. We discuss the consequences of this approximation in some detail. Putting Eq. (67) back into Eq. (66) gives

$$K_{\phi\phi'}^{(d)}(z) = -\frac{i}{W_{\phi'}} \sum_{ij} \frac{\partial}{\partial\phi_i}\frac{\partial}{\partial\phi'_j} \beta^{-1}\Gamma_0^{ij}(\phi)\delta(\phi-\phi')W_\phi$$

$$= -i\sum_{ij} \frac{\partial}{\partial\phi_i}\beta^{-1}\Gamma_0^{ij}(\phi)\left[\frac{1}{W_{\phi'}}\frac{\partial}{\partial\phi'_j}\delta(\phi-\phi')W_{\phi'}\right]$$

$$= -i\sum_{ij} \frac{\partial}{\partial\phi_i}\beta^{-1}\Gamma_0^{ij}(\phi)\frac{1}{W_{\phi'}}\left[W_{\phi'}\frac{\partial}{\partial\phi'_j}\delta(\phi-\phi') + \delta(\phi-\phi')\frac{\partial W_{\phi'}}{\partial\phi'_j}\right]$$

$$= -i\sum_{ij} \frac{\partial}{\partial\phi_i}\beta^{-1}\Gamma_0^{ij}(\phi)\left[-\frac{\partial}{\partial\phi_j}\delta(\phi-\phi') - \beta\frac{\partial\mathcal{H}_\phi}{\partial\phi_j}\delta(\phi-\phi')\right]$$

$$K_{\phi\phi'}^{(d)}(z) = i\sum_{ij} \frac{\partial}{\partial\phi_i}\beta^{-1}\Gamma_0^{ij}(\phi)\left[\frac{\partial}{\partial\phi_j} + \beta\frac{\partial\mathcal{H}_\phi}{\partial\phi_j}\right]\delta(\phi-\phi') \ . \tag{68}$$

9.4.4
Generalized Fokker–Planck Equation

Putting these results for $K^{(s)}$ and $K^{(d)}$ together we obtain, from Eq. (38), the Langevin equation satisfied by $g_\phi(t)$:

$$\frac{\partial}{\partial t}g_\phi(t) + iK^{(s)}_{\phi\bar{\phi}}g_{\bar{\phi}}(t) + i K^{(d)}_{\phi\bar{\phi}}g_{\bar{\phi}}(t) = N_\phi(t) \ . \tag{69}$$

The contribution from the static part of the memory function takes the form:

$$iK^{(s)}_{\phi\bar{\phi}}g_{\bar{\phi}}(t) = \sum_i \frac{\partial}{\partial \phi_i}\left[\mathcal{V}_i[\phi]\delta(\phi - \bar{\phi})g_{\bar{\phi}}(t)\right]$$

$$= \sum_i \frac{\partial}{\partial \phi_i}\left[\mathcal{V}_i[\phi]g_\phi(t)\right] \ , \tag{70}$$

while the contribution from the dynamic part of the memory function can be written as:

$$iK^{(d)}_{\phi\bar{\phi}}g_{\bar{\phi}}(t) = i\sum_{ij}\frac{\partial}{\partial \phi_i}i\beta^{-1}\Gamma^{ij}_0[\phi]\left[\frac{\partial}{\partial \phi_j} + \beta\frac{\partial \mathcal{H}_\phi}{\partial \phi_j}\right]\delta(\phi - \bar{\phi})g_{\bar{\phi}}(t)$$

$$= -\sum_{ij}\frac{\partial}{\partial \phi_i}\beta^{-1}\Gamma^{ij}_0[\phi]\left[\frac{\partial}{\partial \phi_j} + \beta\frac{\partial \mathcal{H}_\phi}{\partial \phi_j}\right]g_\phi(t) \ . \tag{71}$$

Combining the effects of $K^{(s)}$ and $K^{(d)}$ we can define the generalized Fokker–Planck operator [22]:

$$D_\phi = -\sum_i \frac{\partial}{\partial \phi_i}\left[\mathcal{V}_i[\phi] - \sum_j \beta^{-1}\Gamma^{ij}_0[\phi]\left(\frac{\partial}{\partial \phi_j} + \beta\frac{\partial \mathcal{H}_\phi}{\partial \phi_j}\right)\right] \ , \tag{72}$$

then we obtain our primary form for the generalized Fokker–Planck equation (GFPE):

$$\frac{\partial}{\partial t}g_\phi(t) = D_\phi g_\phi(t) + N_\phi(t) \ . \tag{73}$$

This is a fundamental and important equation. Let us check a few of its nice properties. First we require, by integrating over all ϕ' in Eq. (41), that:

$$\langle N_\phi(t)\rangle = 0 \ . \tag{74}$$

However we have *not* required $\langle g_\phi(t)\rangle = 0$; instead we must satisfy:

$$\frac{\partial}{\partial t}\langle g_\phi(t)\rangle = D_\phi \langle g_\phi(t)\rangle \ . \tag{75}$$

Since the equilibrium probability distribution is time independent,

$$\langle g_\phi(t) \rangle = W_\phi \, , \tag{76}$$

we must have:

$$D_\phi W_\phi = 0 \, . \tag{77}$$

We can show this explicitly. We have:

$$\begin{aligned} D_\phi W_\phi &= -\sum_i \frac{\partial}{\partial \phi_i} \left[\mathcal{V}_i[\phi] W_\phi \right] \\ &+ \sum_{ij} \frac{\partial}{\partial \phi_i} \beta^{-1} \Gamma_0^{ij}[\phi] \left[\frac{\partial}{\partial \phi_j} + \beta \frac{\partial \mathcal{H}_\phi}{\partial \phi_j} \right] W_\phi \, . \end{aligned} \tag{78}$$

The first term vanishes due to the *divergence* theorem, Eq. (62), proven earlier. So:

$$D_\phi W_\phi = \sum_{ij} \frac{\partial}{\partial \phi_i} \beta^{-1} \Gamma_0^{ij}[\phi] \left[\frac{\partial W_\phi}{\partial \phi_j} + W_\phi \beta \frac{\partial \mathcal{H}_\phi}{\partial \phi_j} \right] \, . \tag{79}$$

Since:

$$\frac{1}{W_\phi} \frac{\partial W_\phi}{\partial \phi_j} = -\beta \frac{\partial \mathcal{H}_\phi}{\partial \phi_j} \, , \tag{80}$$

Eq. (77) is satisfied. Therefore the equilibrium distribution is stationary under the application of the Fokker–Planck operator. It is also found that when treated as an initial-value problem, then the Fokker–Planck equation, Eq. (75), drives a wide class of initial states, $P_0(\phi) = \langle g_\phi(t_0) \rangle$ to equilibrium, $e^{-\beta \mathcal{H}_\phi}/Z$, as time increases to infinity.

9.4.5
Nonlinear Langevin Equation

The GFPE:

$$\frac{\partial}{\partial t} g_\phi(t) = D_\phi g_\phi(t) + N_\phi(t) \, , \tag{81}$$

can be used to generate an equation of motion for the field $\psi_i(t)$. The associated field $\psi_i(t)$ must now be viewed as a coarse-grained field satisfying an effective dynamics. If we multiply Eq. (81) by ϕ_i and integrate over all ϕ, we immediately obtain the nonlinear Langevin equation:

$$\frac{\partial \psi_i(t)}{\partial t} = \int \mathcal{D}(\phi) \phi_i D_\phi g_\phi(t) + \xi_i(t) \, , \tag{82}$$

where:

$$\psi_i(t) = \int \mathcal{D}(\phi) \phi_i g_\phi(t) \tag{83}$$

and the moment of the noise $N_\phi(t)$ generates the nonlinear Langevin equation noise:

$$\xi_i(t) = \int \mathcal{D}(\phi) \phi_i N_\phi(t) \ . \tag{84}$$

If we look at the middle term in Eq. (8.109), we have:

$$\int \mathcal{D}(\phi) \phi_i D_\phi g_\phi(t)$$
$$= -\int \mathcal{D}(\phi) \phi_i \sum_j \frac{\partial}{\partial \phi_j} \left[\mathcal{V}_j[\phi] - \sum_k \beta^{-1} \Gamma_0^{jk}[\phi] \left(\frac{\partial}{\partial \phi_k} + \beta \frac{\partial \mathcal{H}_\phi}{\partial \phi_k} \right) \right] g_\phi(t) \ . \tag{85}$$

After integrating by parts we obtain:

$$\int \mathcal{D}(\phi) \phi_i D_\phi g_\phi(t) = \int \mathcal{D}(\phi) \left[\mathcal{V}_i[\phi] - \sum_k \Gamma_0^{ik}[\phi] \left(\frac{\partial}{\partial \phi_k} + \frac{\partial \mathcal{H}_\phi}{\partial \phi_k} \right) \right] g_\phi(t)$$
$$= \mathcal{V}_i[\psi(t)] + \sum_k \frac{\partial}{\partial \psi_k(t)} \beta^{-1} \Gamma_0^{ik}[\psi(t)] - \sum_k \Gamma_0^{ik}[\psi] \frac{\partial \mathcal{H}_\psi}{\partial \psi_k(t)} \ , \tag{86}$$

and the nonlinear Langevin equation takes the form:

$$\frac{\partial \psi_i(t)}{\partial t} = \mathcal{V}_i[\psi(t)] + \sum_k \frac{\partial}{\partial \psi_k(t)} \beta^{-1} \Gamma_0^{ik}[\psi(t)]$$
$$- \sum_k \Gamma_0^{ik}[\psi(t)] \frac{\partial}{\partial \psi_k(t)} \mathcal{H}_\psi + \xi_i(t) \ . \tag{87}$$

In a typical case where the bare kinetic coefficients are independent of the fields, then we have:

$$\frac{\partial \psi_i(t)}{\partial t} = \mathcal{V}_i[\psi(t)] - \sum_k \Gamma_0^{ik} \frac{\partial}{\partial \psi_k(t)} \mathcal{H}_\psi + \xi_i(t) \ . \tag{88}$$

This is a key result and we will come back and discuss the various parts of this equation after discussing the noise contributions $\xi_i(t)$ and $N_\phi(t)$.

9.5
Discussion of the Noise

9.5.1
General Discussion

We derived the second fluctuation-dissipation theorem in Chapter 5 and in the present context it takes the form:

$$\langle N_\phi(t) N_{\phi'}(t') \rangle = K^{(d)}_{\phi\phi'}(t-t') S_{\bar\phi\phi'}$$
$$= K^{(d)}_{\phi\phi'}(t-t') W_{\phi'} \quad t > t' \ . \tag{89}$$

We have, from our Markoffian approximation, Eqs. (66) and (67), that:

$$K^{(d)}_{\phi\phi'}(t-t') W_{\phi'} = \sum_{ij} \frac{\partial}{\partial \phi_i} \frac{\partial}{\partial \phi'_j} 2\beta^{-1} \Gamma^{ij}_0(\phi) \delta(\phi - \phi') W_\phi \delta(t-t') \ , \tag{90}$$

so, for self-consistency, we expect:

$$\langle N_\phi(t) N_{\phi'}(t') \rangle = \sum_{ij} \frac{\partial}{\partial \phi_i} \frac{\partial}{\partial \phi'_j} 2\beta^{-1} \Gamma^{ij}_0(\phi) \delta(\phi - \phi') W_\phi \delta(t-t') \ . \tag{91}$$

We have immediately, using Eq. (84), that the autocorrelation function for the noise $\xi_i(t)$ in the nonlinear Langevin equation is given by:

$$\langle \xi_i(t) \xi_j(t') \rangle = \int \mathcal{D}(\phi) \int \mathcal{D}(\phi') \phi_i \phi'_j \langle N_\phi(t) N_{\phi'}(t') \rangle$$
$$= \int \mathcal{D}(\phi) \int \mathcal{D}(\phi') 2\beta^{-1} \Gamma^{ij}_0(\phi) \delta(\phi - \phi') W_\phi \delta(t-t')$$
$$= 2\beta^{-1} \langle \Gamma^{ij}_0[\phi] \rangle \delta(t-t') \ . \tag{92}$$

We see, then, that the noise for the nonlinear Langevin equation has the same autocorrelation as for the linear Langevin equation for the case where Γ^{ij}_0 is independent of ϕ_i.

9.5.2
Gaussian Noise

Let us suppose that the kinetic coefficients $\Gamma^{ij}_0[\phi]$ are independent of ϕ, which is consistent with the assumption that the noise $\xi_i(t)$ is Gaussianly distributed. This means that in an average over the noise of the form:

$$\langle A[\psi] \rangle = \int \mathcal{D}(\xi) P[\xi] A[\psi(\xi)] \ , \tag{93}$$

the probability distribution governing the noise is given by:

$$P[\xi] = \frac{1}{Z_\xi} e^{-\frac{1}{4} \sum_{ij} \int_{-\infty}^{+\infty} dt\, \xi_i(t) \beta (\Gamma_0^{-1})_{ij} \xi_j(t)} \tag{94}$$

and Z_ξ is chosen such that:

$$\int \mathcal{D}(\xi) P[\xi] = 1 \ . \tag{95}$$

The first step in establishing this as a consistent choice is to show that the variance is in agreement with Eq. (92):

$$\langle \xi_i(t)\xi_j(t')\rangle = 2\beta^{-1}\Gamma_0^{ij}\delta(t-t') \ . \tag{96}$$

This follows from the functional identity:

$$\int \mathcal{D}(\xi) \frac{\delta}{\delta\xi_i(t)}[P[\xi]A(\psi,\xi)] = 0 \ . \tag{97}$$

Taking the derivative of the distribution given by Eq. (94),

$$\frac{\delta}{\delta\xi_i(t)} P[\xi] = -\frac{\beta}{2}(\Gamma_0^{-1})_{ik}\xi_k(t)P[\xi] \ , \tag{98}$$

Eq. (97) reduces to:

$$\sum_k \frac{\beta}{2}(\Gamma_0^{-1})_{ik}\langle \xi_k(t)A(\xi)\rangle = \left\langle \frac{\delta A(\xi)}{\delta\xi_i(t)} \right\rangle \ . \tag{99}$$

If we matrix multiply on the left by $2\beta^{-1}\Gamma_0$ we obtain:

$$\langle \xi_i(t)A(\xi)\rangle = 2\beta^{-1}\sum_k \Gamma_0^{ik}\left\langle \frac{\delta A(\xi)}{\delta\xi_k(t)}\right\rangle \ . \tag{100}$$

Let:

$$A(\xi) = \xi_j(t') \ , \tag{101}$$

then:

$$\langle \xi_i(t)\xi_j(t')\rangle = 2\beta^{-1}\sum_k \Gamma_0^{ik}\left\langle \frac{\delta\xi_j(t')}{\delta\xi_k(t)}\right\rangle$$
$$= 2\beta^{-1}\Gamma_0^{ij}\delta(t-t') \ , \tag{102}$$

in agreement with the result derived directly from the second fluctuation-dissipation theorem.

9.5.3
Second Fluctuation-Dissipation Theorem

It is shown in Appendix E that the noise N_ϕ is related to the nonlinear Langevin noise, ξ_i, by:

$$N_\phi(t) = -\sum_{ik}\frac{\partial}{\partial\phi_i}\left[\xi_i(t)\delta_{ik} + \frac{\partial}{\partial\phi_k}\beta^{-1}\Gamma_0^{ik}\right]g_\phi(t) \ . \tag{103}$$

It is further shown in Appendix E, by explicitly doing the average over the Gaussian noise ξ_i, that indeed the second fluctuation-dissipation theorem holds in the form:

$$\langle N_\phi(t) N_{\phi'}(t') \rangle = \sum_{ij} \frac{\partial}{\partial \phi_i} \frac{\partial}{\partial \phi_j} 2\beta^{-1} \Gamma_0^{ij} \delta(\phi - \phi') W_\phi \delta(t - t') \ . \tag{104}$$

9.6
Summary

We have constructed a new self-contained dynamics characterized by the set of Poisson brackets, the bare kinetic coefficients Γ_0^{ij} and a driving effective Hamiltonian. In terms of a generalized nonlinear Langevin equation this is given by:

$$\frac{\partial \psi_i(t)}{\partial t} = \mathcal{V}_i[\psi(t)] - \sum_k \Gamma_0^{ik} \frac{\partial}{\partial \psi_k(t)} \mathcal{H}_\psi + \xi_i(t) \ , \tag{105}$$

where the streaming velocity is given by:

$$\mathcal{V}_i[\psi] = \sum_j \left[Q_{ij}[\psi] \frac{\partial \mathcal{H}_\psi}{\partial \psi_j} - \beta^{-1} \frac{\partial}{\partial \psi_j} Q_{ij}[\psi] \right] \ , \tag{106}$$

where $Q_{ij}[\psi]$ is the set of Poisson brackets among the slow variables. In this case the noise $\xi_i(t)$ is Gaussian with variance:

$$\langle \xi_i(t) \xi_j(t') \rangle = 2 k_B T \Gamma_0^{ij} \delta(t - t') \ . \tag{107}$$

This description is equivalent to working with the Fokker–Planck equation:

$$\frac{\partial}{\partial t} g_\phi(t) = D_\phi g_\phi(t) + N_\phi(t) \ , \tag{108}$$

where the Fokker–Planck operator is given by:

$$D_\phi = -\sum_i \frac{\partial}{\partial \phi_i} \left[\mathcal{V}_i[\phi] - \sum_j \beta^{-1} \Gamma_0^{ij} \left(\frac{\partial}{\partial \phi_j} + \beta \frac{\partial \mathcal{H}_\phi}{\partial \phi_j} \right) \right] \ . \tag{109}$$

Let us move on to discuss the application of these results to various physical systems.

It is key to understand that these models stand on their own as complete dynamical systems. We gave a derivation starting with a microscopic system that has dynamics governed by Liouville operator. If we work with less isolated, more complex systems, it may be more accurate to start with these coarse-grained equations for the slow variables, accepting the crucial role of the noise.

The procedures for specifying a model in a given physical situation are very similar to our linearized Langevin approach. First choose the *slow* variables $\psi_i(t)$. Next address the equilibrium correlations. In linearized hydrodynamics this involves constructing $S_{ij} = \langle \psi_i \psi_j \rangle$, while in *fluctuating nonlinear hydrodynamics* (FNH) one chooses the effective Hamiltonian $\mathcal{H}_\psi$, from which one can determine [23] all equal-time equilibrium correlation functions. Next one determines the Poisson brackets among the slow variables:

$$Q_{ij}[\psi] = \{\psi_i, \psi_j\} . \tag{110}$$

In linearized hydrodynamics, the average of Q_{ij} gives, essentially, the static part of the memory function, while in FNH they determine the streaming velocity and the reversible part of the dynamics. It is clear that it is desirable to choose the set ψ_i such that the Poisson bracket relation Eq. (110) is closed. Finally one must specify the damping matrix Γ_0^{ij} in each case. In linearized hydrodynamics, these are the physical transport coefficients, while in FNH they have the same symmetry but are **bare** transport coefficients that are modified [24] by nonlinear interactions.

9.7 Examples of Nonlinear Models

9.7.1 TDGL Models

Let us turn now to the nonlinear Langevin equation,

$$\frac{\partial \psi_i(t)}{\partial t} = V_i[\psi(t)] - \sum_k \Gamma_0^{ik} \frac{\partial}{\partial \psi_k(t)} \mathcal{H}_\psi + \xi_i(t) , \tag{111}$$

and discuss the various terms. The most straightforward case is where one has a scalar field $\psi(\mathbf{x}, t)$ that does not couple to any other variables. In this example the label i maps onto the continuous spatial label $\mathbf{x}$. In this case we have a field theory that must ultimately be regulated at short distances. One could think in terms of an Ising-like magnetic system where the order parameter is the z-component of the magnetization density or an antiferromagnetic where the order parameter is the z-component of the staggered magnetization density. Since the Poisson bracket of a scalar field with itself is zero, we find that the streaming velocity vanishes. The resulting equation:

$$\frac{\partial \psi_i(t)}{\partial t} = -\sum_k \Gamma_0^{ik} \frac{\partial}{\partial \psi_k(t)} \mathcal{H}_\psi + \xi_i(t) \tag{112}$$

is known as the time-dependent Ginzburg–Landau (TDGL) Model. This model [25] plays a central role in the treatment of dynamic critical phenomena and growth kinetics for dissipative systems where the order parameter is not coupled to other slow modes (like energy). This model also plays a central role in a discussion of ordering systems as discussed in Chapter 11. Such models allow one to treat, for example, temperature quenches and field flips.

The TDGL equations were actually first written down to describe the behavior of superfluids [26] and superconductors [27] near their phase transitions. In these cases, the order parameter ψ is complex and the TDGL equation of motion takes the form:

$$\frac{\partial \psi(\mathbf{x},t)}{\partial t} = -\Gamma_0 \frac{\delta}{\delta \psi^*(\mathbf{x},t)} \mathcal{H}[\psi(t), \psi^*(\mathbf{x},t)] + \xi(\mathbf{x},t) \ , \tag{113}$$

where the noise is complex.

In the area of dynamic critical phenomena these models where a field $\psi(\mathbf{x},t)$ is driven by purely dissipative terms are known, in the classification scheme of Hohenberg and Halperin [28], as models A (nonconserved) and B (conserved) and have been extensively investigated. Other models of this type can be written down for the n-vector model and, for example, a nonconserved order parameter (NCOP) coupled to a conserved field. The conserved-scalar order parameter case is also known as the Cahn-Hilliard model [29] due to its use in treating the ordering kinetics of alloys.

In treating critical phenomena, it is now conventional and reasonable to assume that the effective free-energy or Hamiltonian can be written in the form of the LGW free energy functional:

$$\mathcal{H}[\psi] = \int d^d x \left[\frac{r}{2} \psi^2 + \frac{u}{4} \psi^4 + \frac{c}{2} (\nabla \psi)^2 \right] \ , \tag{114}$$

where u and c are positive constants and a phase transition is characterized by a change of sign in the coefficient $r = a(T - T_c^0)$.

The final step in specifying the model is to choose the matrix of kinetic coefficients that has the form, in this case, of $\Gamma_0[\mathbf{x}, \mathbf{x}']$. There are two important cases. In the simplest case the order parameter is not conserved (NCOP case),

$$\Gamma_0[\mathbf{x}, \mathbf{x}'] = \Gamma_0 \delta(\mathbf{x} - \mathbf{x}') \tag{115}$$

and Γ_0 is a bare kinetic coefficient. The second important case is where the order parameter is conserved (COP case): we write:

$$\begin{aligned}\Gamma_0[\mathbf{x}, \mathbf{x}'] &= D_0 \nabla \cdot \nabla' \delta(\mathbf{x} - \mathbf{x}') \\ &= -D_0 \nabla^2 \delta(\mathbf{x} - \mathbf{x}')\end{aligned} \tag{116}$$

and D_0 is a bare transport coefficient. We can treat both cases by writing:

$$\Gamma_0[\mathbf{x}, \mathbf{x}'] = \hat{\Gamma}_0(\mathbf{x}) \delta(\mathbf{x} - \mathbf{x}') \ , \tag{117}$$

where $\hat{\Gamma}_0$ is a constant for the NCOP case and $-D_0\nabla^2$ for the COP case. The TDGL equation of motion then reads:

$$\frac{\partial \psi(\mathbf{x},t)}{\partial t} = -\hat{\Gamma}_0(\mathbf{x})\frac{\delta \mathcal{H}_\psi}{\delta \psi(\mathbf{x},t)} + \xi(\mathbf{x},t) \ . \tag{118}$$

If we insert the usual LGW effective Hamiltonian into Eq. (118), we obtain the nonlinear Langevin equation:

$$\frac{\partial \psi(\mathbf{x},t)}{\partial t} = -\hat{\Gamma}_0\left[(r-c\nabla^2)\psi(\mathbf{x},t) + u\psi^3(\mathbf{x},t)\right] + \xi(\mathbf{x},t) \ . \tag{119}$$

The only nonlinearity is the ψ^3 term. If we set $u = 0$ we are back to our linearized equation:

$$\frac{\partial \psi(\mathbf{x},t)}{\partial t} = -\hat{\Gamma}_0\left[(r-c\nabla^2)\psi(\mathbf{x},t)\right] + \xi(\mathbf{x},t) \ . \tag{120}$$

We review here the calculation of the associated correlation function. Equation (120) is diagonalized by Fourier transforming over space and time to obtain:

$$\left[-i\omega + \Gamma_0(q)\chi_0^{-1}(q)\right]\psi(\mathbf{q},\omega) = \xi(\mathbf{q},\omega) \ , \tag{121}$$

where $\chi_0^{-1}(q) = r + cq^2$, and $\Gamma_0(q) = \Gamma_0$ for a NCOP and $D_0 q^2$ for a COP. Equation (121) expresses ψ in terms of the noise. Since we know the statistics of the noise, we easily find:

$$\langle \psi(\mathbf{q},\omega)\psi(\mathbf{q}',\omega')\rangle = (2\pi)^{d+1}\delta(\mathbf{q}+\mathbf{q}')\delta(\omega+\omega')C_0(\mathbf{q},\omega) \ , \tag{122}$$

where the multiplicative δ-functions reflect translational invariance in space and time and:

$$C_0(\mathbf{q},\omega) = \frac{2k_B T \Gamma_0(q)}{\omega^2 + [\Gamma_0(q)\chi_0^{-1}(q)]^2} \ . \tag{123}$$

Inverting the frequency Fourier transform we obtain in the time domain:

$$C_0(\mathbf{q},t) = S_0(q)e^{-\Gamma_0(q)\chi_0^{-1}(q)|t|} \ . \tag{124}$$

The equilibrium static structure factor is given by:

$$S_0(q) = \int \frac{d\omega}{2\pi}C_0(\mathbf{q},\omega) = \frac{k_B T}{r+cq^2} = \beta^{-1}\chi_0(q) \ , \tag{125}$$

which is just the expected Ornstein–Zernike result. Notice that the dynamics is relaxational. One has exponential decay of fluctuations.

Nonzero u leads to the corrections in critical phenomena associated with moving from the mean-field universality class to the class associated with physical systems with nontrivial critical indices. We discuss the nonlinear corrections in detail in Chapter 10.

9.7.2
Isotropic Magnets

Let us next consider the case of an isotropic ferromagnet. In this case the order parameter, or slow variable, is just the magnetization density $\mathbf{M}(\mathbf{x}, t)$. We again assume that the effective Hamiltonian or free energy can be written in the LGW form:

$$\mathcal{H}[\mathbf{M}] = \int d^d x \left[\frac{r}{2} \mathbf{M}^2 + \frac{u}{4} (\mathbf{M}^2)^2 + \frac{c}{2} \sum_{i\alpha} (\nabla_i M_\alpha)^2 \right] , \qquad (126)$$

where again c and u are positive and r changes sign at the mean-field transition temperature. In this case, the Poisson brackets close upon themselves,

$$\{ M_\alpha(\mathbf{x}), M_\beta(\mathbf{x}') \} = \sum_\gamma \varepsilon_{\alpha\beta\gamma} M_\gamma(\mathbf{x}) \delta(\mathbf{x} - \mathbf{x}') \qquad (127)$$

and the streaming velocity can be written as:

$$V_\alpha[\mathbf{x}] = \sum_\beta \int d^d x' \left[Q_{\alpha\beta}[x, x'] \frac{\partial \mathcal{H}}{\partial M_\beta(\mathbf{x}')} - \frac{\partial}{\partial M_\beta(\mathbf{x}')} \beta^{-1} Q_{\alpha\beta}[\mathbf{x}, \mathbf{x}'] \right] , \qquad (128)$$

where Q is defined by Eq. (55). Clearly the derivative acting on the Poisson bracket term is zero due to the antisymmetry of $\varepsilon_{\alpha\beta\gamma}$, so:

$$V_\alpha[\mathbf{x}] = \sum_{\beta\gamma} \varepsilon_{\alpha\beta\gamma} M_\gamma(\mathbf{x}) \frac{\delta}{\delta M_\beta(\mathbf{x})} \mathcal{H} . \qquad (129)$$

If we define the *effective* magnetic field:

$$\mathbf{H}(\mathbf{x}) = -\frac{\delta \mathcal{H}}{\delta \mathbf{M}(\mathbf{x})} , \qquad (130)$$

then:

$$V_\alpha(\mathbf{x}) = -\sum_{\beta\gamma} \varepsilon_{\alpha\beta\gamma} M_\gamma(\mathbf{x}) H_\beta(\mathbf{x}) , \qquad (131)$$

or in vector notation:

$$\vec{V} = \mathbf{M} \times \mathbf{H} \qquad (132)$$

and our Langevin equation is of the form [30]:

$$\frac{\partial \mathbf{M}}{\partial t} = \mathbf{M} \times \mathbf{H} - \Gamma_0 \nabla^2 \mathbf{H} + \vec{\zeta} ,\qquad(133)$$

where we have assumed that the magnetization is conserved. In addition to the dissipative terms there is a spin-precession term $\mathbf{M} \times \mathbf{H}$,

If we insert the explicit form for the LGW effective Hamiltonian into this equation we obtain:

$$\mathbf{H} = -\frac{\partial \mathcal{H}}{\partial \mathbf{M}} = -r\mathbf{M} - u(\mathbf{M}^2)\mathbf{M} + c\nabla^2 \mathbf{M} \qquad(134)$$

and:

$$\mathbf{M} \times \mathbf{H} = c\mathbf{M} \times \nabla^2 \mathbf{M} ,\qquad(135)$$

so:

$$\frac{\partial \mathbf{M}}{\partial t} = c\mathbf{M} \times \nabla^2 \mathbf{M} - \Gamma \nabla^2 \mathbf{H} + \vec{\zeta} .\qquad(136)$$

In this equation of motion we have reversible terms reflecting the underlying microscopic equation of motion and dissipative terms reflecting the influence of the **fast** variables in the problem. We discuss this model in detail in Chapter 10. It is important to point out that the reversible terms are nonlinear and are very important in treating the dynamic critical behavior of this model.

In the case of an antiferromagnet [31], where we keep the conserved magnetization $\mathbf{M}$ and the order parameter, the staggered magnetization, $\mathbf{N}$, we obtain the equations of motion:

$$\frac{\partial \mathbf{N}}{\partial t} = \mathbf{M} \times \mathbf{H}_N + \mathbf{N} \times \mathbf{H}_M - \Gamma_N \mathbf{H}_N + \vec{\zeta}_N \qquad(137)$$

$$\frac{\partial \mathbf{M}}{\partial t} = \mathbf{M} \times \mathbf{H}_M + \mathbf{N} \times \mathbf{H}_N + \Gamma_M \nabla^2 \mathbf{H}_M + \vec{\zeta}_M ,\qquad(138)$$

which is written in term of the effective fields:

$$\mathbf{H}_N = -\frac{\delta \mathcal{H}}{\delta \mathbf{N}} \qquad(139)$$

$$\mathbf{H}_M = -\frac{\delta \mathcal{H}}{\delta \mathbf{M}} \qquad(140)$$

and the driving effective Hamiltonian is taken to be of the form:

$$\mathcal{H}[\mathbf{M}, \mathbf{N}] = \int d^d x \left[\frac{r}{2} \mathbf{N}^2 + \frac{u}{4}(\mathbf{N}^2)^2 + \frac{c}{2}(\nabla \mathbf{N})^2 + \frac{r_M}{2} \mathbf{M}^2 \right] ,\qquad(141)$$

where $r_M > 0$. The simple dependence of the effective Hamiltonian on **M** reflects the fact that the magnetization is not the order parameter for an antiferromagnet.

9.7.3
Fluids

Let us turn to the case of a coarse-grained description of fluid dynamics. This, of course, is just *the* fluctuating nonlinear hydrodynamics. This is in some ways more complicated than the magnetic examples. While we can treat the coupled set of variables, the mass density ρ, the momentum density **g** and the heat density q, it is a bit more involved to include the heat mode, so we will not [32] include it for simplicity. Thus we have $\psi_i \to \{\rho(\mathbf{x}), \mathbf{g}(\mathbf{x})\}$.

The first step in constructing our nonlinear model is to make some statements about the effective Hamiltonian governing the equilibrium fluctuations. We want to determine:

$$e^{-\beta \mathcal{H}[\rho, \mathbf{g}]} = \langle g_\phi \rangle \ , \tag{142}$$

where g_ϕ maps the microscopic mass density:

$$\hat{\rho}(\mathbf{x}) = \sum_{i=1}^{N} m\delta(\mathbf{x} - \mathbf{r}_i) \tag{143}$$

and particle current:

$$\hat{\mathbf{g}}(\mathbf{x}) = \sum_{i=1}^{N} \mathbf{p}_i \delta(\mathbf{x} - \mathbf{r}_i) \tag{144}$$

onto classical fields $\rho(\mathbf{x})$ and $\mathbf{g}(\mathbf{x})$. In general this is a complicated procedure because of the interactions in the system. However, with the microscopic Hamiltonian of the form:

$$H = \sum_{i=1}^{N} \frac{\mathbf{p}_i^2}{2m} + V \ , \tag{145}$$

one can coarse grain [15] and obtain Eq. (29) quoted above:

$$\mathcal{H}[\rho, \mathbf{g}] = \int d^d x \left[\frac{1}{2} \frac{\mathbf{g}^2(\mathbf{x})}{\rho(\mathbf{x})} + f[\rho(\mathbf{x})] \right] \ , \tag{146}$$

where $f(\rho)$ is some functional of $\rho(\mathbf{x})$. There should also, in general, be terms depending on the gradient of ρ. The main point here is that the dependence of $\mathcal{H}$ on **g** can be made explicit [15]. This term seems sensible physically if we write:

$$\mathbf{g} = \rho \mathbf{V} \tag{147}$$

and identify:

$$\frac{1}{2}\mathbf{g}^2/\rho = \frac{1}{2}\rho \mathbf{V}^2 \qquad (148)$$

as the kinetic energy density.

The Poisson brackets for the set ρ and $\mathbf{g}$ are worked out in Appendix B, starting with the microscopic definitions of $\rho(\mathbf{x})$ and $\mathbf{g}(\mathbf{x})$. They close on themselves and the nonzero brackets are given by:

$$\{\rho(\mathbf{x}), g^i(\mathbf{x}')\} = -\nabla_x^i[\delta(\mathbf{x}-\mathbf{x}')\rho(\mathbf{x})] \qquad (149)$$

$$\{g_i(\mathbf{x}), g_j(\mathbf{x}')\} = -\nabla_x^j[\delta(\mathbf{x}-\mathbf{x}')g_i(\mathbf{x})] + \nabla_{x'}^i[\delta(\mathbf{x}-\mathbf{x}')g_j(\mathbf{x})] \quad . \qquad (150)$$

The streaming velocities in this case are given by:

$$\mathcal{V}_\rho(\mathbf{x}) = \sum_i \int \{\rho(\mathbf{x}), g_i(\mathbf{x}')\} \frac{\delta \mathcal{H}}{\delta g_i(\mathbf{x}')} d^d x' \qquad (151)$$

and:

$$\mathcal{V}_{g_i}(\mathbf{x}) = \int d^3 x' \left[\{g_i(\mathbf{x}), \rho(\mathbf{x}')\} \frac{\delta \mathcal{H}}{\delta \rho(\mathbf{x}')} + \sum_j \{g_i(\mathbf{x}), g_j(\mathbf{x}')\} \frac{\delta \mathcal{H}}{\delta g_j(\mathbf{x}')} \right] \quad . \qquad (152)$$

$\mathcal{V}_\rho$ and $\mathcal{V}_{g_i}$ are worked out in Problem 9.7 using Eq. (146) with the results:

$$\mathcal{V}_\rho(\mathbf{x}) = -\nabla_x \cdot \mathbf{g}(\mathbf{x}) \qquad (153)$$

$$\mathcal{V}_g^i(\mathbf{x}) = -\rho \nabla^i \frac{\partial f}{\partial \rho} - \sum_j \nabla_j (g_i g_j / \rho) \quad . \qquad (154)$$

$\mathcal{V}_\rho$ is consistent with conservation of mass. We can put the expression for $\mathcal{V}_g^i$ into a more convenient form. We can write:

$$\nabla^i \left[\rho \frac{\delta f}{\delta \rho} - f \right] = (\nabla^i \rho) \frac{\partial f}{\partial \rho} + \rho \nabla^i \frac{\partial f}{\partial \rho} - \frac{\partial f}{\partial \rho} \nabla^i \rho$$

$$= \rho \nabla^i \frac{\partial f}{\partial \rho} \quad , \qquad (155)$$

so we have:

$$\mathcal{V}_g^i = -\nabla_i \left[\rho \frac{\partial f}{\partial \rho} - f \right] - \sum_j \nabla_j (g_i g_j / \rho)$$

$$= -\sum_j \nabla_j \sigma_{ij}^R \qquad (156)$$

and we can identify:

$$\sigma_{ij}^R = \delta_{ij}\left[\rho\frac{\partial f}{\partial \rho} - f\right] + g_i g_j/\rho \tag{157}$$

as the *reversible* part of the stress tensor. Notice that it is symmetric. What is the physical interpretation of the diagonal part? Remember that in thermodynamics that the pressure can be written in the form [33]:

$$p = -\left(\frac{\partial F}{\partial V}\right)_{N,T} . \tag{158}$$

If we express the free energy in terms of the free energy density, $F = fV$, then:

$$p = -f - V\left(\frac{\partial f}{\partial V}\right)_{N,T} , \tag{159}$$

where:

$$\left(\frac{\partial f}{\partial V}\right)_{N,T} = \left(\frac{\partial f}{\partial \rho}\right)_{N,T}\left(\frac{\partial \rho}{\partial V}\right)_{N,T}$$

$$= -\frac{\rho}{V^2}\left(\frac{\partial f}{\partial \rho}\right)_{N,T} , \tag{160}$$

and:

$$p = -f + \rho\frac{\partial f}{\partial \rho} . \tag{161}$$

If we identify f as a fluctuating free-energy density, then p can be identified with the fluctuating pressure. The reversible part of the stress tensor can then be written as:

$$\sigma_{ij}^R = \delta_{ij} p + \rho V_i V_j . \tag{162}$$

We turn next to the specification of the damping matrix Γ_0^{ij} where each index can be associated with ρ or g_i. Clearly, as in the case of normal fluids treated earlier, there is no damping if either index equals ρ. Thus there is no noise or dissipative term in the continuity equation:

$$\frac{\partial \rho}{\partial t} = V_\rho = -\nabla \cdot \mathbf{g} . \tag{163}$$

Thus we can label Γ_0 with the vector index associated with $\mathbf{g}$, $\Gamma_0^{g_i g_j} = \Gamma_0^{ij}$. Since this matrix is symmetric we have:

$$\Gamma_0^{ij} = \sum_{kj}\nabla_k\nabla_l'\eta_{ijkl}\delta(\mathbf{x} - \mathbf{x}') \tag{164}$$

and since the system is isotropic the viscosity tensor η must have the form:

$$\eta_{ijkl} = \eta_1 \delta_{ij}\delta_{kl} + \eta_2 \delta_{ik}\delta_{jl} + \eta_3 \delta_{il}\delta_{jk} \ . \tag{165}$$

Since Γ_0^{ij} is symmetric under exchange of i and j, and the gradient terms should be invariant under exchange of k and l we see that $\eta_2 = \eta_3$ and:

$$\eta_{ijkl} = \eta_1 \delta_{ij}\delta_{kl} + \eta_2 [\delta_{ik}\delta_{jl} + \delta_{il}\delta_{jk}] \ . \tag{166}$$

Putting this result back into the expression for Γ_0 we obtain:

$$\begin{aligned}\Gamma_0^{ij} &= \left[\eta_1 \delta_{ij} \nabla \cdot \nabla' + \eta_2(\nabla_i \nabla_j' + \nabla_j \nabla_i')\right]\delta(\mathbf{x}-\mathbf{x}') \\ &= -\left[\eta_1 \delta_{ij}\nabla^2 + 2\eta_2 \nabla_i \nabla_j\right]\delta(\mathbf{x}-\mathbf{x}') \ . \end{aligned} \tag{167}$$

Inserting this result back into the dissipative contribution to the equation of motion for the momentum density:

$$\sum_j \Gamma_0^{ij}\frac{\partial \mathcal{H}}{\partial g_j} = -\eta_0 \nabla^2 V_i - (\zeta_0 + \eta_0/3)\nabla_i \nabla_j V_j \ , \tag{168}$$

where we have introduced the useful notation for the local velocity:

$$V_i = \frac{\partial \mathcal{H}}{\partial g_i} = g_i/\rho \ . \tag{169}$$

$\eta_1 = \eta_0$ and $2\eta_2 = \eta_0/3 + \zeta_0$ where η_0 and ζ_0 have the interpretation of bare shear and bulk viscosities. Eq. (168) can be interpreted in terms of the derivative of the dissipative part of the stress tensor :

$$\sum_j \Gamma_0^{ij}\frac{\partial \mathcal{H}}{\partial g_j} = \sum_j \nabla_j \sigma_{ij}^D \ . \tag{170}$$

We then find that dissipative part of the stress tensor (which should be symmetric) is given by:

$$\sigma_{ij}^D = -\eta_0 \left[\nabla_i V_j + \nabla_j V_i - \frac{2}{3}\delta_{ij}\nabla \cdot \mathbf{V}\right] - \zeta_0 \delta_{ij} \nabla \cdot \mathbf{V} \ . \tag{171}$$

The momentum equation can then be written:

$$\begin{aligned}\frac{\partial g_i}{\partial t} &= -\nabla_i p - \sum_j \nabla_j(g_i g_j/\rho) \\ &\quad + (\eta_0/3 + \zeta_0)\nabla_i \nabla \cdot (\mathbf{g}/\rho) + \eta_0 \nabla^2(g_i/\rho) + \theta_i\end{aligned}$$

$$= -\nabla_i p - \sum_j \nabla_j (\rho V_i V_j)$$
$$+ (\eta_0/3 + \zeta_0)\nabla_i \nabla \cdot \mathbf{V} + \eta_0 \nabla^2 V_i + \theta_i \ , \tag{172}$$

where the noise satisfies:

$$\langle \theta_i(\mathbf{x},t)\theta_j(\mathbf{x}',t') \rangle = 2k_B T \sum_{kl} \nabla_k \nabla'_l \delta(t-t')\delta(\mathbf{x}-\mathbf{x}')$$
$$\times [\eta_0 \delta_{ij}\delta_{kl} + (\eta_0/3 + \zeta_0)\delta_{ik}\delta_{jl}] \ . \tag{173}$$

In the absence of noise, Eq. (172) is the usual Navier–Stokes equation supplemented with the identification of the fluctuating pressure:

$$p = \rho \frac{\partial f}{\partial \rho} - f \ . \tag{174}$$

The nonlinearities in this case come from the pressure term and the *convective* term $\rho V_i V_j$.

Including the energy in the development is complicated by the need, as explained in Ref. [34], to include multiplicative noise.

We now have several important examples of nonlinear hydrodynamical equations. We could go on and describe more systems. In particular the fluctuating nonlinear hydrodynamics of smectic A liquid crystals is of particular interest [35] because of the strong fluctuations in the smectic A phase. Instead we move on to ask the question: how do we go about analyzing these equations? We discuss here one approach that has been very useful and that gives one some perspective on the structure of the general theory. Another method is discussed in Chapter 10.

9.8
Determination of Correlation Functions

9.8.1
Formal Arrangements

Now, suppose we want to compute the time-correlation function:

$$C_{ij}(t) = \langle \psi_j \psi_i(t) \rangle \tag{175}$$

using our nonlinear models. There are several useful methods. We develop here a method that is somewhat more direct and that builds on our previous work.

Let us start with the generalized Fokker–Planck equation (GFPE):

$$\frac{\partial}{\partial t}g_\phi(t) = D_\phi g_\phi(t) + N_\phi(t) \ . \tag{176}$$

Let us multiply the GFPE by $g_{\phi'} = g_{\phi'}(t=0)$ and average over the Gaussian white noise $\xi_i(t)$. Since, for $t > 0$, due to causality:

$$\langle g_{\phi'} N_\phi(t) \rangle = 0 \ , \tag{177}$$

we obtain:

$$\frac{\partial}{\partial t} G_{\phi\phi'}(t) = D_\phi G_{\phi\phi'}(t) \ , \tag{178}$$

where:

$$G_{\phi\phi'}(t) = \langle g_{\phi'} g_\phi(t) \rangle \ . \tag{179}$$

We have the formal solution to this equation in terms of the GFPE operator D_ϕ,

$$\begin{aligned} G_{\phi\phi'}(t) &= e^{D_\phi t} S_{\phi\phi'} \\ &= e^{D_\phi t} \left(\delta(\phi - \phi') W_\phi \right) \ . \end{aligned} \tag{180}$$

Remember, however, that the order parameter correlation function is given by:

$$\begin{aligned} C_{ij}(t) = \langle \psi_j \psi_i(t) \rangle &= \int \mathcal{D}(\phi') \phi'_j \int \mathcal{D}(\phi) \phi_i \langle g_{\phi'} g_\phi(t) \rangle \\ &= \int \mathcal{D}(\phi') \phi'_j \int \mathcal{D}(\phi) \phi_i \, e^{D_\phi t} [\delta(\phi - \phi') W_\phi] \\ &= \int \mathcal{D}(\phi) \phi_i \, e^{D_\phi t} (\phi_j W_\phi) \ . \end{aligned} \tag{181}$$

Let us define the adjoint operator $\tilde{D}_\phi$ via,

$$\int \mathcal{D}(\phi) A_\phi D_\phi B_\phi = \int \mathcal{D}(\phi) (\tilde{D}_\phi A_\phi) B_\phi \ . \tag{182}$$

We can identify $\tilde{D}_\phi$, after an integration by parts, as:

$$\tilde{D}_\phi = \sum_i \left[\mathcal{V}_i[\phi] - \sum_j \beta^{-1} \Gamma_0^{ij} \left(-\frac{\partial}{\partial \phi_j} + \beta \frac{\partial \mathcal{H}_\phi}{\partial \phi_j} \right) \right] \frac{\partial}{\partial \phi_i} \ . \tag{183}$$

After a sufficient number of integrations by parts, we can rewrite Eq. (181) in the form:

$$C_{ij}(t) = \int \mathcal{D}(\phi) W_\phi \phi_j e^{\tilde{D}_\phi t} \phi_i \ . \tag{184}$$

Looking at this, one should be reminded of the microscopic expression we had for the time-correlation function:

$$C_{ij}(t) = \text{Tr} \frac{e^{-\beta H}}{Z} \psi_j e^{+iLt} \psi_i \ . \tag{185}$$

Indeed they are very similar with $\mathcal{H}$ playing the role of the Hamiltonian and $-i\tilde{D}_\phi$ the role of the Liouville operator. Clearly we can treat this problem using the same memory function methods as used in treating the microscopic case. Since all of this should be rather familiar we need only present the results.

The Laplace transform of Eq. (184) gives the fluctuation function:

$$C_{ij}(z) = -i \int_0^{+\infty} dt\, e^{+izt} C_{ij}(t)$$
$$= \langle \psi_j R(z) \psi_i \rangle \,, \tag{186}$$

where the resolvant operator in this case is given by:

$$R(z) = (z - i\tilde{D}_\psi)^{-1}\,. \tag{187}$$

Again we have that $C_{ij}(z)$ satisfies the kinetic equation:

$$\sum_k (z\delta_{ik} - K_{ik}(z)) C_{kj}(z) = S_{ij}\,, \tag{188}$$

where $S_{ij} = \langle \psi_i \psi_j \rangle$ is the static structure factor and the memory function can be defined by:

$$\sum_k K_{ik} S_{kj} = \Gamma_{ij}\,, \tag{189}$$

with the static part given by (See Problem 9.9):

$$\Gamma_{ij}^{(s)} = i\beta^{-1} \langle Q_{ij} \rangle - i\beta^{-1} \Gamma_0^{ij}\,, \tag{190}$$

while the dynamic part can be expressed (See Problem 9.10) in terms of:

$$\Gamma_{ij}^{(d)}(z) = -\langle I_j^+ R(z) I_i \rangle + \sum_{kl} \langle \psi_l R(z) I_i \rangle C^{-1}(z)_{kl} \langle I_j^+ R(z) \psi_k \rangle\,, \tag{191}$$

where the vertices are defined by:

$$I_i(\phi) = \tilde{D}_\phi \phi_i = V_i[\phi] - \sum_k \Gamma_0^{ik} \frac{\partial \mathcal{H}_\phi}{\partial \phi_k} \tag{192}$$

and:

$$I_i^+(\phi) = -V_i[\phi] - \sum_k \Gamma_0^{ik} \frac{\partial \mathcal{H}_\phi}{\partial \phi_k}\,. \tag{193}$$

We can show, as we found previously, only the nonlinear contributions to I_i and I_i^+ contribute to $\Gamma_{ij}^{(d)}(z)$.

9.8.2
Linearized Theory

The conventional approximation in this approach corresponds to dropping $K^{(d)}(z)$. The resulting theory is identical to linearized hydrodynamics in the Markoffian approximation since, on inspection of Eq. (190), we obtain:

$$\Gamma^{(s)}_{NFH} = \Gamma^{(s)}_{LH} + \Gamma^{(d)}_{LH} \, , \tag{194}$$

where the subscript LH stands for linearized hydrodynamics and where the physical transport coefficients are replaced by the bare transport coefficients.

9.8.3
Mode-Coupling Approximation

In general the analysis of $K^{(d)}_{ij}(z)$ is complicated and its evaluation requires the introduction of sophisticated calculational techniques. In a number of interesting cases the nonlinear interaction of interest is quadratic in the fluctuating fields:

$$I^N_i = \sum_{jk} V_{ijk} \phi_j \phi_k \, . \tag{195}$$

Then, to lowest order in the nonlinear couplings, the dynamic part of the memory function is proportional to:

$$\Gamma^{(d)}_{ij}(z) = -i \int_0^{+\infty} dt \, e^{+izt} \sum_{lm} \sum_{kn} V_{ilm} V^+_{jkn} \langle \delta(\phi_k \phi_n) e^{\tilde{D}^o_\phi t} \delta(\phi_l \phi_m) \rangle_0 \, , \tag{196}$$

where $\tilde{D}^0_\phi$ is the Fokker–Planck operator in the absence of nonlinear interactions and $\langle \rangle_0$ means an average over $e^{-\beta \mathcal{H}_0}$. This assumes that:

$$\sum_{kn} V^+_{jkn} \langle \delta(\phi_k \phi_n) e^{\tilde{D}^o_\phi t} \delta(\phi_i) \rangle_0 = 0 \tag{197}$$

at lowest order and we have:

$$\langle \delta(\phi_k \phi_n) e^{\tilde{D}^o_\phi t} \delta(\phi_l \phi_m) \rangle_0 = \langle \psi_k(0) \psi_n(0) \psi_l(t) \psi_m(t) \rangle_0$$
$$- \langle \psi_k(0) \psi_n(0) \rangle_0 \langle \psi_l(t) \psi_m(t) \rangle_0 \, .$$

In the decoupling approximation we factorize the four-point correlation function into a symmetric product of two-point correlation functions:

$$\langle \psi_k(0) \psi_n(0) \psi_l(t) \psi_m(t) \rangle_0 = C^{(0)}_{kn}(0) C^{(0)}_{lm}(0) + C^{(0)}_{k\ell}(t) C^{(0)}_{nm}(t)$$
$$+ C^{(0)}_{km}(t) C^{(0)}_{\ell n}(t) \, . \tag{198}$$

This approximation is explored in Problem 9.19. Putting this result back into Eq. (196) we find:

$$\Gamma_{ij}^{(d)}(z) = -i \int_0^{+\infty} dt\, e^{+izt} \sum_{lm}\sum_{kn} V_{ilm} V_{jkn}^{+} \left[C_{kl}^0(t) C_{nm}^0(t) + C_{km}^0(t) C_{nl}^0(t) \right] \qquad (199)$$

The evaluation of $\Gamma_{ij}^{(d)}(z)$ involves the sums over the indices l, m, k, and n and the time integral; this is a typical mode-coupling expression.

9.8.4
Long-Time Tails in Fluids

As an example of this development, let us consider a relatively simple but important example: an incompressible fluid. An incompressible fluid is characterized by the condition that the density does not change; there is only flow. Thus we require $\rho = \rho_0$ be constant in time and space:

$$\frac{\partial \rho_0}{\partial t} = 0 \quad \text{and} \quad \nabla_i \rho_0 = 0 \;. \qquad (200)$$

These constraints are used to simplify Eqs. (163) and (172). The continuity equation for the mass density reduces to the condition on the momentum density:

$$\nabla \cdot \mathbf{g} = 0 \;. \qquad (201)$$

Thus the longitudinal degrees of freedom are frozen out. The momentum constraint simplifies the equation of motion given by Eq. (172). This equation simplifies greatly in this case since we can drop all terms proportional to ∇_i:

$$\frac{\partial}{\partial t} g_i = -\nabla_i p - \sum_j \nabla_j (g_i g_j / \rho_0)$$

$$+ (\eta_o/3 + \zeta_0) \nabla_i \nabla \cdot (\mathbf{g}/\rho_0) + \eta_0 \nabla^2 (g_i/\rho_0) + \theta_i \;. \qquad (202)$$

The equation for the transverse part of g_i, g_i^T, is given by:

$$\frac{\partial}{\partial t} g_i^T = \frac{\eta_0}{\rho_0} \nabla^2 g_i^T + \theta_i^T - \sum_{jk} T_{ik} \nabla_j (g_k g_j / \rho_0) \;, \qquad (203)$$

where T_{ij} is a projection operator that picks out the transverse part of a vector depending on the vector index j. Things are a bit more transparent if we Fourier transform Eq. (203) over space:

$$\frac{\partial}{\partial t} g_i^T (\mathbf{k}, t) = -\frac{\eta_0}{\rho_0} k^2 g_i^T (\mathbf{k}, t) + \theta_i^T (\mathbf{k}, t)$$

9.8 Determination of Correlation Functions

$$-\sum_{j,l} T_{il}(\mathbf{k}) \frac{ik_j}{\rho_0} \int \frac{d^d x}{(2\pi)^d} e^{-i\mathbf{k}\cdot\mathbf{x}} g_l(\mathbf{x}) g_j(\mathbf{x}) \ , \qquad (204)$$

and:

$$T_{ij}(\mathbf{k}) = \delta_{ij} - \hat{k}_i \hat{k}_j \qquad (205)$$

is the Fourier transform of the transverse projector.

It is then clear that the zeroth-order solution to our problem is given by the linear equation:

$$\left(\frac{\partial}{\partial t} + \nu_0 k^2\right) g_i^{T,0}(\mathbf{k}) = \theta_i^T(\mathbf{k}) \ , \qquad (206)$$

where $\nu_0 = \eta_0/\rho_0$ is the kinematic viscosity. Then, since:

$$\langle g_j^{T,0}(-\mathbf{k},0) \theta_i^T(\mathbf{k},t) \rangle = 0 \qquad (207)$$

for $t > 0$, we have:

$$\left(\frac{\partial}{\partial t} + \nu_0 k^2\right) C_{ij}^{T,0}(\mathbf{k},t) = 0 \quad t > 0 \qquad (208)$$

and this has the solution:

$$C_{ij}^{T,0}(\mathbf{k},t) = \beta^{-1} \rho_0 T_{ij}(\mathbf{k})_i e^{-\nu_0 k^2 t} = C_T^0(\mathbf{k},t) T_{ij}(\mathbf{k}) \ , \qquad (209)$$

where the initial condition corresponds to the transverse structure factor:

$$S_{ij}(\mathbf{k}) = k_B T \rho_0 T_{ij}(\mathbf{k}) \ . \qquad (210)$$

Notice that $T_{ij}(\mathbf{k})$ insures that $C_{ij}^{T,0}$ is purely transverse. In this case the static part of the memory function is proportional to ν_0.

The nonlinearity that contributes to the nonlinear vertex I_i^N and the dynamic part of the memory function comes from the nonlinear vertex:

$$-\sum_{j,k} \frac{T_{ik}(\mathbf{k})}{\rho_0} (ik_j) \int \frac{d^d x}{(2\pi)^d} e^{-i\mathbf{k}\cdot\mathbf{x}} g_k(\mathbf{x}) g_j(\mathbf{x})$$

$$= -\sum_{j,k} \frac{T_{ik}(\mathbf{k})}{\rho_0} (ik_j) \int \frac{d^d q}{(2\pi)^d} g_k(\mathbf{q}) g_j(\mathbf{k}-\mathbf{q}) \ . \qquad (211)$$

This has the interpretation in terms of the mode-coupling vertex:

$$\sum_{jk} V_{ijk} \phi_j \phi_k \rightarrow \sum_{jk} \int \frac{d^d k_2}{(2\pi)^d} \frac{d^d k_3}{(2\pi)^d} V_{ijk}(\mathbf{k},\mathbf{k}_2,\mathbf{k}_3) g_j(\mathbf{k}_2) g_k(\mathbf{k}_3) \ , \qquad (212)$$

where we can identify:

$$V_{ijk}(\mathbf{k}, \mathbf{k}_2, \mathbf{k}_3) = -T_{ik}(\mathbf{k}) \frac{ik_j}{\rho_0} (2\pi)^d \delta(\mathbf{k} - \mathbf{k}_2 - \mathbf{k}_3) \quad . \tag{213}$$

It is shown in Problem 9.3 that the equilibrium average of the right-hand side of Eq. (212) vanishes. The kinetic equation for the fluctuation function takes the form in this case:

$$\left(z + iv_0 k^2\right) C_{ij}(\mathbf{k}, z) - \sum_l K_{il}^{(d)}(\mathbf{k}, z) C_{lk}(\mathbf{k}, z) = S_{ij}(\mathbf{k}) \tag{214}$$

and the static structure factor is given by Eq. (210) of Chapter 8. We can simplify this equation by looking at the equation for the dynamic part of the memory function:

$$\sum_l K_{il}^{(d)}(\mathbf{k}, z) S_{lj}(\mathbf{k}) = \sum_l K_{il}^{(d)}(\mathbf{k}, z) \rho_0 k_B T T_{lj}(\mathbf{k})$$

$$= \Gamma_{ij}^{(d)}(\mathbf{k}, z) \quad . \tag{215}$$

Since $\sum_j \hat{k}_j T_{lj}(\mathbf{k}) = 0$ we have from Eq. (215) of Chapter 8 that $\sum_j \hat{k}_j \Gamma_{ij}^{(d)}(\mathbf{k}, z) = 0$. Since $\Gamma_{ij}^{(d)}(\mathbf{k}, z)$ is symmetric we can write:

$$\Gamma_{ij}^{(d)}(\mathbf{k}, z) = \Gamma^{(d)}(\mathbf{k}, z) T_{ij}(\mathbf{k}) \quad . \tag{216}$$

Putting this result back into Eq. (215) of Chapter 8 we obtain:

$$K_{ij}^{(d)}(\mathbf{k}, z) = \Gamma^{(d)}(\mathbf{k}, z) T_{ij}(\mathbf{k}) \frac{\beta}{\rho_0} \quad . \tag{217}$$

This means that we can write the correlation function as the product:

$$C_{ij}(\mathbf{k}, z) = C_T(\mathbf{k}, z) T_{ij}(\mathbf{k}) \tag{218}$$

and the kinetic equation can be written in the simplified form:

$$\left(z + iv_0 k^2 - \Gamma^{(d)}(\mathbf{k}, z) \frac{\beta}{\rho_0}\right) = C_T(\mathbf{k}, z) = \rho_0 k_B T \quad . \tag{219}$$

We can then define the renormalized viscosity:

$$v_R(\mathbf{k}, z) = v_0 + i \frac{\beta}{\rho_0 k^2} \Gamma^{(d)}(\mathbf{k}, z) \tag{220}$$

and we can rewrite the kinetic equation in the form:

$$\left(z + iv_R(\mathbf{k}, z) k^2\right) C_T(\mathbf{k}, z) = \rho_0 k_B T \quad . \tag{221}$$

We now focus on the mode-coupling contribution to the dynamic part of the memory function. From Eqs. (199), (212) and (213) we can read off:

$$\Gamma_{ij}^{(d)}(\mathbf{k},z) = -i \int_0^{+\infty} dt\, e^{+izt} \sum_{kslm} T_{ik}(\mathbf{k}) T_{js}(\mathbf{k}) k_l k_m$$
$$\times \int \frac{d^d q}{(2\pi)^d} [C_{ks}^{T,0}(\mathbf{q},t) C_{lm}^{T,0}(\mathbf{k}-\mathbf{q},t)$$
$$+ C_{km}^{T,0}(\mathbf{q}) C_{ls}^{T,0}(\mathbf{k}-\mathbf{q},t)] / \rho_0^2 \quad . \tag{222}$$

Let us further focus on the long-wavelength regime where we can write:

$$\Gamma_{ij}^{(d)}(\mathbf{k},z) = -\frac{i}{\rho_0^2} \sum_{lmks} k_l k_m T_{ik}(\mathbf{k}) T_{js}(\mathbf{k}) \tag{223}$$
$$\times \int_0^{+\infty} dt\, e^{+izt} \int \frac{d^d q}{(2\pi)^d} [C_{ks}^{T,0}(\mathbf{q},t) C_{lm}^{T,0}(-\mathbf{q},t)$$
$$+ C_{km}^{T,0}(\mathbf{q},t) C_{ks}^{T,0}(-\mathbf{q},t)]$$

It is shown in Problem 9.13, using Eq. (218) and the isotropic nature of $C_T^{(0)}(q,t)$, that Eq. (223) can be reduced to:

$$\Gamma_{ij}^{(d)}(\mathbf{k},z) = -ik^2 T_{ij}(\mathbf{k}) \left[1 - \frac{2}{d} + \frac{2}{d(d+2)}\right] \int_0^{+\infty} dt\, \frac{e^{+izt}}{\rho_0^2} I(t) \quad, \tag{224}$$

where we have the remaining integral:

$$I(t) = \int \frac{d^d q}{(2\pi)^d} [C_T^0(q,t)]^2 \quad . \tag{225}$$

This is of the anticipated form and putting this back into Eq. (220) of Chapter 8 we obtain for the renormalized viscosity,

$$\nu_R(z) = \nu_0 + \left(1 - \frac{2}{d} + \frac{2}{d(d+2)}\right) \int_0^{+\infty} dt\, e^{+izt} I(t) \frac{\beta}{\rho_0^3} \quad . \tag{226}$$

Inserting Eq. (210) of Chapter 8 in Eq. (225) for $I(t)$ we obtain explicitly that:

$$\nu_R(z) = \nu_0 + \frac{(d^2-2)}{d(d+2)} \int_0^{+\infty} dt\, e^{+izt} \int \frac{d^d q}{(2\pi)^d} \frac{\beta^{-1}}{\rho_0} e^{-2\nu_0 q^2 t}$$
$$= \nu_0 + \frac{(d^2-2)}{d(d+2)} \frac{\beta^{-1}}{\rho_0} \int \frac{d^d q}{(2\pi)^d} \frac{1}{-iz + 2\nu_0 q^2} \quad . \tag{227}$$

The angular integral can be carried out explicitly [36] with the result:

$$\tilde{K}_d = \int \frac{d\Omega_d}{(2\pi)^d} = \frac{2^{1-d}}{\pi^{d/2} \Gamma(d/2)} \quad, \tag{228}$$

and we are left with:
$$v_R(z) = v_0 + \frac{(d^2-2)}{d(d+2)}\frac{\beta^{-1}}{\rho_0}\tilde{K}_d I_d(z) \tag{229}$$

and the frequency-dependent integral:
$$I_d(z) = \int_0^\Lambda \frac{q^{d-1}dq}{-iz + 2v_0 q^2} , \tag{230}$$

where Λ is a large wavenumber cutoff. This result is most transparent in the time regime where we can write:
$$v_R(t) = 2v_0\delta(t) + \frac{(d^2-2)}{d(d+2)}\frac{\beta^{-1}}{\rho_0}\tilde{K}_d J_d(t) , \tag{231}$$

where:
$$J_d(t) = \int_0^\Lambda q^{d-1}dq\, e^{-2v_0 q^2 t} . \tag{232}$$

At short times:
$$J_d(0) = \frac{\Lambda^d}{d} , \tag{233}$$

which depends on the cutoff. The interpretation of the form for $v_R(t)$ is clear. The leading term, proportional to v_0, corresponds to the contribution to the physical transport coefficient from the Boltzmann equation. The second term is the mode-coupling contribution that dominates at long-times.

If we make the change of variables $q = \sqrt{x/2v_0 t}$ in the integral for J_d, then:
$$J_d(t) = \frac{1}{2(2v_0 t)^{d/2}} \int_0^{2v_0 t \Lambda^2} dx\, x^{d/2-1} e^{-x} . \tag{234}$$

For long times, $2v_0 t \Lambda^2 \to \infty$, the integral reduces to a Γ-function and:
$$J_d(t) = \frac{\Gamma(d/2)}{2(2v_0 t)^{d/2}} \tag{235}$$

independent of the cutoff. We have then that $v_R(t) \approx t^{-d/2}$ at long times. Thus, via the Green–Kubo relation, we have that current–current time-correlation functions decay algebraically in time:
$$\langle \mathbf{J}(t) \cdot \mathbf{J} \rangle \approx t^{-d/2} . \tag{236}$$

These are the long-time tails [37] discussed at the beginning of this chapter.

Going back to the frequency regime we see that the physical transport coefficient is given by:

$$v_R(0) = v_0 + \frac{(d^2-2)}{d(d+2)} \frac{\beta^{-1}}{\rho_0} \tilde{K}_d I_d(0) \qquad (237)$$

and I_d is given at zero frequency by:

$$\begin{aligned} I_d(0) &= \int_0^\Lambda \frac{q^{d-1} dq}{2 v_0 q^2} \\ &= \frac{1}{2v_0} \frac{\Lambda^{d-2}}{d-2} \: . \end{aligned} \qquad (238)$$

This is well behaved only for $d > 2$. The physical transport coefficient does not exist in two or fewer dimensions. It is left as a problem to show in three dimensions for small frequencies that:

$$v_R(z) = v_R(0) + A\sqrt{z} + \ldots \: , \qquad (239)$$

where the constant A is to be determined. In two dimensions for small z:

$$v_R(z) = v_R(0)\left[1 + b_0 \ln(b_1/z)\right] \: , \qquad (240)$$

where the constants b_0 and b_1 are to be determined. Clearly the correction term diverges as $z \to 0$ logarithmically. Conventional hydrodynamics does not exist in two dimensions.

9.9
Mode Coupling and the Glass Transition

The mode-coupling theory of the liquid–glass transition asserts that this is a dynamic ergodic-nonergodic transition. This idea grew out of work in kinetic theory on the divergence of transport coefficients and long-time tails. Building on the work of Dorfmann and Cohen [38], Mazenko [39] showed, using kinetic theory, how one could connect up with the more phenomenological mode coupling calculations of fluctuating nonlinear hydrodynamics. This microscopic work was extended by Sjögren and Sjölander [40] and others. In a dramatic leap, Leutheusser [41], motivated by the kinetic theory work proposed a simple model for dense fluid kinetics that leads to an ergodic-nonergodic as the system becomes more dense. The model describes many of the glassy properties observed experimentally. Leuteussar's model was followed by a more realistic model from Bengtzelius, Götze and Sjölander [42]. Later Das and Mazenko [43] showed that there is a mechanism that cuts off the *ideal* glass transition implying that *the* glass transition is at lower temperatures and higher densities. For reviews and simple introductions see [44–46].

9.10
Mode Coupling and Dynamic Critical Phenomena

Mode-coupling theory also made an important contribution to dynamic critical phenomena before the more rigorous renormalization group method was developed. As reviewed by Hohenberg and Halperin [47], these methods were successful in treating dynamical behavior near critical points of simple fluids and binary fluid mixtures. In particular it explained the divergence of the thermal conductivity near the critical point. The key original references are [48–54].

9.11
References and Notes

1 This work was discussed in Chapters 1–3.

2 Corrections to the Boltzmann equation, introduced by D. Enskog [K. Svensk. Vet.-Akad. Handl. **63**, no.4 (1921)] in the early 1920s allowed the description to be extended to higher densities. However, the basic physics still consisted of the uncorrelated binary collisions discussed in Chapter 7. See the discussion in Chapter 16 in S. Chapman, T. G. Cowling, *The Mathematical Theory of Nonuniform Gases*, 3rd edn., Cambridge University Press, London (1970), Chapter 16.

3 After the establishment of the Green–Kubo equations for transport coefficients, R. W. Zwanzig, Phys. Rev. **129**, 486 (1963) and J. Weinstock, Phys. Rev. **126**, 341 (1962); Phys. Rev. **132**, 454, 470 (1963), proposed a low-density expansion for transport coefficients. Similar work was carried out in K. Kawasaki, I. Oppenheim, Phys. Rev. **136**, 1519 (1964); Phys. Rev. **139**, 689 (1965).

4 The divergence of the density expansion of transport coefficients was discovered by E. G. D. Cohen and J. R. Dorfmann, Phys. Lett. **16**, 124 (1965) and J. Weinstock, Phys. Rev. **140**, 460 (1965). For some history of this discovery see J. R. Dorfmann, in *Lectures in Theoretical Physics, Vol. IXC, Kinetic Theory*, W. E. Brittin (ed), Gordon and Breach, N.Y. (1967), p443. See also M. H. Ernst, L. K. Haines, J. R. Dorfmann, Rev. Mod. Phys. **41**, 296 (1969).

5 The resummation of the expansion was carried out by K. Kawasaki and I. Oppenheim, Phys. Rev. **139**, 1763 (1965).

6 B. J. Alder, T. E. Wainwright, Phys. Rev. Lett. **18**, 988 (1967); Phys. Rev. A **1**, 18 (1970); B. J. Alder, D. M. Gass, T. E. Wainwright, J. Chem. Phys. **53**, 3813 (1970); T. E. Wainwright, B. J. Alder, D. M. Gass, Phys. Rev. A **4**, 233 (1971).

7 The connections between mode-coupling, divergence of transport coefficients and long-time tails were brought together in G. F. Mazenko, Phys. Rev. A **9**, 360 (1974).

8 This choice of variables is discussed by M. S. Green, J. Chem. Phys. **20**, 1281 (1952); J. Chem. Phys. **22**, 398 (1954).

9 There is extensive discussion of effective Hamiltonians or free energies in FOD in Chapters 2 and 3.

10 L. D. Landau and E. M. Lifshitz, *Statistical Physics*, 3rd edn., Pergamon, Oxford (1980).

11 This is discussed in detail in ESM (Appendix O) and FOD (Appendix B).

12 K. Wilson, Phys. Rev. B **4**, 3174 (1971).

13 L. D. Landau, *On the Theory of Phase Transitions*, Phys. Z. Soviet Un. **11**, 26 (1937); reprinted in *Collected Papers of L. D. Landau*, D. ter Haar (ed), Pergamon, New York (1965).

14 This model is treated in mean-field theory in Problem 8.8.

15 J. S. Langer, L. Turski, Phys. Rev. A **8**, 3239 (1973).

16 See Chapter 10 on liquid crystals in FOD.

17. See Section 10.3 on smectic A liquid crystals in FOD.
18. Static elastic theory is studied in Chapter 6 of ESM.
19. V. L. Ginzburg, L. D. Landau, Zh. Eksperim. Teor. Fiz. **20**, 1064 (1950).
20. See the discussion in Chapter 9 on superconductivity in FOD.
21. This development builds on the work of M. S. Green [8] and R. W. Zwanzig, Phys. Rev. **124**, 983 (1961); J. Chem. Phys. **33**, 1330 (1960). A direct use of Mori's method (discussed here for the linear case in Chapter 5) in the nonlinear case is given in H. Mori, H. Fujisaka, Prog. Theor. Phys. **49**, 764 (1973); H. Mori, H. Fujisaka, H. Shigematsu, Prog. Theor. Phys. **51**, 109 (1974). We follow the development of S. Ma and G. F. Mazenko, Phys. Rev. B **11**, 4077 (1975).
22. A. D. Fokker, Ann. Phys. **43**, 812 (1914); M. Planck, Sitzungsber. Preuss. Akad. Wiss., p.324 (1917).
23. $\langle \psi_i(t)\psi_j(t)\psi_k(t)\psi_\ell(t)\rangle = \int D(\phi) W_\phi \phi_i \phi_j \phi_k \phi_\ell$
24. R. W. Zwanzig, K. S. J. Nordholm, W. C. Mitchell, Phys. Rev. A **5**, 2680 (1972), cleared up some early confusion on this point.
25. L. D. Landau, I. M. Khalatnikov, Dokl. Akad. Nauk SSSR **96**, 469 (1954); Reprinted in *Collected Papers of L. D. Landau*, D. ter Haar (ed), Pergamon, New York (1965).
26. This development is the irreversible cousin to the reversible Gross–Pitaevskii equation. E. P. Gross, Nuovo Cimento **20**, 454 (1961); L. P. Pitaevskii, Sov. Phys. JETP **13**, 451 (1961).
27. The TDGL equations for superconductors were first derived by A. Schmid, Phys. Konden. Mater. **5**, 302 (1966); See also M. Cyrot, Rep. Prog. Phys. **36**, 103 (1973). For further discussion and background see A. T. Dorsey, Phys. Rev. B **46**, 8376 (1992).
28. P. C. Hohenberg, B. I. Halperin, Rev. Mod. Phys. **49**, 435 (1977).
29. J. W. Cahn J. E. Hilliard, J. Chem. Phys. **28**, 258 (1958); H. E. Cook, Acta. Metall. **18**, 297 (1970).
30. S. Ma, G. F. Mazenko, Phys. Rev. B **11**, 4077 (1975).
31. R. Freedman, G. F. Mazenko, Phys. Rev. Lett. **34**, 1575 (1975); Phys. Rev. B **13**, 1967 (1976).
32. B. S. Kim, G. F. Mazenko, J. Stat. Phys. **64**, 631 (1991).
33. See Eq. 267 in ESM.
34. The role of multiplicative noise is discussed in Ref. [32].
35. G. Grinstein, R. Pelcovits, Phys. Rev. Lett. **47**, 856 (1981) discussed the equilibrium case, while dynamics were studied by G. F. Mazenko, S. Ramaswamy, J. Toner, Phys. Rev. Lett. **49**, 51 (1982); Phys. Rev. A **28**, 1618 (1983), S. Milner, P. C. Martin, Phys. Rev. Lett. **56**, 77 (1986).
36. This is worked out in Appendix E of ESM.
37. Mode coupling and long-time tails go back to: M. H. Ernst, E. H. Hauge, J. M. J. van Leeuwen, Phys. Rev. Lett. **25**, 1254 (1970); K. Kawasaki, Phys. Lett. A **32**, 379 (1970); Prog. Theor. Phys. **45**, 1691 (1971); Y. Pomeau, P. Resibois, Phys. Rep. C **19**, 64 (1975).
38. J. R. Dorfmann, E. G. D. Cohen, Phys. Rev. Lett. **25**, 1257 (1970); Phys. Rev. A **6**, 776 (1972).
39. Mazenko [7] showed, using kinetic theory, how one could connect up with the more phenomenological mode coupling calculations of fluctuating nonlinear hydrodynamics.
40. L. Sjögren, A. Sjölander, J. Phys. C **12**, 4369 (1979).
41. E. Leutheusser, Phys. Rev. A **29**, 2765 (1984).
42. U. Bengtzelius, W. Götze, A. Sjölander [42], J. Phys. C **17**, 5915 (1984).
43. S. Das, G. F. Mazenko, Phys. Rev. A **34**, 2265 (1986).
44. W. Götze, L. Sjögren, Rep. Prog. Phys. **55**, 241 (1992).
45. S. Das, Rev. Mod. Phys. **76**, 785 (2004).
46. B. Kim, G. F. Mazenko, Phys. Rev. A **45**, 2393 (1992).
47. P. C. Hohenberg, B. I. Halperin, Rev. Mod. Phys. **49**, 435 (2004).
48. M. Fixman, J. Chem. Phys. **36**, 310 (1962).
49. J. Villain, J. Phys. Paris **29**, 321, 687 (1968).
50. L. P. Kadanoff, J. Swift, Phys. Rev. **166**, 89 (1968).
51. K. Kawasaki, Ann. Phys. N.Y. **61**, 1 (1970).
52. R. A. Ferrell, Phys. Rev. Lett. **24**, 1169 (1970).

53 K. Kawasaki, in *Critical Phenomena*, M. S. Green (ed), Academic, N.Y. (1971).
54 K. Kawasaki, in *Statistical Mechanics: New Concepts, New problems, New Applications*, S. A. Rice, K. Freed, J. Light, University of Chicago Press, Chicago (1972).

9.12
Problems for Chapter 9

Problem 9.1: Show that the equilibrium average of the streaming velocity, given by Eq. (61), is zero.

Problem 9.2: If we have a system driven by Gaussian white noise with variance:

$$\langle \xi_i(t)\xi_j(t') \rangle = 2k_B T \Gamma_0^{ij} \delta(t-t') \ ,$$

evaluate the average over the noise $\langle \xi_i(t) g_\phi(t) \rangle$ directly for the case $t > 0$.

Problem 9.3: Show that the equilibrium average of the nonlinear interaction,

$$V_i = -\sum_{jk} T_{ik} \nabla_j (g_k g_j / \rho_0) \ ,$$

is zero.

Problem 9.4: Derive the fluctuating nonlinear hydrodynamical equations for isotropic antiferromagnets using as basic dynamical fields the magnetization **M** and the staggered magnetization **N**. Assume one has the Poisson bracket relations given by Eqs. (8.68), (8.69) and (8.70).

Problem 9.5: An effective Hamiltonian for fluids is of the form:

$$\mathcal{H}[\rho, \mathbf{g}] = \int d^3x \left[\frac{\chi_0^{-1}}{2} (\rho - \rho_0)^2 + \frac{u}{4}\rho^4 + \frac{\mathbf{g}^2}{2\rho} \right] \ .$$

Assuming χ_0^{-1} is large and $\rho \approx \rho_0$ determine the static structure factors:

$$S(\mathbf{q}) = \langle \rho(\mathbf{q})\rho(-\mathbf{q}) \rangle$$

$$S_{ij}(\mathbf{q}) = \langle g_i(\mathbf{q}) g_j(-\mathbf{q}) \rangle \ .$$

Problem 9.6: Consider the GFPE for the case of zero spatial dimensions:

$$\frac{\partial P}{\partial t} = \frac{\partial}{\partial \phi} \beta^{-1} \Gamma \left(\frac{\partial}{\partial \phi} + \beta \frac{\partial \mathcal{H}_\phi}{\partial \phi} \right) P \ ,$$

where the effective Hamiltonian is given by:

$$\mathcal{H}_\phi = \frac{r}{2}\phi^2 \ .$$

Assuming the initial probability distribution function is given by:
$$P_0[\phi] = N e^{-a\phi^2} ,$$
where N is a normalization constant and a is a constant, solve for $P[\phi, t]$. Look in particular at the long-time limit.

Problem 9.7: Starting with Eqs. (151) and (152) for the streaming velocities for a fluid and assuming an effective Hamiltonian of the form given by Eq. (146), show that they reduce to Eqs. (153) and (154).

Problem 9.8: Show that Eq. (168) can be written in the form of Eq. (170) and extract the expression for the dissipative part of the stress tensor given by Eq. (171).

Problem 9.9: Show that the static part of the memory function for the GFPE system is given by:
$$\sum_\ell K_{i\ell}^{(s)} S_{\ell j} = \Gamma_{ij}^{(s)} = \langle \phi_i i \tilde{D}_\phi \phi_i \rangle .$$

Given the GFPE operator, Eq. (183), show:
$$\Gamma_{ij}^{(s)} = i\beta^{-1} \langle Q_{ij} \rangle - i\beta^{-1} \Gamma_0^{ij} ,$$
where the Q_{ij} is the Poisson bracket between ψ_i and ψ_j and the Γ_0^{ij} are the bare kinetic/diffusion coefficients.

Problem 9.10: Show that the dynamic part of the memory function defined by Eqs. (188) and (189) is given by Eq. (191).

Problem 9.11: Consider a zero-dimensional mode-coupling model for a time-correlation function $C(t)$. In terms of Laplace transforms it obeys the kinetic equation:
$$(z + i\Gamma - \Gamma^{(d)}(z))C(z) = 1 ,$$
where Γ is a bare kinetic coefficient and the memory function $\Gamma^{(d)}$ is given by a mode coupling expression:
$$\Gamma^{(d)}(z) = V^2(-i) \int_0^\infty dt e^{izt} 2C^2(t)$$
Determine the physical kinetic coefficient $\Gamma_R^{(d)}(z)$ to second order in the nonlinear coupling V.

Problem 9.12: Show in detail that Eq. (216) is true.

Problem 9.13: Show how to go from Eq. (223) to Eq. (224).

Problem 9.14: Starting with Eq. (227), show in three dimensions for small z that the physical viscosity for a fluid is given by:
$$v_R(z) = v_R(0) + A\sqrt{z} + \ldots$$

Determine the constant A. In two dimensions for small z, the physical viscosity can be written in the form:

$$v_R(z) = v_R(0)\left[1 + b_0 \ln(b_1/z)\right] \; .$$

Determine the constants b_0 and b_1.

Problem 9.15: Consider a dynamic model with two fluctuating scalar fields ψ and ϕ. The system is described by an effective Hamiltonian:

$$\mathcal{H}[\psi,\phi] = \int d^d x \left[\frac{r_\psi}{2} \psi^2 + \frac{r_\phi}{2}\phi^2 + \frac{c_\psi}{2}(\nabla\psi)^2 + \frac{c_\phi}{2}(\nabla\phi)^2 + \lambda\phi\psi \right] \; ,$$

where r_ψ, r_ϕ, c_ψ, and c_ϕ are positive. These fields satisfy the TDGL equations:

$$\frac{\partial \psi}{\partial t} = -\Gamma \frac{\delta \mathcal{H}}{\delta \psi} + \xi$$

$$\frac{\partial \phi}{\partial t} = D\nabla^2 \frac{\delta \mathcal{H}}{\delta \phi} + \eta \; ,$$

where the noise is Gaussian and white with variance:

$$\langle \xi(\mathbf{x},t)\xi(\mathbf{x}',t')\rangle = 2k_B T \Gamma \delta(\mathbf{x}-\mathbf{x}')\delta(t-t')$$

$$\langle \eta(\mathbf{x},t)\eta(\mathbf{x}',t')\rangle = 2k_B TD\nabla_x \cdot \nabla_{x'}\delta(\mathbf{x}-\mathbf{x}')\delta(t-t')$$

$$\langle \xi(\mathbf{x},t)\eta(\mathbf{x}',t')\rangle = 0 \; .$$

Γ is a kinetic coefficient and D a transport coefficient. Find the dynamic structure factors: $C_{\psi\psi}(\mathbf{q},\omega)$, $C_{\phi\phi}(\mathbf{q},\omega)$, and $C_{\psi\phi}(\mathbf{q},\omega)$.

Problem 9.16: Assume we have a system with a complex order parameter $\psi(\mathbf{x},t)$ governed by an effective Hamiltonian:

$$\mathcal{H}[\psi,\psi^*] = \int d^d x \left[\alpha |\psi|^2 + \frac{\beta}{2}|\psi|^4 + c\nabla\psi^* \cdot \nabla\psi \right] \; ,$$

where α, β, and c are real. Assuming a NCOP relaxational dynamics, with complex noise, write down the simplest TDGL equation of motion. Solve this model for the order parameter dynamic structure factor for the case $\beta = 0$.

Problem 9.17: Consider the original Langevin equation for the velocity of a tagged particle (for simplicity in one dimension) in a complex environment:

$$\frac{\partial v(t)}{\partial t} = -\gamma \frac{\partial \mathcal{H}}{\partial v(t)} + \eta \; ,$$

where the effective Hamiltonian is given by:

$$\mathcal{H} = \frac{M}{2}v^2$$

with M the particles mass and we have Gaussian white noise with variance:

$$\langle \eta(t)\eta(t') \rangle = 2k_B T \gamma \delta(t-t') \; .$$

Derive the Fokker–Planck equation satisfied by the velocity probability distribution:

$$P(V,t) = \langle \delta[V - v(t)] \rangle$$

and solve it to obtain the equilibrium velocity probability distribution.

Problem 9.18: Consider the coarse-grained equation of motion for a ferromagnet given by Eq. (133):

$$\frac{\partial \mathbf{M}}{\partial t} = \mathbf{M} \times \mathbf{H} - \Gamma_0 \nabla^2 \mathbf{H} + \vec{\zeta} \; .$$

Assume that we have an applied uniform external field $\mathbf{h} = h\hat{z}$ and $u = c = 0$ in the effective Hamiltonian given by Eq. (126). Using Eq. (130) we have the effective external field:

$$\mathbf{H} = \mathbf{h} - r\mathbf{M}$$

and the equation of motion:

$$\frac{\partial \mathbf{M}}{\partial t} = \mathbf{M} \times \mathbf{h} + \Gamma_0 r \nabla^2 \mathbf{M} + \vec{\zeta} \; .$$

Compute the time evolution of the average magnetization $\mathbf{m}(t) = \langle \mathbf{M}(t) \rangle$, given $\mathbf{m}(t=0) = \mathbf{M}_0$.

Problem 9.19: In treating the mode-coupling approximation we assumed that the four-point correlation function:

$$G_{\ell m, kn}(t) = \langle \delta(\phi_k \phi_n) e^{\tilde{D}_\phi^0 t} \delta(\phi_l \phi_m) \rangle_0$$

is given in the decoupling approximation by:

$$G_{\ell m, kn}(t) = C_{\ell k}^{(0)}(t) C_{mn}^{(0)}(t) + C_{\ell n}^{(0)}(t) C_{mk}^{(0)}(t) \; , \tag{241}$$

where:

$$C_{\ell k}^{(0)}(t) = \langle \delta \phi_k e^{\tilde{D}_\phi^0 t} \delta \phi_l \rangle_0 \; . \tag{242}$$

More specifically, assume a model where:

$$\mathcal{H}_\phi^{(0)} = \sum_i \frac{r_i}{2} \phi_i^2 \; ,$$

$$\Gamma_0^{ij} = \delta_{ij}\Gamma_0 ,$$

and the adjoint Fokker–Planck operator is given by:

$$\tilde{D}_\phi^0 = \sum_i \Gamma_0 \beta^{-1} \left(\frac{\partial}{\partial \phi_i} - \beta r_i \phi_i \right) \frac{\partial}{\partial \phi_i} .$$

1. Evaluate $C_{\ell k}^{(0)}(t)$ explicitly.
2. Evaluate $G_{\ell m, kn}(t)$ explicitly,
3. Show that Eq. (241) is satisfied.
4. Finally show that for this model:

$$\langle \delta(\phi_k \phi_n) e^{\tilde{D}_\phi^0 t} \delta(\phi_i) \rangle_0 = 0 .$$

Problem 9.20: Show in the case of an isotropic ferromagnet that the spin precession term in the equation of motion can be written in terms of a gradient of a current:

$$\mathbf{H} \times \mathbf{M} = -\sum_i \nabla_i \mathbf{J}_i ,$$

where $\mathbf{H}$ is defined by Eq. (130). Identify the current $\mathbf{J}_i$.

10
Perturbation Theory and the Dynamic Renormalization Group

10.1 Perturbation Theory

10.1.1 TDGL Model

In the previous chapter we introduced a set of nonlinear models relevant in dynamic critical phenomena and nonlinear hydrodynamics. We gave one method for analyzing these models at the end of the chapter. Here we introduce some simple perturbation theory methods that have been useful in treating dynamic critical phenomena and hydrodynamical processes. We show that the results for the order parameter time-correlation functions strongly depend on the spatial dimensionality and are not directly useful in three dimensions without interpretation coming from application of the renormalization group method. These methods are developed in the final sections of the chapter.

There now exist a number of sophisticated methods Refs. [1–5] for carrying out perturbation series expansions for nonlinear dynamical models. Some of these methods are discussed in FTMCM. Here we discuss a direct brute-force approach that is sufficient for many purposes. These methods [6] can serve as an introduction to more sophisticated perturbation theory approaches.

We discuss two important model in this chapter: the scalar TDGL model [7] and the isotropic ferromagnet [8]. The methods developed here have been applied to essentially all of the models characterizing dynamic critical phenomena universality classes.

Let us begin with the TDGL model for the scalar order parameter $\psi(\mathbf{x}, t)$:

$$\frac{\partial \psi(\mathbf{x},t)}{\partial t} = -\Gamma_0(\mathbf{x}) \frac{\delta}{\delta \psi(\mathbf{x},t)} \mathcal{H}_E[\psi] + \eta(\mathbf{x},t) \quad , \tag{1}$$

where the Fourier transform of the bare transport coefficient $\Gamma_0(q)$ is a constant, Γ_0, for a NCOP and $D_0 q^2$ for a COP. We choose the LGW form for the effective Hamiltonian:

$$\mathcal{H}_E[\psi] = \int d^d x \int d^d y \frac{1}{2} \psi(\mathbf{x}) \chi_0^{-1}(\mathbf{x}-\mathbf{y}) \psi(\mathbf{y})$$

Nonequilibrium Statistical Mechanics. Gene F. Mazenko
Copyright © 2006 WILEY-VCH Verlag GmbH & Co. KGaA, Weinheim
ISBN: 3-527-40648-4

$$+ \int d^d x \frac{u}{4} \psi^4(\mathbf{x}) - \int d^d x h(\mathbf{x},t)\psi(\mathbf{x}) \quad , \tag{2}$$

where initially we need not specify the zeroth-order static susceptibility χ_0 and $h(\mathbf{x},t)$ is an external time- and space-dependent field. The noise $\eta(\mathbf{x},t)$ in Eq. (1) is assumed to be Gaussian with variance:

$$\langle \eta(\mathbf{x},t)\eta(\mathbf{x}',t') \rangle = 2k_B T \Gamma_0(\mathbf{x})\delta(\mathbf{x}-\mathbf{x}')\delta(t-t') \quad . \tag{3}$$

After taking the functional derivative in Eq. (1) we have the equation of motion:

$$\frac{\partial \psi(\mathbf{x},t)}{\partial t} = -\Gamma_0(\mathbf{x})\left[\int d^d y \chi_0^{-1}(\mathbf{x}-\mathbf{y})\psi(\mathbf{y},t) + u\psi^3(\mathbf{x},t) - h(\mathbf{x},t)\right]$$
$$+ \eta(\mathbf{x},t) \quad . \tag{4}$$

Taking the Fourier transform over space and time we obtain:

$$-i\omega\psi(\mathbf{q},\omega) = -\Gamma_0(\mathbf{q})\left[\chi_0^{-1}(q)\psi(\mathbf{q},\omega) - uN(\mathbf{q},\omega) - h(\mathbf{q},\omega)\right]$$
$$+ \eta(\mathbf{q},\omega) \quad , \tag{5}$$

where the nonlinear contribution can be written as:

$$N(\mathbf{q},\omega) = \int \frac{d^d q'}{(2\pi)^d} \frac{d\omega'}{2\pi} \frac{d^d q''}{(2\pi)^d} \frac{d\omega''}{2\pi} \psi(\mathbf{q}-\mathbf{q}'-\mathbf{q}'', \omega-\omega'-\omega'')$$
$$\times \psi(\mathbf{q}',\omega')\psi(\mathbf{q}'',\omega'') \quad . \tag{6}$$

If we define the zeroth-order response function:

$$G_0(q,\omega) = \left[-i\omega + \Gamma_0(q)\chi_0^{-1}(q)\right]^{-1} \Gamma_0(q) \quad , \tag{7}$$

we can write Eq. (1) as:

$$\psi(\mathbf{q},\omega) = \psi^0(\mathbf{q},\omega) + G_0(\mathbf{q},\omega)h(\mathbf{q},\omega) - G_0(\mathbf{q},\omega)uN(\mathbf{q},\omega) \quad , \tag{8}$$

where:

$$\psi^0(\mathbf{q},\omega) = \frac{G_0(q,\omega)}{\Gamma_0(q)}\eta(\mathbf{q},\omega) \quad . \tag{9}$$

ψ^0 is the zeroth expression for ψ in the absence of an externally applied field.

10.1.2
Zeroth-Order Theory

If we set $u = 0$ in Eq. (8), the equation of motion reduces to:

$$\psi_0(\mathbf{q},\omega) = \psi^0(\mathbf{q},\omega) + G_0(\mathbf{q},\omega)h(\mathbf{q},\omega) \quad . \tag{10}$$

If we average over the noise we obtain the linear response result:

$$\langle \psi_0(\mathbf{q},\omega) \rangle = G_0(\mathbf{q},\omega) h(\mathbf{q},\omega) \tag{11}$$

and we see that G_0 is indeed the linear response function at zero order. If we take the zero frequency limit in Eq. (7) we find:

$$G_0(\mathbf{q},0) = \chi_0(\mathbf{q}) \tag{12}$$

and $G_0(\mathbf{q},0)$ is the static zeroth-order susceptibility.

In zero external field we can compute the zeroth-order correlation function using Eq. (9) twice and the double Fourier transform of Eq. (3):

$$\begin{aligned}
\langle \psi_0(\mathbf{q}_1,\omega_1) \psi_0(\mathbf{q}_2,\omega_2) \rangle &= \langle \psi^0(\mathbf{q}_1,\omega_1) \psi^0(\mathbf{q}_2,\omega_2) \rangle \\
&= \left\langle \frac{G_0(\mathbf{q}_1,\omega_1)}{\Gamma_0(\mathbf{q}_1)} \eta(\mathbf{q}_1,\omega_1) \frac{G_0(\mathbf{q}_2,\omega_2)}{\Gamma_0(\mathbf{q}_2)} \eta(\mathbf{q}_2,\omega_2) \right\rangle \\
&= \frac{G_0(\mathbf{q}_1,\omega_1)}{\Gamma_0(\mathbf{q}_1)} \frac{G_0(\mathbf{q}_2,\omega_2)}{\Gamma_0(\mathbf{q}_2)} \langle \eta(\mathbf{q}_1,\omega_1) \eta(\mathbf{q}_2,\omega_2) \rangle \\
&= \frac{G_0(\mathbf{q}_1,\omega_1)}{\Gamma_0(\mathbf{q}_1)} \frac{G_0(\mathbf{q}_2,\omega_2)}{\Gamma_0(\mathbf{q}_2)} 2k_B T \Gamma_0(\mathbf{q}_1) (2\pi)^{d+1} \delta(\mathbf{q}_1+\mathbf{q}_2) \delta(\omega_1+\omega_2) \\
&= C_0(\mathbf{q}_1,\omega_1) (2\pi)^{d+1} \delta(\mathbf{q}_1+\mathbf{q}_2) \delta(\omega_1+\omega_2) ,
\end{aligned} \tag{13}$$

where:

$$C_0(\mathbf{q},\omega) = \frac{2k_B T \Gamma_0(\mathbf{q})}{\omega^2 + (\Gamma_0(\mathbf{q}) \chi_0^{-1}(\mathbf{q}))^2} . \tag{14}$$

It is left to Problem 10.1 to show that the zeroth-order response and correlation functions are related by:

$$C_0(\mathbf{q},\omega) = k_B T \frac{2}{\omega} \operatorname{Im} G_0(\mathbf{q},\omega) . \tag{15}$$

10.1.3
Bare Perturbation Theory

It is clear that we can generate corrections to the $u = 0$ results by iterating Eq. (8). It is convenient to carry out this iteration graphically. We represent Eq. (8) as shown in Fig. 10.1 where a wavy line stands for ψ, a dashed line for ψ^0; a solid line with an arrow stands for G_0, a line terminating at an external field is represented by a directed line followed by an x, and a dot represents the interaction vertex $(-u)$. Note that at a vertex the wavenumber and frequency are *conserved*. If we assign, for example, a frequency to each line entering a vertex, then it is easy to see that the frequency entering the vertex via G_0 must equal the sum of the frequencies leaving via the wavy lines.

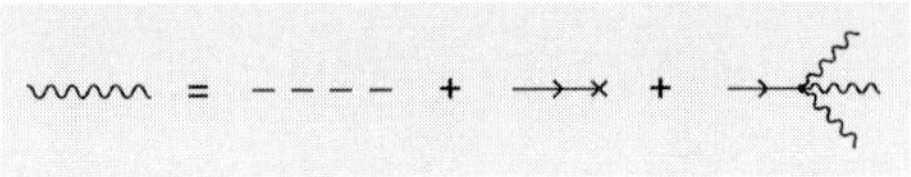

Fig. 10.1 Graphical representation for the equation of motion Eq. (8).

A direct iteration of Eq. (8) (or Fig. 10.1) correct to second order in u and to first order in h, as discussed in Problem 10.2, gives the graphs in Fig. 10.2. In the fourth term on the right-hand side, for example, we get a factor of 3 since the vertex is symmetric under interchange of the wavy lines. We eventually need to carry out averages over the noise. In these averages one has products of dashed lines each of which is proportional to a factor of the noise. We have postulated, however, that the noise is a Gaussian random variable. This means that the average of a product of noise terms is equal to the sum of all possible pairwise averages. Since ψ^0 is proportional to the noise, the average of a product of ψ^0's factors into a sum of all possible pairwise averages. Algebraically we write, with a general index i labeling each field, for a product of four zeroth-order fields,

$$\langle \psi_i^0 \psi_j^0 \psi_k^0 \psi_\ell^0 \rangle = \langle \psi_i^0 \psi_j^0 \rangle \langle \psi_k^0 \psi_\ell^0 \rangle + \langle \psi_i^0 \psi_k^0 \rangle \langle \psi_j^0 \psi_\ell^0 \rangle + \langle \psi_i^0 \psi_\ell^0 \rangle \langle \psi_j^0 \psi_k^0 \rangle. \quad (16)$$

Graphically we introduce the following convention. If before averaging we have four zeroth-order parameter lines as shown in Fig. 10.3, then on averaging we tie these lines together in all possible pairs and obtain the graphs in Fig. 10.4. In these graphs correlation functions are given by:

$$\langle \psi^0(\mathbf{q}_i, \omega_i) \psi^0(\mathbf{q}_j, \omega_j) \rangle = C_0(\mathbf{q}_i, \omega_i)(2\pi)^{d+1} \delta(\mathbf{q_i} + \mathbf{q_j}) \delta(\omega_i + \omega_j) \quad (17)$$

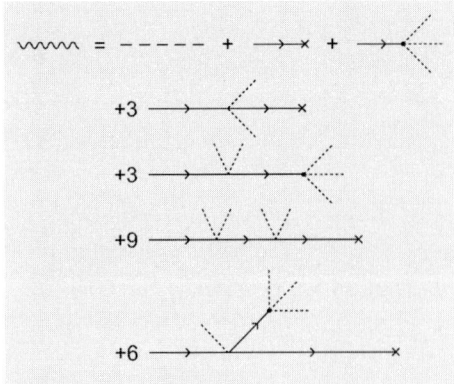

Fig. 10.2 Iterated graphical representation for the equation of motion Eq. (8), keeping terms up to $\mathcal{O}(h^2)$ and $\mathcal{O}(u^3)$.

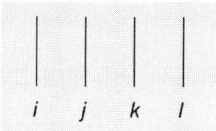

Fig. 10.3 Lines representing ψ^0 with labels i, j, k and ℓ.

and are represented by lines with circles in the middle as shown in Fig. 10.5. With these simple rules we can produce a perturbation theory expansion for any correlation function to any order in the coupling u.

The response function $G(q, \omega)$ is defined in general as the averaged linear response of the field $\psi(\mathbf{x}, t)$ to an external field $h(\mathbf{x}, t)$:

$$G(q, \omega) \equiv \lim_{h \to 0} \frac{\langle \psi(\mathbf{q}, \omega, h) \rangle - \langle \psi(\mathbf{q}, \omega, h = 0) \rangle}{h(q, \omega)} . \tag{18}$$

It is clear from Eq. (2) that G_0 is the linear response function for $u = 0$.

The graphical expansion for G is relatively simple. We take the average over the noise of the graphs in Fig. 10.2. All terms independent of h vanish since they have an odd number of noise lines. We keep only those terms with one external field, x. We therefore obtain (see Problem 10.3), to second order in u, the graphs shown in Fig. 10.5. It is also left to Problem 10.3 to show that these graphs lead to the analytic expressions for the response function given by:

$$\begin{aligned} G(q, \omega) &= G_0(q, \omega) + G_0(q, \omega)\Sigma_1 G_0(q, \omega) + G_0(q, \omega)\Sigma_1 G_0(q, \omega)\Sigma_1 G_0(q, \omega) \\ &\quad + G_0(q, \omega)\Sigma_2 G_0(q, \omega) + G_0(q, \omega)\Sigma_3(q, \omega) G_0(q, \omega) , \end{aligned} \tag{19}$$

where:

$$\Sigma_1 = 3(-u) \int \frac{d^d k}{(2\pi)^d} \frac{d\Omega}{2\pi} C_0(k, \Omega). \tag{20}$$

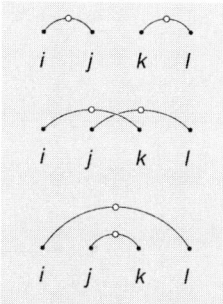

Fig. 10.4 Three ways of tying the lines in Fig. 10.3 together after averaging over the Gaussian noise. *Lines with circles in the middle represent zeroth-order correlation functions.*

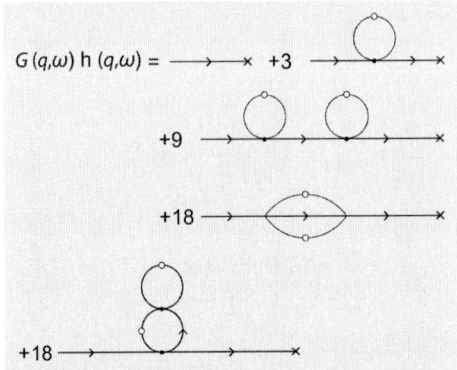

$G(q,\omega)\, h(q,\omega) =$

Fig. 10.5 Graphical perturbation expansion for the response function defined by Eq. (18) including terms of $\mathcal{O}(u^2)$.

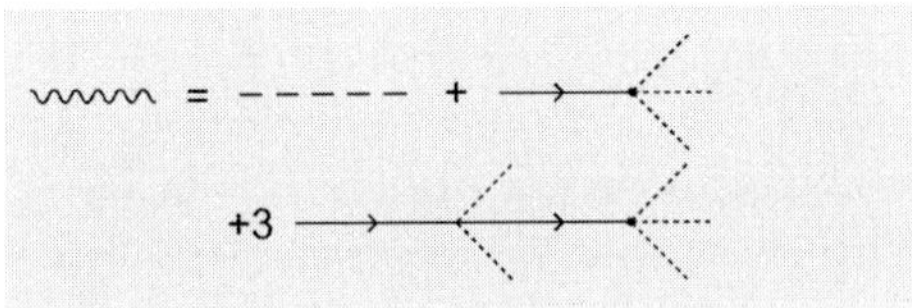

Fig. 10.6 Iterated graphical representation for the order parameter in the absence of an external field up to $\mathcal{O}(u^2)$.

$$\Sigma_2 = 18(-u) \int \frac{d^d k'}{(2\pi)^d} \frac{d\Omega'}{2\pi} C_0(k', \Omega')$$
$$\times (-u) \int \frac{d^d k}{(2\pi)^d} \frac{d\Omega}{2\pi} C_0(k, \Omega) G_0(k, \Omega) \qquad (21)$$

$$\Sigma_3(q, \omega) = 18(-u)^2 \int \frac{d^d k}{(2\pi)^d} \frac{d\Omega}{2\pi} \frac{d^d k'}{(2\pi)^d} \frac{d\Omega'}{2\pi} G_0(\mathbf{q} - \mathbf{k} - \mathbf{k}', \omega - \Omega - \Omega')$$
$$\times C_0(\mathbf{k}, \Omega) C_0(\mathbf{k}', \Omega') \ . \qquad (22)$$

Note that the contributions from Σ_1 and Σ_2 are independent of q and ω. It is left to Problem 10.4 to show that both Σ_1 and Σ_2 are real.

Besides the response function, we can develop the bare perturbation theory for the correlation function:

$$\langle \psi(\mathbf{q}, \omega) \psi(\mathbf{q}', \omega') \rangle = C(q, \omega)(2\pi)^{d+1} \delta(\mathbf{q} + \mathbf{q}') \delta(\omega + \omega') \ . \qquad (23)$$

We do this by setting $h = 0$ in Fig. 10.2 to obtain Fig. 10.6. Then multiply the graphs for ψ times those for ψ' and then average over the noise. It is left to Problem 10.5 to show that one obtains the collection of graphs given in Fig. 10.7. Note that one has many more graphs at second order for C than for G.

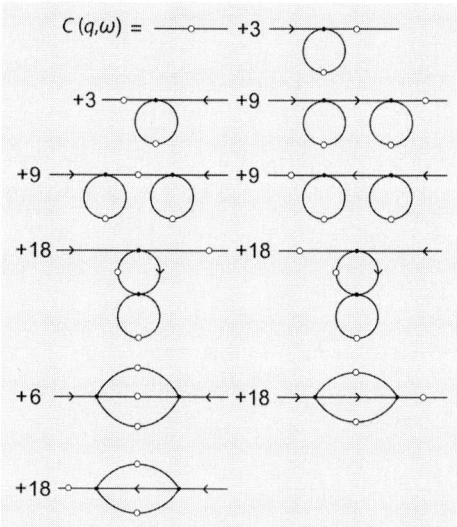

Fig. 10.7 Graphical representation for the perturbation expansion for the correlation function including terms up to $\mathcal{O}(u^2)$.

10.1.4
Fluctuation-Dissipation Theorem

We showed earlier that the zeroth-order correlation function and response function are related by:

$$C^0(q,\omega) = \frac{2}{\omega} k_B T \, \text{Im} \, G_0(q,\omega) \, . \tag{24}$$

It has been shown in Ref. [8] that this *fluctuation-dissipation* theorem (FDT) holds for the class of nonlinear equations defined by Eq. (1) for the full correlation function and response functions:

$$C(q,\omega) = \frac{2}{\omega} k_B T \, \text{Im} \, G(q,\omega). \tag{25}$$

Thus, if we can calculate $G(q,\omega)$, we can obtain $C(q,\omega)$ via the FDT.

We will be satisfied here with showing that the FDT holds up to second order in u in the bare perturbation theory analysis. Higher order terms are discussed by Ma in Ref. [9]. Here we show how the contributions in Fig. 10.5 lead to those in Fig. 10.7. We have already shown the FDT at zeroth order, which we represent graphically in Fig. 10.8. Next we consider the left-hand side of Fig. 10.9. Since Σ_1, the loop given by Eq. (20), is real, we need only consider,

$$\text{Im} \, G^2 = 2G'G'' = (G + G^*)G'' \tag{26}$$

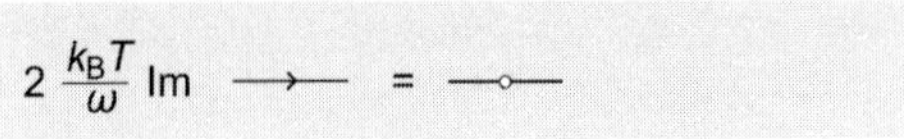

10 Perturbation Theory and the Dynamic Renormalization Group

$2\frac{k_B T}{\omega}$ Im ——→—— = ——o——

Fig. 10.8 Graphical representation of the fluctuation-dissipation theorem (FDT) at zeroth order.

and:

$$\frac{2k_B T}{\omega} \text{Im } G^2 = (G + G^*)C_0 \ . \tag{27}$$

Since $G^*(\omega) = G(-\omega)$ we represent G^* with a line with an arrow pointing to the left. We then find the graphical identity shown in Fig. 10.9 and a matching of contributions between Fig. 10.5 and Fig. 10.7.

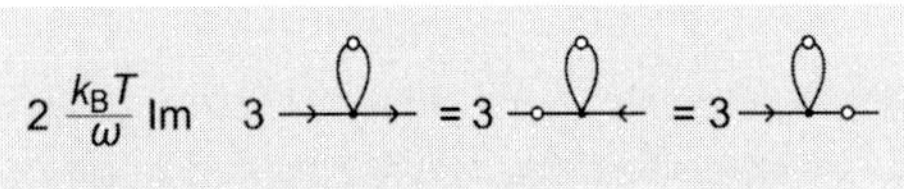

$2\frac{k_B T}{\omega}$ Im 3 ——→◯→—— $= 3$ ——o◯→—— $= 3$ ——→◯o——

Fig. 10.9 Graphical representation of the FDT at first order.

Let us turn to the left-hand side of Fig. 10.10. It is easy to show (see Problem 10.4) that the two-loop insertion given by Σ_2 is real. Therefore we can again use Eq. (27) to obtain the right-hand side of Fig. 10.10 and again map one graph in Fig. 10.5 onto two graphs in Fig. 10.7.

The following identity:

$$\text{Im } ABC = ABC'' + AB''C^* + A''B^*C^* \tag{28}$$

is proven in Problem 10.6. In particular we have:

$$\text{Im } G^3 = G^2 G'' + GG''G^* + G''G^*G^* \tag{29}$$

and:

$$\frac{2k_B T}{\omega} \text{Im } G^3 = GGC_0 + GC_0 G^* + C_0 G^* G^* \ . \tag{30}$$

This leads immediately to the results shown in Fig. 10.11 where one graph in Fig. 10.5 is mapped onto three graphs in Fig. 10.7.

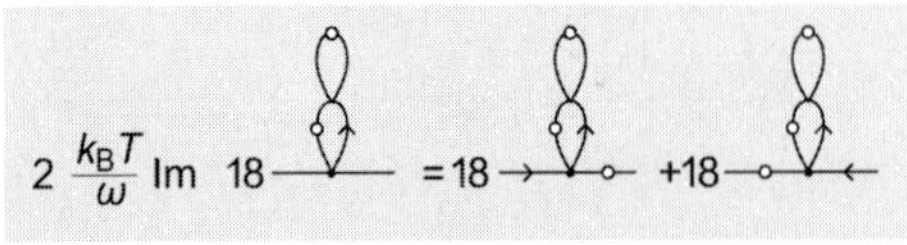

$2\frac{k_B T}{\omega}$ Im 18 ——→◯◯→—— $= 18$ ——→◯o◯→—— $+18$ ——o◯◯→——

Fig. 10.10 One contribution to the graphical representation of the FDT at second order.

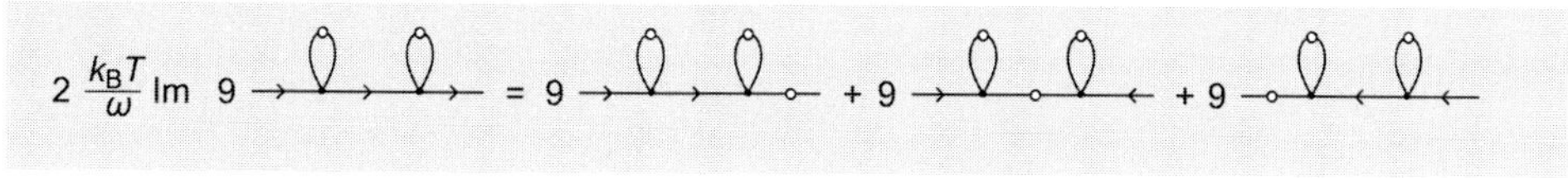

2 $\frac{k_BT}{\omega}$ Im 9
= 9
+ 9
+ 9

Fig. 10.11 Another graphical contribution to the FDT at second order.

2 $\frac{k_BT}{\omega}$ Im
=
+
+ 2 $\frac{k_BT}{\omega}$ Im
G_0''

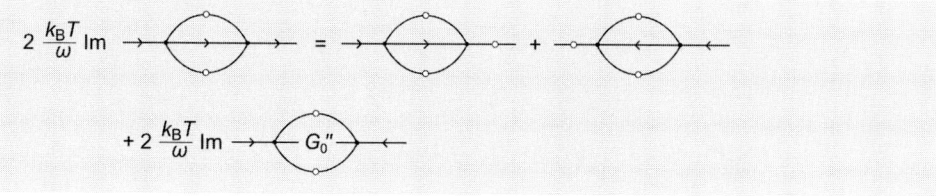

Fig. 10.12 Another graphical contribution to the FDT at second order.

Finally, we turn to the left-hand side of Fig. 10.12. In this case we need Eq. (28) expressed in the form, suppressing the wavenumber dependence,

$$\begin{aligned} \mathrm{Im}\,[G(\omega)\Sigma_3(\omega)G(\omega)] &= G(\omega)\Sigma_3(\omega)G''(\omega) + G(\omega)\Sigma_3''(\omega)G^*(\omega) \\ &\quad + G''(\omega)\Sigma_3^*(\omega)G^*(\omega) \ , \end{aligned} \tag{31}$$

where Σ_3 is given by Eq. (22) and we use the fact that the correlation function lines are real. We then obtain the right-hand side of Fig. 10.12. We still have to look at the last term on the right in Fig. 10.12. We have that the imaginary part of Σ_3 is proportional to:

$$\begin{aligned} &\frac{2k_BT}{\omega}\int_S C_0(\mathbf{q}_1,\omega_1)C_0(\mathbf{q}_2,\omega_2)G_0''(\mathbf{q}_3,\omega_3) \\ &= \frac{2k_BT}{\omega}\int_S C_0(\mathbf{q}_1,\omega_1)C_0(\mathbf{q}_2,\omega_2)C_0(\mathbf{q}_3,\omega_3)\frac{\omega_3}{2k_BT} \\ &= \frac{1}{\omega}\int_S C_0(\mathbf{q}_1,\omega_1)C_0(\mathbf{q}_2,\omega_2)C_0(\mathbf{q}_3,\omega_3)\frac{1}{3}(\omega_3+\omega_1+\omega_2) \\ &= \frac{1}{\omega}\int_S C_0(\mathbf{q}_1,\omega_1)C_0(\mathbf{q}_2,\omega_2)C_0(\mathbf{q}_3,\omega_3)\frac{1}{3}\omega \\ &= \frac{1}{3}\int_S C_0(\mathbf{q}_1,\omega_1)C_0(\mathbf{q}_2,\omega_2)C_0(\mathbf{q}_3,\omega_3) \ , \end{aligned}$$

where we have used the shorthand notation for integrations:

$$\begin{aligned} \int_S &= \int \frac{d^dq_1}{(2\pi)^d}\frac{d^dq_2}{(2\pi)^d}\frac{d^dq_3}{(2\pi)^d}\frac{d\omega_1}{2\pi}\frac{d\omega_2}{2\pi}\frac{d\omega_3}{2\pi}(2\pi)^{d+1} \\ &\quad \times\delta(\mathbf{q}-\mathbf{q}_1-\mathbf{q}_2-\mathbf{q}_3)\delta(\omega-\omega_1-\omega_2-\omega_3) \ . \end{aligned}$$

Clearly we are left with the mapping of the term on the left-hand side of Fig. 10.13 onto the three graphs on the right-hand side of Fig. 10.13. Clearly

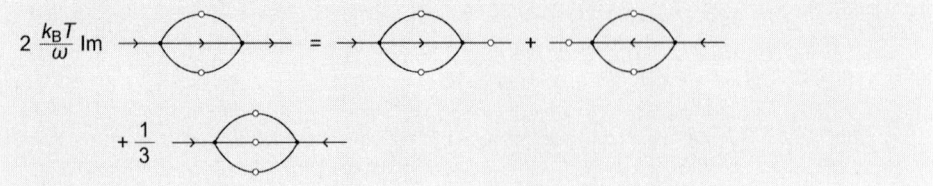

Fig. 10.13 Refined final graphical contribution to the second-order FDT.

this is a mapping of the final terms between Fig. 10.5 and Fig. 10.7 and to this order the FDT is proven.

10.1.5
Static Limit

The linear response function has an interesting feature. If we note the spectral representation [10] for the response function:

$$G(q, \omega) = \int \frac{d\omega'}{\pi} \frac{\text{Im}\, G(q, \omega')}{(\omega' - \omega - i0^+)} , \qquad (32)$$

and use the FDT, we obtain:

$$k_B T G(q, \omega) = \int \frac{d\omega'}{2\pi} \frac{C(q, \omega')\omega'}{(\omega' - \omega - i0^+)} \cdots \qquad (33)$$

If we set $\omega = 0$ we find:

$$k_B T G(q, 0) = \int \frac{d\omega'}{2\pi} C(q, \omega') = C(q) = k_B T \chi(q) . \qquad (34)$$

Therefore we can obtain the static susceptibility from the linear response function simply by setting $\omega = 0$. We have, from Eq. (19), in the analytic expression for the response function, the result at zero frequency:

$$\begin{aligned}\chi(q) &= \chi_0(q) + \chi_0(q)\Sigma_1\chi_0(q) + \chi_0(q)\Sigma_1\chi_0(q)\Sigma_1\chi_0(q) \\ &+ \chi_0(q)\Sigma_2\chi_0(q) + \chi_0(q)\Sigma_3(q, \omega = 0)\chi_0(q) .\end{aligned} \qquad (35)$$

The Σ's appearing here must be purely static quantities that are independent of Γ.

It is clear that Σ_1, defined by Eq. (20), is a purely static quantity given by:

$$\Sigma_1 = 3(-u) \int \frac{d^d k}{(2\pi)^d} k_B T \chi_0(k) . \qquad (36)$$

The quantity Σ_2, defined by Eq. (21), can be written as the product:

$$\Sigma_2 = 6\Sigma_1 J , \qquad (37)$$

where:
$$J = (-u) \int \frac{d^d k}{(2\pi)^d} \frac{d\Omega}{2\pi} C_0(k,\Omega) G_0(k,\Omega) \ . \tag{38}$$

One can perform the frequency integral rather easily. Let us consider the more general frequency integral:
$$K(k,q,\omega) = \int \frac{d\Omega}{2\pi} C_0(k,\Omega) G_0(q,\omega - \Omega) \ . \tag{39}$$

It is convenient to write Eq. (7) for the zeroth order response function as:
$$G_0(q,\omega) = \frac{i\Gamma_0(q)}{\omega + i\gamma(q)} \ , \tag{40}$$

where:
$$\gamma(q) = \Gamma_0(q) \chi_0^{-1}(q) \tag{41}$$

and rewrite Eq. (14) for the correlation function:
$$C_0(q,\omega) = i k_B T \chi_0(q) \left[\frac{1}{\omega + i\gamma(q)} - \frac{1}{\omega - i\gamma(q)} \right] \ . \tag{42}$$

Substituting these results into Eq. (39) gives:
$$K(k,q,\omega) = \int \frac{d\Omega}{2\pi} i k_B T \chi_0(k) \left[\frac{1}{\Omega + i\gamma(k)} - \frac{1}{\Omega - i\gamma(k)} \right] \frac{i\Gamma_0(q)}{\omega - \Omega + i\gamma(q)} \tag{43}$$
$$= \int \frac{d\Omega}{2\pi} k_B T \chi_0(k) \Gamma_0(q) \left[\frac{1}{\Omega + i\gamma(k)} - \frac{1}{\Omega - i\gamma(k)} \right] \frac{1}{\Omega - \omega - i\gamma(q)} \ .$$

One can then carry out the Ω integration using the calculus of residues and, closing the contour in the lower half-plane where there is a single pole at $\Omega = -i\gamma(k)$, we find:
$$K(k,q,\omega) = \frac{-2\pi i}{2\pi} k_B T \chi_0(k) \Gamma_0(q) \frac{1}{-i\gamma(k) - \omega - i\gamma(q)}$$
$$= \frac{i k_B T \chi_0(k) \Gamma_0(q)}{\omega + i[\gamma(k) + \gamma(q)]} \ . \tag{44}$$

Note that if we let $\Omega \to -\Omega$ in Eq. (38) we obtain:
$$J = (-u) \int \frac{d^d k}{(2\pi)^d} K(k,k,0)$$
$$= (-u) \int \frac{d^d k}{(2\pi)^d} \frac{i k_B T \chi_0(k) \Gamma_0(k)}{i 2\gamma(k)}$$

$$= -\frac{u}{2}k_B T \int \frac{d^d k}{(2\pi)^d} \chi_0^2(k) \;, \tag{45}$$

which is purely a static quantity.

We turn finally to Σ_3, given by Eq. (22), evaluated at zero frequency. For general frequency ω we can write:

$$\begin{aligned}\Sigma_3(q,\omega) &= 18(-u)^2 \int \frac{d^d k}{(2\pi)^d} \frac{d\Omega}{2\pi} \frac{d^d k'}{(2\pi)^d} \frac{d\Omega'}{2\pi} G_0(\mathbf{q}-\mathbf{k}-\mathbf{k}',\omega-\Omega-\Omega') \\ &\quad \times C_0(\mathbf{k},\Omega)C_0(\mathbf{k}',\Omega') \\ &= 18(-u)^2 \int \frac{d^d k}{(2\pi)^d} \frac{d^d k'}{(2\pi)^d} \frac{d\Omega'}{2\pi} C_0(\mathbf{k}',\Omega') K(k,\mathbf{q}-\mathbf{k}-\mathbf{k}',\omega-\Omega') \;.\end{aligned}$$

Using Eq. (44) this reduces to:

$$\begin{aligned}\Sigma_3(q,\omega) &= 18u^2 \int \frac{d^d k}{(2\pi)^d} \frac{d^d k'}{(2\pi)^d} \frac{d\Omega'}{2\pi} C_0(\mathbf{k}',\Omega') \\ &\quad \times \frac{ik_B T \chi_0(\mathbf{k}) \Gamma_0(\mathbf{q}-\mathbf{k}-\mathbf{k}')}{\omega-\Omega'+i[\gamma(k)+\gamma(\mathbf{q}-\mathbf{k}-\mathbf{k}')]} \\ &= 18u^2(-i) \int \frac{d^d k}{(2\pi)^d} \frac{d^d k'}{(2\pi)^d} \Gamma_0(\mathbf{q}-\mathbf{k}-\mathbf{k}') k_B T \chi_0(\mathbf{k}) K_1 \;, \end{aligned} \tag{46}$$

where:

$$\begin{aligned}K_1 &= \int \frac{d\Omega'}{2\pi} C_0(\mathbf{k}',\Omega') \frac{1}{\Omega'-\omega-i\gamma(\mathbf{k})-i\gamma(\mathbf{q}-\mathbf{k}-\mathbf{k}')} \\ &= \int \frac{d\Omega'}{2\pi} i k_B T \chi_0(\mathbf{k}') \left[\frac{1}{\Omega'+i\gamma(\mathbf{k}')} - \frac{1}{\Omega'-i\gamma(\mathbf{k}')}\right] \\ &\quad \times \frac{1}{\Omega'-\omega-i\gamma(\mathbf{k})-i\gamma(\mathbf{q}-\mathbf{k}-\mathbf{k}')} \;. \end{aligned} \tag{47}$$

Again, closing the integration contour in the lower half-plane we obtain:

$$K_1 = -i(ik_B T)\chi_0(\mathbf{k}') \frac{1}{-i\gamma(\mathbf{k}')-\omega-i\gamma(\mathbf{k})-i\gamma(\mathbf{q}-\mathbf{k}-\mathbf{k}')}$$

$$= -k_B T \chi_0(\mathbf{k}') \frac{1}{\omega+i[\gamma(\mathbf{k})+\gamma(\mathbf{k}')+\gamma(\mathbf{q}-\mathbf{k}-\mathbf{k}')]} \;. \tag{48}$$

Putting this result back into Eq. (46), we find:

$$\begin{aligned}\Sigma_3(q,\omega) &= 18u^2(i) \int \frac{d^d k}{(2\pi)^d} \frac{d^d k'}{(2\pi)^d} \Gamma_0(\mathbf{q}-\mathbf{k}-\mathbf{k}') k_B T \chi_0(\mathbf{k}) k_B T \chi_0(\mathbf{k}') \\ &\quad \times \frac{1}{\omega+i[\gamma(\mathbf{k})+\gamma(\mathbf{k}')+\gamma(\mathbf{q}-\mathbf{k}-\mathbf{k}')]}\end{aligned}$$

$$
\begin{aligned}
&= 18iu^2(k_BT)^2 \int \frac{d^dk}{(2\pi)^d} \frac{d^dk'}{(2\pi)^d} \chi_0(\mathbf{k})\chi_0(\mathbf{k}') \\
&\quad \times \frac{\Gamma_0(\mathbf{q}-\mathbf{k}-\mathbf{k}')}{\omega + i[\gamma(\mathbf{k}) + \gamma(\mathbf{k}') + \gamma(\mathbf{q}-\mathbf{k}-\mathbf{k}')]} \\
&= 18u^2(k_BT)^2 \int \frac{d^dk}{(2\pi)^d} \frac{d^dk'}{(2\pi)^d} \chi_0(\mathbf{k})\chi_0(\mathbf{k}') \\
&\quad \times \frac{\Gamma_0(\mathbf{q}-\mathbf{k}-\mathbf{k}')}{-i\omega + \gamma(k) + \gamma(k') + \gamma(\mathbf{q}-\mathbf{k}-\mathbf{k}')} \quad .
\end{aligned}
\tag{49}
$$

Notice that this can be written in the more symmetric form:

$$
\begin{aligned}
\Sigma_3(q,\omega) &= 18u^2(k_BT)^2 \int_S \chi_0(k_1)\chi_0(k_2)\chi_0(k_3) \\
&\quad \times \frac{\gamma(k_3)}{-i\omega + \gamma(k_1) + \gamma(k_2) + \gamma(k_3)} \quad ,
\end{aligned}
\tag{50}
$$

where:

$$
\int_S = \int \frac{d^dk_1}{(2\pi)^d} \frac{d^dk_2}{(2\pi)^d} \frac{d^dk_3}{(2\pi)^d} (2\pi)^d \delta(\mathbf{q}-\mathbf{k}_1-\mathbf{k}_2-\mathbf{k}_3) \quad .
\tag{51}
$$

We can then symmetrize with respect to the labels 1, 2 and 3 to obtain:

$$
\begin{aligned}
\Sigma_3(q,\omega) &= 18u^2(k_BT)^2 \int_S \chi_0(k_1)\chi_0(k_2)\chi_0(k_3) \\
&\quad \times \frac{\frac{1}{3}[\gamma(k_1) + \gamma(k_2) + \gamma(k_3)]}{-i\omega + \gamma(k_1) + \gamma(k_2) + \gamma(k_3)} \quad .
\end{aligned}
\tag{52}
$$

If we use the identity:

$$
\frac{a}{-i\omega + a} = 1 + \frac{i\omega}{-i\omega + a} \quad ,
\tag{53}
$$

then we have:

$$
\Sigma_3(q,\omega) = \Sigma_3(q,\omega=0) + i\omega 6u^2(k_BT)^2 J_3(q,\omega) \quad ,
\tag{54}
$$

where:

$$
\Sigma_3(q,\omega=0) = 6u^2(k_BT)^2 \int_S \chi_0(k_1)\chi_0(k_2)\chi_0(k_3) \quad ,
\tag{55}
$$

which is clearly a static quantity. We also have:

$$
J_3(q,\omega) = \int_S \chi_0(k_1)\chi_0(k_2)\chi_0(k_3) \frac{1}{-i\omega + \gamma(k_1) + \gamma(k_2) + \gamma(k_3)} \quad .
\tag{56}
$$

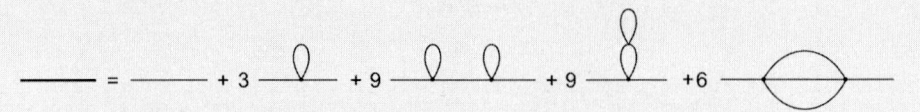

Fig. 10.14 Graphical perturbation theory expansion for the static order parameter correlation function to second order in the quartic coupling.

Thus we have the perturbation theory result, Eq. (35), up to second order in u for the static correlation function. We have the static results for the self-energies:

$$\Sigma_1 = -3u \int \frac{d^d k}{(2\pi)^d} k_B T \chi_0(k) \tag{57}$$

$$\Sigma_2 = 6\Sigma_1 \left(-\frac{u}{2} k_B T\right) \int \frac{d^d k}{(2\pi)^d} \chi_0^2(k) \tag{58}$$

and $\Sigma_3(q, \omega = 0)$ is given by Eq. (55).

This static result can be obtained directly from the function integral over the effective Hamiltonian and is shown graphically in Fig. 10.14. This is an expansion for:

$$C(q) = \langle |\psi(q)|^2 \rangle = k_B T \chi(q) , \tag{59}$$

where the thick line is the full correlation function, thin lines are bare correlation functions, and the four-point vertex in this case (see Problem 10.7) is $-\beta u$.

10.1.6
Temperature Renormalization

The method developed in this chapter will generate all of the terms in the perturbation theory expansion for both the statics and dynamics. One must be careful in interpreting the result of this expansion, however. It is always essential to have some idea of the underlying physics governing the behavior of some quantity before carrying out an expansion. We can demonstrate how problems arise with a simple example in the static limit.

We believe that the static susceptibility at zero wavenumber is given approximately by:

$$\chi(0) = C(T)(T - T_c)^{-\gamma} \tag{60}$$

for $T \geq T_c$, where the constant C is regular as $T \to T_c$. The zeroth-order approximation for $\chi(q)$ is given by the Ornstein–Zernike form:

$$\chi_0(q) = \frac{1}{r_0 + cq^2} , \tag{61}$$

where:

$$\chi_0^{-1}(0) = r_0 = r_1(T - T_c^{(0)}) , \qquad (62)$$

and $\gamma = 1$.

The expansion for $\chi(q)$, including terms up to first order in perturbation theory, is given by:

$$\chi(q) = \chi_0(q) + \chi_0(q)(-3u) \int \frac{d^d k}{(2\pi)^d} k_B T \chi_0(k) \chi_0(q) . \qquad (63)$$

Let us introduce the integral:

$$I_d = \int \frac{d^d k}{(2\pi)^d} \chi_0(k) \qquad (64)$$

and evaluate it first in three dimensions:

$$\begin{aligned} I_3 &= \int \frac{d^3 k}{(2\pi)^3} \frac{1}{r_0 + ck^2} = \frac{4\pi}{8\pi^3} \int_0^\Lambda k^2 dk \frac{1}{r_0 + ck^2} \\ &= \frac{1}{2\pi^2 c} \int_0^\Lambda dk \left[1 - \frac{r_0}{r_0 + ck^2}\right] \\ &= \frac{1}{2\pi^2 c} \left[\Lambda - \frac{r_0}{\sqrt{r_0 c}} \tan^{-1}\left(\sqrt{\frac{c}{r_0}}\Lambda\right)\right] , \end{aligned} \qquad (65)$$

where we have introduced a large wavenumber cutoff, Λ. It is useful to write I_3 in terms of the bare correlation length:

$$\xi_0^2 = \frac{c}{r_0} . \qquad (66)$$

ξ_0 blows up as $r_0 \to 0$. Writing I_3 in terms of ξ_0 we have:

$$I_3 = \frac{1}{2\pi^2 c \xi_0} \left[\Lambda \xi_0 - \tan^{-1}(\Lambda \xi_0)\right] . \qquad (67)$$

Then at zero wavenumber, from Eq. (63), the susceptibility is given by:

$$r^{-1} \equiv \chi(q=0) = r_0^{-1} + r_0^{-1} \frac{(-3uk_B T)}{2\pi^2 c \xi_0} \left[\Lambda \xi_0 - \tan^{-1}(\Lambda \xi_0)\right] r_0^{-1} . \qquad (68)$$

Then as $T \to T_c^{(0)}$, $r_0 \to 0$, $\xi_0 \to \infty$ and our perturbation theory result goes over to:

$$r^{-1} \to -r_0^{-2} \frac{(3uk_B T_c^{(0)} \Lambda)}{2\pi^2 c} . \qquad (69)$$

There are many problems with this result:

- Clearly, in general,
$$C(q) = \langle |\psi(q)|^2 \rangle = k_B T \chi(q) \geq 0 \ , \tag{70}$$
while our result from perturbation theory is negative.

- The *first-order correction* is larger ($r_0^{-2} \gg r_0^{-1}$) than the leading order contributions.

- It appears that $\chi(q=0)$ is blowing up at the *wrong* temperature.

The resolution of this disaster is to realize that while a direct perturbation theory expansion of a *resonant* quantity may be ill advised, the expansion of its inverse may be well behaved [11]. Let us write:

$$\chi^{-1}(q) \equiv \chi_0^{-1}(q) - \Sigma(q) \ , \tag{71}$$

where the *self-energy* $\Sigma(q)$ depends on the nonlinear interaction. We can identify the perturbation theory contributions to $\Sigma(q)$ by assuming $\Sigma(q)$ has a leading term of order u and expanding:

$$\begin{aligned}
\chi(q) &= \frac{1}{\chi_0^{-1}(q) - \Sigma(q)} \\
&= \frac{\chi_0(q)}{1 - \Sigma(q)\chi_0(q)} \\
&= \chi_0(q) + \chi_0(q)\Sigma(q)\chi_0(q) + \chi_0(q)\Sigma(q)\chi_0(q)\Sigma(q)\chi_0(q) + \cdots
\end{aligned} \tag{72}$$

Comparing this result with the expansion in Fig. 10.14, it is easy to see that the self-energy has the expansion shown in Fig. 10.15. The first-order contribution to the self-energy has already been worked out in the case of three dimensions and is given by:

$$\Sigma^{(1)}(q) = \frac{(-3uk_B T)}{2\pi^2 c \xi_0} \left[\Lambda \xi_0 - \tan^{-1}(\Lambda \xi_0) \right] \ . \tag{73}$$

In this case, if we set $q = 0$ in the inverse static susceptibility and use Eq. (71) with $\Sigma = \Sigma^{(1)}$, we have:

$$\chi^{(1)}(q=0) = r = r_0 + \frac{(3uk_B T)}{2\pi^2 c \xi_0} \left[\Lambda \xi_0 - \tan^{-1}(\Lambda \xi_0) \right] \ . \tag{74}$$

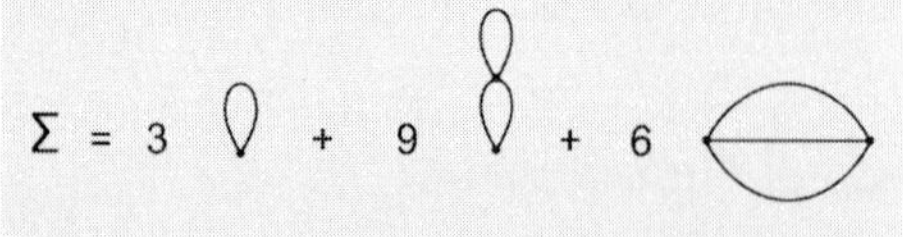

Fig. 10.15 Static self-energy graphs associated with the expansion shown in Fig. 10.14.

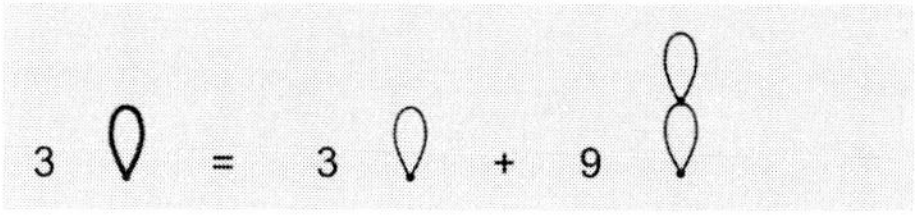

$$3\,\bigcirc = 3\,\bigcirc + 9\,\bigcirc\!\!\!\!\bigcirc$$

Fig. 10.16 Resummation of graphs contributing to the full (*thick line*) static correlation function in the self-energy to first order in u.

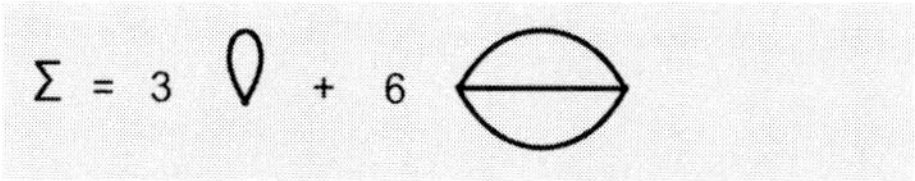

$$\Sigma = 3\,\bigcirc + 6\,\bigcirc\!\!\!\bigcirc$$

Fig. 10.17 Static self-energy expansion in terms of the full static correlation functions.

Suppose $\Lambda \xi_0 \gg 1$, then we have:

$$r = r_1(T - T_c^{(0)}) + \frac{3uk_B T \Lambda}{2\pi^2 c} \ . \tag{75}$$

If r vanishes as $T \to T_c$ this reduces to:

$$\left(r_1 + \frac{3uk_B \Lambda}{2\pi^2 c}\right) T_c = r_1 T_c^{(0)}$$

$$T_c = \frac{r_1 T_c^{(0)}}{r_1 + \frac{3uk_B \Lambda}{2\pi^2 c}} \ . \tag{76}$$

The nonlinear interaction generally acts to reduce the transition temperature from its zeroth-order value. Note that there is still a bit of an inconsistency in our development. We assumed on the right-hand side of Eq. (74) that we could take $\Lambda \xi_0 \gg 1$. However ξ_0 does not become infinite as $T \to T_c$. Self-consistently, it makes sense that the correlation functions appearing in the expansion are the full correlation functions. Thus one would like to rearrange the expansion such that it is χ and not χ_0 that appears in Eq. (20). This involves an additional resummation of graphs. Thus it is shown in Problem 10.8 that the sum of first two graphs on the right-hand side in Fig. 10.15 can be combined to give the single graph as shown in Fig. 10.16. Indeed at second order in perturbation theory we have the expansion for the self-energy given by Fig. 10.17 where all of the lines are full correlation function lines. It is shown how to justify this expansion to all orders in FTMCM.

10.1.7
Self-Consistent Hartree Approximation

After these sets of resummations we have, keeping the first-order contribution to the self-energy, the self-consistent Hartree approximation [12] given by:

$$\chi^{-1}(q) = r_0 + cq^2 + 3uk_BT \int \frac{d^dk}{(2\pi)^d} \chi(k) \ . \tag{77}$$

We want to solve self-consistently for $\chi(q)$. Since the first-order self-energy, the Hartree contribution, is independent of wavenumber, it is pretty clear that we have a solution of the form:

$$\chi^{-1}(q) = r + cq^2 \ , \tag{78}$$

where:

$$r = r_0 + 3uk_BT \int \frac{d^dk}{(2\pi)^d} \frac{1}{r + ck^2} \ , \tag{79}$$

which is a self-consistent equation for r.

The first step in determining the behavior of r as a function of $T - T_c$ is to self-consistently determine T_c as that temperature where r vanishes. At the physical T_c, $r = 0$ and Eq. (79) becomes:

$$0 = r_1(T_c - T_c^{(0)}) + 3uk_BT_c \int \frac{d^dk}{(2\pi)^d} \frac{1}{ck^2} \ . \tag{80}$$

We can carry out the d-dimensional integral:

$$\int \frac{d^dk}{(2\pi)^d} \frac{1}{ck^2} = \frac{\tilde{K}_d}{c} \frac{\Lambda^{d-2}}{d-2} \tag{81}$$

for $d > 2$ where $\tilde{K}_d$ is given by $\tilde{K}_d^{-1} = 2^{d-1}\pi^{d/2}\Gamma(d/2)$. We then obtain the generalization of Eq. (76) to d dimensions:

$$T_c = \frac{r_1 T_c^{(0)}}{\left(r_1 + \frac{3uk_B\tilde{K}_d\Lambda^{d-2}}{(d-2)c}\right)} \ . \tag{82}$$

This result reduces to Eq. (76) for $d = 3$. We now subtract Eq. (80) from Eq. (79) to obtain:

$$r = r_1(T - T_c) + 3uk_BT \int \frac{d^dk}{(2\pi)^d} \frac{1}{r + ck^2} - 3uk_BT_c \int \frac{d^dk}{(2\pi)^d} \frac{1}{ck^2}$$

$$= r_1(T - T_c) + 3uk_B(T - T_c + T_c) \int \frac{d^dk}{(2\pi)^d} \frac{1}{r + ck^2} - 3uk_BT_c \int \frac{d^dk}{(2\pi)^d} \frac{1}{ck^2}$$

$$= r_2(T - T_c) - 3uk_B T_c W_d(r) r \; , \tag{83}$$

where:

$$r_2(T) = r_1 + 3uk_B \int \frac{d^d k}{(2\pi)^d} \frac{1}{r + ck^2} \tag{84}$$

and:

$$-W_d(r)r = \int \frac{d^d k}{(2\pi)^d} \left[\frac{1}{r + ck^2} - \frac{1}{ck^2} \right] . \tag{85}$$

$W_d(r)$ is given by:

$$W_d(r) = \int \frac{d^d k}{(2\pi)^d} \frac{1}{ck^2} \frac{1}{r + ck^2} . \tag{86}$$

Then Eq. (83) is of the form:

$$r \left[1 + 3uk_B T_c W_d(r) \right] = r_2(T - T_c) . \tag{87}$$

It is left to Problem 10.9 to show that in the limit of large $\xi^2 = c/r$, to leading order in this quantity, we have:

$$W_d = \frac{\tilde{K}_d}{c^2} \begin{cases} \frac{\Lambda^{d-4}}{d-4} & d > 4 \\ \ln(\Lambda \xi) & d = 4 \\ \frac{\pi \xi^{4-d}}{2\cos[(d-3)\pi/2]} & d < 4 \end{cases} \tag{88}$$

Thus we have for $d > 4$ and small u, W_d is a small perturbation that corrects the value to the transition temperature and amplitudes. The zeroth-order result is essentially correct. For $d = 4$ W_4 is growing logarithmically as $r \to 0$ and W can no longer be treated as a small perturbation. For smaller dimensionality, W diverges faster and the perturbation theory breaks down badly. In order to handle the regime $d \leq 4$, we need an even more powerful technique beyond perturbation theory. We need the renormalization group method, which will be discussed later in this chapter.

10.1.8
Dynamic Renormalization

The same ideas used to reorganize the static perturbation theory can be generalized to the dynamic case where we can write a Dyson's equation of the form:

$$G^{-1}(\mathbf{q}, \omega) = G_0^{-1}(\mathbf{q}, \omega) - \Sigma(\mathbf{q}, \omega) \; , \tag{89}$$

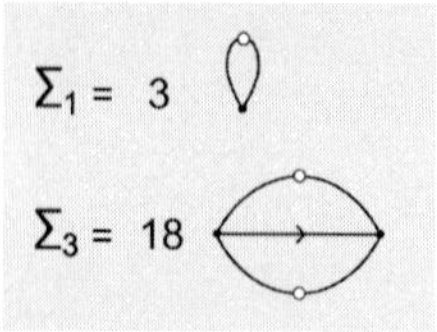

$\Sigma_1 = 3$

$\Sigma_3 = 18$

Fig. 10.18 First-order, Σ_1, and second-order, Σ_3, self-energy contributions.

where the self-energy can be expressed, at this order, in terms of the full correlation functions and response functions as shown in the graphs in Fig. 10.18. These graphs, at the bare level, were treated previously in the static limit. In the renormalized case we have:

$$\Sigma_1(\mathbf{q}, \omega) = 3(-u) \int \frac{d^d q'}{(2\pi)^d} \frac{d\omega'}{2\pi} C(\mathbf{q}', \omega') \tag{90}$$

and:

$$\Sigma_3(\mathbf{q}, \omega) = 18(-u)^2 \int \frac{d^d q'}{(2\pi)^d} \frac{d\omega'}{2\pi} \frac{d^d q''}{(2\pi)^d} \frac{d\omega''}{2\pi} G(\mathbf{q} - \mathbf{q}' - \mathbf{q}'', \omega - \omega' - \omega'')$$
$$\times C(\mathbf{q}', \omega') C(\mathbf{q}'', \omega'') \ . \tag{91}$$

Note that the contribution at $\mathcal{O}(u)$ is independent of q and ω. In particular, this term only serves to change the static $\mathbf{q} = 0$ susceptibility. Suppose we keep only the $\mathcal{O}(u)$ term, then:

$$G^{-1}(\mathbf{q}, \omega) = G_0^{-1}(\mathbf{q}, \omega) - \Sigma_1$$

$$= \frac{-i\omega}{\Gamma_0(\mathbf{q})} + \chi^{-1}(\mathbf{q}) \ , \tag{92}$$

where:

$$\chi^{-1}(\mathbf{q}) = \chi_0^{-1}(\mathbf{q}) - \Sigma_1 \ . \tag{93}$$

Equation (93) can be rewritten as:

$$\chi^{-1}(\mathbf{q}) = r(T) + cq^2 \ , \tag{94}$$

where:

$$r(T) = r_0 - \Sigma_1 \tag{95}$$

was introduced above. The temperature where $r(T) = 0$ is the new transition temperature, correct up to terms of $\mathcal{O}(u^2)$. In this case, the response function can be written in the dynamical scaling form:

$$G^{-1}(\mathbf{q}, \omega) = \chi^{-1}(\mathbf{q}) \left[1 - \frac{i\omega}{\Gamma_0(\mathbf{q}) \chi^{-1}(\mathbf{q})} \right]$$

$$= \chi^{-1}(\mathbf{q})[1 - i\Omega] \;, \qquad (96)$$

where the dimensionless frequency is given by:

$$\Omega = \frac{\omega}{\omega_s(\mathbf{q})} \qquad (97)$$

and the characteristic frequency is given by:

$$\omega_s(\mathbf{q}) = \Gamma_0(\mathbf{q})\chi^{-1}(\mathbf{q}) \;. \qquad (98)$$

The correlation function is given in this case by using the FDT, Eq. (25), and Eq. (96) to obtain:

$$\begin{aligned} C(q,\omega) &= 2\frac{k_B T}{\omega}\text{Im } G(q,\omega) = 2\frac{k_B T}{\omega}\text{Im}\left(\frac{\chi(q)}{1-i\Omega}\right) \\ &= 2\frac{k_B T}{\omega}\chi(q)\text{Im}\left(\frac{1+i\Omega}{1+\Omega^2}\right) = 2\frac{k_B T}{\omega}\chi(q)\frac{\Omega}{1+\Omega^2} \\ &= 2k_B T \frac{\chi(q)}{\omega_s(q)}F(\Omega) \;, \end{aligned} \qquad (99)$$

where the scaling function is given by:

$$F(\Omega) = \frac{1}{1+\Omega^2} \;. \qquad (100)$$

The dynamic scaling index z is determined by writing the characteristic frequency in the form:

$$\omega_s(q) = \Gamma_0(q)\chi^{-1}(q) = q^z f(q\xi) \;.$$

We easily find:

$$\omega_s(q) = \Gamma_0(q)c(q^2 + \xi^2) \qquad (101)$$

and $z = 2$ for the NCOP case, $z = 4$ in the COP case and the scaling function is given by:

$$f(x) = \Gamma_0(q=1)c(1+x^2) \;, \qquad (102)$$

as for the conventional theory.

Let us go back to the more general expression for $G^{-1}(q,\omega)$:

$$G^{-1}(q,\omega) = \frac{-i\omega}{\Gamma_0(q)} + \chi_0^{-1}(q) - \Sigma(q,\omega) \;. \qquad (103)$$

It is convenient to write this in the form:

$$G^{-1}(q,\omega) = \frac{-i\omega}{\Gamma(q,\omega)} + \chi^{-1}(q) \;, \qquad (104)$$

where $\chi(q)$ is the full static susceptibility and $\Gamma(q,\omega)$ is a frequency- and wavenumber-dependent kinetic coefficient. Since, in general,

$$\chi^{-1}(q) = \chi_0^{-1}(q) - \Sigma(q,0) \tag{105}$$

we easily find, after some algebra, that:

$$\Gamma(q,\omega) = \Gamma_0(q)\left[1 + \Gamma_0(q)\left(\Sigma(q,\omega) - \Sigma(q,0)\right)/i\omega\right]^{-1}. \tag{106}$$

In the TDGL model only Σ_3 contributes to $\Gamma(q,\omega)$ to $\mathcal{O}(u^3)$ and we find:

$$\Gamma(q,\omega) = \Gamma_0(q)\left[1 - \frac{\Gamma_0(q)}{i\omega}[\Sigma_3(q,\omega) - \Sigma_3(q,0)] + \mathcal{O}(u^3)\right]. \tag{107}$$

We have from our previous work, Eq. (54), that if we replace the zeroth-order propagators and response functions with those with temperature renormalization $r_0 \to r$, then:

$$\Sigma_3(q,\omega) = \Sigma_3(q,\omega=0) + i\omega 6u^2(k_B T)^2 J_3(q,\omega), \tag{108}$$

where:

$$\Sigma_3(q,\omega=0) = 6u^2(k_B T)^2 \int_S \chi(k_1)\chi(k_2)\chi(k_3), \tag{109}$$

which is clearly a static quantity. We also have Eq. (56):

$$J_3(q,\omega) = \int_S \chi(k_1)\chi(k_2)\chi(k_3)\frac{1}{-i\omega + \gamma(k_1) + \gamma(k_2) + \gamma(k_3)}, \tag{110}$$

where $\chi^{-1}(q) = cq^2 + r$.

Let us consider the case of a NCOP, $\Gamma_0(q) = \Gamma_0$. In this case we can look first at the case of high dimensionality. If we set $r = q = 0$ we obtain:

$$J_3(0,\omega) = \int_S \frac{1}{ck_1^2}\frac{1}{ck_2^2}\frac{1}{ck_3^2}\frac{1}{[-i\omega + \Gamma_0 c(k_1^2 + k_2^2 + k_3^2)]}. \tag{111}$$

If we use simple power counting, we see for $d > 4$ that $J_3(0,0)$ is just some well-defined number that depends on the large wavenumber cutoff. For $d < 4$ we find (see Problem 10.10) that $J_3(0,\omega)$ blows up algebraically in the limit as ω goes to zero. The dividing line is for $d = 4$. As a practical example of perturbation theory calculations, we carry out the determination of $J_3(0,\omega)$ for $d = 4$ in some detail.

In evaluating $\Sigma_3(q,\omega)$ it is easier to work from Eq. (49):

$$\Sigma_3(q,\omega) = 18u^2(k_BT)^2 \int_S \chi_0(k_1)\chi_0(k_2)$$
$$\times \frac{\Gamma_0}{-i\omega + \gamma(k_1) + \gamma(k_2) + \gamma(k_3)} \quad . \tag{112}$$

The interesting case here is for $q = r = 0$ where we focus on the frequency dependence and evaluate:

$$\Sigma_3(0,\omega) = 18 \frac{u^2(k_BT_c)^2}{c^3} S(\Omega) \quad , \tag{113}$$

where:

$$S(\Omega) = \int \frac{d^4k_1}{(2\pi)^4} \int \frac{d^4k_2}{(2\pi)^4} \frac{1}{k_1^2} \frac{1}{k_2^2} \frac{1}{(-i\Omega + k_1^2 + k_2^2 + (\mathbf{k}_1+\mathbf{k}_2)^2)} \quad , \tag{114}$$

where $\Omega = \omega/(c\Gamma_0)$. Wilson [13] has discussed the extension of integrals of this type to d dimensions. In four dimensions we have:

$$S(\Omega) = \int \frac{d^4k_1}{(2\pi)^4} \frac{\tilde{K}_3}{2\pi} \int_0^\Lambda k_2^3 dk_2 \int_0^\pi d\theta \sin^2\theta \frac{1}{k_1^2} \frac{1}{k_2^2} \frac{1}{-i\Omega + 2k_1^2 + 2k_2^2 + 2k_1k_2\cos\theta}$$
$$= \tilde{K}_4 \frac{\tilde{K}_3}{2\pi} \int_0^\Lambda \frac{k_1^3}{k_1^2} dk_1 \int_0^\Lambda \frac{k_2^3}{k_2^2} dk_2 J(k_1,k_2,\Omega) \quad , \tag{115}$$

where:

$$J = \int_0^\pi d\theta \frac{\sin^2\theta}{-i\Omega + 2k_1^2 + 2k_2^2 + 2k_1k_2\cos\theta} \quad . \tag{116}$$

If we let $a = -i\Omega + 2k_1^2 + 2k_2^2$ and $b = 2k_1k_2$ then [14]:

$$J = \int_0^\pi d\theta \frac{\sin^2\theta}{a + b\cos\theta} = \frac{\pi}{a + \sqrt{a^2 - b^2}}$$
$$= \pi \frac{a - \sqrt{a^2 - b^2}}{a^2 - a^2 + b^2} = \frac{\pi}{b^2}\left(a - \sqrt{a^2 - b^2}\right) \quad . \tag{117}$$

We have then:

$$S(\Omega) = \tilde{K}_4 \frac{\tilde{K}_3}{2\pi} \int_0^\Lambda k_1 dk_1 \int_0^\Lambda k_2 dk_2 \frac{\pi}{4k_1^2 k_2^2}\left(-i\Omega + 2k_1^2 + 2k_2^2 - g(\Omega)\right)$$
$$= \frac{\tilde{K}_3 \tilde{K}_4}{8} \int_0^\Lambda \frac{dk_1}{k_1} \int_0^\Lambda \frac{dk_2}{k_2}\left(-i\Omega + 2k_1^2 + 2k_2^2 - g(\Omega)\right) \quad , \tag{118}$$

where we have defined:

$$g(\Omega) = \sqrt{(-i\Omega + 2k_1^2 + 2k_2^2)^2 - 4k_1^2 k_2^2} \quad . \tag{119}$$

Eventually we are interested in subtracting off the $\omega = 0$ contribution and so we consider:

$$\Delta S(\Omega) = S(\Omega) - S(0)$$
$$= \frac{\tilde{K}_3 \tilde{K}_4}{8} \int_0^\Lambda \frac{dk_1}{k_1} \int_0^\Lambda \frac{dk_2}{k_2} [-i\Omega - g(\Omega) + g(0)] \quad . \quad (120)$$

The next step is to let $k_1 = \sqrt{\Omega} y_1$ and $k_2 = \sqrt{\Omega} y_2$ to obtain:

$$\Delta S(\Omega) = \frac{\tilde{K}_3 \tilde{K}_4}{8} \Omega \int_0^{\Lambda/\sqrt{\Omega}} \frac{dy_1}{y_1} \int_0^{\Lambda/\sqrt{\Omega}} \frac{dy_2}{y_2} (-i - g_1 + g_2) \quad , \quad (121)$$

where:

$$g_1 = \sqrt{(-i + 2y_1^2 + 2y_2^2)^2 - 4y_1^2 y_2^2} \quad . \quad (122)$$

and:

$$g_2 = \sqrt{(2y_1^2 + 2y_2^2)^2 - 4y_1^2 y_2^2} \quad . \quad (123)$$

The Ω dependence is confined to the overall factor of Ω and to the large y cutoff. It is left as a problem (Problem 10.11) to show that the small y_1 and y_2 behavior is well defined and the leading behavior comes from large y_1 and y_2. We can then write to leading order:

$$\Delta S(\Omega) = \frac{\tilde{K}_3 \tilde{K}_4}{8} \Omega \int_1^{\Lambda/\sqrt{\Omega}} \frac{dy_1}{y_1} \int_1^{\Lambda/\sqrt{\Omega}} \frac{dy_2}{y_2} F(y_1, y_2) \quad , \quad (124)$$

where:

$$F(y_1, y_2) = -i - g_1 + g_2 \quad . \quad (125)$$

Since any anomalous behavior as a function of frequency is for large y_1 and y_2, we can write:

$$F(y_1, y_2) = F_L(y_1, y_2) + F(y_1, y_2) - F_L(y_1, y_2) \quad , \quad (126)$$

where F_L is the expression for F for large y_1 and y_2 given by:

$$F_L(y_1, y_2) = -i \left[1 - \frac{(y_1^2 + y_2^2)}{[(y_1^2 + y_2^2)^2 - y_1^2 y_2^2]^{1/2}} \right] \quad . \quad (127)$$

We then write:

$$\Delta S(\Omega) = \frac{\tilde{K}_3 \tilde{K}_4}{8} \Omega \int_1^{\Lambda/\sqrt{\Omega}} \frac{dy_1}{y_1} \int_1^{\Lambda/\sqrt{\Omega}} \frac{dy_2}{y_2} F_L(y_1, y_2) \quad , \quad (128)$$

and the remaining integral over $F - F_L$ is regular as $\Omega \to 0$. We have then at leading order:

$$\Delta S(\Omega) = \frac{\tilde{K}_3 \tilde{K}_4}{8}(-i\Omega) \int_1^{\Lambda/\sqrt{\Omega}} \frac{dy_1}{y_1} \int_1^{\Lambda/\sqrt{\Omega}} \frac{dy_2}{y_2}$$
$$\times \left[1 - \frac{(y_1^2 + y_2^2)}{[(y_1^2 + y_2^2)^2 - y_1^2 y_2^2]^{1/2}}\right], \quad (129)$$

plus terms that are linear in Ω as $\Omega \to 0$. Next let $y_2 = u y_1$ in Eq. (129) to obtain:

$$\Delta S(\Omega) = \frac{\tilde{K}_3 \tilde{K}_4}{8}(-i\Omega) K(v), \quad (130)$$

where:

$$K(v) = \int_1^{1/v} \frac{dy_1}{y_1} \int_{1/y_1}^{1/v y_1} du\, f(u), \quad (131)$$

$$v = \frac{\sqrt{\Omega}}{\Lambda}, \quad (132)$$

and:

$$f(u) = \frac{1}{u}\left[1 - \frac{(1+u^2)}{(1+u^2+u^4)^{1/2}}\right]. \quad (133)$$

In going further it is useful to note the property:

$$f(1/u) = u^2 f(u). \quad (134)$$

One integration in Eq. (131) can be carried out by writing:

$$K(v) = \int_1^{1/v} \frac{dy_1}{y_1} \frac{d}{dy_1}(\ln(y_1)) \int_{1/y_1}^{1/v y_1} du\, f(u) \quad (135)$$

and doing an integration by parts. It is shown in Problem 10.12 that:

$$K(v) = \ln(v^{-1}) \int_v^1 du\, f(u) + \int_1^{1/v} du\, \ln(u) [v f(vu) - f(u)]. \quad (136)$$

Further manipulation (see Problem 10.12) gives:

$$K(v) = 2 \ln(v^{-1}) \int_0^1 du\, f(u) + \int_0^1 2du\, \ln(u) f(u) \quad (137)$$

plus terms that vanish as $v \to 0$. Only the first term proportional to $\ln(v^{-1})$ contributes to the anomalous behavior of $K(v)$. It is left to Problem 10.12 to show that:

$$\int_0^1 du\, f(u) = \frac{1}{2} \ln\left(\frac{3}{4}\right) \quad (138)$$

and to leading order:
$$K(v) = \ln(v^{-1}) \ln\left(\frac{3}{4}\right) \tag{139}$$

and:
$$\Delta S(\Omega) = \frac{\tilde{K}_3 \tilde{K}_4}{8}(-i\Omega)\ln(v^{-1})\ln\left(\frac{3}{4}\right) . \tag{140}$$

Since $\tilde{K}_3 = 1/2\pi^2$ and $\tilde{K}_4 = 1/8\pi^2$ [15] we have:
$$\Delta S(\Omega) = \frac{(-i\Omega)}{128\pi^4}\ln(v^{-1})\ln\left(\frac{3}{4}\right) . \tag{141}$$

Putting this back into the appropriate self-energy, Eq. (113) and remembering Eq. (120), gives:
$$\Delta\Sigma_3(0,\omega) = 18c\left(\frac{uk_B T_c}{c^2}\right)^2 \Delta S(\Omega) \tag{142}$$

and the correction to the kinetic coefficient, Eq. (106), can be written in the form:
$$\begin{aligned}
\Gamma^{-1}(0,\omega) &= \Gamma_0^{-1}\left[1 + \frac{\Gamma_0 \Delta\Sigma_3(0,\omega)}{i\omega}\right] = \Gamma_0^{-1}\left[1 + \frac{\Delta\Sigma_3(0,\omega)}{ic\Omega}\right] \\
&= \Gamma_0^{-1}\left[1 + 18\left(\frac{uk_B T_c}{c^2}\right)^2 \frac{\Delta S(\Omega)}{i\Omega}\right] \\
&= \Gamma_0^{-1}\left[1 + 18\left(\frac{uk_B T_c}{c^2}\right)^2 \frac{(-1)}{128\pi^4}\ln(v^{-1})\ln\left(\frac{3}{4}\right)\right] \\
&= \Gamma_0^{-1}\left[1 - u_0^2 b_0 \ln\left(\frac{\Omega}{\Lambda^2}\right)\right] ,
\end{aligned} \tag{143}$$

where:
$$u_0 = \frac{uk_B T_c}{4c^2} \tag{144}$$

and:
$$b_0 = \frac{9}{8\pi^4}\ln\left(\frac{4}{3}\right) . \tag{145}$$

If we compare with Eq. (10) in Ref. [7] we find agreement. Notice that our u_0, defined by Eq. (144), corresponds to their u_0.

Again, as in the static case, we have that $d = 4$ serves as a crossover dimension between regimes where the conventional theory is appropriate ($d > 4$) and where ($d < 4$) there are strong nonlinear corrections.

As it stands we can not complete the analysis for $d = 3$ without further input. The renormalized kinetic coefficient diverges algebraically for $d < 4$ and logarithmically in $d = 4$. To obtain a useful estimate in three dimensions we need to appeal to renormalization group methods. These methods are discussed in Section 10.3 (this chapter) and we return to this problem there.

10.2 Perturbation Theory for the Isotropic Ferromagnet

10.2.1 Equation of Motion

We next want to develop perturbation theory for a slightly more involved system: the isotropic ferromagnet [8]. This case differs from the TDGL case, since one has reversible terms, and, as we shall see, nontrivial corrections to the conventional theory enter at lowest nontrivial order in the expansion. As discussed in Chapter 9, for the isotropic ferromagnet the order parameter M_α is a vector satisfying the Langevin equation:

$$\frac{\partial M_\alpha(\mathbf{x}_1, t_1)}{\partial t_1} = \lambda \sum_{\beta\gamma} \varepsilon_{\alpha\beta\gamma} M_\beta(\mathbf{x}_1, t_1) \frac{\delta \mathcal{H}_E[M]}{\delta M_\gamma(\mathbf{x}_1, t_1)} + D_0 \nabla_1^2 \frac{\delta \mathcal{H}_E[M]}{\delta M_\alpha(\mathbf{x}_1, t_1)}$$
$$+ f_\alpha(\mathbf{x}_1, t_1) \, , \qquad (146)$$

where $\mathcal{H}_E[M]$ is the LGW effective Hamiltonian given by:

$$\mathcal{H}_E[M] = \int d^d x_1 \int d^d x_2 \sum_\alpha \frac{1}{2} M_\alpha(\mathbf{x}_1, t_1) \chi_0^{-1}(\mathbf{x}_1 - \mathbf{x}_2) M_\alpha(\mathbf{x}_2, t_1)$$
$$+ \int d^d x_1 \left[\frac{u}{4} M^4(\mathbf{x}_1, t_1) - \sum_\alpha h_\alpha(\mathbf{x}_1, t_1) M_\alpha(\mathbf{x}_1, t_1) \right] , \qquad (147)$$

where $M^2 = \sum_\alpha M_\alpha^2$, $\mathbf{h}$ is an external magnetic field and the bare susceptibility is assumed to be of the standard Ornstein–Zernike form:

$$\chi_0^{-1}(\mathbf{x}_1 - \mathbf{x}_2) = (r_0 - c\nabla_1^2)\delta(\mathbf{x}_1 - \mathbf{x}_2) \ . \qquad (148)$$

The order parameter is conserved in this case and the noise is assumed to be Gaussian with the variance:

$$\langle f_\alpha(\mathbf{x}_1, t_1) f_\beta(\mathbf{x}_2, t_2) \rangle = 2D_0 k_B T \vec{\nabla}_1 \cdot \vec{\nabla}_2 \delta_{\alpha\beta} \delta(\mathbf{x}_1 - \mathbf{x}_2) \delta(t_1 - t_2) \ . \qquad (149)$$

Finally λ is a nonlinear coupling constant and D_0 a bare diffusion coefficient.

Putting the effective Hamiltonian into the Langevin equation gives the equation of motion:

$$\frac{\partial M_\alpha}{\partial t_1} = \lambda \sum_{\beta\gamma} \varepsilon_{\alpha\beta\gamma} M_\beta \left[(r_0 - c\nabla_1^2) M_\gamma + uM^2 M_\gamma - h_\gamma \right]$$

$$+D_0\nabla_1^2\left[(r_0 - c\nabla_1^2)M_\alpha + uM^2 M_\alpha - h_\alpha\right] + f_\alpha \quad (150)$$

or, in vector notation,

$$\frac{\partial \mathbf{M}}{\partial t_1} = -\lambda \mathbf{M} \times \left(c\nabla_1^2 \mathbf{M} + \mathbf{h}\right)$$
$$+ D_0\nabla_1^2\left[(r_0 - c\nabla_1^2)\mathbf{M} + uM^2\mathbf{M} - \mathbf{h}\right] + \mathbf{f} \quad . \quad (151)$$

There are several comments to be made. First, there are two nonlinear terms: a quadratic term proportional to λ and a cubic term proportional to u that appeared in the TDGL model. The external field also appears in two places and we must be careful to properly treat the $-\lambda \mathbf{M} \times \mathbf{h}$ term. This term generates nonlinear contributions in an expansion where we treat both λ and $\mathbf{h}$ as small.

10.2.2
Graphical Expansion

If we set λ and u equal to zero we obtain the same zeroth-order theory as for the TDGL model with a conserved order parameter. The zeroth-order response function is given again by:

$$G_0^{-1}(q,\omega) = -\frac{i\omega}{\Gamma_0(q)} + \chi_0^{-1}(q) \quad , \quad (152)$$

where $\Gamma_0(q) = D_0 q^2$. We know from our previous work that the conventional theory in this case corresponds to a dynamic index $z = 4$.

Let us move on to treat the contributions of the nonlinear terms. The first step is to set up the appropriate notation. This is quite important if one is to develop an efficient diagrammatic method. Let us look first at the contribution in the Langevin equation due to spin precession. We have the vertex:

$$V(1) = -\lambda \sum_{\beta\gamma} \varepsilon_{\alpha_1\beta\gamma} M_\beta(\mathbf{x}_1, t_1) c \nabla_1^2 M_\gamma(\mathbf{x}_1, t_1) \quad . \quad (153)$$

For reasons that become clear as we go along, we want to introduce the compact notation where a single index includes all of the other indices labeling a field. We write:

$$M(1) = M_{\alpha_1}(\mathbf{x}_1, t_1) \quad . \quad (154)$$

With this in mind we rewrite $V(1)$ in the form:

$$V(1) = -\lambda c \sum_{\alpha_2 \alpha_3} \varepsilon_{\alpha_1\alpha_2\alpha_3} M_{\alpha_2}(\mathbf{x}_1, t_1) \nabla_1^2 M_{\alpha_3}(\mathbf{x}_1, t_1) \quad . \quad (155)$$

Going further, we can insert the identity:

$$\int d^d x_2 dt_2 \delta(\mathbf{x}_2 - \mathbf{x}_1)\delta(t_2 - t_1) = 1 \quad , \quad (156)$$

twice to obtain:

$$V(1) = -\lambda c \sum_{\alpha_2 \alpha_3} \varepsilon_{\alpha_1 \alpha_2 \alpha_3} \int d^d x_2 dt_2 \delta(\mathbf{x}_2 - \mathbf{x}_1) \delta(t_2 - t_1)$$
$$\times \int d^d x_3 dt_3 \delta(\mathbf{x}_3 - \mathbf{x}_1) \delta(t_3 - t_1) M_{\alpha_2}(\mathbf{x}_1, t_1) \nabla_1^2 M_{\alpha_3}(\mathbf{x}_1, t_1)$$
$$= \sum_{\alpha_2 \alpha_3} \int d^d x_2 dt_2 \int d^d x_3 dt_3 V_1(1; 23) M(2) M(3) \tag{157}$$
$$\equiv V_1(1; \bar{2}\bar{3}) M(\bar{2}) M(\bar{3}) \ , \tag{158}$$

where:

$$V_1(1; 23) = -\lambda c\, \varepsilon_{\alpha_1 \alpha_2 \alpha_3} \delta(\mathbf{x}_2 - \mathbf{x}_1)\delta(t_2 - t_1) \nabla_1^2 \delta(\mathbf{x}_3 - \mathbf{x}_1)\delta(t_3 - t_1) \ . \tag{159}$$

In this form we see that we can symmetrize the expression with respect to the summed indices:

$$V(1) = \tilde{V}(1; \bar{2}\bar{3}) M(\bar{2}) M(\bar{3}) \ , \tag{160}$$

where:

$$\tilde{V}(1; \bar{2}\bar{3}) = \frac{1}{2}\left[V_1(1; \bar{2}\bar{3}) + V_1(1; \bar{3}\bar{2})\right]$$
$$= -\frac{\lambda c}{2} \varepsilon_{\alpha_1 \alpha_2 \alpha_3} \delta(t_2 - t_1)\delta(t_3 - t_1)$$
$$\times \left[\delta(\mathbf{x}_2 - \mathbf{x}_1)\nabla_1^2 \delta(\mathbf{x}_3 - \mathbf{x}_1) - \delta(\mathbf{x}_3 - \mathbf{x}_1)\nabla_1^2 \delta(\mathbf{x}_2 - \mathbf{x}_1)\right]$$
$$= \tilde{V}(1; \bar{3}\bar{2}) \ . \tag{161}$$

It is left as an exercise (Problem 10.14), to show that the spatial Fourier transform of this vertex is given by:

$$\tilde{V}_{\alpha_1 \alpha_2 \alpha_3}(\mathbf{k}_1; \mathbf{k}_2, \mathbf{k}_3, t_1, t_2, t_3) = \tilde{V}(\mathbf{k}_1; \mathbf{k}_2, \mathbf{k}_3)\varepsilon_{\alpha_1 \alpha_2 \alpha_3}\delta(t_2 - t_1)\delta(t_3 - t_1) \tag{162}$$

with:

$$\tilde{V}(\mathbf{k}_1; \mathbf{k}_2, \mathbf{k}_3) = -\frac{\lambda c}{2}\left[k_2^2 - k_3^2\right](2\pi)^d \delta(\mathbf{k}_1 + \mathbf{k}_2 + \mathbf{k}_3) \ . \tag{163}$$

We also have the vertex due to the external field:

$$V^h(1) = -\lambda \sum_{jk} \varepsilon_{\alpha_1 jk} M_j(\mathbf{x}_1, t_1) h_k(\mathbf{x}_1, t_1) \ . \tag{164}$$

This can be clearly written in the form:

$$V^h(1) = \tilde{V}^h(1\bar{2}\bar{3}) M(\bar{2}) h(\bar{3}) \ , \tag{165}$$

where:
$$\tilde{V}^h(1;23) = -\lambda\, \varepsilon_{\alpha_1\alpha_2\alpha_3} \delta(\mathbf{x}_2 - \mathbf{x}_1)\delta(t_2 - t_1)\delta(\mathbf{x}_3 - \mathbf{x}_1)\delta(t_3 - t_1) \ . \tag{166}$$

The Fourier transform of this vertex is given by:
$$\begin{aligned}\tilde{V}^h_{\alpha_1\alpha_2\alpha_3}&(\mathbf{k}_1;\mathbf{k}_2,\mathbf{k}_3,t_1,t_2,t_3) \\ &= \tilde{V}^h(\mathbf{k}_1;\mathbf{k}_2,\mathbf{k}_3)\varepsilon_{\alpha_1\alpha_2\alpha_3}\delta(t_2 - t_1)\delta(t_3 - t_1) \ ,\end{aligned} \tag{167}$$

where:
$$\tilde{V}^h(\mathbf{k}_1;\mathbf{k}_2,\mathbf{k}_3) = -\lambda(2\pi)^d \delta(\mathbf{k}_1 + \mathbf{k}_2 + \mathbf{k}_3) \ . \tag{168}$$

We can, of course, include the four-point coupling in the analysis, but for reasons that will become clear, we treat the problem $u = 0$ and focus on the coupling λ. The Langevin equation can then be written in the form:
$$\begin{aligned}\frac{\partial M(1)}{\partial t_1} &= \Gamma_0(1)\chi_0^{-1}(1\bar{2})M(\bar{2}) + \Gamma_0(1)h(1) + f(1) \\ &\quad + \tilde{V}(1;\bar{2}\bar{3})M(\bar{2})M(\bar{3}) + \tilde{V}^h(1;\bar{2}\bar{3})M(\bar{2})h(\bar{3}) \ .\end{aligned} \tag{169}$$

If we multiply by $\Gamma_0^{-1}(1)$ then we can write:
$$\begin{aligned}\Gamma_0^{-1}(1)\frac{\partial M(1)}{\partial t_1} - \chi_0^{-1}(1\bar{2})M(\bar{2}) &= h(1) + \tilde{f}(1) \\ &\quad + V(1;\bar{2}\bar{3})M(\bar{2})M(\bar{3}) + V^h(1;\bar{2}\bar{3})M(\bar{2})h(\bar{3}) \ ,\end{aligned} \tag{170}$$

where:
$$\tilde{f}(1) = \Gamma_0^{-1}(1)f(1) \tag{171}$$

$$V(1;\bar{2}\bar{3}) = \Gamma_0^{-1}(1)\tilde{V}(1;\bar{2}\bar{3}) \tag{172}$$

$$V^h(1;\bar{2}\bar{3}) = \Gamma_0^{-1}(1)\tilde{V}^h(1;\bar{2}\bar{3}) \ . \tag{173}$$

Clearly, a generalization of the zeroth-order response function that carries vector indices is given by:
$$\left[\Gamma_0^{-1}(1)\frac{\partial}{\partial t_1}\delta(1\bar{2}) - \chi_0^{-1}(1\bar{2})\right] G_0(\bar{2}2) = \delta(12) \ , \tag{174}$$

where:
$$\delta(12) = \delta(\mathbf{x}_1 - \mathbf{x}_2)\delta(t_1 - t_2)\delta_{\alpha_1\alpha_2} \tag{175}$$

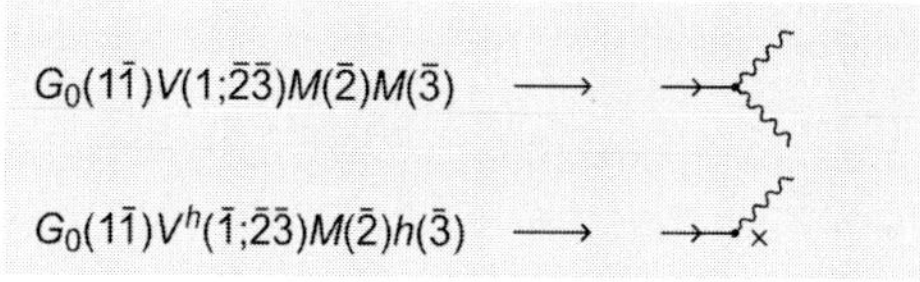

$$G_0(1\bar{1})V(\bar{1};\bar{2}\bar{3})M(\bar{2})M(\bar{3}) \longrightarrow$$

$$G_0(1\bar{1})V^h(\bar{1};\bar{2}\bar{3})M(\bar{2})h(\bar{3}) \longrightarrow$$

Fig. 10.19 Graphical representation for the oriented three-point vertices: $V(1;23)$ and $V^h(1;23)$.

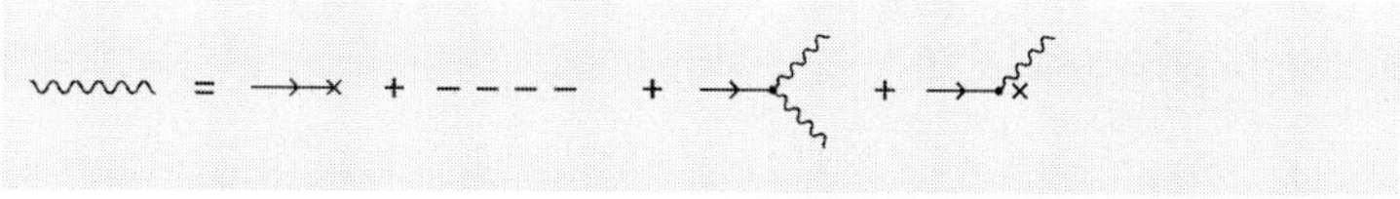

Fig. 10.20 Graphical representation for the order parameter given by Eq. (177).

and:
$$\chi_0^{-1}(12) = (r_0 - c\nabla_1^2)\delta(12) \ . \tag{176}$$

With these definitions the Langevin equation can be written in the form:
$$M(1) = G_0(1\bar{1})h(\bar{1}) + M_0(1) + G_0(1\bar{1})V(\bar{1};\bar{2}\bar{3})M(\bar{2})M(\bar{3})$$
$$+ G_0(1\bar{1})V^h(\bar{1};\bar{2}\bar{3})M(\bar{2})h(\bar{3}) \ , \tag{177}$$

where:
$$M_0(1) = G_0(1\bar{1})\tilde{f}(\bar{1}) \ . \tag{178}$$

This is then the most convenient form to use to develop a graphical approach to perturbation theory. In this case the three-point vertex $V(1;23)$ is represented by a dot as shown in Fig. 10.19. The part of the dot with an entering response line is associated with the first index, and the other two indices labeling the dot are symmetric under interchange. We can then iterate the graph shown in Fig. 10.20 just as for the TDGL case. At first order in λ we have the result for $M(1)$ shown in Fig. 10.21. It is left to Problem 10.13 to show that the graphs contributing to $M(1)$ at second order in λ, and up to first order in h, are given in Fig. 10.22. Next we average over the noise. The surviving contributions up to first order in λ are shown in Fig. 10.23. Clearly the first-order contribution is given by:

$$G^{(1)}(12) = G_0(1\bar{1})V(\bar{1};\bar{2}\bar{3})C_0(\bar{2}\bar{3}) \ . \tag{179}$$

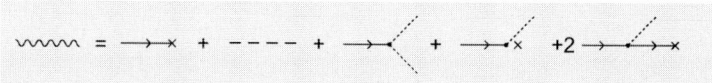

Fig. 10.21 Iterated solution to Fig. 10.20 to first order in the three-point interaction.

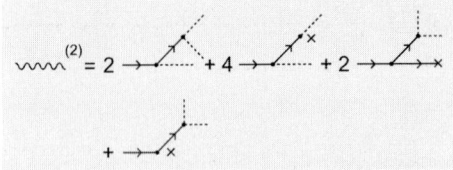

Fig. 10.22 Second-order iterated solution to Fig. 10.20.

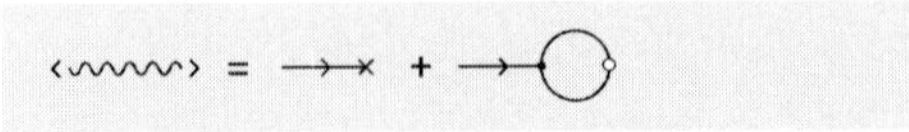

Fig. 10.23 Graphical perturbation theory expansion for the average order parameter to first order.

Since $C_0(\bar{2}\bar{3}) \propto \delta_{\alpha_2 \alpha_3}$ and $V(\bar{1};\bar{2}\bar{3}) \propto \varepsilon_{\alpha_1 \alpha_2 \alpha_3}$ we see that $G^{(1)}(12) = 0$.

10.2.3
Second Order in Perturbation Theory

The contribution to the propagators to second order in λ is given by the graphs in Fig. 10.24. The terms with contributions proportional to $V(\bar{1};\bar{2}\bar{3})C_0(\bar{2}\bar{3})$ vanish as in the first-order contribution. The remaining terms contributing to the propagator are given explicitly by:

$$G(12) = G_0(12) + 4G_0(1\bar{1})V(\bar{1};\bar{2}\bar{3})C_0(\bar{2}\bar{4})G_0(\bar{3}\bar{5})V(\bar{5};\bar{4}\bar{6})G_0(\bar{6}2)$$
$$+ 2G_0(1\bar{1})V(\bar{1};\bar{2}\bar{3})C_0(\bar{2}\bar{4})G_0(\bar{3}\bar{5})V^h(\bar{5};\bar{4}2) \ . \quad (180)$$

As for the TDGL case, we introduce the self-energy. In this case it has contributions:

$$\Sigma^{(i)}(12) = 4V(1;\bar{2}\bar{3})C_0(\bar{2}\bar{4})G_0(\bar{3}\bar{5})V(\bar{5};\bar{4}2) \quad (181)$$

and:

$$\Sigma^{(ii)}(12) = 2V(1;\bar{2}\bar{3})C_0(\bar{2}\bar{4})G_0(\bar{3}\bar{5})V^h(\bar{5};\bar{4}\bar{6})G_0^{-1}(\bar{6},2) \ . \quad (182)$$

If we insert the zeroth-order expressions for the response and correlation functions, we can evaluate these quantities explicitly. The first step in evaluating

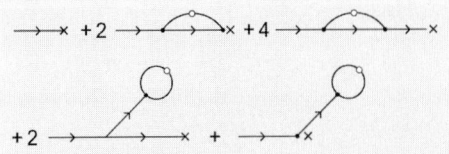

Fig. 10.24 Graphical perturbation theory expansion for the response function to second order.

these quantities is to treat the internal vector sums. Since the response and correlation functions are diagonal in the vector indices we see that we have the vector sums in $\Sigma^{(i)}$ and $\Sigma^{(ii)}$:

$$\sum_{\alpha_2\alpha_3} \varepsilon_{\alpha_1\alpha_2\alpha_3}\varepsilon_{\alpha_3\alpha_2\alpha_2} = -\sum_{\alpha_2\alpha_3} \varepsilon_{\alpha_1\alpha_2\alpha_3}\varepsilon_{\alpha_2\alpha_2\alpha_3} = -2\delta_{\alpha_1\alpha_2} . \tag{183}$$

It is left to Problem 10.15 to write the remaining contributions in terms of Fourier transforms with the result:

$$\Sigma^{(i)}_{\alpha_1\alpha_2}(q,\omega) = \delta_{\alpha_1\alpha_2}\Sigma^{(i)}(q,\omega) , \tag{184}$$

where:

$$\Sigma^{(i)}(q,\omega) = -2\frac{\lambda^2 c^2}{\Gamma_0(q)} \int \frac{d^d k}{(2\pi)^d} \int \frac{d\Omega}{2\pi} \left[k^2 - (\mathbf{q}-\mathbf{k})^2\right]$$
$$\times C_0(k,\Omega)G_0(\mathbf{q}-\mathbf{k},\Omega-\omega)\frac{(k^2-q^2)}{\Gamma_0(\mathbf{q}-\mathbf{k})} , \tag{185}$$

while:

$$\Sigma^{(ii)}_{\alpha_1\alpha_2}(q,\omega) = \delta_{\alpha_1\alpha_2}\Sigma^{(ii)}(q,\omega) , \tag{186}$$

where:

$$\Sigma^{(ii)}(q,\omega) = -2\frac{\lambda^2 c}{\Gamma_0(q)}G_0^{-1}(q,\omega) \int \frac{d^d k}{(2\pi)^d} \int \frac{d\Omega}{2\pi} \left[k^2 - (\mathbf{q}-\mathbf{k})^2\right]$$
$$\times C_0(k,\Omega)G_0(\mathbf{q}-\mathbf{k},\Omega-\omega)\frac{1}{\Gamma_0(\mathbf{q}-\mathbf{k})} . \tag{187}$$

We earlier considered the quantity:

$$K(k,q,\omega) = \int \frac{d\Omega}{2\pi} C_0(k,\Omega)G_0(q,\omega-\Omega)$$
$$= \frac{k_B T \chi_0(k)\Gamma_0(q)}{-i\omega+\gamma(k)+\gamma(q)} , \tag{188}$$

where:

$$\gamma(q) = \Gamma_0(q)\chi_0^{-1}(q) . \tag{189}$$

Equation (185) then takes the form:

$$\Sigma^{(i)}(q,\omega) = -2\frac{\lambda^2 c^2}{\Gamma_0(q)} \int \frac{d^d k}{(2\pi)^d} \left[k^2 - (\mathbf{q}-\mathbf{k})^2\right] K(k,\mathbf{q}-\mathbf{k},\omega)\frac{(k^2-q^2)}{\Gamma_0(\mathbf{q}-\mathbf{k})}$$
$$= -2\frac{\lambda^2 c^2}{\Gamma_0(q)} \int \frac{d^d k}{(2\pi)^d} \left[k^2 - (\mathbf{q}-\mathbf{k})^2\right] \frac{k_B T \chi_0(k)\Gamma_0(\mathbf{q}-\mathbf{k})}{[-i\omega+\gamma(k)+\gamma(\mathbf{q}-\mathbf{k})]} \frac{(k^2-q^2)}{\Gamma_0(\mathbf{q}-\mathbf{k})}$$

$$= -2\frac{\lambda^2 c^2}{\Gamma_0(q)} \int \frac{d^d k}{(2\pi)^d} \left[k^2 - (\mathbf{q}-\mathbf{k})^2\right] \frac{k_B T \chi_0(k)}{-i\omega + \gamma(k) + \gamma(\mathbf{q}-\mathbf{k})} (k^2 - q^2) \quad (190)$$

and:

$$\begin{aligned}
\Sigma^{(ii)}(q,\omega) &= -2\frac{\lambda^2 c}{\Gamma_0(q)} G_0^{-1}(q,\omega) \int \frac{d^d k}{(2\pi)^d} \\
&\quad \times \left[k^2 - (\mathbf{q}-\mathbf{k})^2\right] K(k, \mathbf{q}-\mathbf{k}, \omega) \frac{1}{\Gamma_0(\mathbf{q}-\mathbf{k})} \\
&= -2\frac{\lambda^2 c}{\Gamma_0(q)} G_0^{-1}(q,\omega) \int \frac{d^d k}{(2\pi)^d} \left[k^2 - (\mathbf{q}-\mathbf{k})^2\right] \\
&\quad \times \frac{k_B T \chi_0(k) \Gamma_0(\mathbf{q}-\mathbf{k})}{[-i\omega + \gamma(k) + \gamma(\mathbf{q}-\mathbf{k})]} \frac{1}{\Gamma_0(\mathbf{q}-\mathbf{k})} \\
&= -2\frac{\lambda^2 c}{\Gamma_0(q)} G_0^{-1}(q,\omega) \int \frac{d^d k}{(2\pi)^d} \left[k^2 - (\mathbf{q}-\mathbf{k})^2\right] \\
&\quad \times \frac{k_B T \chi_0(k)}{-i\omega + \gamma(k) + \gamma(\mathbf{q}-\mathbf{k})} \\
&= G_0^{-1}(q,\omega) Q(q,\omega) \; , \quad (191)
\end{aligned}$$

where we define:

$$Q(q,\omega) = -2\frac{\lambda^2 c}{\Gamma_0(q)} \int \frac{d^d k}{(2\pi)^d} \left[k^2 - (\mathbf{q}-\mathbf{k})^2\right] \frac{k_B T \chi_0(k)}{-i\omega + \gamma(k) + \gamma(\mathbf{q}-\mathbf{k})} \; . \quad (192)$$

$$\begin{aligned}
\Sigma^{(i)}(q,\omega) &= -2\frac{\lambda^2 c^2}{\Gamma_0(q)} \int \frac{d^d k}{(2\pi)^d} \left[k^2 - (\mathbf{q}-\mathbf{k})^2\right] \frac{k_B T \chi_0(k)}{-i\omega + \gamma(k) + \gamma(\mathbf{q}-\mathbf{k})} (k^2 - q^2) \\
&= -2\frac{\lambda^2 c}{\Gamma_0(q)} \int \frac{d^d k}{(2\pi)^d} \left[k^2 - (\mathbf{q}-\mathbf{k})^2\right] \frac{k_B T \chi_0(k)}{-i\omega + \gamma(k) + \gamma(\mathbf{q}-\mathbf{k})} (\chi_0^{-1}(k) - \chi_0^{-1}(q)) \\
&= -2\frac{\lambda^2 c}{\Gamma_0(q)} \int \frac{d^d k}{(2\pi)^d} \left[k^2 - (\mathbf{q}-\mathbf{k})^2\right] \frac{k_B T \chi_0(k)}{-i\omega + \gamma(k) + \gamma(\mathbf{q}-\mathbf{k})} \chi_0^{-1}(k) \\
&\quad -\chi_0^{-1}(q) Q(q,\omega) \\
&= -2\frac{\lambda^2 c}{\Gamma_0(q)} \int \frac{d^d k}{(2\pi)^d} \left[k^2 - (\mathbf{q}-\mathbf{k})^2\right] \frac{k_B T}{-i\omega + \gamma(k) + \gamma(\mathbf{q}-\mathbf{k})} \\
&\quad -\chi_0^{-1}(q) Q(q,\omega) \; . \quad (193)
\end{aligned}$$

Notice that the first term changes sign on letting $\mathbf{k} \to \mathbf{q} - \mathbf{k}$ and therefore vanishes and:

$$\Sigma^{(i)}(q,\omega) = -\chi_0^{-1}(q) Q(q,\omega) \; . \quad (194)$$

We can then conveniently combine the two contributions to the second-order self-energy:

$$\Sigma^{(2)}(q,\omega) = \Sigma^{(i)}(q,\omega) + \Sigma^{(ii)}(q,\omega)$$

$$= \left[G_0^{-1}(q,\omega) - \chi_0^{-1}(q)\right] Q(q,\omega)$$
$$= \frac{-i\omega}{\Gamma_0(q)} Q(q,\omega) \ . \tag{195}$$

This means that we have for the response function that:

$$\begin{aligned}G^{-1}(q,\omega) &= G_0^{-1}(q,\omega) - \Sigma^{(2)}(q,\omega) \\ &= -\frac{i\omega}{\Gamma_0(q)}\left[1 - Q(q,\omega)\right] + \chi_0^{-1}(q) \\ &= -\frac{i\omega}{\Gamma(q,\omega)} + \chi_0^{-1}(q) \ , \end{aligned} \tag{196}$$

where the physical damping coefficient is given by:

$$\Gamma(q,\omega) = \Gamma_0(q)\left[1 + Q(q,\omega)\right] \ . \tag{197}$$

Let us focus on the evaluation of $Q(q,\omega)$ given by Eq. (192). The integrand can be put in a more symmetric form if we let $\mathbf{k} \to \mathbf{k} + \mathbf{q}/2$ to obtain:

$$\begin{aligned}Q(q,\omega) &= 2\frac{\lambda^2 c}{\Gamma_0(q)} \int \frac{d^d k}{(2\pi)^d} \left[(\mathbf{k}-\mathbf{q}/2)^2 - (\mathbf{k}+\mathbf{q}/2)^2\right] \\ &\quad \times \frac{k_B T \chi_0(\mathbf{k}+\mathbf{q}/2)}{-i\omega + \gamma(\mathbf{k}+\mathbf{q}/2) + \gamma(\mathbf{k}-\mathbf{q}/2)} \ . \end{aligned} \tag{198}$$

Letting $\mathbf{k} \to -\mathbf{k}$ in the integral, adding the result to the original contribution and dividing by two, we easily find:

$$\begin{aligned}Q(q,\omega) &= \frac{k_B T \lambda^2 c}{\Gamma_0(q)} \int \frac{d^d k}{(2\pi)^d} \left[(\mathbf{k}-\mathbf{q}/2)^2 - (\mathbf{k}+\mathbf{q}/2)^2\right] \\ &\quad \times \frac{\left[\chi_0(\mathbf{k}+\mathbf{q}/2) - \chi_0(\mathbf{k}-\mathbf{q}/2)\right]}{-i\omega + \gamma(\mathbf{k}+\mathbf{q}/2) + \gamma(\mathbf{k}-\mathbf{q}/2)} \ . \end{aligned} \tag{199}$$

It is straightforward to see that this can be written as:

$$\begin{aligned}Q(q,\omega) &= \frac{k_B T \lambda^2 c^2}{\Gamma_0(q)} \int \frac{d^d k}{(2\pi)^d} \left[(\mathbf{k}-\mathbf{q}/2)^2 - (\mathbf{k}+\mathbf{q}/2)^2\right]^2 \\ &\quad \times \frac{\chi_0(\mathbf{k}+\mathbf{q}/2)\chi_0(\mathbf{k}-\mathbf{q}/2)}{-i\omega + \gamma(\mathbf{k}+\mathbf{q}/2) + \gamma(\mathbf{k}-\mathbf{q}/2)} \ . \end{aligned} \tag{200}$$

In this form we see that the limit $\mathbf{q} \to 0$ is a well-behaved constant:

$$\begin{aligned}Q(0,\omega) &= \frac{k_B T \lambda^2 c^2}{D_0} \int \frac{d^d k}{(2\pi)^d} 4(\mathbf{k}\cdot\hat{q})^2 \frac{\chi_0^2(k)}{-i\omega + 2\gamma(k)} \\ &= \frac{k_B T \lambda^2 c^2}{D_0} \frac{4}{d} \int \frac{d^d k}{(2\pi)^d} k^2 \frac{\chi_0^2(k)}{-i\omega + 2\gamma(k)} \ . \end{aligned} \tag{201}$$

Starting with Eq. (197) we see that the temperature dependence of the diffusion coefficient is given by the limits:

$$D = \lim_{q,\omega \to 0} \frac{\Gamma(q,\omega)}{q^2} = D_0[1 + Q(0,0)] \ , \tag{202}$$

where:

$$\begin{aligned} Q(0,0) &= \frac{k_B T \lambda^2 c^2}{D_0} \frac{4}{d} \int \frac{d^d k}{(2\pi)^d} k^2 \frac{\chi_0^2(k)}{2\gamma(k)} \\ &= \frac{2 k_B T \lambda^2 c^2}{d D_0^2} \int \frac{d^d k}{(2\pi)^d} \chi_0^3(k) \ . \end{aligned} \tag{203}$$

If we insert the Ornstein–Zernike form for $\chi_0(k)$ we find in Problem 10.16, for $\Lambda \xi \gg 1$, that:

$$Q(0,0) = \frac{2 k_B T_c \tilde{K}_d \lambda^2}{d D_0^2 c} \begin{cases} \frac{1}{4} \xi^{6-d} \Gamma(d/2) \Gamma(\frac{6-d}{2}) & d < 6 \\ \ell n(\Lambda \xi) - 3/4 & d = 6 \\ \frac{\Lambda^{d-6}}{(d-6)} + \mathcal{O}\left((\Lambda \xi)^{-1}\right) & d > 6 \ . \end{cases} \tag{204}$$

Then, in three dimensions, for example:

$$D = D_0 \left[1 + \frac{k_B T_c}{48\pi c} \left(\frac{\lambda}{D_0}\right)^2 \xi^3 + \mathcal{O}(\lambda^4)\right] \cdots \tag{205}$$

We can see immediately that the usefulness of perturbation theory is strongly coupled to the dimensionality of the system. For dimensions $d > 6$ perturbation theory *works*. For small $f \equiv \sqrt{\frac{k_B T_c}{c}} \frac{\lambda \Lambda^{\frac{d-6}{2}}}{D_0}$ (the dimensionless coupling constant) we have that the expansion,

$$D = D_0 \left(1 + \frac{2 f^2 \tilde{K}_d}{d(d-6)} + \mathcal{O}(f^4)\right) \tag{206}$$

makes sense. In six dimensions the correction is logarithmically divergent:

$$D = D_0 \left(1 + \frac{\tilde{K}_6}{3} f^2 [\ln(\Lambda \xi) - 3/4]\right) \tag{207}$$

and for dimensions fewer than six, the perturbation theory expansion,

$$D = D_0 \left(1 + \frac{2 f^2 \tilde{K}_d}{4d} \Gamma(d/2) \Gamma\left(\frac{6-d}{2}\right) (\Lambda \xi)^{6-d}\right) + \mathcal{O}(f^4) \cdots \tag{208}$$

does not make sense as it stands, since no matter how small f is, eventually, as $T \to T_c$, $(\Lambda\xi)^{6-d}f^2$ will become large. We see from out naive perturbation theory expansion that:

- The reversible terms in the equation of motion are very important near the critical point.

- Six dimensions is the cutoff dimension for a ferromagnet between conventional and nonconventional behavior.

We also see that our direct perturbation theory approach (where the perturbation is large compared to the zeroth-order term) is dubious. The situation is formally identical to the case in treating statics where we carry out a direct perturbation expansion of χ in Eq. (83). The resolution of the problem – a well-defined method for interpreting perturbation theory – is offered by the renormalization group.

10.3
The Dynamic Renormalization Group

10.3.1
Group Structure

Let us turn next to the dynamic renormalization group. This is a rather direct generalization of the static renormalization group developed in Chapters 3 and 4 of FOD. Here, we assume that the reader is familiar with this static development where one determines recursion relations for the parameters characterizing the effective Hamiltonian governing the order parameter. In the TDGL case, one develops recursion relations for the parameters, $\mu = \{r_0, c, u, \Gamma_0\}$, characterizing the associated Langevin equation. Once one has recursion relations one can use all of the renormalization group (RG) phenomenology concerning fixed points, stability, crossover and scaling. We shall return to this later. First we develop the steps in the RG for dynamic problems. We shall see that in some senses the explicit implementation of the dynamic RG is simpler than in the static case.

The implementation of the RG is again a two-step process.

1. Average over the large wavenumber components of the order parameter $\psi(q,t)$ in the wavenumber shell,

$$\Lambda/b < q < \Lambda \, , \quad (209)$$

from the equation of motion.

2. In the resulting equation of motion for $\psi(q,t)$ with $q < \Lambda/b$, replace $\psi(q,t) \to b^{1-\eta/2}\psi(qb, tb^{-z})$, rescale wavenumbers, $q \to q' = bq$, times $t \to$

$t' = tb^{-z}$. This is the same as step 2 in the static RG except that we rescale times with a factor b^{-z}. The new exponent z plays a role in scaling times similar to that played by η in scaling distances. η and z must be adjusted so that we reach a fixed point.

The new equations of motion are then written in the old form. The new parameters are identified as entries in $\mu' = R_b \mu$. We will discuss the implementation of these steps for both the TDGL and ferromagnet cases.

10.3.2
TDGL Case

It is not difficult to see how to carry out the average over the high-wavenumber components of the order parameter. This is most easily done graphically. The TDGL equation of motion can be represented as shown in Fig. 10.25 using the same notation as in Fig. 10.1. The point is that we can then divide the dashed zeroth-order contribution into low- and high-wavenumber components as shown in Fig. 10.26:

$$\psi_0(q,\omega) = \psi_0^<(q,\omega) + \psi_0^>(q,\omega) \,, \tag{210}$$

where:

$$\psi_0^<(q,\omega) = \theta(\Lambda/b - q)\psi_0(q,\omega) \tag{211}$$

and:

$$\psi_0^>(q,\omega) = \theta(q - \Lambda/b)\psi_0(q,\omega) \,. \tag{212}$$

Next observe that an iteration of the equation of motion is shown graphically in Fig. 10.27. This result is exact. If we work to second order in the nonlinear coupling we have the result in Fig. 10.28. Now, in this figure replace ψ_0 by its high- and low-wavenumber components as in Fig. 10.26 and average over the high-momentum components. This is easy to carry out since $\psi_0(q,\omega)$ is a Gaussian variable with independent Fourier components.

Let us take the high-wavenumber average of the three terms in Fig. 10.28 in turn. First we have the result in Fig. 10.29 showing that at zeroth order the low-wavenumber result survives. For the first-order contribution we have the result shown in Fig. 10.30.

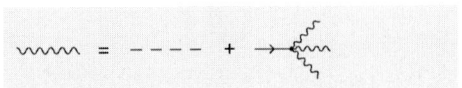

Fig. 10.25 Graphical representation for the order parameter in the time-dependent Ginzburg–Landau (TDGL) case in the absence of an external field.

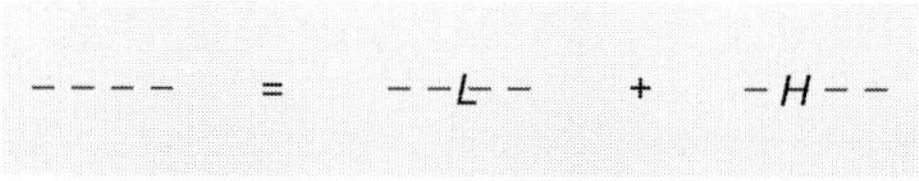

10.3 The Dynamic Renormalization Group

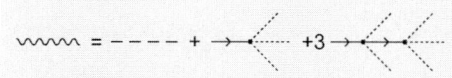

Fig. 10.26 Graphical representation for the zeroth-order contribution to the order parameter divided into high- and low-wavenumber components.

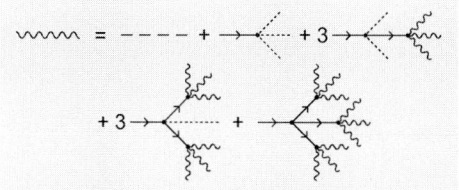

Fig. 10.27 Graphical iteration of Fig. 10.25.

Fig. 10.28 Expansion of order parameter in terms of bare order parameter lines to second order.

Turning to the average over the high-wavenumber components for the third graph in Fig. 10.28, we see that this can be broken up into the averages over the three contributions in Fig. 10.31. These can in turn each be broken up into three contributions shown in Fig. 10.32. Taking the average over the high-wavenumber components gives us the results shown in Fig. 10.33. Putting all of these results together, we have for the coarse-grained average of the order parameter the graphical representation given in Fig. 10.34. These terms can be grouped together in terms of a new equation of motion shown in Fig. 10.35. In this new equation of motion we find that the low-wavenumber *zeroth-order* lines are now given by the sum of graphs shown in Fig. 10.36. The new four-point coupling is shown in Fig. 10.37. A six-point interaction is generated as shown in Fig. 10.38, and the propagator is renormalized as shown in Fig. 10.39.

Let us look at the analytic implications of our coarse-grained equation of motion. Working to first order in the coupling we have for the low-wavenumber order parameter the first two terms on the right-hand side of Fig. 10.36:

$$\psi^{(1)}(q,\omega) = G_0(q,\omega)\frac{1}{2\Gamma_0}\eta(q,\omega) + G_0(q,\omega)\Sigma_H^L G_0(q,\omega)\frac{1}{2\Gamma_0}\eta(q,\omega) \quad (213)$$

and the renormalization of the response function is given by Fig. 10.39:

$$G_0^{(1)}(q,\omega) = G_0(q,\omega) + G_0(q,\omega)\Sigma_H^L G_0(q,\omega) \quad , \quad (214)$$

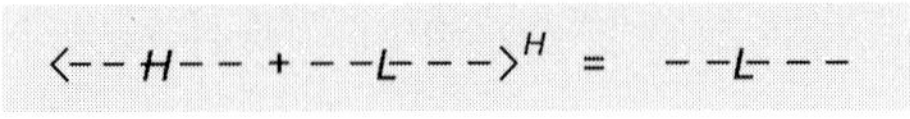

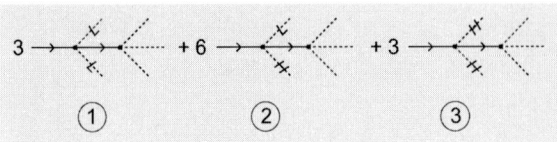

Fig. 10.29 Average over high-wavenumber components of the order parameter at zeroth order.

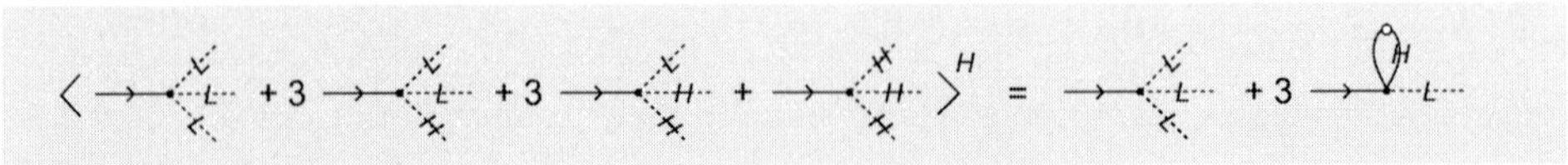

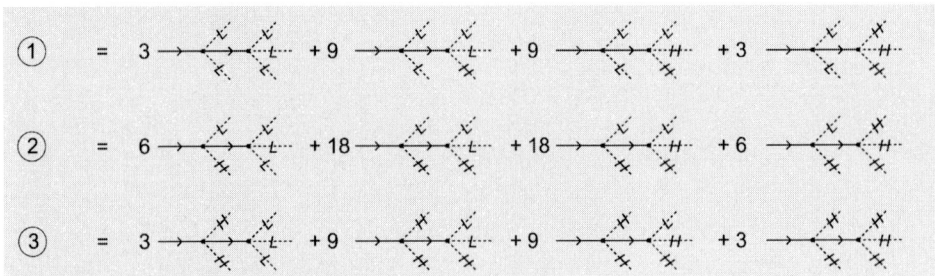

Fig. 10.30 Average over high-wavenumber components of the order parameter at first order.

Fig. 10.31 Setting up the average over high-wavenumber components of the order parameter at second order.

Fig. 10.32 Continuation of the average over high-wavenumber components of the order parameter at second order.

where, in both:

$$\Sigma_H^L = -3u \int_{\Lambda/b}^{\Lambda} \frac{d^d k}{(2\pi)^d} \frac{d\Omega}{2\pi} C_0(k, \Omega) \ . \tag{215}$$

Inspection shows that since:

$$\psi^{(1)}(q, \omega) = G_0^{(1)}(q, \omega) \frac{1}{2\Gamma_0}(q, \omega) \ , \tag{216}$$

the noise is not directly coarse grained. We can then write Eq. (214), consistent to first order in the coupling, as:

$$\begin{aligned}\left(G_0^{(1)}\right)^{-1}(q, \omega) &= G_0^{-1}(q, \omega) - \Sigma_H^L \\ &= -\frac{i\omega}{\Gamma_0} + \chi_0^{-1}(q) - \Sigma_H^L \ . \end{aligned} \tag{217}$$

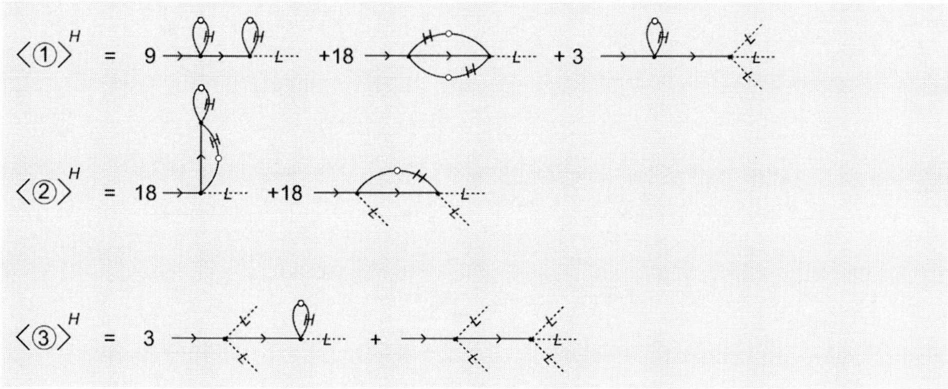

Fig. 10.33 Continuation of the average over the high-wavenumber components of the order parameter at second order.

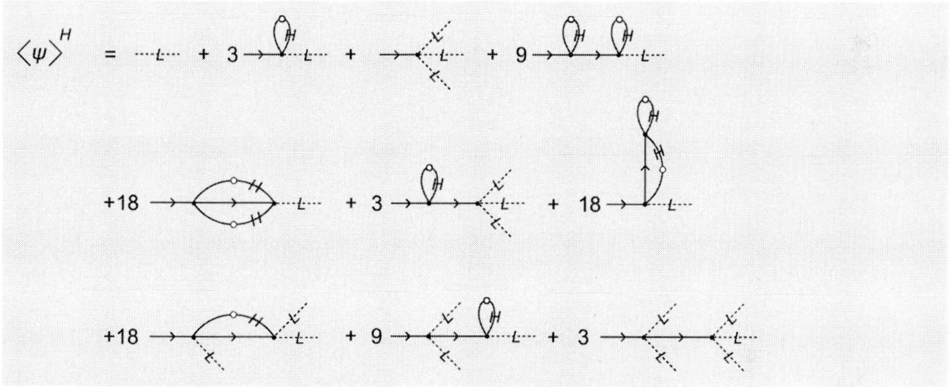

Fig. 10.34 Summary of terms contributing to the average over high-wavenumber components of the order parameter up to second order.

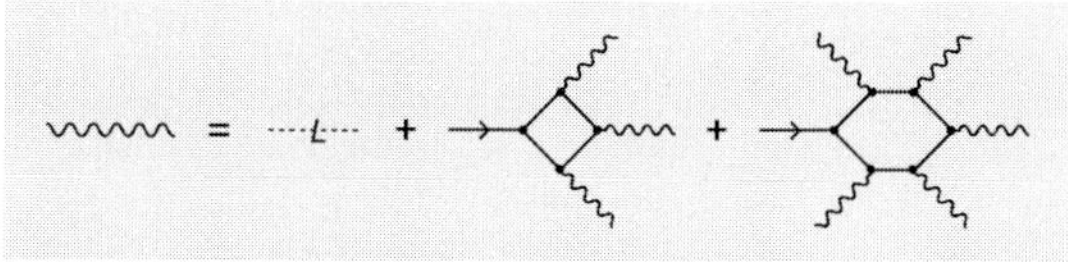

Fig. 10.35 Graphical structure of new equation of motion after averaging over high-wavenumber components.

Then at this order Γ_0 is not renormalized:

$$\Gamma_0^{(1)} = \Gamma_0 \tag{218}$$

and:

$$\left(\chi^{(1)}\right)^{-1}(q) = \chi_0^{-1}(q) - \Sigma_H^L \quad . \tag{219}$$

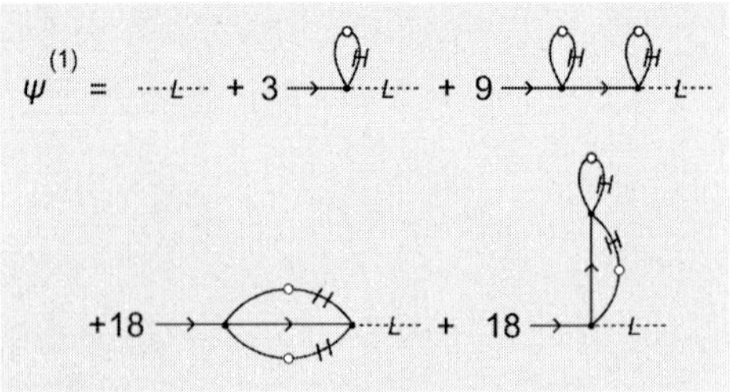

Fig. 10.36 Graphical contributions to the zeroth-order contribution to the equation of motion after averaging over high-wavenumber components.

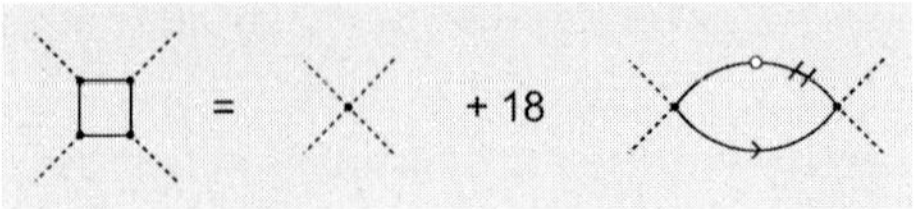

Fig. 10.37 Graphical contributions to the new four-point interaction after averaging over high-wavenumber components.

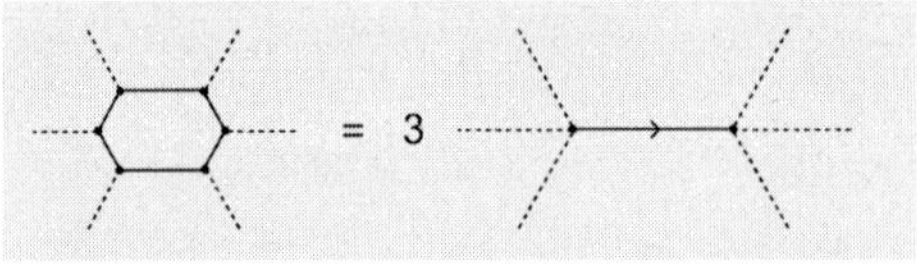

Fig. 10.38 Graphical contributions to the new six-point interaction after averaging over high-wavenumber components.

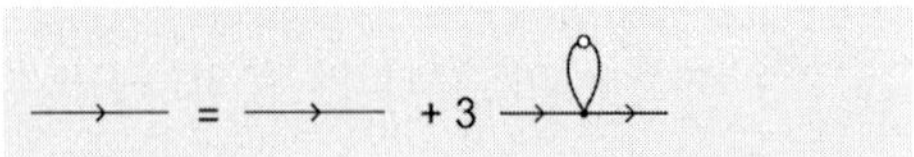

Fig. 10.39 Renormalization of propagator in new equation of motion to first order.

We see that in the long wavelength limit we can still use the LGW parameterization:

$$r^{(1)} + c^{(1)} q^2 = r + cq^2 - \Sigma_H^L \tag{220}$$

and, since Σ_H^L is independent of wavenumber, we can identity:

$$c^{(1)} = c \tag{221}$$

and:

$$\begin{aligned} r^{(1)} &= r - \Sigma_H^L \\ &= r + 3u \int_{\Lambda/b}^{\Lambda} \frac{d^d k}{(2\pi)^d} \chi_0(k) \quad . \end{aligned} \tag{222}$$

Turning to the renormalization of the four-point coupling, shown in Fig. 10.37, we have, taking the small q and ω components:

$$u^{(1)} = u - 18u^2 \int_{\Lambda/b}^{\Lambda} \frac{d^d k}{(2\pi)^d} \frac{d\Omega}{2\pi} C_0(k,\Omega) G_0(k,\Omega) \ . \tag{223}$$

The frequency integral was worked out in Eqs. (38) and (45) with the result:

$$u^{(1)} = u - 18u^2 \int_{\Lambda/b}^{\Lambda} \frac{d^d k}{(2\pi)^d} \frac{1}{2} \chi_0^2(k) \ . \tag{224}$$

It turns out, as expected, that $r^{(1)}$ and $u^{(1)}$ depend only on static parameters.

Next we come to the rescaling step in the RG procedure. We carry this out on the coarse-grained equation of motion given by:

$$\frac{\partial \psi}{\partial t} = -\Gamma_0^{(1)} \left(r^{(1)} \psi - c^{(1)} \nabla^2 \psi + u^{(1)} \psi^3 \right) + \eta \ , \tag{225}$$

where:

$$\psi = \psi^{(1)}(\mathbf{x}, t) \tag{226}$$

and:

$$\eta = \eta^{(1)}(\mathbf{x}, t) \ . \tag{227}$$

The next step in the RG is to rescale space and time coordinates using:

$$\mathbf{x} = b\mathbf{x}' \tag{228}$$

and:

$$t = b^z t' \ . \tag{229}$$

Defining:

$$\tilde{\psi} = \psi^{(1)}(b\mathbf{x}', b^z t') \tag{230}$$

$$\tilde{\eta} = \eta^{(1)}(b\mathbf{x}', b^z t') \ , \tag{231}$$

the equation of motion takes the form:

$$b^{-z} \frac{\partial \tilde{\psi}}{\partial t'} = -\Gamma_0^{(1)} \left(r^{(1)} \tilde{\psi} - c^{(1)} b^{-2} \nabla^2 \tilde{\psi} + u^{(1)} \tilde{\psi}^3 \right) + \tilde{\eta} \ . \tag{232}$$

Next, let:

$$\psi'(\mathbf{x}', t') = b^\alpha \tilde{\psi} \ , \tag{233}$$

then the equation of motion can be put into the form:

$$\frac{\partial \psi'}{\partial t'} = -\Gamma_0^{(1)} b^z \left(r^{(1)} \tilde{\psi} - c^{(1)} b^{-2} \nabla^2 \tilde{\psi} + u^{(1)} b^{-2\alpha} \tilde{\psi}^3 \right) + \eta' \, , \tag{234}$$

where:

$$\eta'(\mathbf{x}', t') = b^{z+\alpha} \tilde{\eta} \, . \tag{235}$$

If we assume that:

$$\Gamma_0' = b^\sigma \Gamma_0^{(1)} \, , \tag{236}$$

then the equation of motion retains its form with the rescaled parameters:

$$r' = b^{z-\sigma} r^{(1)} \tag{237}$$

$$c' = b^{z-\sigma-2} c^{(1)} \tag{238}$$

$$u' = b^{z-\sigma-2\alpha} u^{(1)} \, . \tag{239}$$

We must still consider the renormalization of the noise. We have for its determining variance:

$$\begin{aligned}\langle \eta'(\mathbf{x}'_1, t'_1) \eta'(\mathbf{x}'_2, t'_2) \rangle &= 2 k_B T' \Gamma_0' \delta(\mathbf{x}'_1 - \mathbf{x}'_2) \delta(t'_1 - t'_2) \\ &= b^{2z+2\alpha} \langle \eta(b\mathbf{x}'_1, b^z t'_1) \eta(b\mathbf{x}'_2, b^z t'_2) \rangle \\ &= b^{2z+2\alpha} 2 k_B T \Gamma_0^{(1)} b^{-d-z} \delta(\mathbf{x}'_1 - \mathbf{x}'_2) \delta(t'_1 - t'_2) \, . \end{aligned} \tag{240}$$

From this we can identify:

$$T' \Gamma_0' = b^{2z+2\alpha} b^{-d-z} T \Gamma_0^{(1)} \, . \tag{241}$$

If we choose $T' = T$ then:

$$\Gamma_0' = b^{z+2\alpha-d} \Gamma_0^{(1)} \, . \tag{242}$$

Comparing Eq. (242) with Eq. (236) we identify:

$$\sigma = z + 2\alpha - d \, . \tag{243}$$

It is conventional to choose:

$$\alpha = \frac{1}{2}(d - 2 + \eta) \, , \tag{244}$$

where the η will now turn out to be the usual standard static critical index. With this new notation, we have:

$$\sigma = z - 2 + \eta \, . \tag{245}$$

In summary, the recursion relations are given by:

$$r' = b^{2-\eta} r^{(1)} \tag{246}$$

$$c' = b^{-\eta} c^{(1)} \tag{247}$$

$$u' = b^{4-d-2\eta} u^{(1)} \tag{248}$$

$$\Gamma'_0 = b^{z-2+\eta} \Gamma_0^{(1)} , \tag{249}$$

where $r^{(1)}$ is given by Eq. (222), $c^{(1)}$ by Eq. (221), $u^{(1)}$ by Eq. (224), and $\Gamma_0^{(1)}$ by Eq. (218). We then obtain the final form for the recursion relations:

$$r' = b^{2-\eta} \left(r + 3u \int_{\Lambda/b}^{\Lambda} \frac{d^d k}{(2\pi)^d} \chi_0(k) \right) \tag{250}$$

$$c' = b^{-\eta} c \tag{251}$$

$$u' = b^{4-d-2\eta} \left(u - 18u^2 \int_{\Lambda/b}^{\Lambda} \frac{d^d k}{(2\pi)^d} \frac{1}{2} \chi_0^2(k) \right) . \tag{252}$$

$$\Gamma'_0 = b^{z-2+\eta} \Gamma_0 . \tag{253}$$

Our goal, as discussed in detail in the next section, is now to find the stable fixed-point solutions to these recursion relations. Clearly at this order, to find a fixed point we need to choose,

$$\eta = 0 \tag{254}$$

and:

$$z = 2 . \tag{255}$$

Then the recursion relations reduce precisely to those governing the static critical behavior at first order:

$$r' = b^2 \left(r + 3u \int_{\Lambda/b}^{\Lambda} \frac{d^d k}{(2\pi)^d} \chi_0(k) \right) \tag{256}$$

$$u' = b^{4-d} \left(u - 18u^2 \int_{\Lambda/b}^{\Lambda} \frac{d^d k}{(2\pi)^d} \frac{1}{2} \chi_0^2(k) \right) . \tag{257}$$

The analysis of the static recursion relations for r and u was treated in Section 4.5.3 in FOD. One finds a Gaussian fixed point $u^* = r^* = 0$ for dimensionality greater than four and a nontrivial fixed point for $d < 4$. Expanding in $\varepsilon = 4 - d$ we have to lowest order in ε:

$$\tilde{r}^* = \left(\frac{r}{\Lambda^2}\right)^* = -\frac{\varepsilon}{6} \tag{258}$$

and:

$$\tilde{u}^* = \left(\frac{u}{c}\right)^* = \frac{8\pi^2}{9}\varepsilon \ . \tag{259}$$

The linear stability of this fixed point is given by:

$$\delta\tilde{r}' = \delta\tilde{r}\, b^{y_1} + \delta\tilde{u}\,\frac{3}{16\pi^2}b^2 \tag{260}$$

$$\delta\tilde{u}' = \delta\tilde{u}\, b^{y_2} \ , \tag{261}$$

where the stability exponents are given by:

$$y_1 = 2 - \frac{\varepsilon}{3} > 0 \tag{262}$$

$$y_2 = -\varepsilon < 0 \ . \tag{263}$$

This says that r is the only relevant variable with $y_1 > 0$.

10.3.3
Scaling Results

Next we discuss the scaling implications of the dynamic RG. We have the relation between the original time-correlation function and the coarse-grained version (in terms of the longer wavelength degrees of freedom):

$$\begin{aligned} C(\mathbf{x}_1 - \mathbf{x}_2, t_1 - t_2; \mu) &= \langle \psi^{(1)}(\mathbf{x}_1, t_1)\psi^{(1)}(\mathbf{x}_2, t_2)\rangle \\ &= b^{-2\alpha}\langle \psi'(\mathbf{x}_1/b, t_1/b^z)\psi'(\mathbf{x}_2/b, t_2/b^z)\rangle \\ &= b^{-2\alpha} C\left((\mathbf{x}_1 - \mathbf{x}_2)/b, (t_1 - t_2)/b^z; \mu'\right) \ , \end{aligned} \tag{264}$$

where μ represents the original set of parameters and μ' the RG-transformed parameters. Fourier transforming over space and using Eq. (245) gives:

$$C(q, t; \mu) = b^{2-\eta} C(qb, tb^{-z}; \mu'). \tag{265}$$

If we Fourier transform over time, we obviously obtain:

$$C(q, \omega; \mu) = b^{2-\eta+z} C(q, b, \omega b^z; \mu'). \tag{266}$$

Using the FDT given by Eq. (25), we easily obtain for the response function:

$$G(q,\omega;\mu) = b^{2-\eta}G(qb,\omega b^z;\mu') \ . \tag{267}$$

Suppose we can find a fixed-point solution of our dynamic RG recursion relations:

$$\mu^* = R^b\mu^* \ . \tag{268}$$

We can then write, for situations where we are near the fixed point,

$$\delta\mu' = t_1 b^{y_1} e_1 + \mathcal{O}(b^{y_2}) \tag{269}$$

and $y_2 < 0$. y_2 need not be the same as in the static case. We find then, if we are near the critical surface, that:

$$C(q,\omega;\mu(T)) = b^{2-\eta+z}C\left(qb,\omega b^z;\mu^* + \left(\frac{b}{\xi}\right)^{1/2} e_1 + \mathcal{O}(b^{y_2})\right). \tag{270}$$

If we set $b = \Lambda\xi$, then:

$$C(q,\omega;\mu(T)) = (\Lambda\xi)^{2-\eta+z}C\left(q\xi,\omega(\xi\Lambda)^z;\mu^* + \Lambda^{1/2}e_1 + \mathcal{O}(\Lambda\xi)^{y_2}\right). \tag{271}$$

If terms of $\mathcal{O}[(\Lambda\xi)^{y_2}]$ can be ignored, then we have:

$$C(q,\omega;\mu(T)) = \xi^{2-\eta+z}f(q\xi,\omega\xi^z) \ , \tag{272}$$

which is a statement of dynamic scaling. We also have from Eq. (267) with $b = \Lambda\xi$:

$$G(q,\omega;\mu) = \xi^{2-\eta}f_G(q\xi,\omega\xi^z) \ . \tag{273}$$

We see that while q is naturally scaled by ξ as $T \to T_c$, ω is scaled by ξ^z. Alternatively if we let $T = T_c$ we can write Eq. (270):

$$C(q,\omega;\mu(T_c)) = b^{2-\eta+z}C[qb,\omega b^z;\mu^* + \mathcal{O}(b^{y_2})]. \tag{274}$$

If we then choose $b = (\frac{q}{\Lambda})^{-1}$ we obtain the scaling result:

$$C(q,\omega;\mu(T_c)) = \left(\frac{\Lambda}{q}\right)^{2-\eta+z}C\left(\Lambda,\omega\left(\frac{q}{\Lambda}\right)^{-z};\mu^*\right). \tag{275}$$

If we choose for $b = (\frac{\omega}{\omega_0})^{\frac{1}{z}}$, where ω_0 is some frequency constructed from Γ_0 and Λ, then:

$$C(q,\omega;\mu(T_c)) = \left(\frac{\omega}{\omega_0}\right)^{(2-\eta+z)/z}C\left(k\left(\frac{\omega}{\omega_0}\right)^{\frac{1}{z}},\omega_0;\mu^*\right) \tag{276}$$

and:
$$C(0,\omega;\mu(T_c)) \sim \omega^{(2-\eta+z)/z}. \tag{277}$$

It is convenient to define, as we did in the van Hove theory, a characteristic frequency. One convenient definition is:
$$\omega_c \equiv \chi^{-1}(q)\left[\frac{\partial}{\partial(-i\omega)}G^{-1}(q,\omega)\right]^{-1}_{\omega=0}. \tag{278}$$

If we remember Eq. (7), we are led to identify:
$$\omega_c(q) = \chi^{-1}(q)\Gamma(q,0), \tag{279}$$

and $\Gamma(q,0)$ has the interpretation of a wavenumber-dependent kinetic coefficient. Note that this definition for a characteristic frequency agrees with that used in the conventional theory. Starting with Eq. (267), it is shown in Problem 10.17 that:
$$\omega_c(q;\mu) = b^{-z}\omega_c(qb;\mu'). \tag{280}$$

Following now-familiar arguments, if we let $b = \Lambda\xi$ and ignore terms of $\mathcal{O}[(\Lambda\xi)^{y_2}]$ we obtain the scaling result:
$$\omega_c(q;\mu) = \xi^{-z}f(q\xi). \tag{281}$$

Clearly one of the first orders of business in dynamic critical phenomena is to determine the dynamic index z.

10.3.4
Wilson Matching

In his treatment of the static LGW n-vector model, Wilson [13] argued that the scaling properties of the order parameter structure factor can be obtained directly from the perturbation series in four dimensions, provided one chooses the fixed-point value for the coupling,
$$u_0^* = \frac{2\pi^2\varepsilon}{(n+8)}, \tag{282}$$

where $\varepsilon = 4 - d$, and then exponentiates the logarithmic terms appearing in the expansion. The authors of Ref. [7] assert that the same procedure will yield the correct form for the dynamic correlation functions. Let us see how this works for the scalar TDGL case. We have generally for the response function:
$$G^{-1}(q,\omega) = -\frac{i\omega}{\Gamma(q,\omega)} + \chi^{-1}(q). \tag{283}$$

With $T = T_c$ and $q = 0$ this takes the form:

$$G^{-1}(0, \omega) = -\frac{i\omega}{\Gamma(0, \omega)} \quad . \tag{284}$$

We previously [Eq. (143)] determined, using perturbation theory in $d = 4$, that:

$$\Gamma^{-1}(0, \omega) = \Gamma_0^{-1}\left[1 - u_0^2 b_0 \ln\left(\frac{\Omega}{\Lambda^2}\right)\right] \quad , \tag{285}$$

where $\Omega = \omega/c\Gamma_0$ and:

$$b_0 = \frac{9}{8\pi^4} \ln\left(\frac{4}{3}\right) \quad . \tag{286}$$

Exponentiating the logarithm in Eq. (285) we obtain:

$$\Gamma^{-1}(0, \omega) = \Gamma_0^{-1}\left(\frac{\Omega}{\Lambda^2}\right)^{-u_0^2 b_0} \tag{287}$$

and:

$$G^{-1}(0, \omega) \approx \omega^{1 - u_0^2 b_0} \quad . \tag{288}$$

We can use this result to estimate the dynamic critical index by using scaling arguments. From dynamical scaling we have from Eq. (273):

$$G(q, \omega) = \xi^{2-\eta} f_G(q\xi, \omega\xi^z) \quad . \tag{289}$$

Then as $q \to 0$ and $T \to T_c$ we have:

$$G(0, \omega) \approx \xi^{2-\eta} \frac{1}{(\omega\xi^z)^{(2-\eta)/z}} \approx \omega^{-(2-\eta)/z} \quad . \tag{290}$$

Comparing this to Eq. (288) we identify for $n = 1$:

$$(2 - \eta)/z = 1 - u_0^2 b_0 \quad . \tag{291}$$

Noting from the static calculation [13] $\eta = \varepsilon^2/54$, we can write, using Eqs. (282) and (286),

$$z = 2 + \eta[-1 + 6\ln(4/3)] \tag{292}$$

in agreement with Ref. [7].

10.3.5
Isotropic Ferromagnet

To see how the RG works in the ferromagnetic case let us, for simplicity, set u and r_o equal to zero [16] and investigate how the parameters D_0 and λ

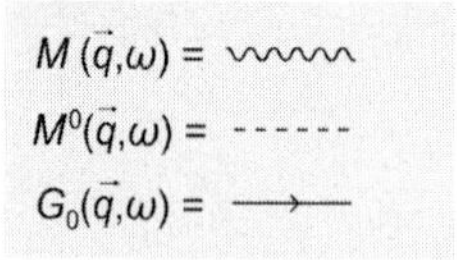

10 Perturbation Theory and the Dynamic Renormalization Group

$M(\vec{q},\omega) =$ ∿∿∿

$M^0(\vec{q},\omega) =$ ------

$G_0(\vec{q},\omega) =$ ⟶

Fig. 10.40 Graphical assignments in ferromagnet case.

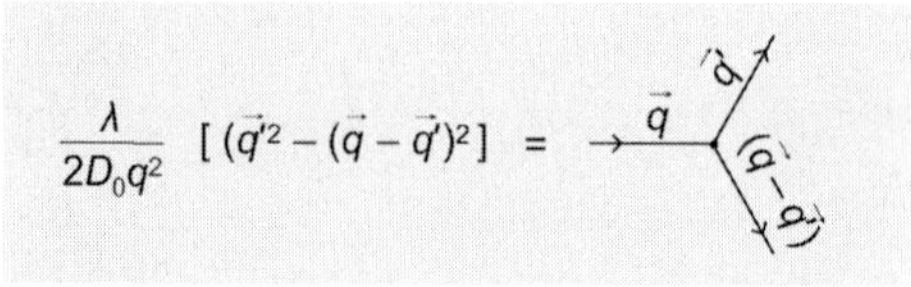

$$\frac{\lambda}{2D_0 q^2}[(\vec{q}'^2 - (\vec{q}-\vec{q}')^2] =$$

Fig. 10.41 Graphical assignment for three-point vertex.

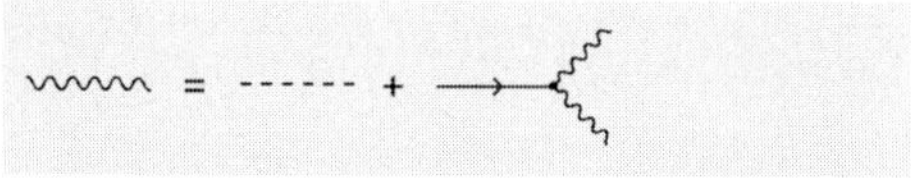

∿∿∿ = ------ + ⟶≺

Fig. 10.42 Graphical representation for the order parameter, Eq. (293).

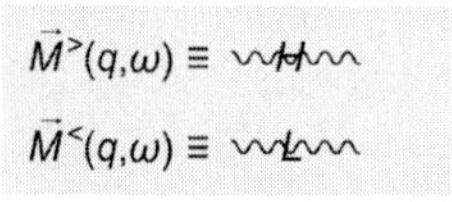

$\vec{M}^{>}(q,\omega) \equiv$ ∿∿∿

$\vec{M}^{<}(q,\omega) \equiv$ ∿∿∿

Fig. 10.43 Graphical representation for high- and low-wavenumber contributions to the order parameter equation of motion.

change under the group transformation. Our Fourier-transformed equation of motion, Eq. (177), with $h = 0$, can then be written as:

$$\mathbf{M}(q,\omega) = \mathbf{M}^0(q,\omega) + G_0(q,\omega) \int \frac{d^d q'}{(2\pi)^d} \frac{d\omega'}{2\pi} \frac{\lambda}{2D_0 q^2}$$
$$\times [(\mathbf{q}')^2 - (\mathbf{q}-\mathbf{q}')^2] \mathbf{M}(\mathbf{q}',\omega') \times \mathbf{M}(\mathbf{q}-\mathbf{q}',\omega-\omega') \quad . \quad (293)$$

It will again be useful to use a graphical technique. If we identify, as in Fig. 10.40, the lines as before, and the three-legged vertex as in Fig. 10.41, then the equation of motion can be written as Fig. 10.42. In carrying out step 1 in the RG we write:

$$\mathbf{M}(\mathbf{q},\omega) = \Theta(|q| > \Lambda)\mathbf{M}(\mathbf{q},\omega) + \Theta(|q| < \Lambda)\mathbf{M}(\mathbf{q},\omega)$$
$$\equiv \mathbf{M}^{>}(\mathbf{q},\omega) + \mathbf{M}^{<}(\mathbf{q},\omega) \quad . \quad (294)$$

We can then break up wavy lines into the high- and low-wavenumber components, (Fig. 10.43) and similarly for response functions (Fig. 10.44). We are then led to the coupled set of equations (Figs. 10.45 and 10.46). We need to *solve* for a wavy line with an H in Fig. 10.45, put it into Fig. 10.46 and then average

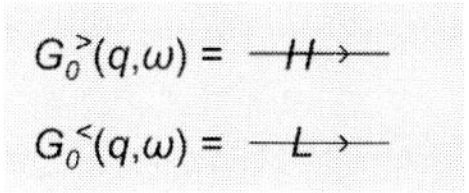

$G_0^>(q,\omega) = $ —H→—

$G_0^<(q,\omega) = $ —L→—

Fig. 10.44 Graphical representation for high- and low-wavenumber contributions to the response function.

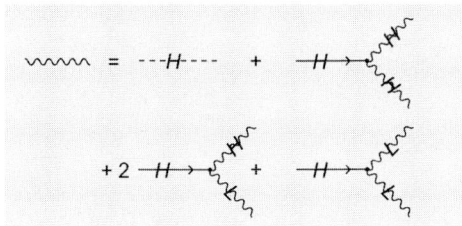

Fig. 10.45 Graphical representation for high-wavenumber contributions to the order parameter.

over the noise associated with high wavenumbers. We will work this out to second order in λ. Note that on iteration, we obtain the graphs in Fig. 10.47. Putting this result for a wavy line with an H in Fig. 10.46 we obtain Fig. 10.48. This looks complicated. However, we must average over all the high-wavenumber components.

All of the graphs with an odd number of $M_0^>$ vanish. When we connect up two $M_0^>$ lines we obtain a factor shown in Fig. 10.49. It is also useful to note that graphs with contributions like Fig. 10.50 vanish due to the symmetry of the three-point interaction. We are then left with the result shown in Fig. 10.51. The first two terms are those we would obtain by ignoring a wavy line with an H altogether. It is shown in Problem 10.20 that the last term in Fig. 10.51 gives no contribution for small enough q due to a noncompatibility of step functions. We are then left with the equation,

$$\mathbf{M}^<(q,\omega) = G_0^<(q,\omega)\frac{1}{D_0 q^2}\mathbf{f}^<(q,\omega) + G_0^<(q,\omega)$$
$$\times \bar{\Sigma}(q,\omega)\mathbf{M}^<(q,\omega) + G_0^<(q,\omega)\int \frac{d^d k}{(2\pi)^d}\frac{d\Omega}{2\pi}\frac{\lambda}{D_0 q^2}$$
$$\times \left[\mathbf{k}^2 - (\mathbf{q}-\mathbf{k})^2\right]\mathbf{M}^<(\mathbf{k},\Omega)\times\mathbf{M}^<(\mathbf{q}-\mathbf{k},\omega-\Omega) \quad, (295)$$

where $\bar{\Sigma}(\mathbf{q},\omega)$, defined graphically in Fig. 10.52, is the same as $\Sigma^{(i)}\mathbf{q},\omega)$, given by Eq. (185), except there are internal wavenumber restrictions. We can bring the $G_0^<\bar{\Sigma}M^<$ term to the left-hand side of Eq. (295) and then solve for $M^<$:

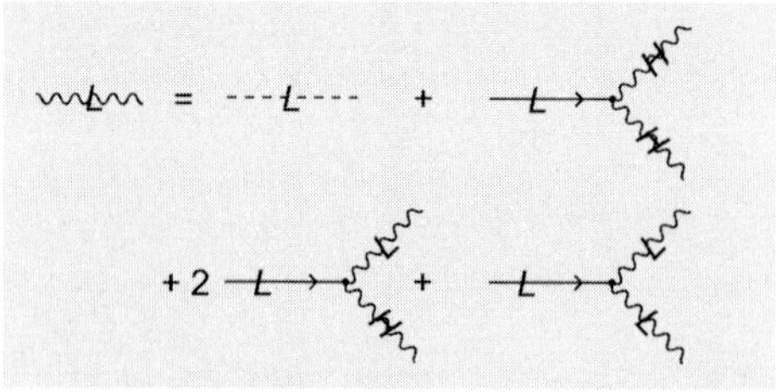

394 | 10 Perturbation Theory and the Dynamic Renormalization Group

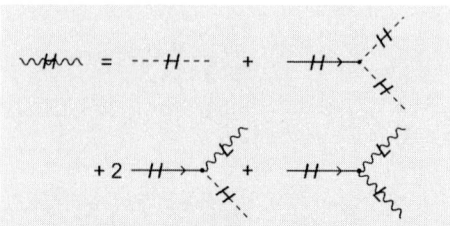

Fig. 10.46 Graphical representation for low-wavenumber contributions to the order parameter.

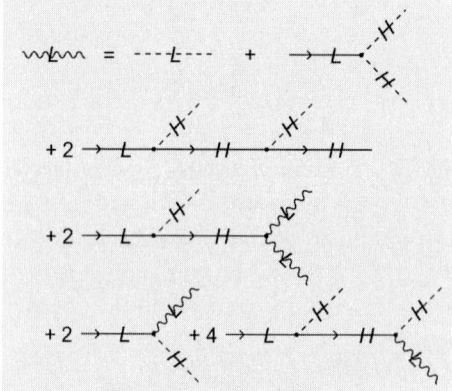

Fig. 10.47 Iterated graphical representation for high-wavenumber contributions to the order parameter.

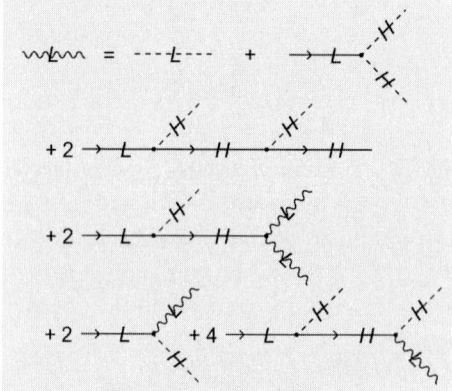

Fig. 10.48 Iterated graphical representation for low-wavenumber contributions to the order parameter.

$$\text{------}H\text{------} \equiv \langle M_0^>(\vec{q},\omega) M_0^>(-\vec{q},-\omega)\rangle = C_0^>(\vec{q},\omega)$$

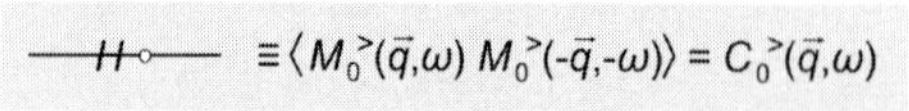

Fig. 10.49 Graphical representation for high-wavenumber contributions to the zeroth-order correlation function.

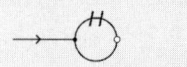

Fig. 10.50 Graphical representation for high-wavenumber contributions to the first-order self-energy.

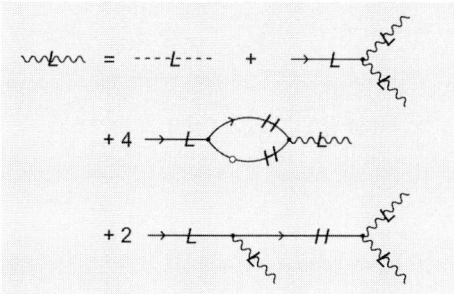

Fig. 10.51 Iterated graphical representation for low-wavenumber contributions to the order parameter to second order.

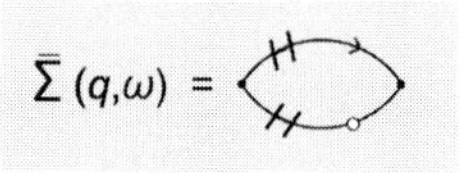

$\bar{\Sigma}(q,\omega) = $ ⟨figure⟩

Fig. 10.52 Graphical representation for high-wavenumber contributions to the second-order self-energy $\bar{\Sigma}(q,\omega)$.

$$\begin{aligned}\mathbf{M}^<(q,\omega) &= \left[1 - G_0^<(q,\omega)\bar{\Sigma}(q,\omega)\right]^{-1} G_0^<(q,\omega)\frac{1}{D_0 q^2}\mathbf{f}^<(\mathbf{q},\omega) \\ &\quad + \left[1 - G_0^<(q,\omega)\bar{\Sigma}(q,\omega)\right]^{-1} G_0^<(q,\omega) \\ &\quad \times \frac{\lambda}{D_0 q^2}\int \frac{d^d k}{(2\pi)^d}\frac{d\Omega}{2\pi}\left[\mathbf{k}^2 - (\mathbf{q}-\mathbf{k})^2\right] \\ &\quad \times \left(\left(\mathbf{M}^<(\mathbf{k},\Omega) \times \mathbf{M}^<(\mathbf{q}-\mathbf{k},\omega-\Omega)\right). \end{aligned} \quad (296)$$

We then note, using Eq. (217), that:

$$\left[1 - G_0^<(q,\omega)\bar{\Sigma}(q,\omega)\right]^{-1} G_0^<(q,\omega)\frac{1}{D_0 q^2} = \frac{1}{\left(G_0^<(q,\omega)\right)^{-1} - \bar{\Sigma}(q,\omega)}\frac{1}{D_0 q^2}$$

$$= \frac{1}{-i\omega + D_0 q^2\left(\chi_0^{-1}(q) - \bar{\Sigma}(q,\omega)\right)} \ . \quad (297)$$

It is left to Problem 10.18 to show in the small q and ω limits that:

$$\bar{\Sigma}(q,\omega) = -\chi_0^{-1}(q)k_B T c^2 (\frac{\lambda}{D_0})^2 I_d \ , \quad (298)$$

where:

$$I_d \equiv \frac{2}{d}\int \frac{d^d k}{(2\pi)^d}\chi_0^3(k)\Theta(k - \frac{\Lambda}{b})\Theta(\Lambda - k). \quad (299)$$

Our new *equation of motion* can then be written as:

$$\mathbf{M}^{(1)}(q,\omega) = \mathbf{M}_0^{(1)}(q,\omega) + G_0^{(1)}(q,\omega)\frac{\lambda}{D_0^{(1)}q^2}\int\frac{d^d k}{(2\pi)^d}\frac{d\Omega}{2\pi}$$
$$\times[\mathbf{k}^2 - (\mathbf{q}-\mathbf{k})^2]\mathbf{M}^{(1)}(\mathbf{k},\omega)\times\mathbf{M}^{(1)}(\mathbf{q}-\mathbf{k},\omega-\Omega)\,, \quad (300)$$

where:

$$\mathbf{M}_0^{(1)}(q,\omega) = \frac{G_0^{(1)}(q,\omega)}{D_0^{(1)}q^2}\mathbf{f}^<(q,\omega) = \frac{\mathbf{f}^<(q,\omega)}{-i\omega + D_0^{(1)}q^2\chi_0^{-1}(q)} \quad (301)$$

$$G_0^{(1)}(q,\omega) = \left[\frac{-i\omega}{D_0^{(1)}q^2} + \chi_0^{-1}(q)\right]^{-1}. \quad (302)$$

Comparing Eqs. (296), (297), (298) and (301), we have:

$$-i\omega + D_0^{(1)}q^2\chi_0^{-1}(q) = -i\omega + D_0 q^2\left(\chi_0^{-1}(q) - \bar{\Sigma}(q,\omega)\right). \quad (303)$$

From this we can identify $D_0^{(1)}$ as:

$$D_0^{(1)} \equiv D_0\left(1 + \frac{k_B T}{c}\left(\frac{\lambda}{D_0}\right)^2 c^3 I_d\right). \quad (304)$$

We see that the full effect of step 1 of the dynamic *RG* is to simply take $D_0 \to D_0^{(1)}$ and λ is unchanged for small q and ω. Higher-order corrections in powers of q and ω are found to be irrelevant [8] in the sense discussed earlier. If we carry out step 2 of the *RG* and rescale we obtain the recursion relations:

$$D_0' = b^{z-4}D_0\left(1 + \frac{k_B T}{c}\left(\frac{\lambda}{D_0}\right)^2 c^3 I_d + \cdots\right) \quad (305)$$

$$\lambda' = b^{(z-1-\frac{d}{2})}\lambda\,.$$

We can use these equations to write a recursion relation for the dimensionless coupling:

$$f = \sqrt{\frac{(k_B T)}{c}}\Lambda^{(d-6)/2}\frac{\lambda}{D_0}\,, \quad (306)$$

given by:

$$f' = b^{3-d/2}f\left(1 - f^2\Lambda^{6-d}c^3 I_d + \mathcal{O}(f^4)\right). \quad (307)$$

Remembering that we have set $r_0 = 0$, we can, as shown in Problem 10.19, evaluate I_d explicitly to obtain for $d \neq 6$:

$$c^3 I_d = \frac{2\tilde{K}_d}{d} \frac{1}{(d-6)} \Lambda^{d-6} \left(1 - b^{6-d}\right) \tag{308}$$

and for $d = 6$:

$$c^3 I_6 = \frac{\tilde{K}_6}{3} \ln b . \tag{309}$$

For $d > 6$ the recursion relation (Eq. 307) can be written for large b as:

$$f' = b^{-(d-6)/2} f \left(1 - \frac{2f^2 \tilde{K}_d}{d(d-6)} + \cdots \right) \tag{310}$$

and f scales to zero for large b. The conventional theory is correct. If $d < 6$ then:

$$f'? = b^{6-d)/2} f \left(1 + \frac{2\tilde{K}_d b^{6-d} f^2}{d(d-6)} + \cdots \right) \tag{311}$$

and we see that for large b the perturbation theory expansion breaks down. If we compare the analysis here with that for the static case we see that we obtain consistency with perturbation theory and a fixed point if we assume that we are very near six dimensions and assume $(f^*)^2 \sim (6-d) \equiv \varepsilon$. Then we can evaluate I_d in six dimensions and write our fixed-point equation, Eq. (307) with $f = f^*$, as:

$$f^* = b^{(6-d)/2} f^* \left(1 - \frac{\tilde{K}_6}{3} (f^*)^2 \ln b + \mathcal{O}(\varepsilon^2)\right) . \tag{312}$$

Since $(f^*)^2$ is very small by assumption, we have after exponentiation:

$$1 = b^{(6-d)/2} b^{-\tilde{K}_6 (f^*)^2 / 3} \tag{313}$$

or:

$$(f^*)^2 = \frac{3}{\tilde{K}_6} \frac{(6-d)}{2} + \mathcal{O}(\varepsilon^2) . \tag{314}$$

Substituting $\tilde{K}_6 = (64\pi^3)^{-1}$ [15], we obtain the fixed-point value [8]:

$$(f^*)^2 = 96(6-d)\pi^3 + \cdots \tag{315}$$

If we then go back to our recursion relation Eq. (305) for D_0, we see that a fixed-point solution requires:

$$D_0^* = b^{z-4} D_0^* \left(1 + \frac{\tilde{K}_6}{3} (f^*)^2 \ln b + \cdots \right) . \tag{316}$$

Exponentiating the logarithm we are led to:

$$z - 4 + \frac{\tilde{K}_6}{3}(f^*)^2 = 0 \tag{317}$$

Thus to first order in $\varepsilon = 6 - d$:

$$\begin{aligned}
z &= 4 - \frac{\tilde{K}_6}{3}(f^*)^2 \\
&= 4 - \frac{\varepsilon}{2} + \mathcal{O}(\varepsilon^2) \\
&= 1 + \frac{d}{2} + \mathcal{O}(\varepsilon^2) \ .
\end{aligned} \tag{318}$$

This is compatible with a fixed-point solution for Eq. (315). If we are bold and let $\varepsilon = 3$ and $d = 3$, we find using Eq. (318) that $z = 5/2$, in agreement with experiment. It has now been shown that this result for z holds to all orders in ε due to certain symmetries of the equation of motion. This does not mean, however, that $(f^*)^2$ is unaffected by high-order terms in ε.

Combining these results with those we found previously for the static parameters, we see that we have found a dynamic fixed point. We could go further and show Ref. [8] that this fixed point is stable. The analysis is along the same lines as in the static case. Instead we note how, once we know the structure of the fixed point, we can make our perturbation theory analysis useful. Let us go back to the expression we had for the transport coefficient resulting from a direct perturbation theory analysis given by Eq. (207). If we assume that our expansion is in powers of f^* and we are near six dimensions, we can write Eq. (207) as:

$$D(0,0) = D_0\left(1 + \frac{(f^*)^2 \tilde{K}_6}{3}\left[\ell n \Lambda \xi - \frac{3}{4}\right] + \mathcal{O}(\varepsilon^2)\right)$$

with $\varepsilon = 6 - d$. Since f^{*2} is small we can rewrite this as:

$$D(0,0) = D_0 (\Lambda \xi)^{(f^*)^2 \tilde{K}_6/3}\left(1 - \frac{f^{*2} \tilde{K}_6}{4} + \mathcal{O}(\varepsilon^2)\right). \tag{319}$$

Remembering Eq. (314) we can rewrite Eq. 319 as:

$$D(0,0) = D_0(\Lambda \xi)^{\frac{\varepsilon}{2}}\left(1 - \frac{3\varepsilon}{8} + \mathcal{O}(\varepsilon^2)\right) \ . \tag{320}$$

This divergence of the transport coefficient as $T \to T_c$ is just the result we need to reconcile the spin-wave definition of the characteristic frequency found in Chapter 8 with the spin-wave damping contribution.

We could go further and calculate $D(0,0)$ to higher powers in ε or, in fact, calculate the complete correlation function Ref. [17], $C(q, \omega)$ using ε as a small parameter. This would take us far afield.

While we can groan over the fact the small parameter for ferromagnets in $\varepsilon = 6 - d$, this is not the case for essentially all other models. It appears that for the dynamics of helium [18], planar ferromagnets [19], isotropic antiferromagnets [19, 20] and fluids [21], $d = 4$ is again the crossover dimension between conventional and nonconventional behavior.

10.4
Final Remarks

In this chapter we have introduced the basic ideas behind the modern theory of dynamic critical phenomena. These are, of course, the ideas of scaling, universality and the nonlinear interaction of long wavelength degrees of freedom. These ideas lead to a renormalization group analysis of semimacroscopic nonlinear equations of motion. There are by now a number of elegant methods for carrying out perturbation theory calculations and these methods will be discussed in FTMCM.

Substantial progress has been made in identifying all of the dynamic universality classes: models that share the same critical surface of a fixed point. We seem to have established that dynamical fixed points are characterized by the underlying static fixed point, the conservation laws governing a system and the Poisson bracket algebra satisfied by the slow variables in the system. Hohenberg and Halperin [22] give an excellent review of the various nonlinear models.

In this chapter we have emphasized dynamical behavior above the phase transition in the disordered state. Very interesting things happen in the ordered state, as indicated at the end of Chapter 8. For example, one has the breakdown in hydrodynamics in the ordered phase of isotropic antiferromagnets [23]. These methods can also be used to explore the breakdown of hydrodynamics in fluid systems [24].

10.5
References and Notes

1 P. C. Martin, E. D. Siggia, H. A. Rose, Phys. Rev. **A8**, 423 (1973).

2 H. K. Janssen, Z. Phys. B **23**, 377 (1976); R. Bausch, H. J. Janssen, H. Wagner, Z. Phys. B **24**, 113 (1976).

3 K. Kawasaki, Prog. Theor. Phys. **52**, 15267 (1974).

4 C. De Dominicis, Nuovo Cimento Lett. **12**, 576 (1975); C. De Dominicis, E. Brezin, J. Zinn-Justin, Phys. Rev. B **12**, 4945 (1975).

5 C. De Dominicis, L. Peliti, Phys. Rev. Lett. **34**, 505 (1977); Phys. Rev. B **18**, 353 (1978).

6 For a more detailed treatment of the method introduced here see S. Ma, *Modern Theory of Critical Phenomena*, Benjamin, N.Y., (1976).

7 B. I. Halperin, P. C. Hohenberg, S. Ma, Phys. Rev. Lett. **29**, 1548 (1972); Phys. Rev. B **10**, 139 (1974) and Phys. Rev. B **13**, 4119 (1976).

8. S. Ma, G. F. Mazenko, Phys. Rev. Lett. **33**, 1383 (1974); Phys. Rev. B **11**, 4077 (1975).
9. S. Ma, *Modern Theory of Critical Phenomena*, Benjamin, N.Y. (1976), p. 510.
10. This integral representation follows from the Plemelj relations discussed in Section 3.3.
11. This is precisely the same logic used in our development of the memory-function formalism.
12. This is the same type approximation as used in calculating the system energy in atomic systems.
13. K. G. Wilson, Phys. Rev. Lett. **28**, 548 (1972). He gives for d-dimensional angular integrals:

$$\int \frac{d^d k}{(2\pi)^d} f(k^2, \mathbf{k}\cdot\mathbf{q}) = \frac{K_{d-1}}{2\pi} \int_0^\infty dk \times \int_0^\pi d\theta\, k^{d-1} (\sin\theta)^{d-2} f(k^2, qk\cos\theta) ,$$

where:

$$\tilde{K}_d^{-1} = 2^{d-1} \pi^{d/2} \Gamma(d/2) .$$

14. This is integral 858.546 in H. B. Dwight, *Tables of Integrals and Other Mathematical Data*, 4th edn., MacMillian, N.Y., (1961).
15. Since $\tilde{K}_d = 1/(\Gamma(d/2)\pi^{d/2} 2^{d-1})$ we have $\tilde{K}_6 = 1/64\pi^3$.
16. We have already investigated the behavior of u and r_0 under the RG in the static case. Since we generate exactly the same statics from our equation of motion we would obtain the same recursion relations.
17. M. J. Nolan, G. F. Mazenko, Phys. Rev. B **15**, 4471 (1977).
18. B. I. Halperin, P. C. Hohenberg, E. D. Siggia, Phys. Rev. Lett. **32**, 1289 (1974); Phys. Rev. B **13**, 1299 (1976); P. C. Hohenberg, B. I. Halperin, E. D. Siggia, Phys. Rev. B **14**, 2865 (1976); E. D. Siggia, Phys. Rev. B **11**, 4736 (1975).
19. B. I. Halperin, P. C. Hohenberg, E. D. Siggia, Phys. Rev. Lett. **32**, 1289 (1974); Phys. Rev. B **13**, 1299 (1976).
20. R. Freedman, G. F. Mazenko, Phys. Rev. Lett. **34**, 1575 (1975); Phys. Rev. B **13**, 1967 (1976).
21. K. Kawasaki, Ann. Phys. N.Y. **61**, 1 (1970); B. I. Halperin, P. C. Hohenberg, E. D. Siggia, Phys. Rev. Lett. **32**, 1289 (1974); E. D. Siggia, B. I. Halperin, P. C. Hohenberg, Phys. Rev. B **13**, 2110 (1976); T. Ohta, K. Kawasaki, Prog. Theor. Phys. (Kyoto) **55**, 1384 (1976).
22. P. C. Hohenberg, B. I. Halperin, Rev. Mod. Phys. **49**, 435 (1977).
23. G. F. Mazenko, M. J. Nolan, R. Freedman, Phys. Rev. B **18**, 2281 (1978); P. Horn, J. M. Hastings, L. M. Corliss, Phys. Rev. Lett. **40**, 126 (1978).
24. D. Forster, D. R. Nelson, M. J. Stephen, Phys. Rev. Lett. **36**, 867 (1976); Phys. Rev. A **16**, 732 (1977); E. D. Siggia, Phys. Rev. A **4**, 1730 (1977).

10.6
Problems for Chapter 10

Problem 10.1: Show that the correlation function, Eq. (14), and the response function, Eq. (7), are related by the FDT Eq. (15).

Problem 10.2: Starting with the graphical expression shown in Fig. 10.1, Eq. (2), iterate in powers of the interaction u, keeping terms up to order u^2 as shown in Fig. 10.2.

Problem 10.3: Starting with the graphs shown in Fig. 10.2, average over the noise and obtain the graphical expression for the response function shown in Fig. 10.5. Show that these graphs lead to the analytic expressions given by Eqs. (19)–(22).

Problem 10.4: Show that the self-energy contributions given by Eqs. (20) and (21) are real.

Problem 10.5: Starting with the graphical expression for ψ shown in Fig. 10.6, find the expansion for $\langle \psi(\mathbf{q},\omega)\psi(\mathbf{q}',\omega')\rangle$, shown in Fig. 10.7.

Problem 10.6: Prove the identity:
$$\text{Im } ABC = ABC'' + AB''C^* + A''B^*C^* \ .$$

Problem 10.7: Starting with the effective Hamiltonian given by Eq. (2), show that to second order in the quartic coupling u, that the static correlation function,
$$C(x-y) = \frac{1}{Z}\int D(\psi) e^{-\beta \mathcal{H}_E} \psi(x)\psi(y) \ ,$$
where Z is the normalizing partition function, is given graphically by Fig. 10.14.

Problem 10.8: Show that the sum of the first two graphs on the right-hand side of Fig. 10.15 can be combined to give the single graph shown in Fig. 10.16.

Problem 10.9: Evaluate the integral:
$$W_d(r) = \int \frac{d^d k}{(2\pi)^d} \frac{1}{ck^2} \frac{1}{r+ck^2}$$
as a function of d in the limit of large $\xi^2 = c/r$.

Problem 10.10: Show that the integral:
$$J_3(0,\omega) = \int_S \frac{1}{ck_1^2} \frac{1}{ck_2^2} \frac{1}{ck_3^2} \frac{1}{[-i\omega + \Gamma_0 c(k_1^2 + k_2^2 + k_3^2)]}$$
diverges as a power of ω and $\omega \to 0$ for $d < 4$. Find the exponent.

Problem 10.11: Show that the small y_1 and y_2 behavior in the integral Eq. (121) is regular and the leading behavior for small Ω is governed by large y_1 and y_2 behavior. Show therefore that $\Delta S(\Omega)$ is given by Eq. (124).

Problem 10.12: Show how one goes from Eq. (131) to Eq. (136) and Eq. (137). Finally do the integral:
$$\int_0^1 du \frac{1}{u}\left[1 - \frac{(1+u^2)}{(1+u^2+u^4)^{1/2}}\right] \ .$$

Problem 10.13: Show that the graphs contributing to $\mathbf{M}$ at second order in λ and up to first order in h are given in Fig. 10.22.

Problem 10.14: Show that the spatial Fourier transform of the three-point vertex defined by Eq. (161) is given by Eqs. (162) and (163).

Problem 10.15: Show that the ferromagnetic self-energies $\Sigma^{(i)}$ and $\Sigma^{(ii)}$, given by Eqs. (181) and (182), reduce to the results given by Eqs. (185) and (187).

Problem 10.16: Evaluate the integral:

$$Q(0,0) = \frac{2k_B T \lambda^2 c^2}{d D_0^2} \int \frac{d^d k}{(2\pi)^d} \chi_0^3(k)$$

as a function of d using the Ornstein–Zernike form for $\chi_0(k)$. Find the leading behavior for $\Lambda \xi \gg 1$.

Problem 10.17: Starting with Eq. (267) and using Eq. (270), show that characteristic frequency satisfies:

$$\omega_c(q; \mu) = b^{-z} \omega_c(qb; \mu') \ .$$

Problem 10.18: Show that Eq. (298) for $\tilde{\Sigma}(q, \omega)$, defined by Fig. 10.52, holds with I_d defined by Eq. (299).

Problem 10.19: Evaluate:

$$I_d \equiv \frac{2}{d} \int \frac{d^d k}{(2\pi)^d} \chi_0^3(k) \Theta\left(k - \frac{\Lambda}{b}\right) \Theta(\Lambda - k)$$

explicitly for $r_0 = 0$ as a function of d.

Problem 10.20: Show that the graph corresponding to the last term in Fig. 10.51 gives no contribution to the low-wavenumber **M** for small enough q due to a noncompatibility of step functions.

Problem 10.21: Starting with Eq. (107), show if $\Delta\Sigma_3(0, \omega)$ exists, then $\Gamma(q, \omega) = \Gamma_0(q)$ to leading order in q in the conserved order parameter case. Therefore there is no correction to the conventional theory result at this order.

Problem 10.22: Consider the anharmonic oscillator governed by the effective Hamiltonian:

$$\mathcal{H}_E[\psi] = \frac{r_0}{2} \psi^2 + \frac{v}{6} \psi^6 - \psi h \ ,$$

where h is an external field that couples to the displacement field ψ. Assume that the dynamics are governed by the Langevin equation:

$$\frac{\partial \psi}{\partial t} = -\Gamma \frac{\partial \mathcal{H}_E[\psi]}{\partial \psi} + \eta \ ,$$

where the Gaussian noise has variance:

$$\langle \eta(t) \eta(t') \rangle = 2 k_B T \Gamma \delta(t - t') \ .$$

Develop perturbation theory for the response function $G(\omega)$. Write down the graphs contributing to $G(\omega)$ to second order in the coupling v. Find the corresponding susceptibility $\chi = G(0)$ to first order in v.

Problem 10.23: Evaluate the zeroth-order response $G_0(q, \omega)$, Eq. (7), and correlation function $C_0(q, \omega)$, Eq. (14), in the q, t domain. What form does the FDT, Eq. (15), take in this regime?

11
Unstable Growth

11.1
Introduction

In this chapter we discuss the problem of the growth of order in unstable thermodynamic systems. For general reviews, see Refs. [1,2]. Let us begin with the conceptually simplest situation. Consider the phase diagram shown in Fig. 11.1 for a ferromagnetic system in the absence of an externally applied magnetic field. At high temperatures, above the Curie temperature T_c, the average magnetization, the order parameter for this system, is zero and the system is in the paramagnetic phase. Below the Curie temperature, for the simplest case of an Ising ferromagnet, one has a nonzero magnetization with two possible orientations: the net magnetization can point in the up-direction or the down-direction. There are two degenerate equilibrium states of the system in the ferromagnetic phase. Now consider [3] the experiment where we first prepare the system in an equilibrium high-temperature state where the average magnetization is zero. We then very rapidly drop the temperature of the thermal bath in contact with the magnet to a temperature well below the Curie temperature. In this case, the magnetic system is rendered thermodynamically unstable. It wants to equilibrate at the new low temperature, but it must choose one of the two degenerate states. Let us look at this problem from the point of view of the TDGL model. Let ψ be the continuous coarse-grained magnetization density and $V(\psi)$ the associated driving potential, shown [4] schematically in Fig. 11.2 before and after the quench. Assume, initially in the paramagnetic phase, ψ is nearly uniform and small in amplitude. Clearly, after the quench, the potential and the associated effective Hamiltonian are minimized by a uniform magnetization with values $+\psi_0$ or $-\psi_0$.

For a quench to zero temperature, [5] the system wants to change the square of the average order parameter ($\langle\psi\rangle^2$) from 0 to ψ_0^2. Right after the temperature quench, the local value of the order parameter will be near zero [6] and the system will be unstable with respect to the two degenerate states. In practice, the system will respond to fluctuations that locally pick out one of the two final states and generate a pattern of ordered domains. Locally, the system

Nonequilibrium Statistical Mechanics. Gene F. Mazenko
Copyright © 2006 WILEY-VCH Verlag GmbH & Co. KGaA, Weinheim
ISBN: 3-527-40648-4

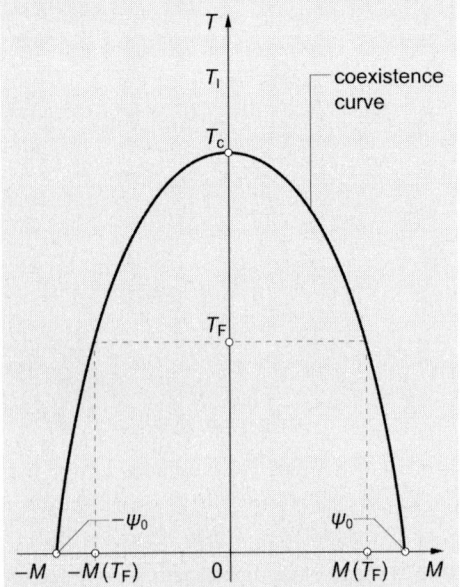

Fig. 11.1 The phase diagram of a ferromagnet. T is the temperature and $M = \langle \psi \rangle$ is the average magnetization, the order parameter in this system. The quench is from T_I in the paramagnetic phase to T_F in the ferromagnetic phase where there are two coexisting degenerate equilibrium states. We assume that the zero temperature value of $M(T = 0)$ is ψ_0.

will take on the values $\pm \psi_0$ while the global average remains zero, $\langle \psi \rangle = 0$. However, as time evolves, competing domains grow inexorably larger, so locally it appears that one has long-range ordering: the longer the time the longer the range.

In order to gain some feeling for the situation (see Problem 11.1) consider the results of a numerical simulations shown in Fig. 11.3. One has a set of Ising spins set on a two-dimensional 64×64 lattice. Initially, panel a, at each site the spin has been randomly chosen to have the values $+1$ or -1. The spins with value $+1$ are represented by a $+$ sign in the figure. The spins with value -1 are left blank. It is then assumed that at time $t = 0$ the system is quenched to zero temperature and then propagated forward in time. Using a standard Monte Carlo algorithm [7, 8], one first selects a spin at random. Then it is flipped if, on average, that flip pushes the system toward thermal equilibrium. After carrying out this process a number of times equal to the number of spins in the system, one has progressed in time one Monte Carlo step per spin (MCS). This choice for a unit of time is sensible, since it is well defined as the number of spins is increased. In Fig. 11.3c one has a typical configuration after 20 MCS where the system is evolving toward the zero temperature state, where the

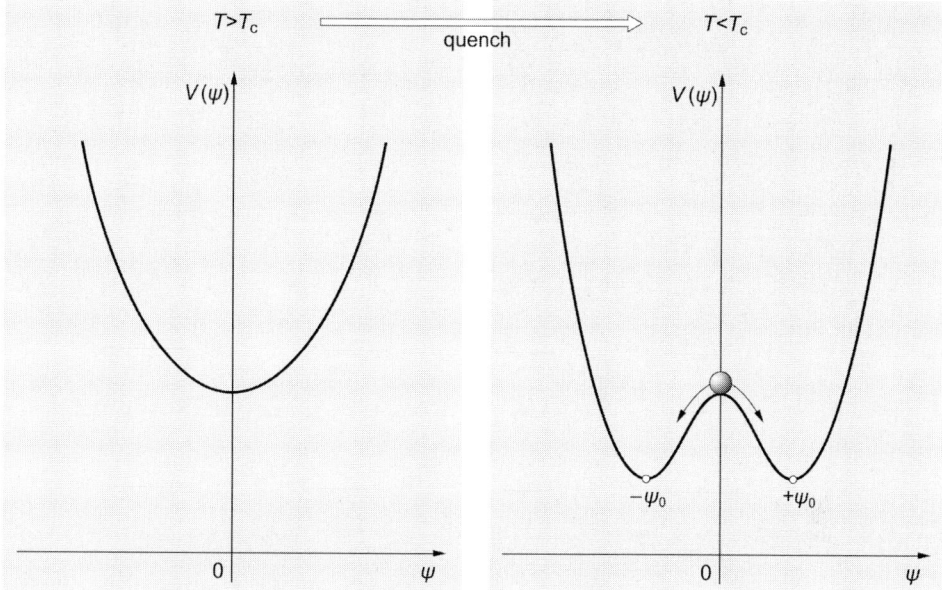

Fig. 11.2 The potential governing the order parameter ψ. As the system is quenched below T_c the potential goes from having a single minimum at $\psi = 0$ to having two degenerate minima at $\psi = \pm\psi_0$. As a result, locally, the system must choose one of these ground states.

spins will all have one or the other of the two possible values. Clearly, as time evolves one can see segregation of like-signed spins into domains. The domains are separated by walls and as *time* proceeds, the curvature of these walls decreases and they tend to become straighter. There is a ramified domain structure where compact objects disappear with time to reveal compact objects on larger scales that, in time, dissolve. The key point is that the size of the domains is growing with time.

There is a close correspondence [9] between the simplest models for Ising magnets and binary alloys. A ferromagnetic transition in an Ising magnet corresponds to phase separation in a binary alloy where like particles *want* to be near one another. An antiferromagnetic transition in an Ising magnet corresponds to an order–disorder transition in a binary alloy. In an antiferromagnet spins alternate signs in adjacent sites in the ordered phase. In a binary alloy that undergoes an order–disorder transition, like Cu_3Au, it is energetically favorable for unlike particles to be close to one another. Time-resolved ordering in alloys is experimentally more accessible than in pure magnet systems because of the large difference in time scales. As indicated below in Fig. 11.9, the phase ordering in alloys takes place on a time scale of minutes. In the case of pure magnetic systems the ordering occurs over microscopic times.

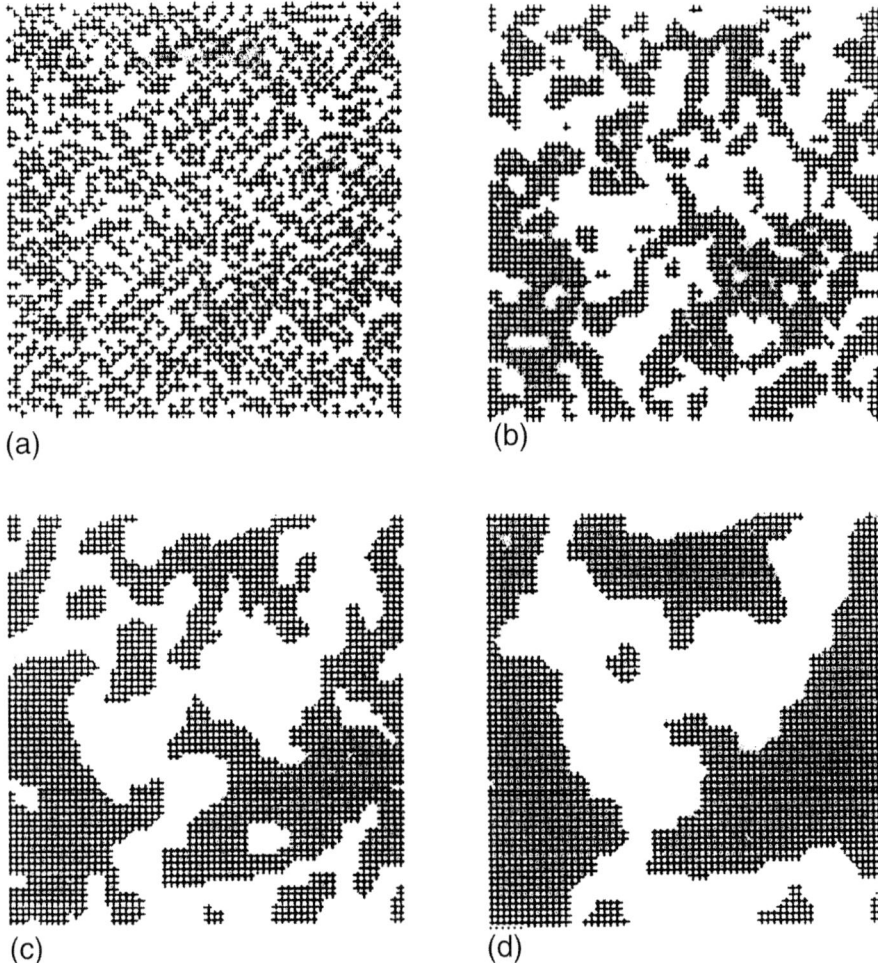

Fig. 11.3a–c. Ising spins evolving under a Monte Carlo algorithm. **a** The initial state with the spin value at each site randomly chosen. **b** The system after 5 Monte Carlo steps (MCS). **c** 20 MCS. **d** 60 MCS.

There are many important physical examples of phase-ordering kinetics beyond simple magnets and alloys. Phase separation in fluids is different from that in solids, since it involves flow that complicates the process. There are a variety of different fluid systems that undergo phase separation, called spinodal decomposition, which are of interest: liquid–gas, binary mixture, polymer mixtures, block copolymers, soap froths, and modulated phases.

We have distinguished between the case of pure magnets and systems with impurities because quenched magnetic impurities can strongly change the ordering properties. It is well known that diluted antiferromagnets in a field and spin glasses [10] order over very long time scales. The discussion here is limited to systems without quenched impurities.

In the examples above, we have emphasized the case of a scalar order parameter where the disordering agents are domain walls. For those cases, where ordering consists of the breaking of a continuous symmetry, [11, 12] the disordering defects are vortices, strings, monopoles, dislocations, disclinations or other more exotic objects. The simplest model for looking at these higher order symmetries is the n-vector model [13]. Important examples include: XY magnets ($n = 2$), superfluid He^4 ($n = 2$), and Heisenberg magnets ($n = 3$). More complicated systems involve superconductors, liquid crystals, the growth of crystals and polymers.

One of the interesting aspects of phase-ordering kinetics is that the simple model descriptions of the problem produce the growth patterns observed experimentally. In Fig. 11.4 we show the phase-ordering pattern from a binary mixture as a function of time after quench. Note the time scale of hundreds of seconds. The early time evolution in this case produces a ramified structure similar to that seen in the simulation shown in Fig. 11.3. At later times, the system phase separates into compact structures.

The ultimate phase-ordering system corresponds to the evolution of the early universe. It seems clear that many of the ideas we have established in a condensed-matter setting will be useful in understanding cosmology [15].

11.2
Langevin Equation Description

Early theoretical work on growth kinetics focused on spin-flip and spin-exchange kinetic Ising models [7], similar to the model used to produce the data in Fig. 11.3. These models were appealing because of the discrete Ising arithmetic, but the associated dynamics is not particularly physical and it is difficult to generalize to more complicated systems like the n-vector model. Most work has centered around the treatment of field theoretic Langevin models [16], like the TDGL equation governing the dynamics of a nonconserved order parameter (NCOP):

$$\frac{\partial}{\partial t}\psi(\mathbf{x},t) = -\Gamma \frac{\delta \mathcal{H}_E}{\delta \psi(\mathbf{x},t)} + \xi(\mathbf{x},t) \quad , \tag{1}$$

where $\psi(\mathbf{x},t)$ is the order parameter field and $\mathcal{H}_E$ is the LGW effective Hamiltonian:

$$\mathcal{H}_E = \int d^d x \left[\frac{c}{2}(\nabla \psi)^2 + V[\psi(\mathbf{x})] \right]. \tag{2}$$

Fig. 11.4 Experimental realization [14] (via a polarizing optical microscope) of phase separation in a polymer mixture. The time after the quench is indicated at the *bottom right* of each picture.

Here, $c > 0$ and V is the potential, typically of the form:

$$V = \frac{r}{2}\psi^2 + \frac{u}{4}\psi^4 \tag{3}$$

with $u > 0$. The simplest nontrivial form for r is $r = r_0(T - T_c^0) < 0$ for unstable growth [17]. $\xi(\mathbf{x}, t)$ is Gaussian thermal noise with variance:

$$\langle \xi(\mathbf{x}, t)\xi(\mathbf{x}', t') \rangle = 2\Gamma k_B T \delta(\mathbf{x} - \mathbf{x}')\delta(t - t') \;, \tag{4}$$

where Γ is a bare kinetic coefficient and T is the final equilibrium temperature.

The physical situation of interest is a quench [18] from a disordered state $T > T_c$ at $t = 0$ where:

$$\langle \psi(\mathbf{x}, 0) \rangle = 0 \;, \tag{5}$$

$$\langle \psi(\mathbf{x}, 0)\psi(\mathbf{x}', 0) \rangle = \varepsilon_I \delta(\mathbf{x} - \mathbf{x}') \tag{6}$$

to a state $T < T_c$. Much of our attention will be focused on quenches to $T = 0$. Quenches to nonzero T ($< T_c$) do not typically lead to qualitative change [19] in the basic findings. If we quench to $T = 0$, we can set the thermal noise to zero and Eq. (1) takes the form:

$$\frac{\partial}{\partial t}\psi(\mathbf{x}, t) = -\Gamma \left[r\psi(\mathbf{x}, t) + u\psi^3(\mathbf{x}, t) - c\nabla^2 \psi(\mathbf{x}, t) \right] \;, \tag{7}$$

where the potential V is given by Eq. (3). The important ingredient is that this nonlinear partial differential equation is supplemented with random initial conditions given by Eqs. (5) and (6). Assuming $r < 0$, which is necessary for unstable growth, we can rescale $\psi \to \sqrt{\frac{|r|}{u}}\psi$, $x \to \sqrt{\frac{c}{|r|}}x$, and $t \to \frac{t}{\Gamma|r|\sqrt{\frac{|r|}{u}}}$, to obtain the dimensionless equation of motion:

$$\frac{\partial}{\partial t}\psi = \psi - \psi^3 + \nabla^2 \psi \;. \tag{8}$$

Notice that there are no parameters left in the problem.

The stationary uniform solutions of this equation of motion are given by:

$$\psi = 0 \text{ and } \pm 1 \;. \tag{9}$$

Since $V[\pm 1] = -\frac{r^2}{4u}$ while $V[0] = 0$, we see that the nonzero values of ψ are the thermodynamically stable solutions.

The equation of motion given by Eq. (8) corresponds to a NCOP. In problems like phase separation one has a conserved order parameter (COP). One

Fig. 11.5a–c. Numerical simulation [20] of the conserved order parameter (COP) TDGL model. **a** $t = 150$ (time in dimensionless units). **b** $t = 600$. **c** $t = 2200$.

then has the dimensionless partial differential equation for quenches to zero temperature given by:

$$\frac{\partial}{\partial t}\psi = \nabla^2 \left[-\psi + \psi^3 - \nabla^2 \psi \right] , \tag{10}$$

which is again supplemented with random initial conditions. An example of the patterns generated by Eq. (10) is shown in Fig. 11.5.

The TDGL model includes only dissipative terms. There are many systems where reversible terms [21] must also be included. In the case of fluids, for example, one must include the coupling of the order parameter, the mass density ρ, to the momentum density **g**. We will focus on the simpler systems without flow here.

11.3
Off-Critical Quenches

In the quench shown in the phase diagram in Fig. 11.1, the average value of the order parameter is zero along the entire path of the quench. One can think of situations where this is difficult to implement in practice or where one is interested in the case where at $t = 0$ $\langle \psi \rangle = m > 0$. In this case, it makes a qualitative difference whether the order parameter is conserved or not. If the order parameter is not conserved, then the initial condition breaks the degeneracy between the final states and the system orders locally in the biased $+\psi_0$ orientation. This can be treated quantitatively, since the average of the order parameter is nonzero and we can write:

$$\psi(\mathbf{x}, t) = m(t) + \phi(\mathbf{x}, t) , \tag{11}$$

where by assumption $\langle \phi(\mathbf{x}, t) \rangle = 0$. The equation of motion for the uniform part of the order parameter is given by substituting $\psi(\mathbf{x}, t) = m(t)$ into Eq. (8):

$$\frac{1}{\Gamma}\frac{d}{dt}m(t) = m(t)\left[1 - m^2(t)\right] , \tag{12}$$

where φ, treated as a perturbation, does not enter at lowest order. It is shown in Problem 11.2 how to solve this equation as an initial-value problem. It is not difficult to show that for long times this equation is satisfied by:

$$1 - m \approx e^{-2\Gamma t}. \tag{13}$$

This case is not a problem of degenerate competing states and the system equilibrates over a rather short-time scale. The situation in the COP case is quite different. The way of thinking about this is to imagine that one has a mixture of A and B particles. For a critical (symmetric) quench the number

Fig. 11.6.a–d Numerical simulation [22] of the COP TDGL model for off-critical quenches. The system with $m = 0$ at $t = 500$ (**a**) and $t = 5000$ (**b**). The system with $m = 0.4$ at $t = 500$ (**c**) and $t = 5000$ (**d**). *Shaded regions* correspond to $\psi < 0$.

of A is equal to the number of B particles. An off-critical quench corresponds to having more, say, A particles than B particles. This breaks the symmetry, but because of the conservation law one can not *eat up* the minority phase of B particles. The system must again break up into domains. There is now a minority and a majority phase and the minority phase may form compact structures, as can be seen in Fig. 11.6, which is again a simulation using Eq. (10) but now with $\langle \psi \rangle \neq 0$.

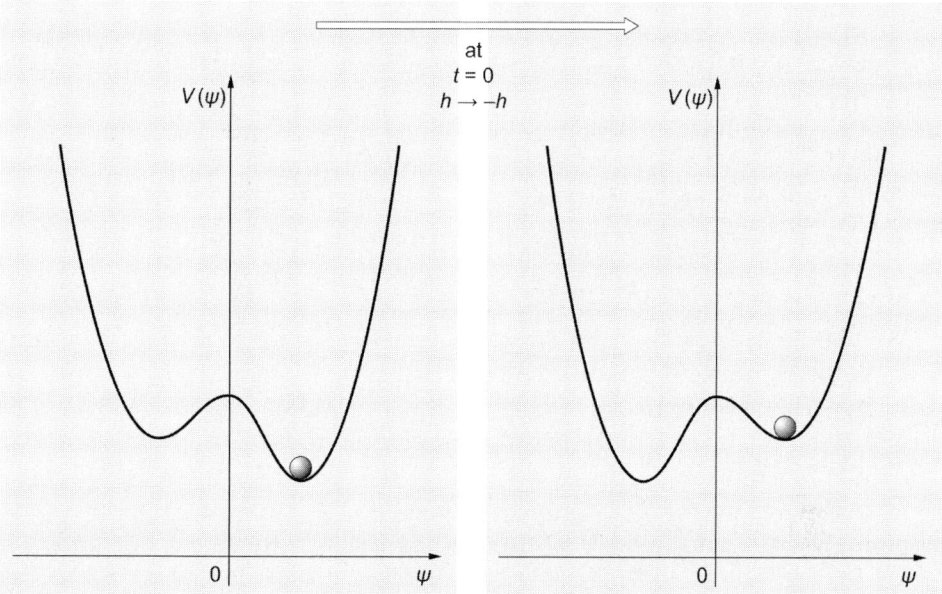

Fig. 11.7 The potential governing the order parameter ψ in the case of nucleation. For $t < 0$ the system is in the lower energy equilibrium state. At $t = 0$ the field h is flipped and the system finds itself in a metastable state. For $t > 0$ fluctuations cause the system to nucleate the state of lower energy.

11.4
Nucleation

Nucleation theory [23] usually refers to a physical situation where one has metastable rather than unstable growth, as shown schematically in Fig. 11.7. One can prepare a magnetic system in a metastable phase by putting the system in an ordered state in the presence of an applied field and then flipping the field as shown in Fig. 11.7. Then the problem is to kinetically surmount a free energy barrier to access the thermodynamically stable state.

The simplest example is where one has a potential:

$$V = -\frac{\psi^2}{2} + \frac{\psi^4}{4} - h\psi \quad (14)$$

and prepares the system in the equilibrium state aligned with h. Then at time $t = 0$, flip $h \to -h$. The system is then in the *wrong* equilibrium state. How does it grow or nucleate the new equilibrium state? This depends strongly on initial or temperature fluctuations.

Examples of metastable systems include supercooled vapor and the ordered magnetic systems in a reversed field mentioned above. With proper care, the vapor can remain supercooled for a long time. Eventually the stable equilibrium liquid phase will show itself via nucleated bubbles of liquid that grow to produce the final liquid phase. These systems are metastable because only droplets larger than some critical size will grow. Droplets smaller than this critical size shrink back into the gas. The question is: given a metastable state, how long on average will it take to produce a growing droplet that leads to the ordering of the system?

The nucleation rate can be written in the form [24]:

$$I = I_0 e^{-\beta \Delta F} , \qquad (15)$$

where ΔF is the free energy barrier that must be overcome in order to nucleate the stable state. This is the dominant factor in I. The prefactor I_0, which is associated with an attempt frequency, is both much less sensitive to the undercooling and much more difficult to compute with precision. The calculation of ΔF is not difficult.

The free energy of a droplet is the sum of a surface term, proportional to the surface tension of the droplet and a bulk term corresponding to the condensation energy [25] of the droplet. These terms balance since the surface term is positive while the bulk term is negative, reflecting the lower energy of the minority phase in the metastable state.

Let us see how we can construct these two contributions to the free energy. Droplets are just compact interfaces. In Section 12.2 of FOD we found the interfacial solution for the Euler–Lagrange equation:

$$\frac{\delta \mathcal{H}_E}{\delta \psi(\mathbf{x})} = 0 , \qquad (16)$$

where $\mathcal{H}_E$ is of LGW form with the potential given by Eq. (14). More explicitly, Eq. (16) takes the form:

$$-\nabla^2 \psi - \psi + \psi^3 - h = 0 , \qquad (17)$$

where, for $h = 0$, we have the interfacial solution:

$$\psi = \psi_0 \tanh\left(\frac{z - z_0}{\sqrt{2}}\right) , \qquad (18)$$

where for our choice of parameters $\psi_0 = 1$ and z_0 is arbitrary. We need to generalize this to the case of spherical interfaces (droplets) for the case of a small applied field h. For small h, the radius of the droplet is large and the *flat* approximation is accurate:

$$\psi_D(r) = \frac{1}{2}(\psi_+ + \psi_-) + \frac{1}{2}(\psi_+ - \psi_-)\tanh\left(\frac{r - R_0}{\sqrt{2}}\right) , \qquad (19)$$

where, to first order in h, the two competing solutions are:

$$\psi_\pm = \pm\psi_0 + \frac{h}{2} \qquad (20)$$

and R_0 is the radius of the droplet. The droplet profile satisfies the boundary conditions:

$$\psi_D(0) = \psi_- \qquad (21)$$

$$\psi_D(\infty) = \psi_+ \ . \qquad (22)$$

The free energy of the droplet is given by:

$$\begin{aligned} F_D &= \mathcal{H}_E(\psi_D) - \mathcal{H}_E(\psi_+) \\ &= F_s + F_B + \mathcal{O}(h^2) \ , \end{aligned} \qquad (23)$$

where F_s is the contribution from the square gradient term in $\mathcal{H}_E$ and F_B is from the potential term in $\mathcal{H}_E$. It is easy to show (Problem 11.4), that:

$$F_s = \sigma \Sigma_d(R_0) \ , \qquad (24)$$

where the surface tension is given by:

$$\sigma = \int_{-\infty}^{\infty} dx \left(\frac{d\psi(x)}{dx} \right)^2 \qquad (25)$$

and $\psi(x)$ is the interfacial profile. The surface area of a droplet is given in d dimensions by:

$$\Sigma_d(R_0) = \frac{2\pi^{d/2}}{\Gamma(d/2)} R_0^{d-1} \ . \qquad (26)$$

The potential contribution can be evaluated analytically if one works to linear order in h. As shown in in Problem 11.5 in the sharp-interface limit we have:

$$F_B = -2\psi_0 V_d |h| \ , \qquad (27)$$

where h is negative if the down state is preferred and the volume of the droplet is given in d dimensions by:

$$V_d = \frac{2\pi^{d/2}}{d\Gamma(d/2)} R_0^d \ . \qquad (28)$$

We have then, remembering that we have flipped the field and $h = -|h|$,

$$F_D = -2\psi_0 V_d |h| + \sigma \Sigma_d \ . \qquad (29)$$

We find the maximum value for F_D, which is the activation energy ΔF, by taking the derivative with respect to R_0 and setting the result to zero:

$$\frac{dF_D}{dR_0} = 0 = -2\psi_0 V_d |h| \frac{d}{R_0} + \sigma \Sigma_d \frac{d-1}{R_0} \ . \qquad (30)$$

In Problem 11.6 we show that this leads to a maximum. The critical droplet radius is given by:

$$R_{0,c} = \frac{\sigma(d-1)}{2\psi_0|h|} \qquad (31)$$

and the associated free energy:

$$\Delta F = F_D(R_{0,c}) = 2\pi^{d/2}\frac{(d-1)^{d-1}}{d\Gamma(d/2)}\frac{\sigma^d}{[2\psi_0|h|]^{d-1}} \qquad (32)$$

thus we find that the activation energy is a very sensitive function of the flipped field h.

To go further and compute the prefactor in the nucleation rate is much more involved [23, 24].

11.5
Observables of Interest in Phase-Ordering Systems

What are the physical observables we should use in order to quantify our analysis of the growth of order in unstable systems? For the case of critical quenches in an infinite system we have:

$$\langle \psi(\mathbf{x},t) \rangle = 0 \qquad (33)$$

for all times. Therefore, the average of the order parameter carries little information except as a check in simulations that finite size effects are not important. Instead we must look at the equal-time order parameter correlation function:

$$C(\mathbf{x},t) \equiv \langle \psi(\mathbf{x},t)\psi(\mathbf{0},t) \rangle , \qquad (34)$$

where the average is over the noise and initial conditions. We are also interested in the Fourier transform:

$$C(\mathbf{q},t) = \langle |\psi(\mathbf{q},t)|^2 \rangle , \qquad (35)$$

which is a time-dependent structure factor. $C(\mathbf{q},t)$ is measured in neutron and x-ray scattering from alloys and in light scattering from fluids. The signature of growth and ordering is the growth of a Bragg peak at ordering [26] wavenumber(s) $\mathbf{q}_0$. The basic phenomenology for the case of a scalar order parameter is that the structure factor is the sum of two pieces:

$$C(\mathbf{q},t) = C_{\text{peak}}(\mathbf{q},t) + \frac{k_B T}{q^2 + \xi^{-2}} . \qquad (36)$$

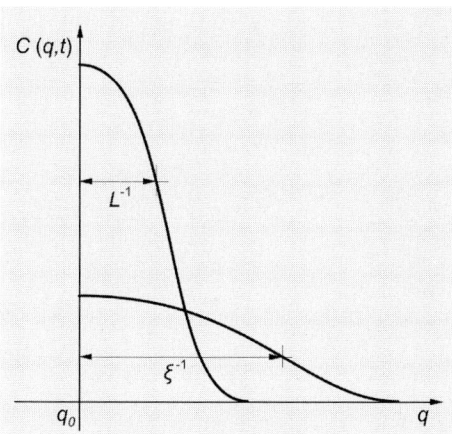

Fig. 11.8 A schematic picture of the structure factor for the nonconserved order parameter (NCOP) case at a late time when there are well-formed domains. There is a Bragg peak contribution whose width is proportional to L^{-1} and an equilibrium Ornstein–Zernike contribution with a width proportional to ξ^{-1}.

Here $C_{\text{peak}}(\mathbf{q}, t)$ is the Bragg peak contribution, which is evolving to the final form:

$$C_{\text{peak}}(\mathbf{q}, t) = \psi_0^2 (2\pi)^d \delta(\mathbf{q} - \mathbf{q_0}) \ . \tag{37}$$

The equilibrated component on the right of Eq. (36) is approximately of the Ornstein–Zernike form. The basic form of the structure factor is shown in Fig. 11.8 for the case of an NCOP. The important point is that there are two peaks characterized by two inverse lengths. The equilibrated piece is characterized by the equilibrium correlation length ξ, while the evolving Bragg peak contribution has a width that is inversely proportional to a characteristic domain size $L(t)$. This can be seen in the x-ray scattering results shown in Fig. 11.9.

In the COP case, $C_{\text{peak}}(\mathbf{q}, t) \to 0$ as $q \to 0$ because of the conservation law. One expects, except for uninteresting terms that depend on the initial conditions, that for small wavenumbers $\psi(\mathbf{q}, t) \sim q^2$, so that:

$$C(\mathbf{q}, t) \sim (q^2)^2. \tag{38}$$

Thus one has a structure factor as shown in Fig. 11.10 for a COP. In this case the position of the maximum and the width of the peak in the structure factor are proportional to L^{-1}.

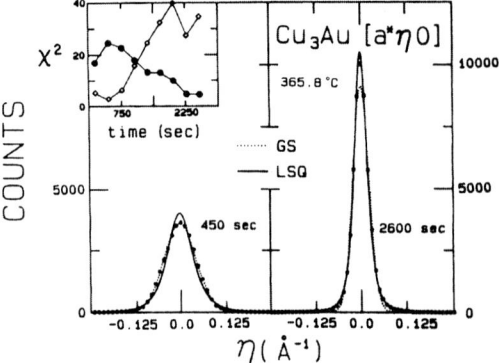

Fig. 11.9 X-ray scattering data [27] for Cu_3Au showing the equilibrium Ornstein–Zernike form and the Bragg peak corresponding to the growing domains. As time passes the Bragg peak becomes taller and sharper.

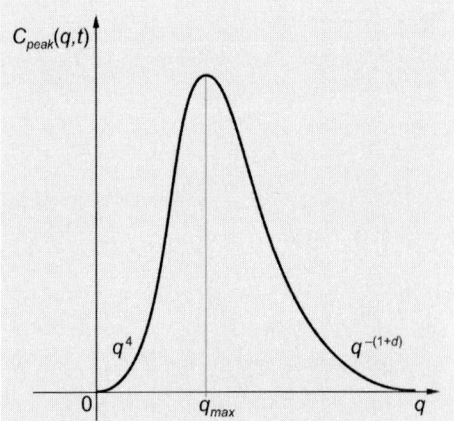

Fig. 11.10 A schematic picture of the structure factor for the COP case. The conservation law is responsible for the q^4 behavior at small q. One has Porod's law $q^{-(1+d)}$ behavior at large q, (see text) a result of scattering from sharp interfaces. The position of the maximum in the structure factor is proportional to L^{-1}. For $L \gg \xi$, the length scale L dominates the problem and one finds that the width of the peak is also proportional to L^{-1}.

11.6
Consequences of Sharp Interfaces

As indicated above, it is important to realize that there are two characteristic lengths in the growth kinetics problem, L and ξ. For long times, $L \gg \xi$ and one has sharp interfaces [28]. Let us consider a method for estimating the contribution of sharp interfaces to the order parameter averages. We assume that the order parameter profile of a domain wall at $z = z_0$ of width ξ can be written in the form:

$$\psi(z) = \psi_0 \tanh[(z-z_0)/\xi] \quad . \tag{39}$$

Then, if there are no other walls around, one can estimate the contribution to the correlation function by averaging over the position of the interface:

$$C(z,z') = \frac{\int_{-L}^{+L} dz_0 \psi(z)\psi(z')}{\int_{-L}^{L} dz_0} \quad . \tag{40}$$

While this average can be worked out explicitly, it is useful to realize that for sharp interfaces with $L \gg \xi$ and for separations $|z-z'| \gg \xi$, we can use the approximation:

$$\psi = \psi_0 sgn(z-z_0) \quad . \tag{41}$$

Inserting Eq. (41) into Eq. (40) we can do the integrations (see Problem 11.7) and obtain:

$$C(z,z') = \psi_0^2 \left[1 - \frac{|z-z'|}{L} + \cdots \right] . \tag{42}$$

While this argument is crude and one-dimensional in nature, it can be generalized [29] to three dimensions and for a collection of sharp interfaces to obtain:

$$C(\mathbf{x},t) = \psi_0^2 \left[1 - \alpha \frac{|\mathbf{x}|}{L} + \cdots \right] , \tag{43}$$

where $\xi \ll x \ll L$ and L can be identified with a typical domain size and α is a number that depends on the precise definition of L. Notice that the term linear in x represents a nonanalytic correction to the leading term. In terms of the structure factor one obtains (see Problem 11.8) for large wavenumbers that:

$$C(\mathbf{q},t) \sim q^{-(1+d)} , \tag{44}$$

which is known as Porod's law [30]. An extension of the result Eq. (43) is that only odd powers of x occur in the expansion of $C(\mathbf{x},t)$ beyond the leading term. The absence of a term of $\mathcal{O}(x^2/L^2)$ is known as the Tomita sum rule [29].

Another simple consequence of sharp interfaces is the estimate [31] one can give for the order parameter autocorrelation function.:

$$S(t) = \langle \psi^2(\mathbf{x},t) \rangle \quad .$$

We know that for long enough times, the order parameter field is equal to $\pm\psi_0$ except where it is equal to zero at an interface. The volume over which the order parameter is zero can be estimated to be proportional to:

$$A\frac{L^{d-1}}{L^d}\xi , \qquad (45)$$

where A is some positive time-independent constant. We can then estimate that $S(t)$ is given by:

$$S(t) = \psi_0^2 \left(1 - \frac{A\xi}{L} + \cdots\right) , \qquad (46)$$

whatever the time dependence of L. This is a nontrivial result from a theoretical point of view that is easily checked numerically.

11.7
Interfacial motion

There is an interplay between the approach that emphasizes the behavior of isolated defects [32] and the statistical approach where one thinks in terms of averaging over many defects. Let us start by looking at the motion of an individual interface. In this case, after specifying the shape of an interface, we are interested in its subsequent motion: an initial value problem. Consider a spherical drop with an interior with an order parameter with value $-\psi_0$ embedded in a sea where the order parameter has the value $+\psi_0$. Let us consider the dynamic evolution of this assumed initial state. It evolves according to the TDGL equation of motion rewritten in the form:

$$\frac{\partial}{\partial t}\psi(\mathbf{x},t) = -\hat{\Gamma}(x)\mu[\psi(x)] , \qquad (47)$$

where the chemical potential is given by:

$$\mu(\mathbf{x},t) = \frac{\delta\mathcal{H}_E}{\delta\psi(\mathbf{x},t)} = V'(\psi) - \nabla^2\psi . \qquad (48)$$

Equation (47) is subject to the boundary condition at time t_0 that there is a sharp spherical drop with profile:

$$\psi(\mathbf{x},t_0) = g[r - R_0(t_0)] , \qquad (49)$$

where $g'(r)$ is peaked at $r \approx 0$ and:

$$g[-R_0(t_0)] \approx -\psi_0 \qquad (50)$$

and:
$$g(\infty) \approx +\psi_0 \ . \tag{51}$$

We are interested in how the radius of the droplet $R_0(t)$ evolves in time. Near the interface, where the order parameter goes through zero as it changes sign,
$$V'(\psi) \approx V''(0)\psi \equiv -V_0 g[r - R_0(t)] \tag{52}$$

and:
$$\nabla^2 \psi = \frac{(d-1)}{r} g'[r - R_0(t)] + g''[r - R_0(t)] \ , \tag{53}$$

so the chemical potential takes the form:
$$\mu(r) = -V_0 g[r - R_0(t)] - \frac{d-1}{r} g'[r - R_0(t)] - g''[r - R_0(t)]. \tag{54}$$

Multiplying this equation by $g'[r - R_0(t)]$ and integrating across the interface we obtain:
$$\begin{aligned}\int_{-\varepsilon}^{+\varepsilon} \mu(z) dz g'(z) &= -V_0 \int_{-\varepsilon}^{+\varepsilon} dz \frac{1}{2} \frac{d}{dz} g^2(z) - \frac{(d-1)}{R_0} \int_{-\varepsilon}^{+\varepsilon} dz [g'(z)]^2 \\ &\quad - \int_{-\varepsilon}^{+\varepsilon} dz \frac{1}{2} \frac{d}{dz} [g'(z)]^2 \ ,\end{aligned} \tag{55}$$

where $R_0(t) \gg \varepsilon \gg \xi$. We can manipulate each of the four terms:
$$\begin{aligned}\int_{-\varepsilon}^{+\varepsilon} \mu(z) dz g'(z) &= \mu(0) \int_{-\varepsilon}^{+\varepsilon} dz g'(z) \\ &= \mu(0)[g(\varepsilon) - g(-\varepsilon)] = 2\psi_0 \mu(0)\end{aligned} \tag{56}$$
$$\int_{-\varepsilon}^{+\varepsilon} dz \frac{1}{2} \frac{d}{dz} g^2(z) = \frac{1}{2}[g^2(\varepsilon) - g^2(-\varepsilon)] = 0 \tag{57}$$
$$\int_{-\varepsilon}^{+\varepsilon} dz [g'(z)]^2 \approx \int_{-\infty}^{+\infty} dz [g'(z)]^2 \tag{58}$$

and finally we have:
$$\int_{-\varepsilon}^{+\varepsilon} dz \frac{1}{2} \frac{d}{dz} [g'(z)]^2 = \frac{1}{2}[(g'(\varepsilon))^2 - (g'(-\varepsilon))^2] = 0. \tag{59}$$

These results together reduce Eq. (55) to the Gibbs–Thomson relation [33]:
$$\mu(0) = -\frac{K\sigma_S}{2\psi_0} \ , \tag{60}$$

where $K = (d-1)/R_0(t)$ is the curvature and:
$$\sigma_S = \int_{-\infty}^{+\infty} dz [g'(z)]^2 \tag{61}$$

is the surface tension.

In the NCOP case the equation of motion, Eq. (47), takes the form:

$$\frac{\partial}{\partial t}\psi(\mathbf{x},t) = g'[R - R_0(t)]\left(-\frac{d}{dt}R_0(t)\right) = \Gamma\mu \ . \tag{62}$$

Multiplying by g', integrating across the interface, and using Eq. (60) we obtain:

$$\sigma_S \frac{d}{dt}R_0(t) = -\frac{(d-1)\Gamma}{R_0(t)}\sigma_S \ . \tag{63}$$

Canceling the common factor of σ_S, we can easily integrate to obtain:

$$R^2(t) = R^2(0) - 2(d-1)\Gamma t. \tag{64}$$

Thus, if we have a spherical droplet it will shrink with a radius decreasing with time as given by Eq. (64). This result was checked numerically in Ref. [34].

Let us turn to the COP case, which is more involved. The equation of motion is given by:

$$\frac{\partial}{\partial t}\psi(\mathbf{x},t) = \nabla^2 \mu[\psi(\mathbf{x},t)] \tag{65}$$

and assumes that all of the time evolution of the order parameter occurs near the interface. Therefore we assume we have a boundary value problem where, away from the interface,

$$\nabla^2 \mu = 0 \tag{66}$$

and $\mu(R)$ is specified on the surface of the sphere by the Gibbs–Thomson relation Eq. (60). The solution is shown in Problem 11.9. to be given by:

$$\mu(R) = -\frac{2\sigma_S}{2\psi_0 R_0(t)}\Theta[R_0(t) - R] - \frac{2\sigma_S}{2\psi_0 R}\Theta[R - R_0(t)]. \tag{67}$$

This yields the result:

$$\nabla^2 \mu = \frac{2\sigma_S}{2\psi_0 R_0^2(t)}\delta[R - R_0(t)] \ , \tag{68}$$

which can be substituted back into the equation of motion to obtain:

$$g'[R - R_0(t)]\left(-\frac{d}{dt}R_0(t)\right) = D_0 \frac{2\sigma_S}{2\psi_0 R_0^2(t)}\delta[R - R_0(t)]. \tag{69}$$

Again, integrating this equation across the interface results in the equation for the radius:

$$\frac{d}{dt}R_0(t) = -\frac{2D_0\sigma_S}{(2\psi_0)^2 R_0^2(t)} \ , \tag{70}$$

which has the solution:

$$R_0^3(t) = R_0^3(0) - \frac{6 D_0 \sigma_s t}{(2\psi_0)^2} \quad . \tag{71}$$

Thus the droplet radius decays as $t^{1/3}$ for the conserved case.

Ideas of this type can be used to develop a theory [35–38] for the evolution of a low-density set of minority phase droplets immersed in a sea of a majority phase. One finds a scaling result and that the characteristic size of the minority phase droplets increases as $t^{1/3}$.

11.8
Scaling

Scaling is the most important statistical property of coarsening domain structures. Scaling was established by Marro, Lebowitz, and Kalos [39] in their numerical simulations of evolving alloy systems and asserts for the ordering contribution to the structure factor that:

$$C(\mathbf{q}, t) = L^d(t) \psi_0^2 \tilde{F}[qL(t)] \tag{72}$$

$$C(\mathbf{x}, t) = \psi_0^2 F[|\mathbf{x}|/L(t)] \tag{73}$$

where $L(t)$ is the growing characteristic length or growth law. Experimentally, confirmation of scaling in a phase-separating binary alloy is shown in the neutron-scattering results of Hennion et al. [40] in Fig. 11.11.

This scaling is a manifestation of the self-similarity of the morphological structure of the system under a spatial rescaling. This reflects the comment above that compact structures appear at every length scale as time evolves.

As first emphasized by Furukawa [41], there is an interesting two-time scaling:

$$\begin{aligned} C(\mathbf{x}, t_1, t_2) &= \langle \psi(\mathbf{x}, t_1) \psi(\vec{0}, t_2) \rangle \\ &= \psi_0^2 F[\mathbf{x}/L(t_1), L(t_1)/L(t_2)] \quad , \end{aligned} \tag{74}$$

and one finds that the autocorrelation function is governed [42] by the nonequilibrium exponent λ:

$$C(\mathbf{0}, t_1, t_2) \sim \left[\frac{\sqrt{t_1 t_2}}{T}\right]^\lambda \quad , \tag{75}$$

where $T = \frac{1}{2}(t_1 + t_2)$ and one of the times t_1 or t_2 is much larger than the other.

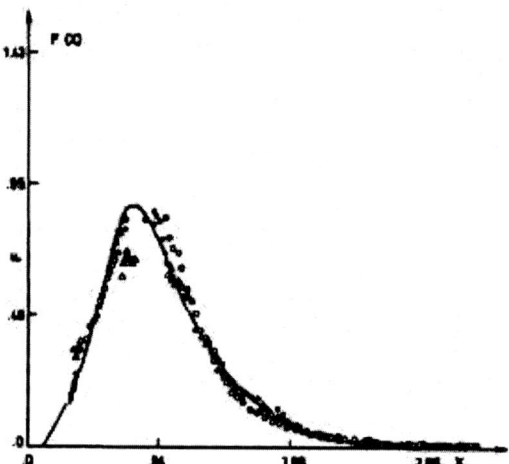

Fig. 11.11 Neutron-scattering data [40] from a binary alloy supports scaling. Here the scaling function for the structure factor, $\tilde{F}(x)$, is plotted against scaled wavenumber x. See Eq. (72).

One of the consequences of scaling is that we can combine it with the droplet calculation described above. In the NCOP case we typically do not have a collection of droplets, but we can talk about the local curvature near an interface with a characteristic inverse curvature R_0, which we showed in Eq. (63) satisfies:

$$\dot{R}_0 = -\frac{\Gamma(d-1)}{R_0} = -\Gamma K. \tag{76}$$

If there is a single characteristic length, $L(t)$ then we argue, essentially by dimensional analysis, that:

$$\dot{L}(t) \sim \Gamma \bar{K} \sim \Gamma/L \tag{77}$$

or:

$$L\dot{L} \sim \text{constant}, \tag{78}$$

which leads to the result for the growth law:

$$L \sim t^{1/2}, \tag{79}$$

which is known as the Lifshitz–Cahn–Allen growth law [43].

In the COP case, using similar arguments, we estimate:

$$\dot{R}_0 \sim \frac{1}{(R_0)^2}, \tag{80}$$

which gives the Lifshitz–Slyozov–Wagner growth law [35]:

$$L \sim t^{1/3}. \tag{81}$$

To summarize, for a given system undergoing phase order kinetics, the questions of importance include:

- What is the time dependence of the growth law $L(t)$?

- What is the nonequilibrium exponent λ?

- Is there universality? That is, do different physical systems share the same $L(t), \lambda, F(x)$, and $F(x,y)$?

11.9 Theoretical Developments

11.9.1 Linear Theory

We now need to construct a theory to answer the questions posed at the end of the previous section. Let us start with a brief review of the early work on this problem, beginning with the Cahn–Hilliard–Cook theory [44]. We work with the TDGL equation of motion:

$$\frac{\partial \psi}{\partial t} = -\hat{\Gamma}\left[-\psi + \psi^3 - \nabla^2 \psi\right] + \eta \, , \tag{82}$$

where $\hat{\Gamma} = 1$ for an NCOP and $\hat{\Gamma} = -\nabla^2$ for a COP. In this simplest theory the nonlinear term in the equation of motion is dropped because it can be assumed to be small at early times. The resulting equation of motion (including the thermal noise) [45],

$$\frac{\partial}{\partial t}\psi = -\hat{\Gamma}\left[-\psi + \nabla^2 \psi\right] + \eta \, , \tag{83}$$

is linear and can be solved by first Fourier transforming over space:

$$\frac{\partial}{\partial t}\psi(q,t) = -\Gamma(q)\left[-1 + q^2\right]\psi(q,t) + \eta(q,t) \, . \tag{84}$$

Next multiply by $\psi(-q,t)$ and average, using the fact (see Problem 11.10) that $\langle \eta(q,t)\psi(-q,t) \rangle = \Gamma(q)k_B T$, to obtain:

$$\frac{1}{2}\frac{\partial}{\partial t}C(\mathbf{q},t) = -\Gamma(q)\left[-1 + q^2\right]C(q,t) + \Gamma(q)k_B T. \tag{85}$$

The solution to this equation is:

$$C(q,t) = e^{-\Lambda(q)t}C(q,0) + \frac{2\Gamma(q)k_B T}{\Lambda(q)}\left[1 - e^{-\Lambda(q)t}\right] , \qquad (86)$$

where $\Lambda(q) = 2\Gamma(q)[-1 + q^2]$ and $C(q,0)$ is the initial value of the order parameter structure factor. This treatment gives exponential growth for small q, which is almost never seen except for very early times. This theory basically allows one to see that the system is unstable on the longest length scales. One must do better if one is to understand scaling, obtain sharp interfaces, and physical ordering.

11.9.2
Mean-Field Theory

A more sophisticated approximation is mean-field theory [46]. In this theory the nonlinear term in the equation of motion, Eq. (82), is *linearized* by making a Gaussian approximation for ψ by writing:

$$\psi^3 \to 3\langle\psi^2\rangle\psi . \qquad (87)$$

The equation of motion, Eq. (82), then reads:

$$\frac{\partial}{\partial t}\psi = -\hat{\Gamma}\left[A(t)\psi - \nabla^2\psi\right] + \eta , \qquad (88)$$

where:

$$A(t) = -1 + 3\langle\psi^2\rangle. \qquad (89)$$

We must determine $\langle\psi^2\rangle$ self-consistently. Let us focus first on the case of a quench to $T = 0$. Following the same steps used in the Cahn?Hilliard–Cook case and integrating the resulting first-order differential equation in time we obtain:

$$C(q,t) = C(q,0)e^{-2\int_0^t d\tau \Gamma(q)[q^2 + A(\tau)]} . \qquad (90)$$

We then need to specify the initial condition. It is convenient to choose:

$$C(q,0) = C_I e^{-\frac{1}{2}\alpha q^2} . \qquad (91)$$

If we assume that the system has a NCOP so $\Gamma(q) = \Gamma$, then Eqs. (89) and (90) lead to the self-consistent equation for $\langle\psi^2\rangle$:

$$\langle\psi^2\rangle = \int \frac{d^d q}{(2\pi)^d} C(q,t)$$

$$= C_I e^{-2\Gamma \int_0^t d\tau A(\tau)} \int \frac{d^d q}{(2\pi)^d} e^{-2\Gamma q^2 t} e^{-\frac{1}{2}\alpha q^2}$$

$$\equiv \frac{B}{(2\Gamma[t+t_0])^{d/2}} e^{-2\Gamma \int_0^t d\tau A(\tau)}, \qquad (92)$$

where $B = \tilde{K}_d \Gamma(d/2) C_I$,

$$\tilde{K}_d = \int \frac{d^d q}{(2\pi)^d} \delta(q-1) \qquad (93)$$

is proportional to the surface area of a unit sphere and $t_0 = \alpha/4\Gamma$. We then have, after using Eq. (89) in Eq. (92), the self-consistent equation for $A(t)$:

$$\frac{1}{3}(A+1) = \frac{B}{(2\Gamma[t+t_0])^{d/2}} e^{-2\Gamma \int_0^t d\tau A(\tau)}. \qquad (94)$$

Taking the derivative with respect to time, we find an equation for A as a function of time:

$$\frac{1}{3}\dot{A} = -\frac{d}{2(t+t_0)} \frac{(A+1)}{3} - 2\Gamma A \frac{1}{3}(A+1), \qquad (95)$$

which reduces to:

$$\dot{A} = -(A+1)\left(\frac{d}{2(t+t_0)} + 2\Gamma A\right). \qquad (96)$$

It is shown in Problem 11.11 that for large times Eq. (96) has a solution of the form:

$$A = \frac{A_1}{(t+t_0)} + \frac{A_2}{(t+t_0)^2} + \cdots, \qquad (97)$$

where:

$$A_1 = -\frac{d}{4\Gamma} \qquad (98)$$

$$A_2 = \frac{d(1-d)}{8\Gamma^2}. \qquad (99)$$

We can solve Eq. (96) for A numerically for different initial conditions for A. These initial values are constrained by Eq. (94), which reads:

$$\frac{1}{3}(A_0+1) = \frac{B}{(2\Gamma t_0)^{d/2}}. \qquad (100)$$

One finds that each initial condition drives the system to the asymptotic state given by Eq. (97). Using Eq. (97) in Eq. (89) gives:

$$\langle \psi^2 \rangle = \frac{1}{3} - \frac{d}{12\Gamma t}. \qquad (101)$$

The structure factor given by Eq. (90) can then be written in the form:

$$C(q,t) = C(q,0)\kappa L^d(t) e^{-[qL(t)]^2} \qquad (102)$$

with $L(t) = (2\Gamma t)^{1/2}$ and (see Problem 11.12):

$$\kappa = \lim_{t \to \infty} \frac{e^{-2\Gamma \int_0^t d\tau A(\tau)}}{(2\Gamma(t+t_0))^{d/2}} \qquad (103)$$

is a constant. We see that the structure factor does obey scaling. However, the prefactor depends on the initial conditions as does the constant factor κ.

At first sight this theory is appealing but, unfortunately there are many problems with this approach:

- It gives the wrong ordering values:

$$\langle \psi^2 \rangle \to \frac{1}{3} \text{ not } 1 \qquad (104)$$

- It does not include sharp interfaces since one has:

$$\langle \psi^2 \rangle - \frac{1}{3} \sim \mathcal{O}\left(\frac{1}{L^2}\right) \text{ not } \mathcal{O}\left(\frac{1}{L}\right) \qquad (105)$$

as suggested by Eq. (46).

- One does not obtain Porod's law, $C(q) \approx q^{-(d+1)}$.

- The scaling function depends multiplicatively on the initial conditions.

- If we include the noise in the development we find the equation of motion takes the form:

$$\frac{\partial}{\partial t} C(\mathbf{q},t) = -2\Gamma(q)\left[A(t) + q^2\right]C(\mathbf{q},t) + 2\Gamma(q)kT \quad . \qquad (106)$$

Looking for the long time stationary solution, $A \to 0$, we find:

$$C(\mathbf{q},t) - C_{\text{peak}}(\mathbf{q},t) = \frac{k_B T}{q^2} + \mathcal{O}\left(\frac{1}{2\Gamma t}\right) \quad . \qquad (107)$$

This is not the correct Ornstein–Zernike form given by Eq. (36).

A theory that resolves these problems requires a rather different approach.

11.9.3
Auxiliary Field Methods

Significant theoretical progress has been made in the theory of phase-ordering kinetics [2,47,48] using methods that introduce auxiliary fields. These theories well describe the qualitative scaling features of ordering in unstable systems. We develop one such approach [49] here for the case of a scalar order parameter. These ideas can be generalized [50] to the case of the n-vector model and systems with continuous symmetry in the disordered state.

Our goal is to solve for the order parameter scaling properties for an unstable dynamics governed by the equation of motion in dimensionless units:

$$\frac{\partial \psi}{\partial t} = -V'[\psi] + \nabla^2 \psi \ , \tag{108}$$

with initial conditions where ψ is Gaussian with $\langle \psi \rangle = 0$ and:

$$\langle \psi(\mathbf{r}_1, t_0) \psi(\mathbf{r}_2, t_0) \rangle = g_0(\mathbf{r}_1 - \mathbf{r}_2) \ . \tag{109}$$

We can put Eq. (108) in the form:

$$\Lambda(1)\psi(1) = -V'[\psi(1)] \ , \tag{110}$$

where the diffusion operator:

$$\Lambda(1) = \frac{\partial}{\partial t_1} - \nabla_1^2 \tag{111}$$

is introduced along with the shorthand notation that 1 denotes $(\mathbf{r}_1, t_1)$.

Suppose we decompose the order parameter into a sum of two parts:

$$\psi = \sigma + u \ . \tag{112}$$

The idea [51] is that σ represents the ordering degrees of freedom and give rise to the Bragg peak in Fig. 11.8, while u is associated with the equilibrated degrees of freedom, and, in the presence of noise, leads to the Ornstein–Zernike peak in Fig. 11.8.

We then need to construct equations of motion separately for σ and u. We first show that we can consistently construct equations of motion for σ that are ultimately compatible with Eq. (108). The first step [48] is to assume that σ is a local function of an auxiliary field m and $\sigma[m]$ is the solution to the Euler–Lagrange equation for the associated stationary interface problem:

$$\frac{d^2\sigma}{dm^2} = V'(\sigma[m]) \ . \tag{113}$$

In this equation m is taken to be the coordinate. It will generally be useful to introduce the notation:

$$\sigma_\ell \equiv \frac{d^\ell \sigma}{dm^\ell} \ . \tag{114}$$

A key point in the introduction of the field m is that the zeros of m locate the zeros of the order parameter and give the positions of interfaces in the system. Locally, the magnitude of m gives the distance to the nearest defect. As a system coarsens and the distance between defects increases, the typical value of m increases linearly with $L(t)$. From a scaling point of view we can estimate $m \approx L$. For the usual double-well potential:

$$V(x) = -\frac{1}{2}x^2 + \frac{1}{4}x^4 , \tag{115}$$

we have an analytic solution to Eq. (113) given by:

$$\sigma = \tanh\left(\frac{m}{\sqrt{2}}\right) . \tag{116}$$

More generally σ, as a function of m, is defined by Eq. (113) with the boundary condition associated with an interface. The field u is defined by Eq. (112). We still must assign a dynamics to both σ and u, starting with Eq. (108)

Let us substitute Eq. (112) into the equation of motion, Eq. (108), for the order parameter, use the chain-rule for differentiation, and find:

$$\Lambda(1)u(1) + \sigma_1(1)\Lambda(1)m(1) = -V'[\sigma(1) + u(1)] + \sigma_2(1)(\nabla m(1))^2 . \tag{117}$$

This can be regarded as an equation for the field u. We then have the freedom to assume that m is driven by an equation of motion of the form:

$$\Lambda(1)m(1) = \Xi(m(1), t_1) . \tag{118}$$

Ξ is undetermined at this point except that it must be chosen such that the auxiliary field equation of motion, Eq. (118) leads to scaling. The key idea is that m should scale with $L(t)$ for large t. We return to this point later.

Using Eq. (118) in Eq. (117) for u gives:

$$\Lambda(1)u(1) = -V'[\sigma(1) + u(1)] + \sigma_2(1)[\nabla m(1)]^2 - \sigma_1(1)\Xi(1) . \tag{119}$$

Consider the special case where we have the potential given by Eq. (115). The equation for u is given then by:

$$\Lambda u + (3\sigma^2 - 1)u + 3\sigma u^2 + u^3 = -\sigma_2\left[1 - (\nabla m)^2\right] - \sigma_1\Xi . \tag{120}$$

In the limit of large $|m|$ the derivatives of σ go exponentially to zero and the right-hand side of Eq. (120) is exponentially small in $|m|$. Clearly we can construct a solution for u where it is small and we can linearize the left-hand side of Eq. (120). Remembering that $\sigma^2 = \psi_0^2 = 1$ away from interfaces in the bulk of the ordered material we have:

$$\Lambda u + 2u = -\sigma_2(1 - (\nabla m)^2) - \sigma_1\Xi . \tag{121}$$

In this regime we can use dimensional analysis to make the following estimates: $(\nabla m)^2 \approx \mathcal{O}(1)$, $\sigma_2 \approx \mathcal{O}(L^{-2})$, $\Xi \approx \mathcal{O}(L^{-1})$, $\sigma_1 \Xi \approx \mathcal{O}(L^{-2})$, and $\Lambda u \approx \mathcal{O}(L^{-4})$. Notice on the left-hand side of Eq. (121) that u has acquired a *mass* (= 2) and in the long-time long-distance limit, the term where u is multiplied by a constant dominates the derivative terms:

$$2u = -\sigma_2[1 - (\nabla m)^2] - \sigma_1 \Xi \ . \tag{122}$$

That the u field picks up a mass in the scaling limit can easily be seen to be a general feature of a wide class of potentials where $q_0^2 = V''[\sigma = \pm \psi_0] > 0$. In the case of the potential given by Eq. (115), $q_0^2 = 2$. We have then, on rather general principles, that the field u must vanish rapidly as one moves into the bulk away from interfaces. From a scaling analysis we see that $u \approx L^{-2}$.

The explicit construction of correlations involving the field u is rather involved and depends on the details of the potential chosen. If we restrict our analysis to investigating universal properties associated with bulk ordering, we will not need to know the statistics of u explicitly. If we are interested in determining interfacial properties then we need to know u in some detail. Thus, for example, if we want to determine the correlation function:

$$C_{\psi^2}(12) = \langle [\psi^2(1) - \psi_0^2][\psi^2(2) - \psi_0^2] \rangle \ , \tag{123}$$

we will need to know the statistics of the field u. However if we are interested in quantities like:

$$C(12...n) = \langle \psi(1)\psi(2) \cdots \psi(n) \rangle \ , \tag{124}$$

where the points 12...n are not constrained to be close together, then we do not need to know u in detail. Why is this? Consider, for example,

$$C(12) = \langle \psi(1)\psi(2) \rangle = \langle [\sigma(1) + u(1)][\sigma(2) + u(2)] \rangle.$$

Since $u(1)$ vanishes exponentially for large $|m(1)|$, the averages over the u's are down by a factor of L^{-2} relative to the averages over the field $\sigma(1)$. Thus $\langle \sigma(1)\sigma(2) \rangle \approx \mathcal{O}(1)$ as $L(t) \to \infty$, while $\langle \sigma(1)u(2) \rangle$ and $\langle u(1)\sigma(2) \rangle$ are of $\mathcal{O}(L^{-2})$ and $\langle u(1)u(2) \rangle$ of $\mathcal{O}(L^{-4})$. Henceforth we focus on the ordering correlations where:

$$C(12) = \langle \sigma(1)\sigma(2) \rangle \ . \tag{125}$$

The ordering component of the order parameter correlation function is given by Eq. (125) where $\sigma(1) = \sigma[m(1)]$. The strategy is to assume that the statistics of the field m are simpler than for ψ and the equation of motion for m given by Eq. (118) can be developed using simple assumptions. We show this in the next section. For a fuller account see Ref. [49].

11.9.4
Auxiliary Field Dynamics

The dynamics of the auxiliary field are governed by a nonlinear diffusion equation of the general form given by Eq. (118). The simplest choice for Ξ is that it is linear in m. This assumption leads to the equation of motion for the auxiliary field:

$$\Lambda(1)m(1) = \Omega(t_1)m(1) \; , \tag{126}$$

where $\Omega(t_1)$ is to be determined. From dimensional analysis $m \approx L$, $\Lambda \approx L^{-2}$, so $\Omega(t_1) \approx L^{-2}(t_1)$. The auxiliary field correlation function,

$$C_0(12) = \langle m(1)m(2) \rangle \; , \tag{127}$$

satisfies the equation of motion:

$$\left(\frac{\partial}{\partial t_1} - \nabla_1^2\right) C_0(12) = \Omega(1)C_0(12) \; . \tag{128}$$

Fourier transforming over space we find, assuming Ω is a function only of time,

$$\left(\frac{\partial}{\partial t_1} + q^2\right) C_0(q, t_1 t_2) = \Omega(t_1) C_0(q, t_1 t_2) \; . \tag{129}$$

We can integrate this equation to obtain:

$$\begin{aligned} C_0(q, t_1 t_2) &= \exp\left(\int_{t_2}^{t_1} d\tau(-q^2 + \Omega(\tau))\right) C_0(q, t_2 t_2) \\ &= R(t_1, t_2) e^{-q^2(t_1 - t_2)} C_0(q, t_2 t_2) \; , \end{aligned} \tag{130}$$

where:

$$R(t_1, t_2) = \exp\left(\int_{t_1}^{t_2} d\tau \, \Omega(\tau)\right) \; . \tag{131}$$

To go further we must evaluate the equal-time correlation function. One can easily show, using Eq. (129), that this quantity satisfies the equation of motion:

$$\left(\frac{\partial}{\partial t} + 2q^2\right) C_0(\mathbf{q}, tt) = 2\Omega(t) C_0(\mathbf{q}, tt) \; . \tag{132}$$

This can be integrated much as for Eq. (129) and we find:

$$C_0(q, t_2 t_2) = R^2(t_2, t_0) e^{-2q^2(t_2 - t_0)} C_0(q, t_0 t_0) \; , \tag{133}$$

where we must then give the initial condition $C_0(q, t_0 t_0) = \tilde{g}(q)$. Putting Eq. (133) back into the two-time expression, Eq. (130), we obtain:

$$C_0(q, t_1 t_2) = R(t_1, t_2) e^{-q^2(t_1 - t_2)} R^2(t_2, t_0) e^{-2q^2(t_2 - t_0)} \tilde{g}(q) \; . \tag{134}$$

Since $R(t_1,t_2)R(t_2,t_1) = 1$, this takes the form:

$$C_0(q,t_1 t_2) = R(t_1,t_0)R(t_2,t_0)e^{-q^2(t_1+t_2-2t_0)}\tilde{g}(q) \quad , \tag{135}$$

which is properly symmetric. Inverting the Fourier transform we obtain:

$$C_0(r,t_1 t_2) = R(t_1,t_0)R(t_2,t_0)\int \frac{d^d q}{(2\pi)^d} e^{i\vec{q}\cdot\vec{r}}\tilde{g}(q)e^{-2q^2 T} \quad , \tag{136}$$

where it is convenient to introduce:

$$T = \frac{t_1+t_2}{2} - t_0 \quad . \tag{137}$$

While we are primarily interested in the long-time scaling properties of our system, we can retain some control over the influence of initial conditions and still be able to carry out the analysis analytically if we introduce the initial condition:

$$\tilde{g}(q) = g_0 e^{-\frac{1}{2}(q\ell)^2} \tag{138}$$

or:

$$g(r) = g_0 \frac{e^{-\frac{1}{2}(r/\ell)^2}}{(2\pi\ell^2)^{d/2}} \quad . \tag{139}$$

Inserting this form into Eq. (136) and doing the wavenumber integration we obtain:

$$C_0(r,t_1 t_2) = R(t_1,t_0)R(t_2,t_0)\frac{g_0}{[2\pi(\ell^2+4T)]^{d/2}} e^{-\frac{1}{2}r^2/(\ell^2+4T)}.$$

In the long-time limit this reduces to:

$$C_0(r,t_1 t_2) = R(t_1,t_0)R(t_2,t_0)g_0 \frac{e^{-r^2/8T}}{(8\pi T)^{d/2}} \quad . \tag{140}$$

Let us turn now to the quantity $R(t_1,t_2)$ defined by Eq. (131). For consistency, we must assume for long times:

$$\Omega(t) = \frac{\omega}{t_c+t} \quad , \tag{141}$$

where ω is a constant we will determine and t_c is a short-time cutoff that depends on details of the earlier-time evolution. Evaluating the integral:

$$\int_{t_2}^{t_1} d\tau\, \Omega(\tau) = \int_{t_2}^{t_1} d\tau \frac{\omega}{t_c+\tau} = \omega \ln\left(\frac{t_1+t_c}{t_2+t_c}\right) \quad , \tag{142}$$

we obtain:
$$R(t_1, t_2) = \left(\frac{t_1 + t_c}{t_2 + t_c}\right)^\omega . \tag{143}$$

Inserting this result back into Eq. (140) leads to the expression for the correlation function:
$$C_0(r, t_1 t_2) = g_0 \left(\frac{t_1 + t_c}{t_0 + t_c}\right)^\omega \left(\frac{t_2 + t_c}{t_0 + t_c}\right)^\omega \frac{e^{-r^2/8T}}{(8\pi T)^{d/2}} . \tag{144}$$

If we are to have a self-consistent scaling equation then the autocorrelation function ($r = 0$), at large equal times $t_1 = t_2 = t$, must show the behavior $\langle m^2(t) \rangle = S_0(t) \approx L^2 \approx t$. Thus Eq. (144) gives the result:
$$S_0(t) = C_0(0, t_1, t_1) = g_0 \left(\frac{t}{t_0 + t_c}\right)^{2\omega} \frac{1}{(8\pi t)^{d/2}}$$
$$= t^{2\omega - d/2} \frac{1}{(t_0 + t_c)^{2\omega}} \frac{g_0}{(8\pi)^{d/2}} = A_0 t , \tag{145}$$

where A_0 is a constant to be determined. For a solution to Eq. (145) we see the exponent ω must be given by:
$$\omega = \frac{1}{2}\left(1 + \frac{d}{2}\right) \tag{146}$$

and the amplitude by:
$$A_0 = \frac{1}{(t_0 + t_c)^{2\omega}} \frac{g_0}{(8\pi)^{d/2}} . \tag{147}$$

The general expression for the auxiliary correlation function, Eq. (144), can be rewritten in the convenient form:
$$C_0(r, t_1 t_2) = \sqrt{S_0(t_1) S_0(t_2)} \Phi_0(t_1 t_2) e^{-\frac{1}{2}r^2/(\ell^2 + 4T)} \tag{148}$$

where:
$$\Phi_0(t_1 t_2) = \left(\frac{\sqrt{(t_1 + t_c)(t_2 + t_c)}}{T + t_c + t_0}\right)^{d/2} . \tag{149}$$

The nonequilibrium exponent is defined in the long-time limit by:
$$\frac{C_0(0, t_1 t_2)}{\sqrt{S_0(t_1) S_0(t_2)}} = \left(\frac{\sqrt{(t_1 + t_c)(t_2 + t_c)}}{T + t_0 + t_c}\right)^{\lambda_m} \tag{150}$$

and we obtain the result [2] for the exponent:

$$\lambda_m = \frac{d}{2} . \tag{151}$$

We put the subscript m on λ to indicate the exponent associated with the m field. In general $C_0(12)$ and the order parameter correlation function $C(12)$ need not share the same exponents.

Looking at equal times, Eq. (148) reduces to:

$$f_0(x) = \frac{C_0(r,tt)}{S_0(t)} = e^{-x^2/2} , \tag{152}$$

where the scaled length is defined by $x = r/2\sqrt{t}$. This result for $f_0(x)$, the scaled auxiliary correlation function, is originally due to Ohta, Jasnow and Kawasaki [47].

11.9.5
The Order Parameter Correlation Function

We still must relate $C(12)$ to $C_0(12)$. In the scaling regime the order parameter correlation function is given by:

$$C(12) = \langle \sigma[m(1)]\sigma[m(2)]\rangle . \tag{153}$$

$C(12)$ can be written in terms of the two-point probability distribution:

$$P_0(x_1, x_2, 1, 2) = \langle \delta[x_1 - m(1)]\delta[x_2 - m(2)]\rangle , \tag{154}$$

$$C(12) = \int dx_1 \int dx_2\, \sigma(x_1)\sigma(x_2) P_0(x_1 x_2, 12) . \tag{155}$$

We can evaluate P_0 by using the integral representation for the δ function:

$$P_0(x_1, x_2, 12) = \int \frac{dk_1}{2\pi} \int \frac{dk_2}{2\pi} e^{-ik_1 x_1} e^{-ik_2 x_2} \langle e^{\mathcal{H}(12)}\rangle , \tag{156}$$

where $\mathcal{H}(12) \equiv ik_1 m(1) + ik_2 m(2)$. Equation (126) is consistent with m being a Gaussian variable. We can then show in Problem 11.13 that [52]:

$$\langle e^{\mathcal{H}(12)}\rangle = e^{\frac{1}{2}\langle \mathcal{H}^2(12)\rangle} = e^{-\frac{1}{2}[k_1^2 S_0(1) + k_2^2 S_0(2) + 2k_1 k_2 C_0(12)]} , \tag{157}$$

where:

$$S_0(1) = C(11) . \tag{158}$$

Putting Eq. (157) back into Eq. (156), and doing (see Problem 11.14) the Gaussian integrals over k_1 and k_2 we obtain:

$$P_0(x_1, x_2, 12) = \frac{\gamma}{2\pi\sqrt{S_0(1)S_0(2)}} e^{\left\{-\frac{\gamma^2}{2}\left[\frac{x_1^2}{S_0(1)} + \frac{x_2^2}{S_0(2)} - \frac{2x_1 x_2 f}{\sqrt{S_0(1)S_0(2)}}\right]\right\}} , \tag{159}$$

where:

$$f(12) = \frac{C_0(12)}{\sqrt{S_0(1)S_0(2)}} \tag{160}$$

and:

$$\gamma = \frac{1}{\sqrt{1-f^2}} \ . \tag{161}$$

Putting Eq. (159) back into Eq. (155) and letting:

$$x_1 = \sqrt{S_0(1)}\gamma^{-1}y_1 \tag{162}$$

$$x_2 = \sqrt{S_0(2)}\gamma^{-1}y_2 \ , \tag{163}$$

the structure factor is then given by:

$$C(12) = \frac{1}{2\pi\gamma}\int dy_1 dy_2\, \sigma\left(\sqrt{S_0(1)}\gamma^{-1}y_1\right)\sigma\left(\sqrt{S_0(2)}\gamma^{-1}y_2\right)$$
$$\times e^{-\frac{1}{2}[y_1^2+y_2^2-2y_1y_2 f]} \ . \tag{164}$$

For late times, where S_0 is arbitrarily large, we can replace:

$$\sigma\left(\sqrt{S_0(1)}\gamma^{-1}y_1\right) \to \psi_0 \mathrm{sgn}(y_1) \tag{165}$$

and obtain:

$$C(12) = \frac{\psi_0^2}{2\pi\gamma}\int dy_1 \int dy_2\, \mathrm{sgn}(y_1)\mathrm{sgn}(y_2) e^{-\frac{1}{2}[y_1^2+y_2^2-2y_1y_2 f]} \ . \tag{166}$$

Evaluation of the remaining integrals is left as a problem (Problem 11.15). One has the simple final result:

$$C(12) = \frac{2}{\pi}\psi_0^2 \sin^{-1} f(12) \ . \tag{167}$$

Combining Eq. (148) with Eq. (167), we obtain the scaling result [47]:

$$F(x,t_1/t_2) = \frac{2}{\pi}\sin^{-1}\left[\Phi_0(t_1,t_2)e^{-\frac{1}{2}x^2}\right] \ , \tag{168}$$

where:

$$\Phi_0(t_1,t_2) = \left(\frac{\sqrt{t_1 t_2}}{T}\right)^{\lambda_0} \tag{169}$$

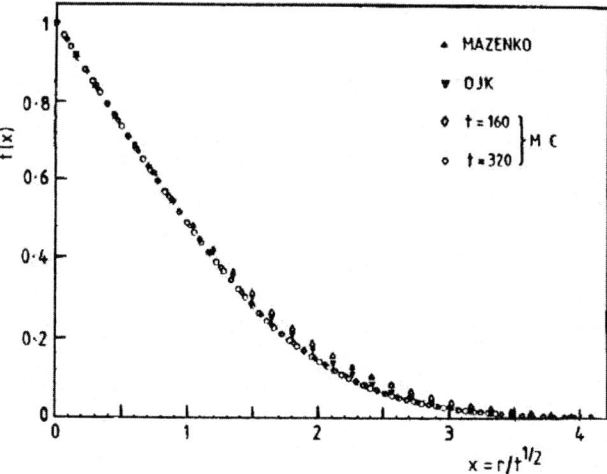

Fig. 11.12 Numerical test labeled MC, of the equal-time scaling function given by Eq. (167), labeled OJK. See Fig. 12 in Ref. [2].

and $\lambda_0 = d/2$. Note that the final result is independent of the initial conditions and of the form of the potential V. This result for the equal-time scaling function has been checked against numerical results in Fig. 11.12.

Looking at these results for $C(12)$ one can see that the autocorrelation function $C(\mathbf{x}_1, \mathbf{x}_1, t_1, t_2)$, for $t_1 \gg t_2$ is proportional to $C_0(\mathbf{x}_1, \mathbf{x}_1, t_1, t_2)$ and gives the nonequilibrium exponent $\lambda = \lambda_0 = d/2$ in this simplest nontrivial approximation. Corrections to this Gaussian approximation are developed in Ref. [49].

11.9.6
Extension to n-Vector Model

Thus far in this chapter there has been an emphasis on systems with scalar order parameters and order disrupting interfaces in two and three dimensions. For the case of an n-vector order parameter, $\vec{\psi}$, as discussed in some detail in Chapter 12 of FOD, one has different defect structures depending on the value of n and the spatial dimension d. Thus for the case of $n = d$ one has point defects, for $n = d - 1$ one has line defects and for $n = d - 2$ wall defects. Examples of point defects correspond to kinks for $n = d = 1$, point vortices for $n = d = 2$, and point monopoles for $n = d = 3$. For line defects one has line interfaces for $d = 2$ and $n = 1$ as in the two-dimensional Ising model, line vortices for $d = 3$ and $n = 2$ in superfluids, and domain walls for $d = 3$ and $n = 1$. In terms of examples, one has vortices disrupting the pathway to ordering in superfluids [53–55], superconductors [56–59], and XY magnets [60–66]. Also of interest are dislocations in solids [67], disclinations and monopoles in nematic liquid crystals [68–73], and vortices in Bose–Einstein condensed systems [74–76]. There has been considerable exchange of ideas between the study of defects in phase-ordering kinetics and the role of defects as density seeds [77–80] in galaxy formation in the evolution of the early universe [81].

The dynamics of this set of defects can be understood in terms of dynamics generated by the n-vector nonconserved TDGL model driving the vector order parameter $\vec{\psi}$:

$$\frac{\partial}{\partial t}\vec{\psi} = \nabla^2\vec{\psi} + \vec{\psi} - (\vec{\psi})^2\vec{\psi} = -\frac{\delta\mathcal{H}_E}{\delta\vec{\psi}} \ . \tag{170}$$

The TDGL model generates a dissipative dynamics. There is also interest in reversible dynamics driven by the nonlinear Schroedinger equation (NLSE) [82]:

$$-i\frac{\partial}{\partial t}\psi = \nabla^2\psi + \psi - |\psi|^2\psi = -\frac{\delta\mathcal{H}_E}{\delta\psi^*} \ , \tag{171}$$

where the order parameter ψ is complex. The NLSE generates a quite different dynamics compared to the dissipative TDGL case, which drives a system to equilibrium. Note that the quantity $\int d^dx\,\psi^*(\mathbf{x})\psi(\mathbf{x})$ is conserved for the NLSE (see Problem 11.16).

The theory for the order parameter correlation function given above for a scalar order parameter can be generalized [83, 84] to the n-vector case. One again has a mapping in terms of an auxiliary field:

$$\vec{\psi} = \vec{\sigma}(\mathbf{m}) \ , \tag{172}$$

where $\vec{\psi}$ represents the ordering part of the order parameter. The mapping $\vec{\sigma}(\mathbf{m})$ satisfies the Euler–Lagrange equation determining the defect profile for isolated defects:

$$\sum_{\alpha=1}^{n}\frac{\partial^2}{\partial m_\alpha \partial m_\alpha}\sigma_\mu = \frac{\partial}{\partial \sigma_\mu}V[\vec{\sigma}] \ , \tag{173}$$

where V is the usual wine-bottle potential. Thus for the vortex case (see Section 12.6.1 in FOD) the single vortex profile has the form:

$$\sigma_\mu = f(m)\hat{m}_\mu \ , \tag{174}$$

where $f(\infty) = 1$ and for small m f is linear in m. Then, in close analogy with the development in the scalar case, we have for the scaling contribution:

$$C(12) = \langle \vec{\psi}(1) \cdot \vec{\psi}(2) \rangle \tag{175}$$
$$= \langle \vec{\sigma}[\mathbf{m}(1)] \cdot \vec{\sigma}[\mathbf{m}(2)] \rangle \ , \tag{176}$$

which, as for Eq. (155), is a function of:

$$C_0(12) = \langle \mathbf{m}(1) \cdot \mathbf{m}(2) \rangle \ . \tag{177}$$

Much as for the scalar case one can show [50] in the simplest approximation that $C_0(12)$ has the form given by Eq. (152). In the limit where $S_0(1) = C_0(11)$ is large, where the auxiliary field is taken to be Gaussian, one again has scaling:

$$C(r,t) = \psi_0^2 F[r/L(t)] \ , \tag{178}$$

where the scaling function is given by [84]:

$$F(x) = \frac{nf_0}{\pi} B\left[\frac{1}{2}, \frac{n+1}{2}\right] \int_0^1 \frac{(1-z^2)^{(n-1)/2}}{(1-z^2 f_0^2)^{1/2}} dz \ , \tag{179}$$

where $B(a,b) = \Gamma(a)\Gamma(b)/\Gamma(a+b)$, $f_0(x)$ is given by Eq. (152), and the growth law obeys $L(t) \approx t^{1/2}$. An interesting result, due to Bray and Puri [84], is that the generalization of Porod's law to $n > 1$ is simply that the Fourier transform of the order parameter scaling function for large $Q = qL$ goes as:

$$f(Q) \approx Q^{-(d+n)} \tag{180}$$

for general n.

11.10
Defect Dynamics

Theoretical descriptions of defect structures go back, at least, to Lord Kelvin [85] and his theorems concerning conservation of vorticity in fluids. The inviscid and incompressible equations of hydrodynamics are the simplest model for producing vortices. From a more modern point of view, our understanding of defects is carried out in terms of an order parameter [11, 12] field. The nature of the defect in a given system depends on the symmetry of the order parameter (n) and the spatial dimensionality (d).

Our understanding of defect structures is developed at three different levels of description. These levels are associated with three characteristic lengths: (a). L is the typical distance between defects, (b) ξ is the length characteristic of a vortex core or interfacial width, and (c) a is some microscopic length like a lattice spacing. At the longest length scale $L(\gg \xi \gg a)$, the *hydrodynamic* length scale, one can formulate a coarse-grained continuum theory that is largely independent of the model details and where defect cores are treated as point or line singularities. Examples are the elastic theory of dislocations [86] in solids and disclinations in nematic liquid crystals, the London theory [87] applied to vortices in superconductors, the phase-field models for superfluids [88] and the Landau–Lifshitz equations [89] for magnetic systems. In those cases where one has topological defects there has been a prejudice [90] that the behavior at large distances from the core is independent of the details of the core. We will refer to such descriptions as *phase-field* theories.

There is a vast literature on the equations of motion satisfied by defects. Much of it, for obvious reasons, is devoted to the phase-field equation description, which holds very nicely for a system of well-separated defects. For the case of classical fluids the inviscid and incompressible equations of hydrodynamics offer a good example. The nature of the dynamics of the vortices described by these equations has been understood for some time. Consider the comments of Sommerfeld [91] about the strangeness of vortex dynamics in fluids: "The dynamics of vortices...is indeed a very peculiar one and deviates decisively from the dynamics of mass points." To begin with, Newton's first law is altered. "The isolated vortex... remains in a state of rest. A uniform rectilinear motion can only be acquired by association with a second vortex of equal strength but opposite sense of rotation or under the action of a wall at rest." The modification of the second law is even more remarkable. The external action originating in a second vortex does not determine the acceleration but the velocity. The content of the law of motion of the mass center is shifted accordingly: "not the acceleration, but the velocity of the mass center vanishes."

The conventional pathway from the more microscopic field equations to the phase-field equations is well established. The first step is to work out the stationary single-defect solutions for the field equations that give the classical defect profile solutions for points, lines and walls as discussed above. These solutions are parameterized by defect positions. In the phase-field approximation, working for example with a complex order parameter $\psi = Re^{i\theta}$, we can establish [92] that the influence of the vortex at large distances is governed by the associated Nambu–Goldstone mode, the phase θ. Thus the amplitude R is assumed to be constant. The equations satisfied by θ in this regime are typically linear and one can use superposition to look at the phase of a set of N vortices in two dimensions at positions $\mathbf{r}_i = x_i \hat{x} + y_i \hat{y}$ with charge or circulation m_i:

$$\theta(\mathbf{r}, t) = \sum_{i=1}^{N} m_i \tan^{-1}\left(\frac{y - y_i}{x - x_i}\right) . \tag{181}$$

It is then not difficult to work out the associated equations of motion satisfied by the $\mathbf{r}_i(t)$ generated by the original equation of motion. For the case where one has the inviscid equations of hydrodynamics one is led to the conclusions of Sommerfeld discussed above. There is a substantial literature [35, 37, 43] of work along these lines in the case of phase-ordering kinetics. As long as the defects are well separated this description is sensible and one can look at the motion of isolated vortices and the interactions among small sets of vortices [38].

There are several major drawbacks to this phase-field approach. The first is that it does not allow one to follow the setting up of the evolving defect structures and therefore the initial conditions are not known. This is nontrivial because the correlations that are associated with properties like scaling may already be present at the time when a well-defined defect structure has developed. Thus, one has a serious problem knowing how to take averages and this could clearly affect the results at the longest length scales, which involves many defects.

The phase-field method is not a complete description since it breaks down when the cores of defects begin to overlap. Traditionally, the approach has been to assume that one can introduce some short-distance cutoff into the theory that can be handled phenomenologically. More recent work shows that this approach can be misleading and miss some of the general physics of the problem. As an example, consider the relationship between the vortex velocity $\mathbf{v}$ and a constant driving phase gradient $\mathbf{k} =< \nabla\theta >$ in the case of the TDGL model for $n = d = 2$. It is now understood [93–96] that to obtain the *mobility* connecting these two quantities, one must match up the inner phase-field solution with the outer core solution to the field equations. This leads to the nontrivial result, for weak driving gradients, that:

$$k_\alpha = \frac{m}{2} \ln\left(\frac{v_0}{v}\right) \sum_\beta \varepsilon_{\alpha\beta} v_\beta , \qquad (182)$$

where $\varepsilon_{\alpha\alpha} = 0$ and $\varepsilon_{xy} = -\varepsilon_{yx} = 1$, and m is the charge of the vortex and for $|m| = 1$, and $v_0 =\approx 3.29\ldots$ in dimensionless units.

In work on phase-ordering kinetics one [97] is interested in answering the question: What is the probability of finding a vortex a distance r from an antivortex? Or, what is the probability of a vortex having a velocity, $\mathbf{v}$? A motivating factor in the development is the realization that in treating statistical properties of the defects, we do not want to go over to an explicit treatment in terms of the defect positions. That would lead back to the problem of specification of initial conditions. Instead, one can look for a way of implicitly finding the positions of the defects using the field $\vec{\psi}$ itself.

Let us consider the case of point defects where $n = d$. The basic idea is that the positions of defects are located by the zeros [98–100] of the order parameter field $\vec{\psi}$. Suppose, instead of the positions $\mathbf{r}_i(t)$ we want to write our description in terms of the zeros of $\vec{\psi}(\mathbf{r}, t)$. It is not difficult to see that we can write:

$$\delta(\vec{\psi}) = \frac{1}{|\mathcal{D}(\mathbf{r})|} \sum_{i=1}^{N} \delta[\mathbf{r} - \mathbf{r}_i(t)] , \qquad (183)$$

where in the denominator on the right-hand side one has the Jacobian associated with the change of variables from the set of vortex positions to the field $\vec{\psi}$:

$$\mathcal{D} = \frac{1}{n!} \varepsilon_{\mu_1\mu_2\ldots\mu_n} \varepsilon_{\nu_1\nu_2\ldots\nu_n} \nabla_{\mu_1}\psi_{\nu_1} \nabla_{\mu_2}\psi_{\nu_2} \cdots \nabla_{\mu_n}\psi_{\nu_n} , \qquad (184)$$

where $\varepsilon_{\mu_1\mu_2\ldots\mu_n}$ is the n-dimensional fully antisymmetric tensor and summation over repeated indices is implied. The quantities on which we will focus here are the signed or charge density:

$$\rho(\mathbf{r},t) = \delta[\vec{\psi}(\mathbf{r},t)]\mathcal{D}(\mathbf{r},t) = \sum_{i=1}^{N} q_i \delta[\mathbf{r} - \mathbf{r}_i(t)] , \qquad (185)$$

where $q_i = \mathcal{D}(\mathbf{r}_i)/|\mathcal{D}(\mathbf{r}_i)|$, and the unsigned scalar density $n(\mathbf{r},t) = |\rho(\mathbf{r},t)|$. In order to interpret the quantity q_i, we can use a simple model for a vortex with charge m near its core for the case $n = d = 2$. In this case, the order parameter is given by:

$$\psi_x = \psi_0 r^{|m|} \cos(m\phi + \phi_0), \quad \psi_y = \psi_0 r^{|m|} \sin(m\phi + \phi_0) , \qquad (186)$$

where r and ϕ are the cylindrical coordinates relative to the core at the origin. It is then a simple calculation (see Problem 11.17) to work out $\mathcal{D} = m|m|r^{2(|m|-1)}\psi_0^2$, and $q_i = sgn(m_i)$. For systems where only unit charges are present $m_i = \pm 1$, then ρ is the topological charge density. Notice that $q_i = sgn(m_i)$ is well defined even for system like classical fluids where m_i corresponds to the circulation associated with vortex i.

The dynamical implications of this approach are simple. If topological charge is indeed conserved then we would expect the charge density, ρ, to obey a continuity equation. It was shown in Ref.([101]) that ρ satisfies a continuity equation of the form:

$$\partial_t \rho = -\vec{\nabla} \cdot (\rho \mathbf{v}) , \qquad (187)$$

where the vortex velocity field $\mathbf{v}$ is given explicitly by:

$$\mathcal{D} v_\alpha = -\frac{1}{(n-1)!} \varepsilon_{\alpha\mu_2\ldots\mu_n} \varepsilon_{\nu_1\nu_2\ldots\nu_n} \dot{\psi}_{\nu_1} \nabla_{\mu_2}\psi_{\nu_2} \cdots \nabla_{\mu_n}\psi_{\nu_n} , \qquad (188)$$

where $\mathcal{D}$ is defined by Eq. (184) and we must remember that $\mathbf{v}$ is multiplied by the vortex core-locating δ function. Equation (188) gives one an explicit expression for the defect velocity field expressed in terms of derivatives of the order parameter. This expression for the defect velocity seems to be very general. Notice that we have not specified the form of the equation of motion for the order parameter only that the order parameter be a vector and $n = d$.

Does the expression for the velocity agree with our expectations for known cases? To answer this question we need to restrict the analysis to a particular set of field equations. A highly nontrivial test is to look at the complex Ginzburg–Landau equation:

$$\partial_t \psi = b\nabla^2\psi + (1 - u\psi^*\psi)\psi \ , \tag{189}$$

where b and u are complex. Our expression for the velocity reduces in this case, using complex notation ($n = 2$), to the result:

$$\mathcal{D}v_\alpha = -\frac{i}{2}\sum_\beta \varepsilon_{\alpha\beta}\left(b\nabla^2\psi\nabla_\beta\psi^* - b^*\nabla^2\psi^*\nabla_\beta\psi\right) \ . \tag{190}$$

Let us assume that we have a vortex of charge m at the origin of our two-dimensional system and write the order parameter in the form: $\psi = Re^{i\theta}$, $R = r^{|m|}e^w$ and $\theta = m\phi + \theta_B$, where again r and ϕ are the cylindrical coordinates relative to the core at the origin. It is then a straightforward bit of calculus to show that the velocity given by Eq. (190) reduces to:

$$v_\alpha = 2b''\left(\nabla_\alpha\theta_B + \frac{m}{|m|}\sum_\beta \varepsilon_{\alpha\beta}\nabla_\beta w\right)$$
$$-2b'\left(\nabla_\alpha w - \frac{m}{|m|}\sum_\beta \varepsilon_{\alpha\beta}\nabla_\beta\theta_B\right) \ . \tag{191}$$

If we ignore the contributions due to the variation in the amplitude, w, this equation reduces to $v_\alpha = 2b''\nabla_\alpha\theta_B + b'\frac{m}{|m|}\sum_\beta \varepsilon_{\alpha\beta}\nabla_\beta\theta_B$. The first term is the only contribution in the NLSE case and states that a vortex moves with the local superfluid velocity [102]. The second term is the Peach-Koehler [103] term first found in this context by Kawasaki [104]. These are the results from the phase-field approach and lead, for example, to the interaction between two vortices of the type discussed above by Sommerfeld [91]. If one looks at the velocity of a single isolated vortex it is zero. For a set of two isolated vortices one has the expected behavior for the TDGL and NLSE cases. Thus, the velocity given by Eq. (191) reproduces the most sophisticated results obtained using other methods [105].

The expression for the defect velocity, Eq. (188), reduces to conventional forms in the case of a few defects and reproduces results for the phase-field models in the appropriate limit. It can be used to determine the point-defect velocity probability distribution function defined by:

$$\langle n \rangle P(\mathbf{V}, t) \equiv \langle n\delta(\mathbf{V} - \mathbf{v}[\vec{\psi}])\rangle \ , \tag{192}$$

where $n = |\rho|$ is the unsigned defect density. $P(\mathbf{V}, t)$ gives the probability that a vortex has a velocity $\mathbf{V}$ at the time t after a quench.

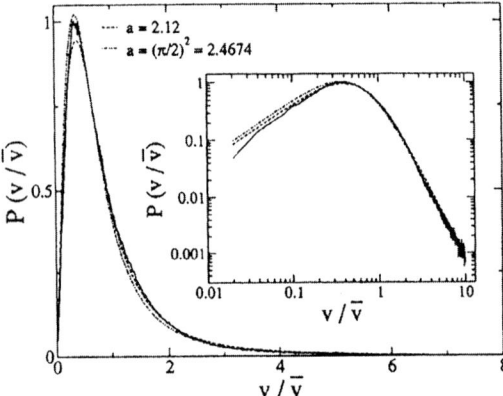

Fig. 11.13 The vortex speed distribution probability density $VP(\mathbf{V})$ from numerical simulation (*solid line*) versus scaled speed $x = V/\bar{v}$. The *dashed lines* are fits to the form $VP(\mathbf{V}) = 2ax/(1+ax)^2$. The *inset* shows the same data in the logarithmic scale. See Fig. 3 in Ref. [107].

While the analytic evaluation of $P(\mathbf{V}, t)$ looks improbable because of the complicated dependence of $\mathbf{v}$ on ψ in Eq. (188), it turns out that there are several factors working to help one in this calculation. If one makes a local mapping from the order parameter $\vec{\psi}$ onto some auxiliary field $\mathbf{m}$, $\vec{\psi} = \vec{\psi}(\vec{m})$, and the two fields share the same zeros, it is easy to see that one can evaluate ρ using $\vec{\psi}$ or $\mathbf{m}$, similarly (see Problem 11.18) $\mathbf{v}[\psi] = \mathbf{v}[m]$. If we can make a transformation to a field $\mathbf{m}$, which can be treated as Gaussian, then the average in Eq. (192) can be worked out explicitly [101] in the case of the TDGL model with the result:

$$P(\mathbf{V}, t) = \Gamma\left(\frac{n+1}{2}\right)\left(\frac{1}{\pi\bar{v}^2(t)}\right)^{n/2}\left(1 + \mathbf{V}^2/\bar{v}^2(t)\right)^{-(n+2)/2} \qquad (193)$$

where the characteristic velocity $\bar{v}(t)$ is expressed in terms of derivatives of the normalized auxiliary field correlation function $f(r,t)$ ($f(0,t) = 1$) evaluated at $r = 0$. One way of determining this auxiliary field correlation function [106] $f(r,t)$ is to satisfy the continuity equation Eq. (187) on average. This leads to the remarkably simple final result: $f(r,t) = e^{-\frac{x^2}{2}}$ where $x = r/L(t)$ is the scaled length and $L(t) \approx \sqrt{t}$ where t is the time. The characteristic velocity (in dimensionless units) is given then by $\bar{v}^2 = 2n/t$. This result given by Eq. (193) has been tested numerically in Ref. [107] for the case $n = d = 2$. The excellent agreement between theory and simulation are shown in Fig. 11.13.

The result for $P(\mathbf{V}, t)$ given by Eq. (193), indicates that the probability of finding a vortex with a large velocity decreases with time. However, since this distribution falls off only as $V^{-(n+2)}$ for large V, only the first moment beyond the normalization integral exists. This seems to imply the existence of a source of large velocities. Assuming the large velocities of defects can be associated with the final collapse of a defect structure (vortex–antivortex pair annihilation for point defects), Bray [108] used general scaling arguments to obtain the same large velocity tail given by Eq. (193).

One can also look at the spatial correlations between defects [97] and the analysis for point defects can be generalized [109] to the case of line defects $n = d - 1$.

11.11
Pattern Forming Systems

In another class of equilibrating systems one has pattern formers. The equilibrium structures of some such systems are discussed in Chapter 13 of FOD. Examples of such systems are magnetic systems that form patterns of Bloch walls [110], including arrays of magnetic bubbles [111, 112]. In superconductors we have the domain walls separating the normal and superconducting phase in the intermediate phase [113, 114] and the flux lattice formed in type II superconductors [56–58]. There is considerable current interest [115–119] in the variety of structures formed by diblock copolymers.

The simplest Langevin model generating such patterns is the Swift–Hohenberg model [120]. This is a model describing the growth of stripes or layers in a 50/50 composition of a diblock copolymers in the weak segregation limit. The equilibrium properties of this model were discussed in Chapter 13 in FOD. In treating the dynamics we have the Langevin equation:

$$\partial_t \psi = -\Gamma \frac{\delta \mathcal{H}_E}{\delta \psi} + \eta \, , \tag{194}$$

where the effective Hamiltonian $\mathcal{H}_E$ is given by Eq. (13.86) in FOD. Here we use the standard notation for dimensionless variables:

$$\mathcal{H}_E = \int d^d x \left[\frac{1}{2} [(\nabla^2 + 1)\psi]^2 - \frac{\varepsilon}{2} \psi^2 + \frac{1}{4} \psi^4 \right] \, , \tag{195}$$

where the wavelength of the generated stripes is $\lambda = 2\pi$ and ε is a control parameter. For quenches to zero temperature and in time units of Γ we have the equation of motion:

$$\partial_t \psi = \varepsilon \psi - \psi^3 - (\nabla^2 + 1)^2 \psi \, . \tag{196}$$

Starting this system with random initial conditions one can look at the set of defects generated as the system evolves toward an ordered striped system. As shown in Fig. 11.14, in two dimensions one finds [121] a complicated defect structure including disclinations, dislocations, and grain boundaries.

Fig. 11.14 Typical configuration for the Swift–Hohenberg model for a quench to zero temperature. The *black points* correspond to ψ > 0. See Fig. 5 in Ref. [121].

Experimentally there has been significant progress on this problem. In Ref. [122] they have observed the ordering of diblock copolymers with a characteristic growth law $L(t) \approx t^{1/4}$ and the eventual disordering agents, consistent with this growth law, are annihilating quadruples of disclinations. The various models studied numerically have not yet seen the annihilating quadrupoles. Analytic theory is far from giving a satisfactory description of this system.

11.12
References and Notes

1 J. D. Gunton, M. San Miguel, P. S. Sahni, in *Phase Transitions and Critical Phenomena*, Vol. 8, C. Domb, J. L. Lebowitz (eds), Academic, N.Y. (1983), p. 267.

2 A. J. Bray, Adv. Phys. **43**, 357 (1994).

3 This problem was discussed in FOD at the end of Section 12.2.1.

4 This figure is over simplified. The potential can be thought of as effective temperature-dependent potential.

5 For quenches to temperatures $T_F > 0$ the average magnitude of the order parameter $M = \langle \psi \rangle$ will be reduced to a value $M(T_F) < \psi_0$.

6 In the field theory model one can choose initial conditions $\psi(\mathbf{x}) = \varepsilon$ where ε is a random variable in the range $-\varepsilon_0 \leq \varepsilon \leq \varepsilon_0$ where ε_0 can be arbitrarily small. For an Ising model the coarse-grained value of the spins will have a spatial average near zero.

7 The Monte Carlo method is discussed in detail in K. Binder, D. Heermann, *Monte Carlo Methods in Statistical Physics: An Introduction*, Springer-Verlag. Heidelberg (1988).

8 See also the discussion in ESM in Chapter 5.

9. Here we use the lattice-gas description of a binary alloy where at each lattice site **R** there is a variable $n(\mathbf{R})$, which has the value $n(\mathbf{R}) = +1$ if the site is occupied by a particle of type A and $n(\mathbf{R}) = -1$ if the site is occupied by a particle of type B. Clearly this model is mathematically identical to an Ising model where the external magnetic field for the magnet is related to the chemical potential of the alloy. For a discussion see K. Huang, *Statistical Mechanics*, Wiley, N.Y. (1963), p. 331.

10. J. A. Mydosh, *Spin Glasses: An Experimental Introduction*, Taylor and Francis, London, (1993); K. H. Fischer, J. A. Hertz, *Spin Glasses*, Cambridge University Press, Cambridge (1991).

11. N. D. Mermin, Rev. Mod. Phys. **51**, 591 (1979).

12. M. Kleman, *Points, Lines and Walls in Liquid Crystals, Magnetic Systems and Various Ordered Media*, Wiley, N.Y. (1983).

13. There is a full discussion of the *n*-vector model in FOD in Chapters 1 and 4.

14. T. Hashimoto *et al.*, in *Dynamics of Ordering Processes in Condensed Matter*, S. Komura, M. Furukawa (eds), Plenum, N.Y. (1988).

15. To see a discussion from this point of view consult G. F. Mazenko, Phys. Rev. D **34**, 2223 (1986); Phys. Rev. Lett. **54**, 2163 (1985).

16. For early numerical treatments see R. Petschek, H. Metiu, J. Chem. Phys. **79**, 3443 (1983); O. T. Valls, G. F. Mazenko, Phys. Rev. B **34**, 7941 (1986).

17. It is T_c^0 and not the physical transition temperature T_c that appears in r since the quartic terms in V self-consistently reduce T_c^0 to T_c. T_c^0 is the transition temperature for the case $u = 0$.

18. Quenches from $T > T_c$ to $T = T_c$ are special. See the discussion in Bray [2].

19. Exceptions to this statement occur for systems like the spin-exchange Kawasaki model, where the system freezes following quenches to zero temperature.

20. T. M. Rogers, K. R. Elder, R. C. Desai, Phys. Rev. B **37**, 9638 (1988).

21. Reversible or mode-coupling terms were discussed in Chapter 10.

22. T. M. Rogers, R. C. Desai, Phys. Rev. B **39**, 11 956 (1989).

23. J. D. Gunton, M. Droze, in *Introduction to the Dynamics of Metastable and Unstable States, Vol. 183, Lecture Notes in Physics*, J. Zittartz (ed), Springer-Verlag, Berlin (1983).

24. Classical nucleation theory was developed by R. Becker and W. Döring, Ann. Phys. **24**, 719 (1935); see F. F. Abraham, *Homogeneous Nucleation Theory*, Academic, N.Y. (1974). The field theoretical description was developed by J. S. Langer, Ann. Phys. N.Y. **54**, 258 (1969); in *Systems Far From Equilibrium, Vol. 132, Lecture Notes in Physics*, L. Garrido (ed)., Springer-Verlag, Berlin (1980).

25. The condensation energy is the energy gained in going from the higher-free-energy phase to the more stable lower-free-energy phase.

26. $\mathbf{q}_0 = 0$ for fluids, but for magnetic solids one has a set of ordering wavenumbers reflecting the underlying periodicity of the lattice. One can distinguish between ferromagnetic and antiferromagnetic points.

27. S. Nagler, R. F. Shannon, Jr., C. R. Harkness, M. A. Singh, R. M. Nicklow, Phys. Rev. Lett. **61**, 718 (1988).

28. We assume that the interfacial width is proportional to ξ.

29. H. Tomita, Prog. Theor. Phys. **72**, 656 (1984); **75**, 482 (1986).

30. G. Porod, Kolloid Z. **124**, 83 (1951); **125**, 51 (1952). See also G. Porod, in *Small Angle X-Ray Scattering*, O. Glatter, L. Kratky (eds), Academic, N.Y. (1982).

31. This approach was suggested by Professor P. Nozieres.

32. These ideas are fully discussed in the excellent review by Alan Bray [2].

33. W. Thomson, Proc. R. Soc. Edinburgh **7**, 63 (1870); Philos. Mag. **42**, 448 (1871).

34. A numerical study of droplet decay is given by S. A. Safron, P. S. Sahni, G. S. Grest, Phys. Rev. B **28**, 2693 (1983).

35. I. M. Lifshitz, V. V. Slyosov, J. Phys. Chem. Solids **19**, 35 (1961). C. Wagner, Z. Elektrochem. **65**, 581 (1961).

36. E. M. Lifshitz, L. P. Pitaevskii, *Physical Kinetics*, Pergamon, Oxford, (1981), Chapter XII.

37 K. Kawasaki, Ann. Phys. N.Y. **154**, 319 (1984).

38 M. Tokuyama, K. Kawasaki, Physica A **123**, 386 (1984); M. Tokuyama, Y. Enomoto, K. Kawasaki, Physica A **143**, 183 (1987).

39 J. Marro, J. L. Lebowitz, M. Kalos, Phys. Rev. Lett. **43**, 282 (1979).

40 M. Hennion, D. Ronzaud, P. Guyot, Acta. Metall. **30**, 599 (1982).

41 H. Furukawa, J. Phys. Soc. Jpn. **58**, 216 (1989).

42 D. S. Fisher, D. A. Huse, Phys. Rev. B **38**, 373 (1988).

43 I. M. Lifshitz, Zh. Eksp. Teor. Fiz. **42**, 1354 (1962) [Sov. Phys. JETP **15**, 939 (1962)]; S. M. Allen, J. W. Cahn, Acta. Metall. **27**, 1085 (1979).

44 J. W. Cahn, J. E. Hilliard, J. Chem. Phys. **28**, 258 (1958); H. E. Cook, Acta Metall. **18**, 297 (1970).

45 The autocorrelation function for the noise is given by Eq. (4).

46 J. S. Langer, M. Bar-on, H. D. Miller, Phys. Rev. A **11**, 1417 (1975).

47 T. Ohta, D. Jasnow, K. Kawasaki, Phys. Rev. Lett. **49**, 1223 (1983).

48 G. F. Mazenko, Phys. Rev. B **42**, 4487 (1990).

49 We follow the development in G. F. Mazenko, Phys. Rev. E **58**, 1543 (1998).

50 G. F. Mazenko, Phys. Rev. E **61**, 1088 (2000).

51 G. F. Mazenko, O. T. Valls, M. Zannetti, Phys. Rev. B **38**, 520 (1988).

52 For a discussion of Gaussian integrals see Appendix O in ESM and Appendix B in FOD.

53 J. Wilks, *The Properties of Liquid and Solid Helium*, Oxford, London (1967).

54 R. J. Donnelly, *Quantized Vortices in Helium II*, Cambridge University Press, Cambridge (1991).

55 L. M. Pismen, *Vortices in Nonlinear Fields*, Oxford University Press, London (1999), Chapter 4.

56 P. G. de Gennes, *Superconductivity of Metals and Alloys*, Addison-Wesley, Redwood City (1989).

57 A. L. Fetter, P. C. Hohenberg, ?Theory of Type II Superconductors?, in *Superconductivity, Vol. 2*, R. D. Parks (ed), Dekker, N.Y. (1969), p817.

58 M. Tinkham, *Introduction to Superconductivity*, McGraw-Hill, New York (1994).

59 L. M. Pismen, *Vortices in Nonlinear Fields*, Oxford University Press, London (1999), Chapter 6.

60 B. Yurke, A. Pargellis, T. Kovacs, D. Huse, Phys. Rev. E **47**, 1525 (1993).

61 M. E. Gouvea, G. M. Wysin, A. R. Bishop and F. G. Mertens, Phys. Rev. B **39**, 11840 (1989).

62 N. Papanicolaou, Phys. Lett. A **186**, 119 (1994).

63 N. Papanicolaou, cond-mat/9711035.

64 S. Komineas, N. Papanicolaou, Nonlinearity **11**, 265 (1998). cond-mat/9612043.

65 F. G. Mertens, H. J. Schnitzer, A. R. Bishop, Phys. Rev. B **56**, 2510 (1997).

66 F. G. Mertens, A. R. Bishop, cond-mat/9903037.

67 F. R. N. Nabarro, *Theory of Crystal Dislocations*, Oxford University Press, London (1967).

68 P. G. de Gennes, J. Prost, *The Physics of Liquid Crystals*, 2nd edn., (Clarendon, Oxford (1993).

69 I. Chuang, B. Yurke, N. Pargellis, N. Turok, Phys. Rev. E **47** 3343 (1993).

70 H. Orihara, Y. Ishibashi, J. Phys. Soc. Jpn. **55**, 2151 (1986).

71 I. Chuang, B. Yurke, N. Turok, Phys. Rev. Lett. **66** 2472 (1991)

72 A. N. Pargellis, N. Turok, B. Yurke, Phys. Rev. Lett. **67**, 1570 (1991).

73 A. N. Pargellis, I. Chuang, B. Yurke, Physica B **56**, 1780 (1992).

74 A. L. Fetter, J. Low. Temp. Phys. **113**, 189 (1998).

75 E. Lundh, P. Ao, Phys. Rev. A **61**, 2000.

76 A. A. Svidzinsky, A. L. Fetter, Phys. Rev. A **62**, 063617 (2000).

77 J. Dziarmagai, M. Sadzikowski, Phys. Rev. Lett. **82**, 4192 (1999).

78 G. J. Stephens, E. A. Calzetta, B. L. Hu, Phys. Rev. D **59**, 5009 (1999).

79 M. Hindmarsh, Phys. Rev. Lett. **77**, 4495 (1996).

80 N. D. Antunes, L. M. A. Bettencourt, Phys Rev D **55**, 925 (1997).

81. See articles by Kibble and Zurek in *Formation and Interactions of Topological Defects*, A. C. Davis, R. Brandenberger (eds), Plenum, N.Y. (1995).
82. E. P. Gross, Nuovo Cimento **20**, 454 (1961); L. P. Pitaevski, Sov. Phys. JETP **13**, 451 (1961).
83. S. Puri, C. Roland, Phys. Lett. **151**, 500 (1990); H. Toyoki, K. Honda, Prog. Theor. Phys. **78**, 237 (1987); F. Liu, G. F. Mazenko, Phys. Rev. B **45**, 6989 (1992).
84. A. J. Bray, S. Puri, Phys. Rev. Lett. **67**, 2760 (1991).
85. W. Thomson, (Lord Kelvin), *Mathematical and Physical Papers, Vol. IV*, Cambridge University Press, Cambridge (1910).
86. See Section 12.10.2 in FOD.
87. See Section 12.7.2 in FOD.
88. See Section 12.6.3 in FOD.
89. A. Aharoni, *Introduction to the Theory of Ferromagnetism*, 2nd edn., Oxford University Press, Oxford (2000).
90. For some discussion of the lore see M. R. Geller, D. J. Thouless, S. W. Rhee, W. F. Vinen, J. Low Temp. Phys. **121**, 411 (2000).
91. A. Sommerfeld, *Mechanics of Deformable Bodies*, Academic, N.Y. (1950), p. 160.
92. See Section 12.6.3 in FOD.
93. J. C. Neu, Physica D **43**, 385 (1990).
94. L. M. Pismen, J. D. Rodriguez, Phys. Rev. A **42**, 2471 (1990).
95. M. C. Cross, P. C. Hohenberg, Rev. Mod. Phys. **65**, 851 (1993), p. 921.
96. L. M. Pismen, Oxford, London (1999), p. 37.
97. Fong Liu, Gene F. Mazenko, Phys. Rev. B **46**, 5963 (1992).
98. S. O. Rice, Mathematical Analysis of Random Noise, in *Bell System Journal Vols. 23 and 24*, reprinted in *Selected Papers on Noise and Stochastic Processes*, N. Wax (ed), Dover, N.Y. (1954).
99. B. I. Halperin, M. Lax, Phys. Rev. **148**, 722 (1966).
100. B. I. Halperin, Statistical Mechanics of Topological Defects, in *Physics of Defects*, R. Balian et al., North-Holland, N.Y. (1981).
101. G. F. Mazenko, Phys. Rev. Lett. **78**, 401 (1997).
102. A. L. Fetter, Phys. Rev. **151**, 100 (1966).
103. M. Peach, J. S. Koehler, Phys. Rev. **80**, 436 (1950).
104. K. Kawasaki, Prog. Theor. Phys. Suppl. **79**, 161 (1984).
105. O. Törnkvist, E. Schröder, Phys. Rev. Lett. **78**, 1908 (1997).
106. G. F. Mazenko, R. Wickham, Ordering Kinetics of Defect Structures, Phys. Rev. E **57**, 2539 (1998).
107. H. Qian, G. F. Mazenko, Phys. Rev. **68**, 021109 (2003).
108. A. Bray, Phys. Rev. E **55**, 5297 (1997).
109. G. F. Mazenko, Phys. Rev. E **59**, 1574 (1999).
110. Soshin Chikazumi, *Physics of Ferromagnetism*, Oxford University Press, New York (1997), Chapters 15–17.
111. A. P. Malozemoff, J. C. Slonczewski, *Magnetic Domain Walls in Bubble Material*, Academic, N.Y. (1979)
112. A. H. Eschenfelder, *Magnetic Bubble Technology*, Springer-Verlag, Berlin (1981).
113. R. P. Huebener, *Magnetic Flux Structures in Superconductors*, Springer-Verlag, N.Y. (1979).
114. L. D. Landau, Sov. Phys. JETP **7**, 371 (1937); Phys. Z. Sowjet, **11**, 129 (1937).
115. Y. Oono, M. Bahiana, Phys. Rev. Lett. **61**, 1109 (1988).
116. F. Liu, N. Goldenfeld, Phys. Rev. A **39**, 4805 (1989).
117. J. Christensen, A. J. Bray, Phys. Rev. E **58**, 5364 (1998).
118. Q. Hou, S. Sasa, N. Goldenfeld, Physica A **239**, 219 (1997)
119. D. Boyer, J. Viñals, Phys. Rev. E **64**, 050101 (2001).
120. J. B. Swift, P. C. Hohenberg, Phys. Rev. A **15**, 319 (1977).
121. H. Qian, G. F. Mazenko, Phys. Rev. **67**, 036102 (2003).
122. C. Harrison, D. H. Adamson, Z. Cheng, J. M. Sebastian, S. Sethuraman, D. A. Huse, R. A. Register, P. M. Chaikin, Science **290**, 1558 (2000); C. Harrison, Z. Cheng, S. Sethuraman, D. A. Huse, P. M. Chaikin, D. A. Vega, J. M. Sebastian, R. A. Register, D. H. Adamson, Phys. Rev. E **66**, 011706 (2002).

11.13
Problems for Chapter 11

Problem 11.1: Write a Monte Carlo program for evaluating the phase-ordering kinetics of a one-dimensional nearest-neighbor Ising model. Assume the Hamiltonian for the system is given by:

$$H = -J \sum_{i=1}^{N} \sigma_i \sigma_{i+1}$$

with periodic boundary conditions $\sigma_{N+1} = \sigma_1$ and $J > 0$.

Construct the kinetics in the following fashion: repeatedly carry out the following steps.

1. Randomly select a site j.

2. If $\sigma_{j+1}\sigma_{j-1} = 1$, set $\sigma_j = \sigma_{j+1} = \sigma_{j-1}$

3. If $\sigma_{j+1}\sigma_{j-1} = -1$, generate a random number w between 0 and 1. If $w < 0.5$, flip σ_j, otherwise do not flip σ_j.

Starting with a random initial distribution ($\sigma_i = \pm 1$ at random) compute the number of kinks ($\sigma_i \sigma_{i+1} = -1$) in the system per site as a function of Monte Carlo steps per spin. Average your results over 40 runs.

Problem 11.2: Solve as an initial-value problem the equation of motion:

$$\frac{1}{\Gamma}\frac{d}{dt}m(t) = m(t)\left[1 - m^2(t)\right]$$

governing the time evolution of the average magnetization after an off-critical quench.

Problem 11.3: Assume a potential $V(\psi)$ given by Eq. (14). Show, to first order in the applied field h, that the minima:

$$V'(\psi_\pm) = 0$$

are given by Eq. (20).

Problem 11.4: Show that the square gradient term in the effective Hamiltonian in the nucleation problem leads to the surface energy given by Eq. (24).

Problem 11.5: Show that the coefficient of h in the activation energy, evaluated in the sharp interface limit, is given by:

$$F_B^{(1)} = -2\psi_0 V_d |h| \quad ,$$

where the volume of the droplet is given by:

$$V_d = \frac{2\pi^{d/2}}{d\Gamma(d/2)} R_0^d \quad .$$

Problem 11.6: Show that the extremum as a function of droplet radius R_0 of the droplet free energy difference:

$$F_D = -2\psi_0 V_d |h| + \sigma \Sigma_d$$

is a maximum.

Problem 11.7: Insert the interfacial ansatz:

$$\psi = \psi_0 sgn(z - z_0)$$

into the correlation function:

$$C(z, z') = \frac{\int_{-L}^{L} dz_0 \psi(z) \psi(z')}{\int_{-L}^{L} dz_0}$$

and do the integrations.

Problem 11.8: Show that the Fourier transform of the short-distance expansion of the equal-time order parameter correlation function:

$$C(\mathbf{x}, t) = \psi_0^2 \left[1 - \alpha \frac{|\mathbf{x}|}{L} + \cdots \right] \ ,$$

leads to Porod's law, given by Eq. (44), for large wavenumbers.

Problem 11.9: Solve Laplace's equation:

$$\nabla^2 \mu = 0 \ ,$$

together with the condition that μ is given by the Gibbs–Thomson relation:

$$\mu = -\frac{K\sigma_s}{2\psi_0}$$

on the surface of a sphere separating two phases as described in Section 11.7.

Problem 11.10: Show, for the TDGL model, remembering from Chapter 9 the result:

$$\frac{\delta \psi(\mathbf{x}', t)}{\delta \eta(\mathbf{x}, t)} = \frac{1}{2} \delta(\mathbf{x} - \mathbf{x}') \ ,$$

that:

$$\langle \eta(\mathbf{q}, t) \psi(-\mathbf{q}, t) \rangle = k_B T \Gamma(q) \ ,$$

where $\eta(\mathbf{x}, t)$ is gaussianly distributed white noise.

Problem 11.11: Consider the equation:

$$\dot{A} = -(A + 1)\left(\frac{d}{2(t + t_0)} + 2\Gamma A \right) \ ,$$

which arises in the mean-field theory treatment of phase-ordering kinetics. Show that for large times this has a solution of the form:

$$A = \frac{A_1}{(t+t_0)} + \frac{A_2}{(t+t_0)^2} + \cdots ,$$

where:

$$A_1 = -\frac{d}{4\Gamma}$$

$$A_2 = \frac{d(1-d)}{8\Gamma^2} .$$

Problem 11.12: As part of mean-field theory treatment of phase-ordering kinetics investigate numerically the dependence of the constant:

$$\kappa = \lim_{t\to\infty} \frac{e^{-2\Gamma \int_0^t d\tau A(\tau)}}{(2\Gamma(t+t_0))^{d/2}}$$

on the initial conditions.

Problem 11.13: Assuming m is a Gaussian field and:

$$\mathcal{H}(12) = ik_1 m(1) + ik_2 im(2) ,$$

show that the average over m gives:

$$\langle e^{\mathcal{H}(12)} \rangle = e^{-\frac{1}{2}[k_1^2 S_0(1) + k_2^2 S_0(2) + 2k_1 k_2 C_0(12)]} ,$$

where:

$$C_0(12) = \langle m(1)m(2) \rangle$$

and:

$$S_0(1) = C_0(11) .$$

Problem 11.14: Carry out the integrations over k_1 and k_2 giving the joint probability distribution:

$$P_0(x_1, x_2, 12) = \int \frac{dk_1}{2\pi} \int \frac{dk_2}{2\pi} e^{-ik_1 x_1} e^{-ik_2 x_2} e^{-\frac{1}{2}[k_1^2 S_0(1) + k_2^2 S_0(2) + 2k_1 k_2 C_0(12)]} .$$

Problem 11.15: Evaluate the integrals defined in Eq. (166) and verify Eq. (167). Hint: Use the integral representation for the sgn function:

$$sgn\, y = \frac{y}{|y|} = y \int \frac{ds}{\sqrt{2\pi}} e^{-\frac{1}{2}s^2 y^2} .$$

Problem 11.16: Show that the quantity $\int d^d x\, \psi^*(\mathbf{x})\psi(\mathbf{x})$ is conserved in the NLSE.

Problem 11.17: Show that the Jacobian defined by Eq. (184) is given by $\mathcal{D} = m|m|r^{2(|m|-1)}\psi_0^2$, and $q_i = \mathrm{sgn}(m_i)$ for an order parameter given by Eq. (188) near a vortex core.

Problem 11.18: Consider the defect velocity defined by Eq. (188). If the order parameter is a local function of an auxiliary field $\mathbf{m}$, which shares the same zeros, show that:

$$\mathbf{v}[\psi] = \mathbf{v}[m] \ .$$

Problem 11.19: Evaluate the ordered energy of the Swift–Hohenberg effective Hamiltonian in the single mode approximation:

$$\psi = A\cos q_0 z \ .$$

Choose the amplitude A such that this energy is a minimum.

Problem 11.20: Solve the self-consistent equations:

$$A(t) = -1 + 3\langle \psi^2 \rangle.$$

$$C(q,t) = C(q,0) e^{-2\int_0^t d\tau \Gamma(q)[q^2 + A(\tau)]}$$

with the initial condition:

$$C(q,0) = C_I e^{-\frac{1}{2}\alpha q^2} \ .$$

in the case of a COP. First determine $\langle \psi^2 \rangle$.

Problem 11.21: Evaluate the integral:

$$I = \int \frac{dz_1 dz_2}{2\pi} e^{-\frac{1}{2}[z_1^2 + z_2^2 - 2z_1 z_2 \tilde{f}]} \ .$$

Appendix A
Time-Reversal Symmetry

The reversal in time of a state $|\psi\rangle$ changes it into a state $|\psi'\rangle$ that develops in accordance with the opposite sense of progression of time. For the new state the signs of all linear and angular momenta are reversed but other quantities are unchanged. Time reversal is effected by a time-independent operator T:

$$T|\psi\rangle = |\psi'\rangle \ .$$

Time-reversal invariance means that if $|\psi_i\rangle$ is an energy eigenstate:

$$H|\psi_i\rangle = E_i|\psi_i\rangle \ ,$$

so also $|\psi'\rangle$ is an eigenstate of H with eigenvalue E_i.

Consider now a pair of evolution operations that we can perform on $|\psi_i\rangle$ that are expected to lead to the same physical state. In case A we allow the state to propagate to time t and then let $t \to -t$. In case B we reverse time at $t = 0$ and then allow the reversed state to propagate with the opposite sense of progression of time, that is, to time $-t$.

In operation A we allow the state $|\psi_i\rangle$ to first evolve forward in time:

$$e^{-iHt/\hbar}|\psi_i\rangle = e^{-iE_i t/\hbar}|\psi_i\rangle \tag{A.1}$$

and then we apply the time-reversal operator:

$$T\, e^{-iE_i t/\hbar}|\psi_i\rangle = |\psi_i\rangle_A \ . \tag{A.2}$$

For operation B, we first apply T and then let the system evolve backward in time:

$$e^{iE_i t/\hbar}T|\psi_i\rangle = |\psi_i\rangle_B \ . \tag{A.3}$$

We require that these operations prepare the system in the same quantum state:

$$|\psi_i\rangle_A = |\psi_i\rangle_B \ , \tag{A.4}$$

so:

$$Te^{-iE_i t/\hbar}|\psi_i\rangle = e^{iE_i t/\hbar}T|\psi_i\rangle \ . \tag{A.5}$$

Nonequilibrium Statistical Mechanics. Gene F. Mazenko
Copyright © 2006 WILEY-VCH Verlag GmbH & Co. KGaA, Weinheim
ISBN: 3-527-40648-4

We see from this relation that T is *not* a linear operator. The time-reversal operator is: *antilinear*

$$T(a_1|\psi_1\rangle + a_2|\psi_2\rangle) = a_1^* T|\psi_1\rangle + a_2^* T|\psi_2\rangle \ . \tag{A.6}$$

Suppose we construct a complete and orthonormal set of energy eigenstates $|n\rangle$,

$$H|n\rangle = E_n|n\rangle \tag{A.7}$$

$$\langle m|n\rangle = \delta_{m,n} \tag{A.8}$$

$$\sum_n |n\rangle\langle n| = 1 \tag{A.9}$$

and we assume that these states remain orthonormal after time reversal:

$$|n'\rangle = T|n\rangle \tag{A.10}$$

$$\langle m'|n'\rangle = \delta_{m,n} \ . \tag{A.11}$$

We can then expand two states $|\psi\rangle$ and $|\phi\rangle$ as:

$$|\psi\rangle = \sum_n |n\rangle\langle n|\psi\rangle \ , \tag{A.12}$$

and:

$$|\phi\rangle = \sum_n |n\rangle\langle n|\phi\rangle \ . \tag{A.13}$$

When we apply T to Eq. (A.12) we obtain:

$$\begin{aligned}|\psi'\rangle &= T|\psi\rangle = \sum_n \langle n|\psi\rangle^* T|n\rangle \\ &= \sum_n \langle\psi|n\rangle T|n\rangle = \sum_n \langle\psi|n\rangle|n'\rangle \ .\end{aligned} \tag{A.14}$$

Similarly:

$$|\phi'\rangle = T|\phi\rangle = \sum_m \langle\phi|m\rangle|m'\rangle \ . \tag{A.15}$$

Taking the inner product:

$$\langle\phi'|\psi'\rangle = \sum_{m,n} \langle m|\phi\rangle\langle m'|n'\rangle\langle\psi|n\rangle$$

$$\begin{aligned}&= \sum_{m,n}\langle m|\phi\rangle\delta_{m,n}\langle\psi|n\rangle\\&= \sum_{n}\langle\psi|n\rangle\langle n|\phi\rangle\end{aligned}$$
$$\langle\phi'|\psi'\rangle = \langle\psi|\phi\rangle \ . \tag{A.16}$$

The two properties:
$$T(a_1|\psi_1\rangle + a_2|\psi_2\rangle) = a_1^* T|\psi_1\rangle + a_2^* T|\psi_2\rangle \tag{A.17}$$

$$\langle\phi'|\psi'\rangle = \langle\psi|\phi\rangle = \langle\phi|\psi\rangle^* \ , \tag{A.18}$$

where $|\psi'\rangle = T|\psi\rangle$, $|\phi'\rangle = T|\phi\rangle$, specify that T is an *antiunitary operator*.

We can also derive the appropriate transformation law for matrix elements of operators. Consider the states $|\alpha\rangle$ and $|\beta\rangle = \hat{B}|\mu\rangle$, where $\hat{B}$ is some operator. We have in general from Eq. (A.18) that:

$$\langle\alpha|\beta\rangle = \langle\beta'|\alpha'\rangle \ , \tag{A.19}$$

where:
$$|\alpha'\rangle = T|\alpha\rangle \tag{A.20}$$

and:
$$\begin{aligned}|\beta'\rangle &= T\hat{B}|\mu\rangle\\&= T\hat{B}T^{-1}T|\mu\rangle\\&= \hat{B}'|\mu'\rangle \ ,\end{aligned} \tag{A.21}$$

where we see that operators transform as $\hat{B}' = T\hat{B}T^{-1}$. We have then the relationship between matrix elements:

$$\begin{aligned}\langle\alpha|\beta\rangle &= \langle\alpha|B|\mu\rangle\\&= \langle\mu'|(B')^\dagger|\alpha'\rangle\\&= \langle\mu'|(TBT^{-1})^\dagger|\alpha'\rangle \ .\end{aligned} \tag{A.22}$$

If we look at an average of an operator $\hat{A}$ over some probability operator, $\hat{\rho}$,

$$\langle\hat{A}\rangle = Tr\hat{\rho}\hat{A} \ , \tag{A.23}$$

then we can write:
$$\langle\hat{A}\rangle = \sum_i \langle i|\hat{\rho}\hat{A}|i\rangle \ . \tag{A.24}$$

Using Eq. (A.22) for the matrix element gives:

$$\langle \hat{A} \rangle = \sum_i \langle i'|(T\hat{\rho}\hat{A}T^{-1})^\dagger |i'\rangle \quad , \tag{A.25}$$

where $|i'\rangle = T|i\rangle$. Since the set $|i'\rangle$ is also complete,

$$\langle \hat{A} \rangle = Tr\, (T\hat{\rho}T^{-1}T\hat{A}T^{-1})^\dagger \quad . \tag{A.26}$$

If the probability operator is invariant under time reversal:

$$T\hat{\rho}T^{-1} = \hat{\rho} \quad , \tag{A.27}$$

we have:

$$\langle \hat{A} \rangle = Tr\, (\hat{\rho}T\hat{A}T^{-1})^\dagger \quad . \tag{A.28}$$

For two operators $\hat{A}$ and $\hat{B}$, $(\hat{A}\hat{B})^\dagger = \hat{B}^\dagger \hat{A}^\dagger$, and:

$$\langle \hat{A} \rangle = Tr\, (T\hat{A}T^{-1})^\dagger \hat{\rho}^\dagger \quad . \tag{A.29}$$

Since the probability operator is hermitian, $\hat{\rho}^\dagger = \hat{\rho}$, we can use the cyclic invariance of the trace to obtain the invariance principle:

$$\langle \hat{A} \rangle = \left\langle \left(T\hat{A}T^{-1}\right)^\dagger \right\rangle \quad . \tag{A.30}$$

We can build up the properties of various quantities under time reversal using the fundamental relations: if **p** is a momentum operator,

$$T\mathbf{p}(t)T^{-1} = -\mathbf{p}(-t) \quad , \tag{A.31}$$

while if **x** is a position operator:

$$T\mathbf{x}(t)T^{-1} = \mathbf{x}(-t) \quad . \tag{A.32}$$

Clearly, from this one has, if **L** is an angular momentum operator:

$$T\mathbf{L}(t)T^{-1} = -\mathbf{L}(-t) \tag{A.33}$$

We see then that Hamiltonians that depend on $\mathbf{p}^2$ and **x** are time-reversal invariant:

$$THT^{-1} = H \quad . \tag{A.34}$$

Consider Maxwell's equations, :

$$\nabla \cdot \mathbf{E} = 4\pi \rho_e \tag{A.35}$$

$$c\nabla \times \mathbf{B} = \frac{\partial \mathbf{E}}{\partial t} + \mathbf{J} \; , \tag{A.36}$$

where ρ_e is the electric charge density and $\mathbf{J}$ is the charge-current density. Since ρ_e is even under time reversal and $\mathbf{J}$ is odd under time reversal, the electric and magnetic fields transform as:

$$T\mathbf{E}(\mathbf{x},t)T^{-1} = \mathbf{E}(\mathbf{x},-t) \tag{A.37}$$

while:

$$T\mathbf{B}(\mathbf{x},t)T^{-1} = -\mathbf{B}(\mathbf{x},-t) \; . \tag{A.38}$$

Observables $A(\mathbf{x},t)$ typically have a definite signature ε_A under time reversal:

$$A'(\mathbf{x},t) = TA(\mathbf{x},t)T^{-1} = \varepsilon_A A(\mathbf{x},-t) \; , \tag{A.39}$$

where $\varepsilon = +1$ for positions, electric fields, and $\varepsilon = -1$ for momenta, magnetic fields and angular momenta.

Appendix B
Fluid Poisson Bracket Relations

We are interested in the Poisson brackets:

$$\{A, B\} = \sum_i \left(\frac{\partial A}{\partial \mathbf{r}_i} \cdot \frac{\partial B}{\partial \mathbf{p}_i} - \frac{\partial A}{\partial \mathbf{p}_i} \cdot \frac{\partial B}{\partial \mathbf{r}_i} \right) , \quad (B.1)$$

where A and B are certain fluid densities. In particular A and B are taken to be the particle density,

$$n(\mathbf{x}) = \sum_i \delta(\mathbf{x} - \mathbf{r}_i) , \quad (B.2)$$

and the momentum density:

$$g_\alpha(\mathbf{x}) = \sum_i p_i^\alpha \delta(\mathbf{x} - \mathbf{r}_i) . \quad (B.3)$$

Using the results:

$$\frac{\partial}{\partial r_i^\beta} n(\mathbf{x}) = -\nabla_x^\beta \delta(\mathbf{x} - \mathbf{r}_i) \quad (B.4)$$

$$\frac{\partial}{\partial p_i^\beta} n(\mathbf{x}) = 0 \quad (B.5)$$

$$\frac{\partial}{\partial r_i^\beta} g^\alpha(\mathbf{x}) = -\nabla_x^\beta \delta(\mathbf{x} - \mathbf{r}_i) p_i^\alpha \quad (B.6)$$

$$\frac{\partial}{\partial p_i^\beta} g^\alpha(\mathbf{x}) = \delta_{\alpha,\beta} \delta(\mathbf{x} - \mathbf{r}_i) , \quad (B.7)$$

we easily evaluate the Poisson brackets among n and $\mathbf{g}$:

$$\{n(\mathbf{x}), g^\alpha(\mathbf{y})\} = \sum_i \sum_\beta [-\nabla_x^\beta \delta(\mathbf{x} - \mathbf{r}_i) \delta(\mathbf{y} - \mathbf{r}_i) \delta_{\alpha,\beta}]$$

Nonequilibrium Statistical Mechanics. Gene F. Mazenko
Copyright © 2006 WILEY-VCH Verlag GmbH & Co. KGaA, Weinheim
ISBN: 3-527-40648-4

$$= -\nabla_x^\alpha [\delta(\mathbf{x} - \mathbf{y}) n(\mathbf{x})] \ . \tag{B.8}$$

$$\{n(\mathbf{x}), n(\mathbf{y})\} = 0 \tag{B.9}$$

$$\{g_\alpha(\mathbf{x}), g_\beta(\mathbf{y})\} = \sum_{i,\gamma} \left(\frac{\partial g_\alpha(\mathbf{x})}{\partial r_i^\gamma} \frac{\partial g_\beta(\mathbf{x}')}{\partial p_i^\gamma} - \frac{\partial g_\alpha(\mathbf{x})}{\partial p_i^\gamma} \frac{\partial g_\beta(\mathbf{x}')}{\partial r_i^\gamma} \right)$$

$$= \sum_{i,\gamma} \left(-\nabla_x^\gamma \delta(\mathbf{x} - \mathbf{r}_i) p_i^\alpha \delta(\mathbf{y} - \mathbf{r}_i) \delta_{\beta,\gamma} + \nabla_y^\gamma \delta(\mathbf{y} - \mathbf{r}_i) p_i^\beta \delta(\mathbf{x} - \mathbf{r}_i) \delta_{\alpha,\gamma} \right)$$

$$= - \left(\nabla_x^\beta [\delta(\mathbf{x} - \mathbf{y}) g_\alpha(\mathbf{x})] \right) + \nabla_y^\alpha \left(\delta(\mathbf{x} - \mathbf{y}) g_\beta(\mathbf{y}) \right) \ . \tag{B.10}$$

We can also explicitly work out the Poisson bracket between the energy density,

$$\varepsilon(\mathbf{x}) = K(\mathbf{x}) + V(\mathbf{x}) \tag{B.11}$$

$$K(\mathbf{x}) = \sum_i \frac{p_i^2}{2m} \delta(\mathbf{x} - \mathbf{r}_i) \tag{B.12}$$

$$V(\mathbf{x}) = \tfrac{1}{2} \sum_{ij} V(\mathbf{r}_i - \mathbf{r}_j) \delta(\mathbf{x} - \mathbf{r}_i) \ , \tag{B.13}$$

and the particle density:

$$\{n(\mathbf{x}), \varepsilon(\mathbf{y})\} = \{n(\mathbf{x}), K(\mathbf{y})\}$$

$$= \sum_{i,\beta} \left(\frac{\partial n(\mathbf{x})}{\partial r_i^\beta} \frac{\partial K(\mathbf{y})}{\partial p_i^\beta} - \frac{\partial n(\mathbf{x})}{\partial p_i^\beta} \frac{\partial K(\mathbf{y})}{\partial r_i^\beta} \right) \tag{B.14}$$

$$= \sum_{i,\beta} \left(-\nabla_x^\beta \delta(\mathbf{x} - \mathbf{r}_i^\beta) \right) \frac{p_i^\beta}{m} \delta(\mathbf{y} - \mathbf{r}_i)$$

$$= -\nabla_x \cdot (\delta(\mathbf{x} - \mathbf{y}) \mathbf{g}(\mathbf{x})/m) \ . \tag{B.15}$$

The other Poisson brackets between the energy density and the momentum density are not very illuminating.

Appendix C
Equilibrium Average of the Phase-Space Density

Here we compute the canonical ensemble average of the phase-space density $f(\mathbf{x},\mathbf{p})$:

$$f_0(\mathbf{p}) = \langle \hat{f}(\mathbf{x},\mathbf{p}) \rangle_{EQ}$$
$$= \frac{\left(\prod_{i=1}^{N} \int d^3 r_i d^3 p_i\right) e^{-\beta H} \hat{f}(\mathbf{x},\mathbf{p})}{\left(\prod_{i=1}^{N} \int d^3 r_i d^3 p_i\right) e^{-\beta H}} = \frac{Num}{Z} \quad , \tag{C.1}$$

where the Hamiltonian is assumed to be of the form:

$$H = K + V = \sum_{i=1}^{N} \frac{\mathbf{p}_i^2}{2m} + V(\mathbf{r}_1, \mathbf{r}_2, \ldots, \mathbf{r}_N) \quad . \tag{C.2}$$

Taking the partition function first:

$$Z = \left(\prod_{i=1}^{N} \int d^3 r_i d^3 p_i\right) e^{-\beta H}$$
$$= I^N \int d^3 r_1 \ldots d^3 r_N e^{-\beta V}$$
$$\equiv I^N Q_N \quad , \tag{C.3}$$

where:

$$I = \int d^3 p_i \, e^{-\beta \mathbf{p}_i^2/2m} = (2\pi k_B T m)^{3/2} \quad . \tag{C.4}$$

The numerator is then given by:

$$Num = \sum_{j=1}^{N} \left(\prod_{i=1}^{N} \int d^3 p_i \, e^{-\beta \mathbf{p}_i^2/2m}\right) \int d^3 r_1 \ldots d^3 r_N e^{-\beta V} \delta(\mathbf{x}-\mathbf{r}_j) \delta(\mathbf{p}-\mathbf{p}_j)$$
$$= \sum_{j=1}^{N} I^{N-1} e^{-\beta \mathbf{p}^2/2m} \int d^3 r_1 \ldots d^3 r_N e^{-\beta V} \delta(\mathbf{x}-\mathbf{r}_j)$$
$$= I^{N-1} e^{-\beta \mathbf{p}^2/2m} Q_N \langle \sum_{j=1}^{N} \delta(\mathbf{x}-\mathbf{r}_j) \rangle_{EQ}$$

Nonequilibrium Statistical Mechanics. Gene F. Mazenko
Copyright © 2006 WILEY-VCH Verlag GmbH & Co. KGaA, Weinheim
ISBN: 3-527-40648-4

$$= I^{N-1} e^{-\beta \mathbf{p}^2/2m} Q_N \langle n(\mathbf{x}) \rangle_{EQ} = I^{N-1} e^{-\beta \mathbf{p}^2/2m} Q_N n \qquad (C.5)$$

and we have finally, dividing Eq. (C.5) by Eq. (C.3),

$$f_0(\mathbf{p}) = n \frac{e^{-\beta \mathbf{p}^2/2m}}{I} = n \frac{e^{-\beta \mathbf{p}^2/2m}}{(2\pi k_B T m)^{3/2}} \;, \qquad (C.6)$$

which is the Maxwell velocity distribution.

Appendix D
Magnetic Poisson Bracket Relations

Let us consider the Poisson bracket relations for a magnetic system. Suppose our fields of interest are the magnetization density:

$$\mathbf{M}(\mathbf{x}) = \sum_{\mathbf{R}} \delta(\mathbf{x} - \mathbf{R})\mathbf{S}(\mathbf{R}) \tag{D.1}$$

and the staggered magnetization density.:

$$\mathbf{N}(\mathbf{x}) = \sum_{\mathbf{R}} \delta(\mathbf{x} - \mathbf{R})\eta(\mathbf{R})\mathbf{S}(\mathbf{R}) \;, \tag{D.2}$$

where the $\mathbf{S}(\mathbf{R})$ are magnetic moments or spins that satisfy the fundamental Poisson brackets for a set of classical spins:

$$\{S_i(\mathbf{R}), S_j(\mathbf{R}')\} = \delta_{\mathbf{R},\mathbf{R}'} \sum_k \varepsilon_{ijk} S_k(\mathbf{R}) \quad . \tag{D.3}$$

We can then work out the Poisson brackets among the components of the densities. For the magnetization density we find:

$$\begin{aligned}\{M_i(\mathbf{x}), M_j(\mathbf{y})\} &= \sum_{\mathbf{R},\mathbf{R}'} \delta(\mathbf{x} - \mathbf{R})\delta(\mathbf{y} - \mathbf{R'})\{S_i(\mathbf{R}), S_j(\mathbf{R}')\} \\ &= \sum_{\mathbf{R},\mathbf{R}'} \delta(\mathbf{x} - \mathbf{R})\delta(\mathbf{y} - \mathbf{R'})\delta_{\mathbf{R},\mathbf{R}'} \sum_k \varepsilon_{ijk} S_k(\mathbf{R}) \;, \end{aligned} \tag{D.4}$$

or:

$$\{M_i(\mathbf{x}), M_j(\mathbf{y})\} = \delta(\mathbf{x} - \mathbf{y}) \sum_k \varepsilon_{ijk} M_k(\mathbf{x}) \quad . \tag{D.5}$$

The Poisson brackets among the components of the staggered magnetization are given by:

$$\begin{aligned}\{N_i(\mathbf{x}), N_j(\mathbf{y})\} &= \sum_{\mathbf{R},\mathbf{R}'} \delta(\mathbf{x} - \mathbf{R})\delta(\mathbf{y} - \mathbf{R'})\eta(\mathbf{R})\eta(\mathbf{R}')\{S_i(\mathbf{R}), S_j(\mathbf{R}')\} \\ &= \sum_{\mathbf{R},\mathbf{R}'} \delta(\mathbf{x} - \mathbf{R})\delta(\mathbf{y} - \mathbf{R'})\eta(\mathbf{R})\eta(\mathbf{R}')\delta_{\mathbf{R},\mathbf{R}'} \sum_k \varepsilon_{ijk} S_k(\mathbf{R}) \;. \end{aligned} \tag{D.6}$$

Nonequilibrium Statistical Mechanics. Gene F. Mazenko
Copyright © 2006 WILEY-VCH Verlag GmbH & Co. KGaA, Weinheim
ISBN: 3-527-40648-4

Since $\eta(\mathbf{R})^2 = 1$, this reduces to:

$$\{N_i(\mathbf{x}), N_j(\mathbf{y})\} = \delta(\mathbf{x} - \mathbf{y}) \sum_k \varepsilon_{ijk} M_k(\mathbf{x}) \quad . \tag{D.7}$$

Finally we have the cross Poisson bracket:

$$\begin{aligned}
\{M_i(\mathbf{x}), N_j(\mathbf{y})\} &= \sum_{\mathbf{R},\mathbf{R}'} \eta(\mathbf{R}')\delta(\mathbf{x} - \mathbf{R})\delta(\mathbf{y} - \mathbf{R}')\{S_i(\mathbf{R}), S_j(\mathbf{R}')\} \\
&= \sum_{\mathbf{R},\mathbf{R}'} \eta(\mathbf{R}')\delta(\mathbf{x} - \mathbf{R})\delta(\mathbf{y} - \mathbf{R}')\delta_{\mathbf{R},\mathbf{R}'} \sum_k \varepsilon_{ijk} S_k(\mathbf{R}) \\
&= \sum_{\mathbf{R}} \eta(\mathbf{R})\delta(\mathbf{x} - \mathbf{R})\delta(\mathbf{y} - \mathbf{R}) \sum_k \varepsilon_{ijk} S_k(\mathbf{R}) \\
&= \delta(\mathbf{x} - \mathbf{y}) \sum_k \varepsilon_{ijk} N_k(\mathbf{x}) \quad . \tag{D.8}
\end{aligned}$$

Appendix E
Noise and the Nonlinear Langevin Equation

The nonlinear Langevin equation,

$$\frac{\partial \psi_i(t)}{\partial t} = \mathcal{V}_i[\psi(t)] - \sum_k \Gamma_0^{ik}[\psi(t)] \frac{\partial}{\partial \psi_k(t)} \mathcal{H}_\psi + \xi_i(t) \;, \tag{E.1}$$

where $\xi_i(t)$ is white Gaussian noise with variance given by Eq. (9.77), forms a complete dynamical system. We show here, starting from Eq. (E.1), that we can derive the GFPE, Eq. (9.81), with an expression for the noise $N_\phi(t)$ in terms of $\xi_i(t)$. We show that the noise $N_\phi(t)$ satisfies the usual properties, $\langle g_{\phi'} N_\phi(t) \rangle = 0$ and the second FDT Eq. (9.104)

We can regain the GFPE and a direct relationship between N_ϕ and ξ_i by using the chain-rule for differentiation:

$$\frac{\partial}{\partial t} g_\phi(t) = - \sum_i \frac{\partial}{\partial \phi_i} g_\phi(t) \frac{\partial \psi_i(t)}{\partial t} \tag{E.2}$$

and the nonlinear Langevin equation, Eq. (E.1), for $\dot{\psi}_i$:

$$\frac{\partial}{\partial t} g_\phi(t) = - \sum_i \frac{\partial}{\partial \phi_i} g_\phi(t) \Bigg[\mathcal{V}_i[\psi(t)]$$

$$+ \sum_k \frac{\partial}{\partial \psi_k(t)} \beta^{-1} \Gamma_0^{ik}[\psi(t)] - \sum_k \Gamma_0^{ik}[\psi] \frac{\partial \mathcal{H}[\psi(t)]}{\partial \psi_k(t)} + \xi_i(t) \Bigg] \tag{E.3}$$

$$= - \sum_i \frac{\partial}{\partial \phi_i} \Bigg[\mathcal{V}_i[\phi] + \sum_k \beta^{-1} \frac{\partial}{\partial \phi_k} \Gamma_0^{ik}[\phi] - \sum_k \Gamma_0^{ik}[\phi] \frac{\partial \mathcal{H}_\phi}{\partial \phi_k} + \xi_i(t) \Bigg] g_\phi(t) \;.$$

Comparing this with the GFPE, Eq. (9.73), written out:

$$\frac{\partial}{\partial t} g_\phi(t) = - \sum_i \frac{\partial}{\partial \phi_i} \Bigg[\mathcal{V}_i[\phi]$$

$$- \sum_k \beta^{-1} \Gamma_0^{ik}[\phi] \Big(\frac{\partial}{\partial \phi_k} + \beta \frac{\partial \mathcal{H}_\phi}{\partial \phi_k} \Big) \Bigg] g_\phi(t) + N_\phi(t) \;, \tag{E.4}$$

we find:

$$-\sum_i \frac{\partial}{\partial \phi_i} \sum_k \left[\left(\frac{\partial}{\partial \phi_k} \beta^{-1} \Gamma_0^{ik}[\phi] \right) g_\phi(t) + \sum_i \frac{\partial}{\partial \phi_i} \left(\xi_i(t) g_\phi(t) \right) \right]$$

$$= \sum_{ik} \frac{\partial}{\partial \phi_i} \beta^{-1} \Gamma_0^{ik}[\phi] \frac{\partial}{\partial \phi_k} g_\phi(t) + N_\phi(t) \qquad (E.5)$$

Solving for $N_\phi(t)$, we obtain the basic relation between ξ_i and N_ϕ:

$$N_\phi(t) = -\sum_{ik} \frac{\partial}{\partial \phi_i} \left[\xi_i(t) \delta_{ik} + \beta^{-1} \left(\frac{\partial \Gamma_0^{ik}[\phi]}{\partial \phi_k} \right) + \beta^{-1} \Gamma_0^{ik}[\phi] \frac{\partial}{\partial \phi_k} \right] g_\phi(t)$$

$$= -\sum_{ik} \frac{\partial}{\partial \phi_i} \left[\xi_i(t) \delta_{ik} + \frac{\partial}{\partial \phi_k} \beta^{-1} \Gamma_0^{ik}[\phi] \right] g_\phi(t) \ . \qquad (E.6)$$

Let us restrict ourselves to the simple Gaussian noise case where Γ_0^{ik} is independent of ϕ. We can then check some of the statistical properties of the noise $N_\phi(t)$.

We have first:

$$\langle N_\phi(t) \rangle = -\sum_{ik} \frac{\partial}{\partial \phi_i} \left\langle \left[\xi_i(t) \delta_{ik} + \frac{\partial}{\partial \phi_k} \beta^{-1} \Gamma_0^{ik} \right] g_\phi(t) \right\rangle \ . \qquad (E.7)$$

We need to evaluate:

$$\langle \xi_i(t) g_\phi(t) \rangle = 2\beta^{-1} \sum_k \Gamma_0^{ik} \langle \frac{\delta}{\delta \xi_k(t)} g_\phi(t) \rangle \ . \qquad (E.8)$$

We have, using the chain-rule:

$$\frac{\delta}{\delta \xi_k(t)} g_\phi(t) = -\sum_\ell \frac{\partial}{\partial \phi_\ell} g_\phi(t) \frac{\delta \psi_\ell(t)}{\delta \xi_k(t)} \ . \qquad (E.9)$$

It is self-consistent to assume:

$$\frac{\delta \psi_\ell(t)}{\delta \xi_k(t)} = \frac{1}{2} \delta_{\ell k} \ , \qquad (E.10)$$

then:

$$\frac{\delta}{\delta \xi_k(t)} g_\phi(t) = -\frac{1}{2} \frac{\partial}{\partial \phi_k} g_\phi(t) \qquad (E.11)$$

and:

$$\langle \xi_i(t) g_\phi(t) \rangle = 2\beta^{-1} \sum_k \Gamma_0^{ik} \langle -\sum_\ell \frac{\partial}{\partial \phi_\ell} g_\phi(t) \frac{1}{2} \delta_{\ell k} \rangle$$

$$= -\beta^{-1} \sum_k \Gamma_0^{ik} \frac{\partial}{\partial \phi_k} \langle g_\phi(t) \rangle \ . \tag{E.12}$$

Putting this back into Eq. (E.7), we find as required that:

$$\langle N_\phi(t) \rangle = 0 \ . \tag{E.13}$$

Next we consider:

$$\langle g_{\phi'} N_\phi(t) \rangle = -\sum_{ik} \frac{\partial}{\partial \phi_i} \langle g_{\phi'} \left[\xi_i(t) \delta_{ik} + \frac{\partial}{\partial \phi_k} \beta^{-1} \Gamma_0^{ik} \right] g_\phi(t) \rangle \ . \tag{E.14}$$

This requires that we evaluate the average over the noise:

$$\langle \xi_i(t) g_{\phi'} g_\phi(t) \rangle = 2\beta^{-1} \sum_k \Gamma_0^{ik} \langle \frac{\delta}{\delta \xi_k(t)} \left[g_{\phi'} g_\phi(t) \right] \rangle \ . \tag{E.15}$$

Consider first:

$$\frac{\delta}{\delta \xi_k(t)} g_{\phi'} = -\sum_j \frac{\partial}{\partial \phi'_j} g_{\phi'} \frac{\delta \psi_j(0)}{\delta \xi_k(t)} \ , \tag{E.16}$$

then clearly, because of causality,

$$\frac{\delta \psi_j(0)}{\delta \xi_k(t)} = 0 \tag{E.17}$$

if $t > 0$. Then $\delta g_{\phi'} / \delta \xi_k(t)$ vanishes for $t > 0$, and:

$$\langle \xi_i(t) g_{\phi'} g_\phi(t) \rangle = 2\beta^{-1} \sum_k \Gamma_0^{ik} \langle g_{\phi'} (-\sum_j \frac{\partial}{\partial \phi_j} g_\phi(t) \frac{\delta \psi_j(t)}{\delta \xi_k(t)}) \rangle$$

$$= -\sum_{jk} \frac{\partial}{\partial \phi_j} 2\beta^{-1} \Gamma_0^{ik} \langle g_{\phi'} g_\phi(t) \frac{\delta \psi_j(t)}{\delta \xi_k(t)} \rangle \ . \tag{E.18}$$

$$= -\beta^{-1} \sum_k \Gamma_0^{ik} \frac{\partial}{\partial \phi_k} \langle g_{\phi'} g_\phi(t) \rangle \ . \tag{E.19}$$

Putting this back into Eq. (E.14) we find:

$$\langle g_{\phi'} N_\phi(t) \rangle = 0 \quad \text{for } t > 0 \tag{E.20}$$

Next we want to evaluate $\langle N_\phi(t) N_{\phi'}(t') \rangle$ where the average is over the noise ξ_i. Using Eq. (E.6) twice and assuming $t \geq t'$, we have:

$$\langle N_\phi(t) N_{\phi'}(t') \rangle = \sum_{ijkl} \frac{\partial}{\partial \phi_i} \frac{\partial}{\partial \phi'_k} \langle \left[\xi_i(t) \delta_{ij} \right. $$

$$\left. + \frac{\partial}{\partial \phi_j} \beta^{-1} \Gamma_0^{ij} \right] g_\phi(t) \left[\xi_k(t') \delta_{kl} + \frac{\partial}{\partial \phi'_l} \beta^{-1} \Gamma_0^{kl} \right] g_{\phi'}(t') \right\rangle$$

$$= \sum_{ik} \frac{\partial}{\partial \phi_i} \frac{\partial}{\partial \phi'_k} \left[\langle \xi_i(t) \xi_k(t') g_\phi(t) g_{\phi'}(t') \rangle \right.$$

$$+ \langle \xi_i(t) g_\phi(t) \sum_l \frac{\partial}{\partial \phi'_l} \beta^{-1} \Gamma_0^{kl} g_{\phi'}(t') \rangle$$

$$+ \langle \xi_k(t') g_{\phi'}(t') \sum_j \frac{\partial}{\partial \phi_j} \beta^{-1} \Gamma_0^{ij} g_\phi(t) \rangle$$

$$\left. + \sum_{jl} \frac{\partial}{\partial \phi'_l} \frac{\partial}{\partial \phi_j} \beta^{-2} \Gamma_0^{ij} \Gamma_0^{kl} \langle g_{\phi'}(t') g_\phi(t) \rangle \right] . \quad (E.21)$$

Using Eq. (E.19) we see that the second and fourth terms in Eq. (E.21) cancel. We note that the first term in Eq. (E.21) has a factor that can be written as:

$$\langle \xi_i(t) \xi_k(t') g_\phi(t) g_{\phi'}(t') \rangle = 2\beta^{-1} \sum_l \Gamma_0^{il} \langle \frac{\delta}{\delta \xi_l(t)} \xi_k(t') g_\phi g_{\phi'}(t') \rangle$$

$$= 2\beta^{-1} \sum_l \Gamma_0^{il} \left[\delta_{lk} \delta(t - t') \langle g_\phi(t) g_{\phi'}(t') \rangle + \langle \xi_k(t') g_{\phi'}(t') \frac{\delta g_\phi(t)}{\delta \xi_l(t)} \rangle \right] .$$

The derivative of $g_{\phi'}(t')$ with respect to $\xi_l(t)$ vanishes unless $t = t'$, in which case it is swamped by the term proportional to $\delta(t - t')$ and can be dropped. Next we can use Eq. (E.11) to obtain:

$$\langle \xi_i(t) \xi_k(t') g_\phi(t) g_{\phi'}(t') \rangle = 2\beta^{-1} \sum_l \Gamma_0^{il}$$

$$\left[\delta_{lk} \delta(t - t') \delta(\phi - \phi') W_\phi - \frac{1}{2} \frac{\partial}{\partial \phi_l} \langle \xi_k(t') g_{\phi'}(t') g_\phi(t) \rangle \right] . \quad (E.22)$$

Notice that the second term on the right in Eq. (E.22) cancels the third term on the right hand side of Eq. (E.21) and we have the final result:

$$\langle N_\phi(t) N_{\phi'}(t') \rangle = \sum_{ik} \frac{\partial}{\partial \phi_i} \frac{\partial}{\partial \phi'_k} 2\beta^{-1} \Gamma_{ik} \delta(t - t') \delta(\phi - \phi') W_\phi ,$$

which agrees with Eq. (104) of Chap. 9.

Index

a
Adamson, D. H. 449
adiabatic speed of sound 191
Aharoni, A. 449
Akcasu, A. Z. 266
Alder, B. J. 170, 300, 336
Als-Nielsen, J. 296
Ampere's law 114
analytic properties 70
Anderson, H. C. 171, 266
Andrews, T. 178, 202
angular momenta commutation relations 33
anticommutator 126
antiferromagnet 274, 321
antilinear operator 456
antiunitary operator 457
Antunes, N.D. 448
Ao, P. 448
Arfken, G. 96
argon gas 43
Ashcroft, N. 120, 266
auxiliary field methods 429

b
Bahiana, M. 449
Bar-on, M. 448
bare kinetic coefficient 318
bare transport coefficient 318
Bausch, R. 399
Baym, G. 59
Becker, R. 447
Bengtzelius, U. 335, 337
Bennett, H. S. 170
Berne, B. J. 59
Bettencourt, L.M.A. 448
Bhatnagar, P. L. 265
binary alloy 405
Binder, K. 446
Bishop, A.R. 448
Bloch equations of motion 33
Bloch walls 445
Bloch, F. 33, 58
Bloembergen, N. 170

Boltzmann H-function 221
Boltzmann collision integral 218
Boltzmann equation 130, 205, 219
Boltzmann paradigm 14
Boltzmann, L. 265
Boon, J. P. 170, 202
Born approximation 43
Boyer, D. 449
Bragg peak 416, 429
Bray, A.J. 439, 445–447, 449
Brezin, E. 399
Brillouin doublet 191
Brillouin, L. 202
broken continuous symmetry 15
Brown, R. 15
Brownian motion 6, 142, 143
Brownian particle 156
Brownian particles 8
Brush, S.G. 202, 265
bulk viscosity 200, 325
Burbury, S. H. 265

c
Cahn, J. W. 337, 448
Cahn-Hilliard model 318
Cahn–Hilliard–Cook theory 425
Callen, H. B. 96, 266
Calzetta, E.A. 448
canonical ensemble 64
Carrier, G. 96
causal response function 72
Chaikin, P. M. 449
Chapman, S. 265, 336
characteristic frequency 281
Cheng, Z. 449
Chikazumi, S. 449
Christensen, J. 449
Chuang, I. 448
Clausius, R. 210, 265
coarse graining 303
coarse-grained field 312
Cohen, E. G. D. 266, 335–337
coherent neutron scattering 52
collisional invariants 220

complex Ginzburg–Landau   443
complex Kubo function   80
complex relaxation function   78
complex response function   72
conservation laws   15, 123
conservation of energy   141
conservation of momentum   141
conserved variables   134, 301
constitutive relations   2, 11, 127
constrained equilibrium   39, 79
continuity equation   2, 125, 141
conventional theory   271, 280
Cook, H. E.   337, 448
COP   318, 422
Corliss, L. M.   400
correlation length   278
Cowling, T. G.   265, 336
critical droplet   416
critical dynamics   5
critical indices   279
critical opalescence   178
critical phenomena   304
critical quench   411
critical slowing down   124, 281
Cross, M.C.   15, 449
Curie temperature   403
Curie-Weiss theory   279
current–current correlation function   132
cyclotron frequency   246
Cyrot, M.   337

d
Döring, W.   447
damping matrix   317
Das, S.   335, 337
De Dominicis, C.   399
de Gennes, P.G.   448
De Leener, M.   146, 171
defect density   443
defect profile   438
defect structures   439
degenerate equilibrium states   403
density expansion   300
density matrix   79
Desai, R. C.   447
diblock copolymers   445
dielectric constant   56
dielectric function   32, 113
dielectric material   32
Dietrich, O. W.   296
diffusion coefficient   11
diffusion equation   128
diffusion limited aggregation   15
diluted antiferromagnets   407
Dirac, P. A. M.   135, 170
direct correlation function   254
director field   305

disclinations   407, 439
dislocations   407, 439
disordering defects   407
displacement field   113
dissipation   89
dissipative systems   318
divergence of transport coefficients   335
Donnelly, R.J.   448
Dorfmann, J. R.   265, 266, 335–337
Dorsey, A. T.   337
driven harmonic oscillator   75
droplet profile   415
Droze, M.   447
Drude formula   235
Drude model   116
Drude theory   32
Drude, P.   58, 120
Duderstadt, J. J.   266
dynamic critical index   281
dynamic critical phenomena   271, 318, 343
dynamic fixed point   398
dynamic renormalization group   379
dynamic scaling   281, 389
dynamic structure factor   47, 63
dynamic susceptibility   27, 63, 129
dynamical part of the memory function   146
dynamical scaling   362
Dziarmagai, J.   448

e
early universe   407
easy plane   273
effective Hamiltonian   303, 304, 307
effective magnetic field   320
Einstein, A.   6, 12, 15, 178, 202
elastic scattering   48
elastic theory   439
Elder, K. R.   447
electrical conductivity   31, 83
electron scattering   49
electron spin resonance   32
energy current   142
energy density   175
energy density current   176
Enomoto, Y.   448
Enskog, D.   336
equipartition theorem   9, 77, 151
ergodic theory   14
Ernst, M. H.   336, 337
Eschenfelder, A.H.   449
Euler relation   179
Euler–Lagrange equation   429, 438
evolution of the early universe   438
exchange interaction   136, 273

f

Fabelinskii, I. L. 202
FDT 349
Fermi scattering length 50
Fermi's golden rule 43
Ferrell, R. A. 296, 337
Fetter, A.L. 448, 449
Fick's law 11
Fick, A. 15, 170
field operators 40
field theories 302
finite size effects 416
first law of thermodynamics 180
Fischer, K. H. 447
Fisher, D. S. 448
Fixman, M. 337
Fleury, P. S. 202
fluctuating nonlinear hydrodynamics 317, 326
fluctuation function 66, 70
fluctuation-dissipation theorem 5, 63, 66, 349
fluctuations 6
fluid dynamics 322
flux lattice 445
Fokker, A. D. 337
Fokker–Planck operator 316
Forster, D. 19, 58, 96, 266, 400
Fourier's law 2, 239
Fourier, J 15
Frank effective Hamiltonian 305
free energy barrier 413
Freedman, R. 337, 400
Frenkel, J. 202
Frey, E. 15
friction 142
friction constant 7
Fujisaka, H. 337
Furukawa, H. 423, 448

g

Götze, W. 335, 337
galaxy formation 438
Gass, D. M. 170, 336
Geller, M.R. 449
generalized Fokker–Planck equation 311
generalized Fokker–Planck operator 311
generalized Langevin equation 142, 143, 306
Gibbs–Duhem relation 179
Gibbs–Thomson relation 421
Ginsburg-Landau free energy 305
Ginzburg, V. L. 337
Goldenfeld, N. 449
Gould, H. 266
Gouvea, M.E. 448
grand canonical ensemble 65

Green, M. 201
Green, M. S. 19, 58, 130, 170, 202, 336, 337
Green–Kubo equation 132
Green–Kubo formula 85, 130, 299
Green–Kubo relation 334
Grest, G. S. 447
Grinstein, G. 337
Gross, E. P. 202, 265, 337, 449
growth kinetics 318
growth law 423
growth patterns 407
Gubbins, K. E. 266
Gunton, J. D. 446
Guyot, P. 448

h

Haines, L. K. 336
Hall coefficient 245
Hall effect 245
Hall, E. H. 266
Halperin, B. I. 296, 318, 336, 337, 399, 400, 449
Hamiltonian 138, 149
harmonic oscillator 149
harmonic solid 49
Harrison, C. 449
Hartree approximation 360
Hashimoto, T. 447
Hastings, J. M. 400
Hauge, E. H. 337
heat current 2
heat density 176
heat equation 2, 128
Heermann, D. 446
Heisenberg equations of motion 20, 91, 134
Heisenberg ferromagnet 136
Heisenberg model 273
Heisenberg representation 21, 79
Hennion, M. 423, 448
Hertz, J. A. 447
Hilliard, J. E. 337, 448
Hindmarsh, M. 448
Hohenberg, P. C. 15, 296, 318, 336, 337, 399, 400, 448, 449
hole function 249
homogeneous function 278
Honda, K. 449
Horn, P. 400
Hou, Q. 449
Hu, B. L. 448
Huang, K. 447
Huebener, R.P. 449
Huse, D. A. 448, 449
hydrodynamic modes 128
hydrodynamic pole 128

i

ideal gas law   209
incoherent neutron scattering   52
incompressible fluid   330
induced charge density   107
induced current density   109
induced polarization   113
interaction representation   24
interfacial growth   5
interfacial profile   415
intermediate phase   445
intrinsic fluctuations   5
invariance principle   458
irreversibility   14
Ishibashi, Y.   448
Ising model   274
isotropic ferromagnet   320, 343

j

Jackson, E. A.   265
Janssen, H. K.   399
Jasnow, D.   448
Johnson, J. B.   96

k

Kadanoff, L. P.   19, 58, 278, 296, 337
Kalos, M.   423, 448
Kaufmann, B.   296
Kawasaki, K.   146, 171, 336, 337, 399, 400, 443, 448, 449
Kestin, J.   265
Khalatnikov, I. M.   337
Kibble, T.   449
Kim, B. S.   337
kinetic coefficients   310, 364
kinetic energy current density   206
kinetic energy density   139
kinetic equation   328
kinetic modeling   226
kinetic theory   141, 205, 246
kinetics of first-order phase transitions   5
Kirkwood, J. G.   171
Kittel, C.   87, 96, 120
Kleman, M.   447
Kloy, K.   15
Koehler, J.S.   449
Komineas, S.   448
Koopman's operator   136
Koopman, B.O.   170
Kovacs, T.   448
Kramers–Kronig relations   72
Krech, M.   170
Krook, M.   96, 265
Kubo function   129
Kubo response function   80
Kubo, R.   19, 58, 79, 96, 130, 170

l

λ-transition   277
Landau theory   279
Landau, D. P.   170
Landau, L. D.   202, 303, 304, 336, 337, 449
Landau–Ginzburg–Wilson model   318
Landau–Placzek ratio   190
Landau-Lifshitz equations   439
Langer, J. S.   336, 447, 448
Langevin equation   7, 134, 156
Langevin models   407
Langevin paradigm   14
Langevin, P.   15
Laplace   191
Laplace transform   72, 136
lattice gas   447
Lax, M.   449
layer-field displacement   305
Lebowitz, J.L.   423, 448
Leutheusser, E.   335, 337
LGW   304, 343, 407
LGW model   318
Lifshitz, E. M.   336, 447
Lifshitz, I. M.   303, 447, 448
Lifshitz–Cahn–Allen growth law   424
Lifshitz–Slyozov–Wagner growth law   425
light scattering   55, 56, 416
line defects   437, 445
linear response   3, 5, 63
linear response function   129
linear response theory   19, 32, 247
linearized Boltzmann collision integral   223
linearized Boltzmann collision operator   259
linearized hydrodynamics   301
Liouville operator   134, 135, 139, 150
liquid-glass transition   335
Liu, F.   449
local equilibrium   237
localization   15
London theory   439
long-time tails   300, 330, 334
Lord Kelvin   439
Lorentz force   244
Lorentz number   240
Lundh, E.   448

m

Ma, S.   337, 349, 399
magnetic bubbles   445
magnetic neutron scattering   52
magnetic susceptibility   91
magnetization   321
magnetization density   124, 137
magnetization-density current   126
Mahan, G. D.   120

Malozemoff, A.P.   449
Markoffian approximation   155, 310
Markov, A. A.   171
Marro, J.   423, 448
Martin, P. C.   19, 58, 96, 120, 170, 266, 337, 399
Mash, D. I.   202
mass density   322
material deposition   4
Maxwell molecules   264, 266
Maxwell velocity distribution   205, 207
Maxwell, J. C.   265, 266
Mazenko, G. F.   171, 258, 265, 266, 335–337, 400, 447–449
Mazo, R.   15
Mc Coy, B. M.   296
MCS   404
mean-field theory   279, 426
mean-free path   205, 210, 213
mean-free time   205
memory function   143, 310, 328
memory-function formalism   144
Menyhárd, N.   296
Mermin, N. D.   120, 266, 447
Mertens, F.G.   448
Messiah, A.   58
Miller, H. D.   448
Milner, S.   337
Mitchell, W. C.   337
mode-coupling   329
mode-coupling approximation   146
mode-coupling effects   301
mode-coupling theory   335, 336
modulated phases   406
molecular dynamics   123, 132
momentum density   93, 139, 175, 206
monopoles   407
Monte Carlo   404
Monte Carlo dynamics   123
Montgomery, D. C.   266
Mori, H.   143, 170, 171, 201, 202, 337
multiplicative noise   326
Muskhelishvili, N. I.   96
Mydosh, J. A.   447

n

n-vector model   318, 390, 407
Néel transition   275
Nabarro, F.R.N.   448
Nagler, S.   447
Nambu–Goldstone modes   15, 123, 283, 301, 305, 440
Navier–Stokes equation   326
NCOP   318
Nelkin, M.   266
Nelson, D. R.   400
nematic liquid crystals   305, 439

Neu, J. C.   449
neutron scattering   42, 416
Newton, I.   191
NLSE   438, 443
noise   7, 142, 143, 307, 313
Nolan, M. J.   400
nonequilibrium exponent   423
nonlinear diffusion   432
nonlinear hydrodynamics   5, 343
nonlinear Langevin equation   312, 317
nonlinear models   304
nonlinear response   299
nonlinear Schroedinger equation   438
Nordholm, K. S. J.   337
normal fluid   303
Nozieres, P.   447
nuclear magnetic resonance   32
nucleation   4
nucleation rate   414
nucleation theory   413
number density   175, 206
Nyquist theorem   85
Nyquist, H.   96

o

Ohm's law   235, 246
Ohm, G.   170
Ohta, Jasnow and Kawasaki   435
Ohta, T.   400, 448
one body irreducible   147
Onsager   66
Onsager, L.   96, 266, 296
Oono, Y.   449
Oppenheim, I.   336
optical frequency   119
order parameter   271, 301
order parameter correlation function   416
order–disorder transition   405
Orihara, H.   448
Ornstein, L. S.   178, 202, 266, 295
Ornstein–Zernike form   272, 356
Ostrowski, G.   52, 58

p

pair potential   139
Pake, G.   32
Pake, G. E.   58
Papanicolaou, N.   448
paramagnet   2, 32
paramagnetic relaxation   3
Pargellis, A.   448
Pargellis, N.   448
particle density   139
Passell, L.   296
pattern formers   445
Peach, M.   449
Peach-Koehler   443

Pearson, C. 96
Pecora, R. 59
Pelcovits, R. 337
Peliti, L. 399
Perrin, J. 6, 15, 265
perturbation theory 343
phase separation 406
phase-field theories 440
phase-ordering kinetics 406
phase-space density 139, 206
Pismen, L.M. 448, 449
Pitaevskii, L. P. 337, 447, 449
Placzek, G. 202
Planck, M. 337
plasma frequency 108, 255
Plemelj relations 71
Poincare-Bertrand identity 71
point defects 437
Poisson 191
Poisson bracket 135
Pomeau, Y. 337
Porod's law 419, 439
Porod, G. 447
pressure 207
principle-value part 71
projection operators 143
Prost, J. 448
Puri, S. 449

q
Qian, H. 449
quench 409
quenched impurities. 15
quenched magnetic impurities 406

r
radial distribution function 248
Ramaswamy, S. 337
ramified structure 405
Randolph, P. D. 52, 58
random initial conditions 411
random walk 6, 15
Ranganthan, S. 266
rate of effusion 210
Rayleigh peak 191
Rayleigh–Benard experiment 5
recursion relations 379
Reed, T. M. 266
relaxation function 129
relaxational function 39
relaxational process 156
renormalization group 304, 343
renormalization group theory 279
Resibois, P. 146, 171, 337
resolvant operator 136, 306, 328
response function 27, 344, 347
Rhee, S.W. 449

Rice, S.O. 449
Robert Brown 6
Rodriguez, J.D. 449
Rogers, T. M. 447
Roland, C. 449
Ronzaud, D. 448
Rose, H. A. 399
rotational invariance 137
Rowe, J. 52
Rowe, J. M. 58
Ruelle, D. 15

s
Sadzikowski, M. 448
Safron, S. A. 447
Sahni, P. S. 446, 447
Salpeter, E. E. 266
San Miguel, M. 446
Sander, L. 15
Sasa, S. 449
scaling 277, 423
scaling hypothesis 278
Schiff, L. 58
Schmid, A. 337
Schmidt, H. 296
Schnitzer, H.J. 448
Schröder, E. 449
Schroedinger representation 22, 87
Schwable, F. 296
screened response function 115
Sebastian, J. M. 449
second fluctuation-dissipation theorem 147, 314
second sound 295
Seebeck effect 241
self energy 358
self-similarity 423
Sethuraman, S. 449
sharp interfaces 418
shear viscosity 325
Shigematsu, H. 337
Siggia, E. D. 399, 400
simple fluids 305
single relaxation time approximation 205
singlet probability distribution 206
Sjögren, L. 171, 335, 337
Sjölander, A. 171, 335, 337
Sköld, K. 52, 58
Slichter, C. P. 32, 58
Slonczewski, J.C. 449
slow variables 123, 134, 301
Slyosov, V. V. 447
smectic A liquid crystals 305, 326
Smoluchowski, M. Von 178
soap froths 406
solids 305
Sommerfeld, A. 440, 443, 449

sound attenuation   3, 201
sound velocity   3
spin diffusion   124
spin glasses   407
spin-commutation relations   137
spin-diffusion coefficient   128
spin-exchange kinetic Ising model   407
spin-flip kinetic Ising model   407
spin-wave   289
spin-wave frequency   289
spinodal decomposition   4, 406
spontaneous symmetry breaking   285
staggered magnetization   274, 321
Starunov, V. S.   202
static part of the memory function   146
static susceptibility   32, 38
Stephens, G.J.   448
Stokes–Einstein relation   12
strain fields   305
streaming velocity   309
stress tensor   141, 176, 324
strings   407
sublattice   275
superconductors   305, 318, 437
supercooled vapor   413
superfluid helium   277
superfluids   318, 437
surface tension   415, 422
Svidzinsky, A.A.   448
Swift, J. B.   337, 449
Swift–Hohenberg model   445
symmetry principle   123
symmetry properties of correlation functions   67
Szépfalusy, P.   296

t
TDGL model   318, 343, 403, 407, 438
thermal conductivity   2, 199
thermal neutrons   42
thermal noise   409
thermodynamic derivatives   38
thermodynamic stability   129
thermodynamic state variables   124
Thomson, W.   447, 449
Thouless, D.J.   449
Tidman, D. A.   266
time reversal   455
time-correlation functions   27, 144
time-dependent Ginzburg–Landau Model   318
time-reversal operator   455, 456
time-reversal symmetry   68, 455
time-translational invariance   27, 143
Tinkham, M.   448
Törnkvist, O.   449
Tokuyama, M.   448

Tomita sum rule   419
Tomita, H.   447
Toner, J.   337
topological charge density   442
topological defects   440
Toyoki, H.   449
transition temperature   359
translational invariance   48, 137
transport coefficient.   300
turbulence   4, 5
Turok, N.   448
Turski, L.   336
type II superconductors   445

u
universality   277
unstable thermodynamic systems   403

v
Valls, O. T.   447, 448
Van der Waals, J. D.   296
van der Waals' theory   279
van Hove self-correlation function   52
van Hove, L.   47, 58, 296
van Leeuwen, J. M. J.   337
van Hove approximation   280
velocity autocorrelation function   10, 158
Viñals, J.   449
Villain, J.   337
Vinen, W.F.   449
viscosity tensor   325
Vlasov approximation   253
Von Smoluchowski, M.   202
vortex core   439
vortices   407
vorticity   439

w
Wagner, C.   447
Wagner, H.   399
Wainwright, T. E.   170, 300, 336
wall defects   437
Wei, T.   265
Weinstock, J.   336
Weiss, P.   296
Welton, T. R.   96
white noise   8
Wickham, R.   449
Widom, B.   278, 296
Wiedemann-Franz   240
Wilks, J.   448
Wilson matching   390
Wilson, K. G.   304, 336, 390, 400
Witten, T.   15
Wu, T. T.   296
Wysin, G.M.   448

x

X-Ray scattering 55
x-ray scattering 56, 416
XY model 273, 437

y

Yip, S. 96, 170, 265
Yurke, B. 448

z

Zannetti, M. 448
Zeeman coupling 20
Zernike, F. 178, 202, 266, 295
zero-point motion 77
Zinn-Justin, J. 399
Zwanzig, R. W. 143, 170, 336, 337